Advances in Sustainable and Digitalized Factories: Manufacturing, Measuring Technologies and Systems

Advances in Sustainable and Digitalized Factories: Manufacturing, Measuring Technologies and Systems

Editors

Roque Calvo
José A. Yagüe-Fabra
Guido Tosello

MDPI • Basel • Beijing • Wuhan • Barcelona • Belgrade • Manchester • Tokyo • Cluj • Tianjin

Editors

Roque Calvo
Department of Mechanical, Chemical and Industrial Design Engineering
Universidad Politécnica de Madrid
Madrid
Spain

José A. Yagüe-Fabra
Engineering Research Institute of Aragon
Universidad de Zaragoza
Zaragoza
Spain

Guido Tosello
Department of Civil and Mechanical Engineering
Technical University of Denmark
Kgs. Lyngby
Denmark

Editorial Office
MDPI
St. Alban-Anlage 66
4052 Basel, Switzerland

This is a reprint of articles from the Special Issue published online in the open access journal *Applied Sciences* (ISSN 2076-3417) (available at: www.mdpi.com/journal/applsci/special_issues/sustainable_digitalized_factories).

For citation purposes, cite each article independently as indicated on the article page online and as indicated below:

LastName, A.A.; LastName, B.B.; LastName, C.C. Article Title. *Journal Name* **Year**, *Volume Number*, Page Range.

ISBN 978-3-0365-7659-6 (Hbk)
ISBN 978-3-0365-7658-9 (PDF)

Cover image courtesy of wbk Institute of Production Science
Karlsruhe Institute of Technology (KIT), Karlsruhe, Germany

Contents

About the Editors

Roque Calvo

Roque Calvo is currently an Associate Professor at Universidad Politécnica de Madrid (UPM), Department of Mechanical, Chemical and Industrial Design Engineering, sharing teaching and research activities with academic duties since 2018 as Vice-Dean Head of Studies of Escuela Técnica Superior de Ingeniería y Diseño Industrial. His teaching activity includes lectures and thesis supervision in manufacturing engineering in bachelor, master, and doctoral studies.

He has developed formerly managerial and manufacturing operation responsibilities for 15+ years in the aeronautical, automotive, and metal industries.

His research interests are in the broad field of manufacturing processes and systems, including dimensional metrology, with over 30 publications in those fields. At the same time, he serves as a peer reviewer of many JCR journals on a regular basis.

His ongoing research includes point coordinate metrology and stochastic modeling of manufacturing systems.

Roque has been a member of several technical committees of the Spanish Association for Standardization and Certification (AENOR). He has a Ph.D. degree in Industrial Engineering, a MSc degree as Aeronautical Engineer from UPM, and a MBA degree from the IE Business School.

José A. Yagüe-Fabra

José Yagüe-Fabra is Full Professor at the Department of Design and Manufacturing Engineering at the University of Zaragoza (Spain), and since 2019 he serves as Dean of the School of Engineering and Architecture.

His research work is focused on precision design, micro-nano-metrology, machine tool metrology, large scale precision engineering, computed tomography and precision manufacturing. He has published over 130 peer-reviewed articles in international scientific journals and conferences. Jose has been the project coordinator or member of the research team in more than 30 research projects and has supervised over 80 MSc and PhD theses.

In 2021, he became a Fellow of the International Academy for Production Engineering (CIRP). He is also a member of the European Society for Precision Engineering and Nanotechnology (euspen) since 2011 and belongs to its council since 2022. He is also a member of the Manufacturing Engineering Society. He has been a member of international scientific committees for euspen, Manufacturing Engineering Society International Conference, CIRP Global Web Conference (CIRPe), CIRP CAT Conference, CIRP Conference on Manufacturing Systems, and International Conference on Industrial Computed Tomography.

José is an editorial board member of the journals *Nanomanufacturing and Metrology* (Springer), *Manufacturing Letters* (Elsevier) and *Metrology* (MDPI). He serves as reviewer for peer-reviewed journals on a regular basis. He holds a MSc degree in Mechanical Engineering (2000) and a PhD in Machine Tool Metrology (2005), both from the University of Zaragoza (Spain).

Guido Tosello

Guido Tosello is currently Associate Professor at the Technical University of Denmark, Department of Civil and Mechanical Engineering, Section of Manufacturing Engineering, where he is also Head of Studies of the MSc program in Materials and Manufacturing Engineering. He is a senior lecturer, research manager, supervisor of PhD, MSc, and BSc projects, industrial and management consultant, and start-up board member.

He has 15+ years of research experience in the analysis, characterization, monitoring, control, optimization and simulation of precision molding processes of thermoplastic materials from conventional size down to micro- and nano-scales. He is currently the coordinator of the European project DIGIMAN4.0 "Digital Manufacturing Technologies for Zero-defect Industry 4.0 Production" (2019–2024). His latest research interests are in the field of digitalization of polymer processing technologies, recycling of plastic materials, artificial intelligence and digital twin for the optimization of manufacturing processes. He has supervised 70+ MSc thesis, 20+ PhD thesis, and published 200+ articles in peer-reviewed scientific journals and international conference proceedings.

Guido Tosello is the Editor of the book 'Micro Injection Molding', published by Hanser in 2018. He is Associate Member of the International Academy for Production Engineering (CIRP), Board Member of the CIRP Scientific Technical Committee 'Surfaces' (STC S), Council Member of the European Society for Precision Engineering and Nanotechnology (EUSPEN). He is currently Associate Editor of *Advances in Industrial and Manufacturing Engineering* (Elsevier).

Guido holds a MSc degree in Mechanical and Production Engineering from the University of Padova (Italy), a PhD degree in the field of Micro Manufacturing from the Technical University of Denmark, and a MBA degree from the DTU Executive School of Business.

applied sciences

MDPI

Editorial

Advances in Sustainable and Digitalized Factories: Manufacturing, Measuring Technologies and Systems

Roque Calvo [1,*], José A. Yagüe-Fabra [2] and Guido Tosello [3]

[1] Department of Mechanical, Chemical and Industrial Design Engineering, Universidad Politécnica de Madrid, Ronda de Valencia, 3, 28012 Madrid, Spain

[2] Engineering Research Institute of Aragon—I3A, Universidad de Zaragoza, María de Luna 3, 50018 Zaragoza, Spain; jyague@unizar.es

[3] Department of Civil and Mechanical Engineering, Technical University of Denmark, Produktionstorvet Building 427A, 2800 Kgs. Lyngby, Denmark; guto@dtu.dk

* Correspondence: roque.calvo@upm.es

Citation: Calvo, R.; Yagüe-Fabra, J.A.; Tosello, G. Advances in Sustainable and Digitalized Factories: Manufacturing, Measuring Technologies and Systems. *Appl. Sci.* **2023**, *13*, 5570. https://doi.org/10.3390/app13095570

Received: 25 April 2023
Revised: 27 April 2023
Accepted: 28 April 2023
Published: 30 April 2023

The evolution from current to future factories is supported by research contributions in many fields of technology. While lean manufacturing techniques represent a main improvement paradigm, the integration of new processes and technologies is a breakthrough for step-change improvements and systems evolution. The dominant paradigm of Industry 4.0 is a common framework under development. Accordingly, this Special Issue presents research results focused on the latest technologies and the design/operation of manufacturing systems contributing to this evolution towards the digitalization of production as well as the factory itself. The Special Issues consists of 14 original research papers and two review papers, which cover both fundamental and key enabling digital technologies, as well as the design and application of these technologies for the digitalization of production, covering six main domains: (1) Lean Manufacturing and Industry 4.0; (2) Internet of Things in Manufacturing; (3) Virtual Reality/Augmented Reality in Manufacturing; (4) Digital Technologies for Production Planning; (5) Machine Learning in Manufacturing; (6) Digitalization in Handling and Assembly.

(1) Lean Manufacturing and Industry 4.0. Peças et al. [1] proposed a conceptual approach for the continuous improvement cycle Plan-Do-Check-Act, depicting how Industry 4.0 technological concepts should be used for its enhancement. Gil et al. [2], using a systematic literature review process, clarified the origin, evolution, and diversification of the lean concept and concluded that the use of "lean" as systemic thinking is likely to be further extended to new research fields.

(2) Internet of Things in Manufacturing. Hosseinifard et al. [3] developed a simple wireless and passive RFID system for analogic pressure manometers, showing an affordable digitalization of analogic industrial systems. Vuković et al. [4] presented a low-cost prototyping solution for the legacy machine retrofit through the Industrial Internet of Things (IoT), facilitating process monitoring through hardware digitalization.

(3) Virtual Reality (VR)/Augmented Reality (AR) in Manufacturing. Ho et al. [5] used a systematic literature review methodology to draw conclusions concerning the emerging interest in using AR as an assisting technology for the quality sector in an Industry 4.0 context and proposed a design procedure for the development of long-term AR-based solutions in the future. Kim and Jeong [6] proposed a VR/AR collaboration model that increases the efficiency of manufacturing environments simultaneously with participants through 3D rendering virtualization of facilities or robots.

(4) Digital Technologies for Production Planning. Chiacchio et al. [7] accomplished a performance analysis of a repairable production line using a hybrid dependability queuing model based on Monte Carlo simulation. Rodríguez-Padial et al. [8] presented the assisted-driven design of customized maintenance plans for industrial plants. Liu

and Urgo [9] showed the robust scheduling framework for re-manufacturing activities of turbine blades. Moshiri et al. [10] presented an injection molding industrial case study for high-volume production in which the value chains for the production of additively and conventionally manufactured multi-cavity tool steel inserts are compared. May et al. [11] discussed the simulation of ontology-based production using the commercially available software OntologySim. Kubalíc et al. [12] solved the facility layout problem by applying alternative facility variants modeling.

(5) Machine Learning in Manufacturing. Chen et al. [13] critically discussed how machine learning is applied today in manufacturing towards Industry 4.0 at Product, Process, Machine, System levels; Mollaei et al. [14] presented an application of Machine Learning for defect prediction and process optimization in plastic injection molding.

(6) Digitalization in Handling and Assembly. Bianchi et al. [15] presented a modular under-actuated surface to perform handling tasks, such as sorting, stopping, or slowing down material flows, by exploiting the previous speed or gravity. Bencak et al. [16] proposed a simulation model based on ADAMS/MATLAB co-simulation for robotic pick-point evaluation using a two-fingered robotic gripper.

This variety of technologies, new methodologies, and novel approaches anticipate a near future with plenty of fruitful contributions to research and professional practice in sustainable and digitalized factories.

Funding: The Special Issue is partially supported and funded by the European Commission Horizon2020 Framework Programme for Research and Innovation through the Marie Skłodowska-Curie Innovative Training Network DIGIMAN4.0 ("DIGItal MANufacturing Technologies for Zero-defect Industry 4.0 Production", http://www.digiman4-0.mek.dtu.dk, accessed on 15 April 2023, 2019–2024, Project ID: 814225) is acknowledged.

Institutional Review Board Statement: Not applicable.

Informed Consent Statement: Not applicable.

Data Availability Statement: Not applicable.

Acknowledgments: We wish to thank all the authors who contributed with their papers to this Special Issue. We would also like to acknowledge all the reviewers whose careful and timely reviews have ensured the high quality of this Special Issue, "Advances in Sustainable and Digitalized Factories: Manufacturing, Measuring Technologies and Systems".

Conflicts of Interest: The authors declare no conflict of interest.

References

1. Peças, P.; Encarnação, J.; Gambôa, M.; Sampayo, M.; Jorge, D. PDCA 4.0: A New Conceptual Approach for Continuous Improvement in the Industry 4.0 Paradigm. *Appl. Sci.* **2021**, *11*, 7671. [CrossRef]
2. Gil-Vilda, F.; Yagüe-Fabra, J.; Sunyer, A. From Lean Production to Lean 4.0: A Systematic Literature Review with a Historical Perspective. *Appl. Sci.* **2021**, *11*, 10318. [CrossRef]
3. Hosseinifard, M.; Alzubaidi, S.; Michel, A.; Fantoni, G. RFID Technology as a Low-Cost and Passive Way to Digitize Industrial Analogic Indicators. *Appl. Sci.* **2022**, *12*, 1451. [CrossRef]
4. Vuković, M.; Jorg, O.; Hosseinifard, M.; Fantoni, G. Low-Cost Digitalization Solution through Scalable IIoT Prototypes. *Appl. Sci.* **2022**, *12*, 8571. [CrossRef]
5. Ho, P.; Albajez, J.; Santolaria, J.; Yagüe-Fabra, J. Study of Augmented Reality Based Manufacturing for Further Integration of Quality Control 4.0: A Systematic Literature Review. *Appl. Sci.* **2022**, *12*, 1961. [CrossRef]
6. Kim, J.; Jeong, J. Design and Implementation of OPC UA-Based VR/AR Collaboration Model Using CPS Server for VR Engineering Process. *Appl. Sci.* **2022**, *12*, 7534. [CrossRef]
7. Chiacchio, F.; Oliveri, L.; Khodayee, S.; D'Urso, D. Performance Analysis of a Repairable Production Line Using a Hybrid Dependability Queueing Model Based on Monte Carlo Simulation. *Appl. Sci.* **2023**, *13*, 271. [CrossRef]
8. Rodríguez-Padial, N.; Marín, M.; Domingo, R. Assisted-Driven Design of Customized Maintenance Plans for Industrial Plants. *Appl. Sci.* **2022**, *12*, 7144. [CrossRef]
9. Liu, L.; Urgo, M. A Robust Scheduling Framework for Re-Manufacturing Activities of Turbine Blades. *Appl. Sci.* **2022**, *12*, 3034. [CrossRef]

10. Moshiri, M.; Raza, M.; Sahlab, M.; Malik, A.; Bilberg, A.; Tosello, G. Value Chain Comparison of Additively and Conventionally Manufactured Multi-Cavity Tool Steel Inserts: An Injection Molding Industrial Case Study for High-Volume Production. *Appl. Sci.* **2022**, *12*, 10410. [CrossRef]
11. May, M.; Kiefer, L.; Kuhnle, A.; Lanza, G. Ontology-Based Production Simulation with OntologySim. *Appl. Sci.* **2022**, *12*, 1608. [CrossRef]
12. Kubalík, J.; Kurilla, L.; Kadera, P. Facility layout problem with alternative facility variants. *Appl. Sci.* **2023**, *13*, 5032. [CrossRef]
13. Chen, T.; Sampath, V.; May, M.; Shan, S.; Jorg, O.; Aguilar Martín, J.; Stamer, F.; Fantoni, G.; Tosello, G.; Calaon, M. Machine Learning in Manufacturing towards Industry 4.0: From for Now to Four-Know. *Appl. Sci.* **2023**, *13*, 1903. [CrossRef]
14. Mollaei Ardestani, A.; Azamirad, G.; Shokrollahi, Y.; Calaon, M.; Hattel, J.; Kulahci, M.; Soltani, R.; Tosello, G. Application of Machine Learning for Prediction and Process Optimization—Case Study of Blush Defect in Plastic Injection Molding. *Appl. Sci.* **2023**, *13*, 2617. [CrossRef]
15. Bianchi, E.; Jorg, O.; Fantoni, G.; Brosed Dueso, F.; Yagüe-Fabra, J. Study and Simulation of an Under-Actuated Smart Surface for Material Flow Handling. *Appl. Sci.* **2023**, *13*, 1937. [CrossRef]
16. Bencak, P.; Hercog, D.; Lerher, T. Simulation Model for Robotic Pick-Point Evaluation for 2-F Robotic Gripper. *Appl. Sci.* **2023**, *13*, 2599. [CrossRef]

applied sciences

MDPI

Article

PDCA 4.0: A New Conceptual Approach for Continuous Improvement in the Industry 4.0 Paradigm

Paulo Peças [1,*], João Encarnação [2], Manuel Gambôa [2], Manuel Sampayo [2] and Diogo Jorge [3]

[1] IDMEC, Instituto Superior Técnico, Universidade de Lisboa, 1049-001 Lisbon, Portugal
[2] Instituto Superior Técnico, Universidade de Lisboa, 1049-001 Lisbon, Portugal; joao.encarnacao@tecnico.ulisboa.pt (J.E.); manuel.gamboa@tecnico.ulisboa.pt (M.G.); manuel.sampaio@tecnico.ulisboa.pt (M.S.)
[3] EfficiencyRising, Lda. (Erising), 1800-082 Lisboa, Portugal; diogojorge@erising.pt
* Correspondence: ppecas@tecnico.ulisboa.pt

Featured Application: The proposal of a conceptual approach towards the application and evolution of continuous improvement in the context of Industry 4.0.

Abstract: Continuous improvement (CI) is a key component of lean manufacturing (LM), which is fundamental for organizations to remain competitive in an ever more challenging market. At present, the new industrial revolution, Industry 4.0 (I4.0), is taking place in the manufacturing and service markets, allowing more intelligent and automated processes to become a reality through innovative technologies. Not much research was found regarding a holistic application of I4.0′s technological concepts towards CI, which clarifies the potential for improving its effectiveness. This clearly indicates that research is needed regarding this subject. The present publication intends to close this research gap by studying the main I4.0 technological concepts and their possible application towards a typical CI process, establishing the requirements for such an approach. Based on that study, a conceptual approach is proposed (PDCA 4.0), depicting how I4.0 technological concepts should be used for CI enhancement, while aiming to satisfy the identified requirements. By outlining the PDCA 4.0 approach, this paper contributes to increasing the knowledge available regarding the CI realm on how to support the CI shift towards a I4.0 industrial paradigm.

Keywords: Industry 4.0; continuous improvement; lean manufacturing

Citation: Peças, P.; Encarnação, J.; Gambôa, M.; Sampayo, M.; Jorge, D. PDCA 4.0: A New Conceptual Approach for Continuous Improvement in the Industry 4.0 Paradigm. *Appl. Sci.* **2021**, *11*, 7671. https://doi.org/10.3390/app11167671

Academic Editors: José A. Yagüe-Fabra, Guido Tosello and Roque Calvo

Received: 19 July 2021
Accepted: 16 August 2021
Published: 20 August 2021

1. Introduction

Continuous improvement (CI) is a key component of lean manufacturing (LM) [1], being generally defined as a culture of sustained improvement targeting the elimination of waste in all systems and processes of an organization [2]. In highly dynamic and demanding markets, the CI of production processes and other value chain activities is crucial for organizations to remain competitive [3]. In this regard, the current fourth industrial revolution, Industry 4.0 (I4.0), is taking place in manufacturing companies, causing the shifting, or at least the adaptation, of the LM and CI paradigms [4]. Recently, several approaches on the integration between the LM realm and I4.0 were formulated, and authors reached important conclusions on how both paradigms can work together to enhance manufacturing performance and flexibility [5,6].

Approaching LM in its purest form does not require information technology [7]. However, both LM and I4.0 paradigms aim to solve present and future challenges in manufacturing [8]. Among the publications that study the applicability of I4.0, several mention CI as a part of LM (e.g., [4,8,9]), whereas others focus exclusively on CI (e.g., [10–12]). The existing contributions in the literature show, directly or indirectly, the potential of CI enhancement under a I4.0 context, referring to various I4.0 technological concepts to support this transformation. However, they do not propose a holistic methodology or a complete strategy for the CI shifting towards I4.0. Therefore, the present article proposes PDCA 4.0: a new conceptual approach for CI in the I4.0 environment, aiming to cover the knowledge gap found in the literature.

A thorough literature review was performed to craft a complete and reality-adapted conceptual approach. For the systematization of the review and to organize the conceptual approach, the application of CI was formalized as a project-based activity, with the following eight subsequent actions: CI's documentation management, problem identification, problem mapping, and the problem-solving sequence, with the Plan-Do-Check-Act (PDCA) cycle at its core, i.e., diagnosis, root cause analysis, countermeasures, implementation, follow-up, and standardization. To support the build-up of the approach, the work begins with the study of the roots of traditional CI practices in order to understand their purpose and identify their current limitations. Secondly, the design principles of I4.0 were studied and 10 of I4.0's technological concepts were considered for the analysis. Thirdly, conceptual, empirical, and practical approaches to the application of I4.0 technologies on LM/CI tools and methods were studied, and their potential was discussed. This analysis allowed for the identification and characterization of the challenges and limitations of conventional CI practices, based on published literature, which, analyzed together with the potential of the I4.0 concept, allowed for the statement of eleven functional requirements for the implementation of CI/LM in the Industry 4.0 paradigm.

The article concludes by proposing a new conceptual PDCA 4.0 approach, including how the technological solutions should be used and explaining the mechanisms of interaction and data management (satisfying the identified requirements). By outlining the PDCA 4.0 approach, this paper contributes to increasing the knowledge available regarding the CI realm on how to support the CI shift towards a I4.0 industrial paradigm.

2. Lean Manufacturing and Continuous Improvement

LM is rooted in the Toyota production system (TPS) [11,13]. TPS integrates a set of methods and tools with a management philosophy, aiming at the constant identification and elimination of waste [14]. TPS principles follow the logic of a house, with CI at its core [13]. Bhuiyan and Baghel [2] define CI as a culture of sustained improvement that aims at eliminating waste in all organizational systems and processes involving people. CI consists of solving problems that were previously identified. Therefore, preliminary tasks of identifying opportunities for improvement are essential to this matter. This section presents a summary of conventional CI practices and how a typical management process of CI projects works. An analysis on their challenges and limitations is the other main objective of this section.

2.1. Problem Identification and Mapping

At an early stage of a CI project, several methods can be used to enhance the identification of improvement opportunities. Using tools for key process indicators (KPI) analysis, mapping the value chain, or "simply" considering workers' suggestions are typical practices at the beginning of a CI project. KPIs are defined as a set of indicators aiming to analyze and control the process under investigation [15]. Dashboards are typically used to represent them [15,16]. The mapping activity can be performed using tools such as SIPOC (suppliers, input, process, output, customers) and value stream mapping (VSM) [17]. Problems and waste identified in this phase result in opportunities for improvement that can be prioritized through a matrix of effort vs. impact [18].

2.2. Problem Solving

With an efficient identification process, several problems will be solved in the problem-solving phase. CI has its origins in the PDCA cycle: a problem-solving method consisting of a four-step iterative cycle: Plan, Do, Check, and Act [11,19]. The PDCA cycle's logic is patent in several problem-solving methodologies, such as the eight disciplines (8D) [20] and the A3 problem solving [20–22]. This last one is a visual tool in an A3 sheet format that enhances the communication of complex problems existing in the production system and stands out for being one of the most complete tools for LM and CI-specific problems [22]. Based on [23,24], a PDCA problem-solving structure can be systematized in the following steps: diagnosis (including problem description and problem analysis), root cause analysis, countermeasures' definition (which includes the definition of the target value), implementation, follow-up, and standardization.

Thus, as referred to in Section 2.1 before assigning a problem for the problem-solving approach, two activities are necessary: the KPI analysis and the value chain mapping for the identification of potential problems. These are a crucial part of the planning phase of the PDCA cycle or culture [15,17]. One activity that is usually not mentioned in problem-solving strategies is the information management, representing the tasks of managing the PDCA projects and controlling its development, usually carried out in an *obeya* room [25]. Therefore, based on these PDCA related activities, the sequence depicted in Table 1 was used in this research to systematize the literature analysis and the proposed approach explanation.

Table 1. Typical process for managing and executing CI projects.

Phase	General	Problem Identification		Problem Solving				
Activity	Information Management	KPI analysis	Mapping	Diagnosis	Root Causes	Countermeasures	Implementation	Follow-up and Standardization
PDCA	P D C A	P	P	P	P	P	D	C A

2.3. Challenges and Limitations

An analysis about the challenges and limitations of the conventional CI practices is important in order to identify gaps that can be mitigated by the proposed PDCA 4.0 approach. A comprehensive survey was carried out through the use of the Google Scholar search engine, using the keywords "Limitations of Continuous Improvement", "Limitations of Problem Solving", "Limitations of Kaizen", and "Limitations of Lean Manufacturing". In total, 16 related publications were found. Three major aspects were identified among the selected publications regarding the challenges and limitations of CI practices (Figure 1).

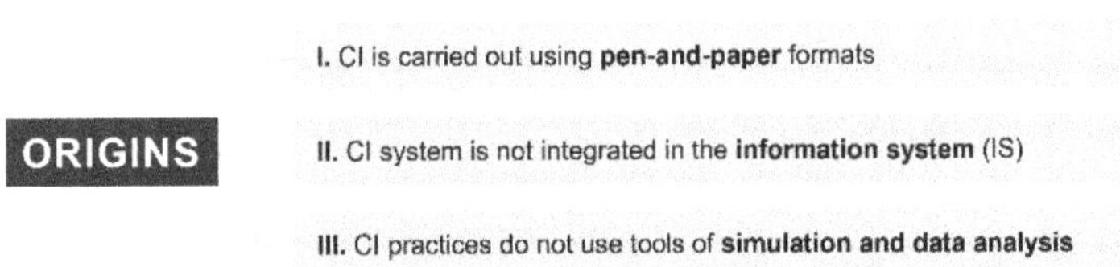

Figure 1. Typical origins of the challenges and limitations of conventional CI practices.

The typical pen-and-paper format of documentation [7,10,22] implies several constraints in CI projects and is not ideal to achieve an effective documentation management, as is further described. Despite their team building usefulness, the existence of Obeya rooms made the access to CI information by all the elements involved in the CI projects difficult [25]. Some authors describe the limitations of the traditional pen-and-paper value stream mapping (VSM) [7,17], indicating that this aspect is also relevant in the mapping activity. According to Hambach et al. [10], due to the physical format of CI documentation, the lack of aggregated and simplified information about the status of problem-solving projects may constrain the implementation and follow-up activities. In addition, pen-and-paper formats do not allow for the storage of CI information in a computer system [22], resulting in the inexistence of a database with previous problem-solving projects, which is a need addressed by some authors [10,26]. Therefore, the inexistence of an effective information system (IS) and its regular documentation management leads to not using knowledge acquired in previous CI initiatives, which might result in reworks in finding root causes and countermeasures.

The inexistence of an IS, or its lack of application for CI purposes, leads to more challenges and limitations. In addition, the absence of an automatic data collection and result analysis platform also contributes to limit and constrain the CI project's impact and assertiveness. Regarding the KPI analysis activity, production planning using ERP combined with manual Excel sheets is a conventional practice in organizations that do not support reliable and real-time data collection to the dashboards [27]. This difficulty is also felt for the VSM because it offers only a "photograph" of the system, and a small change in the real situation would change its validity [7]. Concerning the diagnosis, various authors point out that data are typically collected and analyzed manually, which is very time-consuming [11,28]. Other authors even mention the lack of access to process the data, which makes them impossible to measure, control, and improve [11,26,29]. This indicates the lack of connection between physical production objects and virtual IS [11,28]. This limitation also spans the follow-up and standardization activities [11]. Multiple authors also address the lack of an IS as a cause for an inefficient communication system that constrains the dissemination of improvements [11,30], which is relevant to the follow-up and standardization. According to Vo et al. [30], the cellular way in which traditional businesses operate, together with the absence of an IS system linked to CI, contributes to minimizing collaboration and knowledge sharing, with the best practices being contained only in their corresponding departments. As a consequence, best practices are not used for subsequent improvements [11].

Using basic data analysis tools, instead of data analytics software integrated in the IS to process Big Data, corresponds to another limitation for CI practices [11,26]. Relevant to the diagnosis, Meister et al. [26] state that advanced analytics techniques are necessary due to the increasing number of production parameters, mentioning that conventional tools, such as Excel, present limitations and are not sufficient to solve new complex problems. In fact, the problem-solving process is positively benefited by the advanced manufacturing analytics (MA) techniques, as they boost execution speed [26]. The same authors also point out the inability of basic MA practices to correlate variables and determine the root causes of a given problem. This results in the waste of time and resources by the "firefighting" approach, or simply because it takes longer to solve the problems [26]. Moreover, the existence of a limitation arising from the lack of the use of simulation and optimization techniques in order to provide assistance for improving decision-making is evident in the literature, with multiple authors addressing the combination of these methods with LM [12,31–33]. This implies a possible difficulty in the virtual testing of countermeasures before their physical implementation. Associated with the lack of these technologies in order to predict process anomalies, Rittberger et al. [11] refer to the challenge of problem prediction and prevention in advance, which is something that is not conventionally possible. Thus, the KPI analysis and mapping activities are also affected by this limitation.

Table 2 synthesizes these literature findings, where the CI activities defined in Table 1 are matched with their respective challenges and limitations. Clearly, the conventional CI practices face several obstacles, which the article in hand addresses with the development of a holistic PDCA 4.0 approach towards CI using I4.0 principles and technologies.

Table 2. Challenges and limitations of conventional practices for each CI activity.

	Challenges	Information Management	KPI Analysis	Mapping	Diagnosis	Root Causes	Countermeasures	Implementation	Follow-Up and Standardization
I	Project status information can only be consulted in *Obeya Rooms*	X		X					
	Insufficient information on the current status of problem-solving projects							X	X
	Lack of access to previous problem-solving projects	X				X	X		
II	Manual data collection and analysis		X	X	X				X
	Inefficient system for communicating best practices								X
III	Use of basic analytics tools				X	X			
	Lack of use of simulation and optimization techniques		X	X			X		

3. Industry 4.0 and Continuous Improvement

Several authors propose the use of I4.0 principles and associated technologies as a way to overcome some of the challenges and limitations of CI. In this chapter, a brief overview of the I4.0 design principles and technological concepts is given in Section 3.1. (For a more detailed analysis, refer to Appendix A) and, after that, in Section 3.2, a discussion about the existing publications on CI in the context of I4.0 is presented.

3.1. Industry 4.0 Design Principles and Technological Concepts

I4.0 denotes an unprecedented transformation in both industry flexibility and agility. Nowadays, the business world recognizes the huge opportunities for growth offered by this innovation stream [34]. Several authors point out the advantages of integrating I4.0 technologies with LM [1,4,7,12] and, more specifically, with CI [10,11,35]. In order to understand why the I4.0 trend is so important to CI projects, knowledge about I4.0 design principles and I4.0 technological concepts must first be acquired. In order to systematize the Industry 4.0 knowledge and describe its elementary constituents, Hermann et al. [36] conducted an extensive study resulting in four design principles of Industry 4.0, summarized in Figure 2.

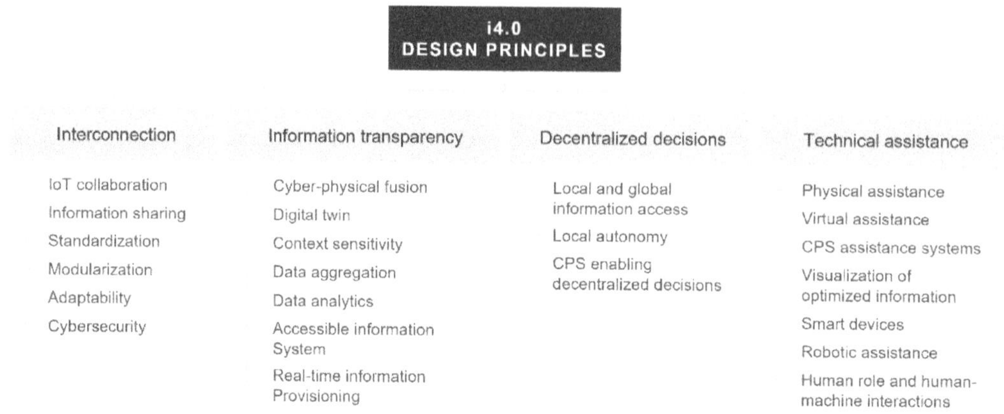

Figure 2. I4.0 design principles, adapted from [36].

Along with I4.0 transition comes the availability of new or improved technological concepts to foster the business models transition [37,38]. There is a proliferation of publications about these technologies addressing their identification, association, and relationships between [39–43]. Bibby et al. [44] summarized I4.0′s technologies into eight different technological concepts: additive manufacturing, Cloud, manufacturing executing systems, Internet of Things and cyber-physical systems, Big Data, sensors, e-value chains, and autonomous robots. This clustering is simple, objective, and well-grounded by previous work. Nevertheless, other authors propose an organization of technologies driven by the type of use [41,43,45,46], which is very useful for process-based analysis, such as the one made in this study. Based on these two approaches, an organization of the technological concepts just for the practical depiction of the proposed approach PDCA 4.0 is used (Figure 3). In total, 10 technological concepts are considered in this study: Internet of Things (IoT), cyber-physical systems (CPS), Big Data (BigData), Cloud (Cloud), sensors and actuators (Sens&Act), autonomous robotics (AutRob), simulation and virtualization (Sim&Virt), additive manufacturing (3DP), manufacturing execution systems (MES), and e-value chains (eVC). The description and justification of each technological concept is presented in Appendix A.

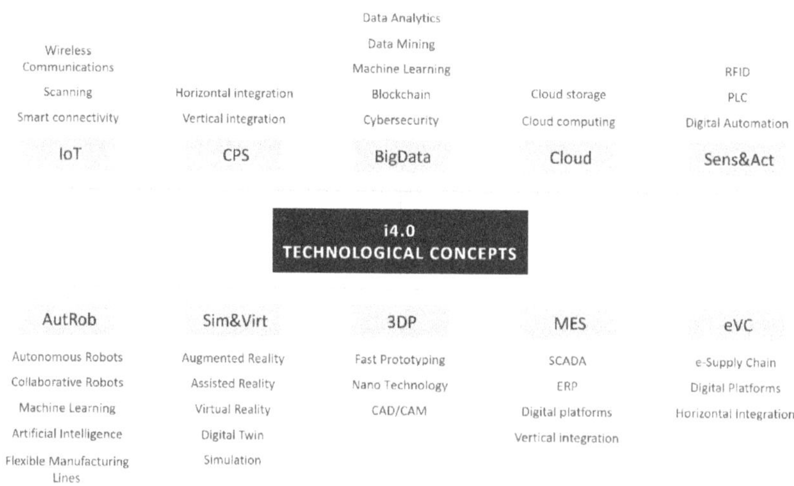

Figure 3. Coverage of each technological concept regarding other I4.0 concepts and technologies considered in this study.

3.2. Approaches towards the Applicability of I4.0 to LM/CI

There are several published documents fostering the use of LM/CI in the context of I4.0 that were considered in the design of the proposed PDCA 4.0 approach. It should be noted that there is no clear distinction between CI and LM tools, as both are a part of the same management philosophy [13,14]. In fact, LM is supported by CI practices, implying the use of lean tools. The documents were selected by using the following list of keywords in the Google Scholar search engine: "Continuous Improvement 4.0", "Kaizen 4.0", "Lean Manufacturing 4.0", "Industry 4.0 Continuous Improvement", "Industry 4.0 Kaizen", and "Industry 4.0 Lean". In total, 47 documents were found; those that contribute to this area of knowledge are discussed in this section. Three main types of approaches were identified: (i) conceptual approaches that study the applicability and impact of I4.0 towards LM/CI; (ii) empirical approaches that are based on perceptions extracted from the industry regarding the same subject; (iii) and practical approaches in which the application of I4.0's technologies to LM/CI is demonstrated through use cases.

3.2.1. Conceptual Approaches

Among the 23 publications on the conceptual application of I4.0 technologies to LM/CI found in the literature, 19 of them do not mention their applicability towards CI in a direct and structured way. After analyzing the content of these 23 papers, the findings were assigned according to their relevance to each of the CI activities. As an example, Sim&Virt, through augmented/assisted reality (AR/AsR), enables real-time remote support in manual operations [8,12], relevant to the documentation management, implementation, and follow-up and standardization activities. The result is exposed in Tables A1 and A2 (Appendix B).

From this study, it can be stated that IoT and Cloud have the potential to cover all CI activities, with IoT having the main role of enabling data transmission and access [9,47], and Cloud the sharing of information [9,12] and cloud-based data storage [12,48], as well as cloud computing capabilities [35,48,49]. Sim&Virt is also quite overarching, allowing the use of AR/AsR to aid in manual operations [8,12] and observe the current state of a process [11], and the use of virtual reality (VR) to facilitate training [8]. Additionally, simulation technologies can be used to test countermeasures before their real life implementation [12,31,50]. For this, a factory digital twin coupled with simulations can also be used for CI [51]. Big Data also possess a high degree of applicability, associated with data analytics [9,12], predictive analysis [11,12,52,53], data mining [53], correlation analysis [47], root cause analysis [11,53], and machine learning [11,43,53], as well as advanced analytics for planning [8]. CPS, on the other hand, has relevancies to the collection and access to real-time data [9,54], and can predict machine failures [52,55]. AutRob, besides improving manufacturing flexibility and productivity [9] through the automation of routine tasks [11], also makes the automatic detection of machine failures possible [1,9], as well as automated logistics systems [8,9,12]. It can also be used in collaboration with operators [9]. Sens&Act is a fundamental technological concept, as it is used to collect production data, including machine performance and object location [12]. MES is mainly used to collect data, as well as to display KPIs and data charts [56,57]. On the other hand, eVC allows for the connectivity between stakeholders of the value chain and the information exchange along the supply chain, with on-demand access to value chain information through digital platforms [43]. 3DP is essentially connected with mass customization [12] and smart product development [9], permitting the test of product designs.

3.2.2. Empirical Approaches

The empirical approaches to the application of I4.0 technologies to LM/CI are carried out mainly through surveys for data collection regarding perceptions extracted from the industry. Some of these insights are listed in Table A3 (Appendix C). It can be stated that there are difficulties associated with I4.0's concepts; namely the lack of knowledge about their impacts, as well as high cost factors [58–60]. A simultaneous approach for the adoption of I4.0′s technologies and LM (and CI) is needed [58,59,61]. Furthermore, several benefits of I4.0′s integration with LM are indicated [61].

Regarding the greater synergies between LM and I4.0 technologies, cloud computing/machine learning are related with waste prevention/increased productivity, Big Data are interconnected with the concept of zero defects, and AR/VR with visual management [59,60]. Tortorella et al. [62] studied the correlations between LM principles and I4.0 technologies. As examples, for the principle of "digitally controlled processes", digital sensors/interfaces and the remote control of production are considered as facilitators for the identification of abnormal product/process conditions [62]. The authors also mention technologies such as AR/VR with low correlation values, presenting a low adoption level. However, they state that these applications cannot be disregarded, as the topic of I4.0 is recent, and new relationships may arise as the manufacturer's awareness escalates [62]. Dombrowski et al. [59] cover CI-related interdependencies, concluding that CI has the biggest correlation with Big Data, followed by cloud computing, RFID/identification, and sensors/actuators. Some additional references are made towards AR/VR, automated guided vehicles, and smart glasses [59]. Regarding technologies with greater consideration for implementation, IoT, big data analytics, and cloud computing received greater consideration for implementation, followed by additive manufacturing, AsR/AR, and robotics [61].

3.2.3. Practical Approaches

In the literature, multiple practical approaches were also found, in which the application of I4.0′s technologies toward LM/CI are demonstrated through use cases. The corollary of this study is presented through Tables A4 and A5 (Appendix D). This table summarizes various use cases of I4.0′s technologies, connecting them to the different CI activities in which they are relevant. As an example, the use case referring to a data collection system for an actual machine using sensors and a CPS, allowing for the real-time visualization of KPIs [63], can be considered as more relevant to KPI analysis and the diagnostic activities.

Only one publication [30] references a more complete description of a CI process, including most of CI's activities. The authors mention the use of web-based monitoring tools to collect data, MES to monitor manufacturing processes in real-time, and root cause analysis with the aid of digital boards and 3D printing to test product designs [30]. Other use cases cover fewer CI stages, such as a data collection system for an actual machine through the use of sensors and a CPS, allowing for the real-time visualization of KPIs [63], a Big Data tool stack that processes a high volume of data, feeding predictive models based on machine learning, and a descriptive analysis module that, through graphs, aids in accessing recent problems and their root causes [64]. An approach that is more closely related to problem identification through mapping refers to an RFID (Sens&Act)-based system that can collect data, such as the quantity of items in a given place and cycle times [65], having the potential to be adapted for a real-time VSM [66]. Additionally, a value stream analysis based on a Big Data model [67] and the combination of VSM with a simulation [68] are described in the literature. More associated with the development and testing of countermeasures, a publication mentions scheduling solutions based on real-time simulations [69]. Regarding implementation activities, multiple publications report developments in task organization through Kanban boards; namely a web-based board [70], a board operated with a smartphone [71], and a computer-aided task board that tracks a physical panel in real-time [72]. Other use cases, such as a highly flexible measurement-aided welding [73] and a safe human–robot collaborative assembly cell based on a CPS [74], were also found. Lastly, some use cases of AR/AsR (Sim&Virt) were deemed relevant for follow-up and standardization regarding the production [1,75], maintenance [76], and quality control [77] activities, while also facilitating the management of documents regarding work instructions.

Many of the practical approaches that were found are specifically related to one LM tool, covering different CI activities. Use cases related to TPM, namely an online root cause analysis [78] and condition monitoring for machines [8,79], essentially cover the KPI analysis, diagnosis, and root cause stages. Other studies describing the incorporation of CPS in *Jidoka* [80,81] are also associated with these activities in a similar fashion.

3.2.4. Discussion

In summary, a body of knowledge exists regarding the application of I4.0′s technologies to CI, although many of the studied approaches relate to CI in an indirect way. In fact, most of the conceptual approaches found in the literature directly reference LM tools, which means that their functionalities were posteriorly analyzed in terms of their applicability to CI, which is a more specific practice of the same management philosophy. Regarding practical approaches, SMED is only briefly mentioned within this subject, and only a few use cases relating to mapping were found, although the only suitable use case towards VSM is not purposefully applied for that method. Few approaches refer to documentation management and improvement implementation activities, although the technologies to do so are readily available. In terms of general CI practices, the existing knowledge regarding conceptual, empirical, and practical approaches is very limited, as only a few publications that directly study this subject were found. Nonetheless, they do not offer comprehensive and extensive maps of the applicability of I4.0 towards typical CI processes. These findings further confirm the need for a more holistic and direct approach towards CI.

4. Functional Requirements for PDCA 4.0

Based on the compilation of several conceptual, empirical, and practical approaches regarding the application of I4.0 technologies to LM/CI (Tables A1–A5) and the challenges and limitations of conventional CI practices (Table 1 of Section 2), the functional requirements of a new conceptual PDCA 4.0 approach are defined. In total, 11 requirements are stated. Each one of them is briefly presented and substantiated with examples taken from the three types of approaches mentioned above. Table 3 presents the match of the technological concepts with each of the 11 requirements for PDCA 4.0.

Table 3. Technological concepts for PDCA 4.0′s functional requirements.

		PDCA 4.0′S Functional Requirements										
		R1	**R2**	**R3**	**R4**	**R5**	**R6**	**R7**	**R8**	**R9**	**R10**	**R11**
I4.0 Technological concepts	IoT	X	X	X	X	X	X	X	X	X	X	X
	CPS			X	X						X	
	Big Data	X	X	X		X	X	X				
	Cloud	X	X	X	X	X	X	X	X	X		X
	Sens&Act	X										
	AutRob										X	
	Sim&Virt	X			X	X	X			X		
	3DP											X
	MES	X			X			X				
	eVC	X			X							

4.1. *Automatic Data Collection System (R1)*

As conventional practices depend heavily on the manual collection of data, which implies serious limitations in the process, an automatic data collection system is proposed for the CI team in order to have readily available data. This requirement is patent in conceptual approaches that mention that such systems are enabled by the application of Sens&Act to collect data [82], as well as IoT for the intelligent monitoring of the production and supply chain management functions [47]. Real-time data collection from the production system is also enabled by using MES [56,57], this being the primary reason for its deployment [57].

Regarding data generated outside of the factory, eVC can increase connectivity and allow information exchange along the supply chain [43,45]. After further refinement, the collected data can be stored in Cloud databases that constitute the heart of a factory digital twin [51]. Lastly, the concept of Big Data, as it is associated with the gathering of data from sensor readings [43], and because it is needed in order to establish a digital twin [43], is also associated with this requirement. Based on an empirical study, the use of digital sensors/interfaces facilitates the identification of abnormal product/operating conditions [62]. Additionally, IoT can be used as a supporting mechanism to interconnect products and processes [62]. Finally, this requirement is also present in practical use cases, such as a sensor-based data collection system that measures different machine parameters [63] and data collection by MES to support root cause problem solving [30].

4.2. Advanced Analysis Tool (R2)

When data are collected automatically from sensors, an advanced analysis tool capable of handling a high volume of data is required. This way, it is possible to use advanced data analytics tools in an automatic fashion, avoiding Excel sheets and the limitations of simpler analysis tools. This requirement is evident in conceptual approaches that point out that the concept of Big Data, and therefore big data analytics, can process a high volume of data into information that can be used to improve the system's performance [8,12]. This technology can be used to perform simple data analysis [9,12] as well as advanced data analytics, such as machine learning [11,43,53], data mining [43,53], root cause analysis [11,53], correlation analysis [47], and predictive analysis [11,12,52,53]. As an example, Rittberger et al. [11] suggest that a machine learning algorithm can be used in conjunction with a problem-solving database in order to infer cause–problem relationships. On a support point of view, other authors mention that IoT enables data transmission from machines to end user software [9], which is pertinent to this requirement. The Cloud concept is also relevant, as it can provide cloud computing capacity for data analytics [35,42,83] as well as enable data transfer from cloud storage to analysis tools [12]. Empirically, a study mentions that Big Data, namely big data analytics, and IoT, were given higher consideration for implementation [61]. Other authors present Big Data as the most interdependent technological concept towards CI [59]. Regarding practical approaches, as described earlier, a Big Data tool stack that allows data to feed predictive and descriptive analysis modules already exists [64], as well as a value stream analysis based on a Big Data model [67].

4.3. Problem Prediction System (R3)

A system that predicts problematic situations before they happen is required to anticipate problems, which is something that is not possible in conventional practices. This requirement is present in conceptual approaches that mention the potential of Big Data, namely big data analytics, for helping employees to determine cause-and-effect correlations, as well as trends to predict problems that are occurring in a process [11]. In a more specific approach towards machines, CPS architectures with embedded analytics (machine learning) can be applied to monitor, predict, and diagnose machine failures [52,55]. IoT is a relevant support technology for this requirement, as it enables real-time operation/machine monitoring [47,55]. Multiple practical use cases that demonstrate this requirement also exist, such as CPS-based systems with the ability to predict equipment failure [8,80,81,84], aided by machine learning algorithms and cloud computing [8,84].

4.4. Real-Time Visualization System to Consult Production Data (R4)

A real-time visualization system is needed so that relevant information is displayed in real-time, enabling the user to have knowledge about the current state of the production system, and avoiding decision making based on obsolete data. This requirement is present in conceptual studies that indicate the role of MES to determine KPIs, create reports, and provide user interfaces to visualize and manage shop floor operations [56,57]. Data from the value chain can also be visualized through digital platforms, enabled by eVC [43]. A digital twin, integrating different types of data from the manufacturing site and recreating a production line in a digital space, can also be used as the basis for the display of production and product information [51]. Other authors refer to CPS as useful for real-time data collection, which allows for effective KPI monitoring [9]. Lastly, AR/AsR devices (Sim&Virt) can also be used to display relevant process information, aiding users in problem-solving actions [12]. From a support perspective, IoT may be applied in order to provide real-time visualizations of information [47], and Cloud may be applied in order to access the collected data [9,12]. This requirement is further present in practical use cases; namely a data collection system that uses a CPS in order to monitor and visualize KPIs in real-time [63], and a RFID (Sens&Act)-based system used to collect data from the shop floor and present process data, with the potential to be integrated with VSM [65].

4.5. System That Analyzes Countermeasures' Impact before Their Implementation (R5)

The analysis of the countermeasures' impact before their real-life implementation is also defined as a requirement for PDCA 4.0, as it can aid in developing viable and feasible countermeasures towards existing problems. Conceptually, this requirement is evidenced as the future of modelling, and the simulation will allow for the creation of near to real-time models with low building cycles, providing a tool for decision making and semi-autonomous problem solving [31]. Therefore, when connected to Big Data sources (stored in the Cloud), this technology can be used to test improvements to the production system, evaluating their impact in a virtual environment [12,31]. In this context, a digital twin of the manufacturing site that uses a simulation can be used towards CI [51]. This requirement can be further supported by on-demand cloud computing resources that allow high-speed simulation analytics [85], as well as IoT to transmit data between machines and sensors to software tools [9]. Regarding practical use cases, this requirement is present in scheduling solutions based on both real-time simulations [69] and the combination of VSM with simulation models in order to validate current and future states, aiding in decision-making processes [68].

4.6. System That Prioritizes CI Projects (R6)

The prioritization of CI projects, constituting a hierarchy of problems to be solved, is also important because it guarantees the awareness of the impact vs. the effort of a CI project, allowing for the implementation of the most effective improvements. This requirement is justified by the difficulty in establishing consensus regarding the starting points towards process improvement [86]. Furthermore, the lack of resource availability for CI [87] reinforces the need to focus efforts correctly. I4.0 can play a role in the scope of this requirement, as simulations can be used to analyze the possible impact of countermeasures [12,31,51]. Big Data (big data analytics) has the potential to improve upon conceptual methodologies for the prioritization of projects and the allocation of resources, such as the one proposed by Allan et al. [86]. Similarly to the previous requirement, cloud computing and IoT can be used to support this need.

4.7. Dynamic Planning of Improvement Activities with Alarmistic (R7)

In order to assist the implementation process of CI projects, a dynamic planning tool that displays the activities to be carried out, those that are being carried out, and those already carried out is needed. Regarding conceptual approaches, Krishnaiyer et al. [49] propose a cloud Kanban framework in order to monitor and control the resource consumption and production of an enterprise. In the scope of planning, Mayr et al. [8] mention advanced data analytics towards planning. For this, MES also plays an important role, as it supports advanced production planning (Gantt charts, for example) and resource allocation [56,57]. This requirement is also present in practical case studies that mention the advantages of digital Kanban boards in terms of its flexibility for visualizing assigned tasks [70–72]. As examples, Nakazawa et al. [71] developed a digital Kanban board that is controlled by a smartphone, and Bacea et al. [72] developed a virtual Kanban board that tracks physical task cards through a camera. These use cases demonstrate the importance of device interconnectivity, relating to the IoT technological concept.

4.8. Digital Support for CI Documentation (R8)

As it enables easy access to information regardless of the users' location, the digital support for CI documentation is also a requirement for PDCA 4.0, along with a database for its storage and future access. This requirement is patent in the conceptual approach by Hambach et al. [10], in which the authors mention the advantages of a digital CI system in terms of data storage and access through document management systems, digital communication that is independent of space and time, and data visualization. For this, cloud-based data storage is essential in order to allow for more effective data sharing within departments [48], a concept that is aligned with the life cycle of technical documentation in the context of I4.0 [88]. Data exchange through IoT is therefore also fundamental for this requirement, being supported by authors who argue that technical documentation should be connected to IoT in order to facilitate communication between machines and humans [89]. Related to this requirement, a database for CI documentation storage is necessary. This way, through a library search, knowledge from past situations (state of the project, diagnostic, root causes, countermeasures, and their degree of success) can be accessed and used for present CI projects. This is present in conceptual approaches that mention the need for problem-solving databases to aid future CI projects [10,26]. In fact, Rittberger et al. [11] propose that successful improvement initiatives should be filed as standards in a database that contains problem information; namely its root causes and countermeasures. Once again, this is supported through the use of Cloud data storage [12,48], as well as IoT, to ensure data exchange between different devices [9], making documentation accessible everywhere.

4.9. Organizationally Transversal System for Consulting Best Practices (R9)

A transversal system for consulting best practices is defined as a requirement for PDCA 4.0, as it will enable knowledge sharing across the company regardless of the user's location, avoiding information retention in organizational silos, and facilitating the standardization of new countermeasures. This requirement is present in conceptual approaches that mention that although AR/AsR (Sim&Virt) enables users to access real-time information about process data [11], it can also be used to remotely support employees in performing manual operations, such as maintenance or production tasks [8,12]. This way, work instructions can be shared with the operator [8], guaranteeing that best practices are shared across the organization, facilitating their standardization. In a similar context, VR can also facilitate the employees' training [8]. According to the literature, the use of these technologies is supported by Cloud and IoT [90,91]. Empirically, AR and VR are considered as interdependent with CI, although to a lesser extent when compared with visual management [59]. This requirement is further demonstrated in practical use cases. Several publications cover the application of these technologies through use cases in the context of production, maintenance, and quality control activities [1,75–77]. As an example, Kolberg et al. [1] mention the use of AR on manual workstations for the identification of both tasks and relevant information.

4.10. Automatic and Intelligent Work System (R10)

An automatic and intelligent work system is also defined as a requirement for PDCA 4.0, as it can ensure a more efficient process of implementing countermeasures regarding new machine parameters and work sequences, guaranteeing their immediate standardization. This requirement is justified in conceptual publications, as one study mentions that employees may lack the capacity to implement countermeasures during the implementation ("Do") phase, suggesting the automation of routine tasks [11]. In this scope, authors argue that autonomous robots both increase manufacturing flexibility [9], easily adapting to changes and thus allowing for automated logistics systems [9,12], and communicate failures automatically, calling other systems for fault-repair actions [1]. In fact, the connectivity achieved through IoT enables a new level of automation [92]. From a practical point of view, this requirement is demonstrated through a use case by Tuominen [73], who describes a measurement-aided welding cell that is able to adjust itself in order to produce different products, and also has the ability to inspect weld beads. Nikolakis et al. [74] describe a CPS system for enabling human–robot collaboration based on a safety distance evaluation.

4.11. Rapid Prototyping System (R11)

A rapid prototyping system is specified as a requirement, as this will enable a faster process of testing countermeasures. This requirement is justified, as conceptual studies mention the additive manufacturing's ability to produce customized products, being a highly flexible manufacturing process [12], potentiating smart product development processes [9]. In this context, IoT is useful to directly connect printers to the Cloud, enabling their remote control and monitoring [73]. Regarding practical use cases, one publication in the context of root cause problem solving describes the utilization of 3D printing to test improved product solutions [30].

In a final analysis, observing Table 3 it can be stated that IoT, Cloud, and Big Data are the most wide-ranging technological concepts regarding the 11 requirements for PDCA 4.0. This result is aligned with the base technologies for I4.0 identified by Frank et al. [43]. In fact, IoT is relevant to all the requirements, being the medium by which connectivity is achieved between various objects and software. Cloud, with cloud-based services such as data storage and computing power, is also applicable to the majority of the requirements. Big Data, referring to advanced data analytics, is pertinent to the PDCA 4.0′s functionalities that make use of such technologies. Sim&Virt is also a very ample concept, supporting decision making through simulation and optimization techniques, and also virtualizing the physical documentation and the physical processes that take place in the shopfloor. Sens&Act, despite directly covering only one requirement, is a base technological concept to PDCA 4.0, as it is the means through which data are collected, feeding the other technological concepts as well as most of the CI activities. The least overarching technology corresponds to 3DP and AutRob, each matching to a specific requirement.

In the following section, the PDCA 4.0 approach is proposed based on satisfying these 11 requirements, where the match between the requirements and the CI activities is carried out as the approach design is specified.

5. PDCA 4.0′s Framework Dynamics

As previously stated in Section 3.2.4, a more holistic and direct approach towards the application of I4.0 to CI is needed. Supported by both the previous analyses of the applicability of I4.0′s technological concepts and the definition of the requirements for a new CI approach (matching visible in Figure 4), the vision for PDCA 4.0 is defined. Figure 5 visually represents its framework.

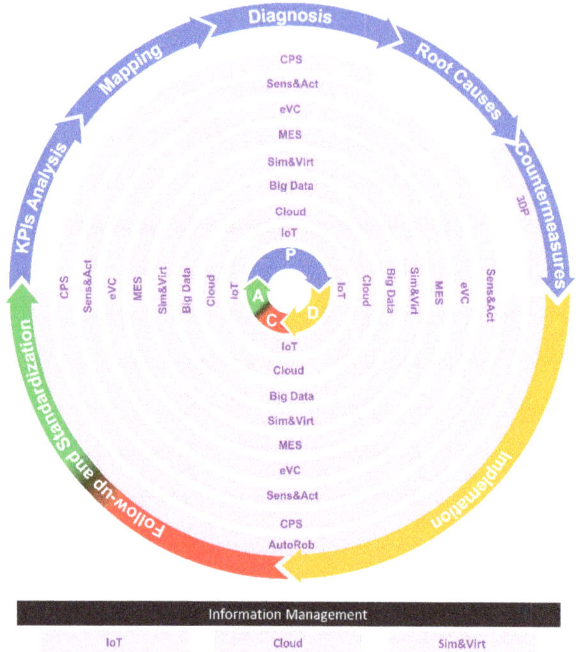

Figure 4. I4.0 technological concepts for the PDCA 4.0 methodology.

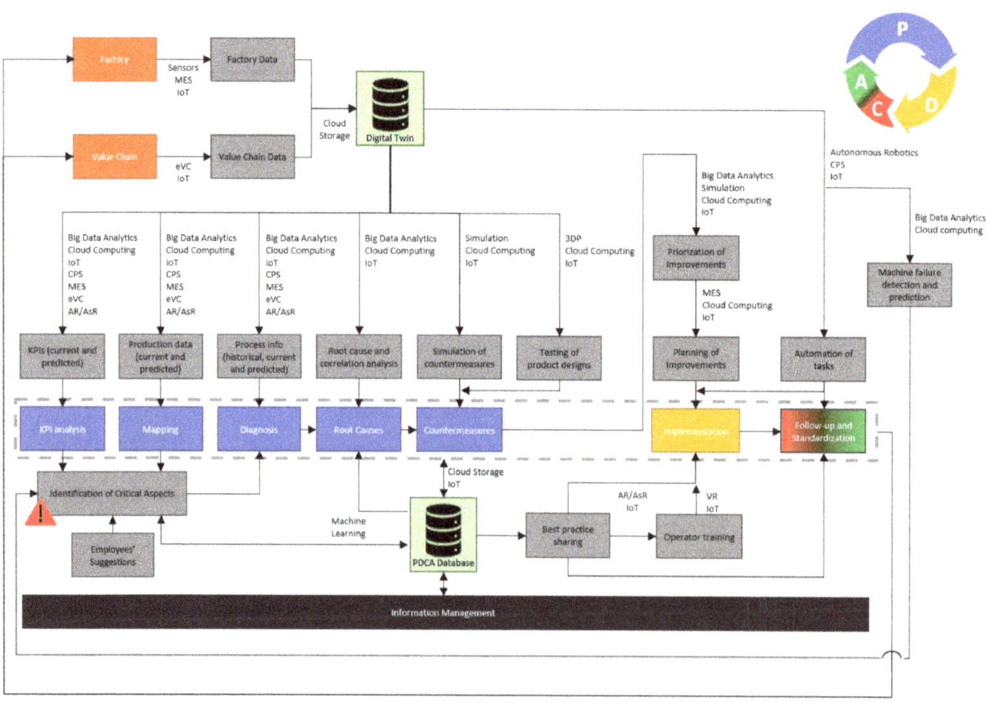

Figure 5. PDCA 4.0 framework.

In the beginning of the CI process, namely the activities of KPI analysis, mapping, and diagnosis, Big Data collected automatically from the production system (R1) by Sens&Act and MES systems are stored in Cloud databases (Cloud), which also contain data regarding the whole value chain via eVC. The various types of data can then be integrated, forming the heart of a factory digital twin (Sim&Virt) that represents a virtual model of the physical world. The role of IoT is the same throughout this framework, ensuring connectivity and data transmission between machines and software. With these technologies, data coming from the shop floor, as well as from the value chain, can be used by CI teams for data-driven PDCA, avoiding situations where data are insufficient, obsolete, or inexistent.

The stored data can then be processed by advanced big data analytics (Big Data) (R2), generating information regarding KPI analysis and mapping activities, through which the current state of the production system is known. A prediction of the future state may also be possible via predictive analytics (Big Data) (R3). For analytics (Big Data) and simulation services (Sim&Virt), cloud computing (Cloud) can be a supporting technology for providing the necessary computing ability. The information can be displayed in real-time to employees (to assist those responsible for the improvement to make decisions) through CPS, MES interfaces, eVC digital platforms, and AR/AsR devices (Sim&Virt) (R4), with its goal being the identification of negative critical aspects existing (or that may exist) in the organization. More specifically towards equipment, machine fault detection and prediction is possible through AutRob and CPS with embedded machine learning algorithms (R10). The PDCA database (R8) allows access to digitalized information regarding already identified problems, supporting the process of finding critical parameters and factors. This can also correspond to the employee's suggestions that are stored in the database (Cloud). All of these advanced data processing systems must have a user-friendly and simple-to-use interface, where the user (who is not an expert in data analytics) simply has to ask for a correlation analysis to the interface, and receives the output of several data analytics algorithms in the form of degree of influence and/or level of correlation (allowing an informed root cause analysis).

When the critical aspect is identified, an automatic diagnosis can be achieved via an advanced data analysis tool (Big Data) (R2) that retrieves historical and current data from the factory digital twin (Sim&Virt) in order to construct a diagnosis regarding the critical aspect. Predictions for the future state (Sim&Virt) (R3) can also be used in order to give emphasis to the hindered performance of the production system. This type of information could also be accessed and visualized through the same technologies stated in the previous paragraph (R4). Additionally, for the next stage of the CI process, advanced big data analytics, such as correlation analysis and root cause analysis, or a machine learning algorithm that infers cause–effect relationships based on a historical problem-solving database (R8), can be used to help the CI team determine the root causes of a problem. These functionalities ensure that manual, simple, and inefficient analysis tools are avoided, allowing the CI team to save time in carrying out these tasks. In addition, they make sure that the real source of the identified problems is addressed, avoiding superficial quick fixes.

For the next activity, the development of countermeasures, simulation models with low building cycles (Sim&Virt), can be used to predict the effect of improvement solutions in a virtual factory environment through software (R5), as well as 3DP, in order to test solutions regarding product designs (R11). These technologies aid the decision-making process, allowing the CI team to find the most effective and feasible solutions to problems, all while offering a low-risk and faster way to test changes to the system/product.

The results of the simulations (Sim&Virt) can also be used to further better the decision-making process by allowing to prioritize CI projects according to their impact and perceived effort (R6), ensuring a more effective implementation of countermeasures. Big Data analysis techniques can also be used in conjunction with the results of simulations, enabling the automatic hierarchization of projects (R6).

Regarding the planning of activities, MES can be used to support the planning process through Gantt charts and resource allocation information. Additionally, cloud Kanban boards (Cloud) can be used to enable remote access to assigned tasks (R7). A system to consult best practices (Sim&Virt) (R9) further enhances the implementation of countermeasures by giving employees access to important information regarding work procedures, aiding them in performing new tasks. VR (Sim&Virt), namely AsR, can also help in this regard, as it facilitates employee training in a virtual environment. The automation of tasks (AutRob) (R10) is also pertinent to the implementation stage, aiding people in implementing the developed countermeasures, which accelerates the process. This implies an automatic and intelligent work system that easily adapts to changes in manufacturing (R10).

For the follow-up and standardization activities, data collected from the production system (R1), regarding the new system performance, can once again be used to identify critical aspects, allowing to evaluate whether or not the improvement initiatives were successful and if new problems arose. This constitutes a CI cycle over which successive problem-solving needs are identified and undertaken. More related to standardization, best practice sharing through AR/AsR devices (Sim&Virt) (R9) facilitates the normalization of improvement solutions through the remote access to documentation such as work instructions by the employees in the shopfloor. Furthermore, autonomous and collaborative robotics (AutRob) allow for a more flexible and intelligent work system (R10), easily adapting to changes. This potentiates the immediate standardization of new work procedures and work parameters, project impact and effort, and thus its hierarchization, as well as implementation planning information.

Lastly, information management is essential to the success of the improvement. Contrary to the usual pen-and-paper practices, having all the information digitized is crucial to store it in a database (Cloud) (R9) that enables everyone involved in the project to easily access important data.

Regarding the connection of the technological concepts with the CI activities, Figure 4 highlights the embracing way that they are interconnected. It is possible to see that most technologies relate to all CI activities by analyzing the large amount of complete circular crowns around the center. Only some technological concepts are less overarching. AutRob is present only in the check and act phases, being important for the implementation, follow-up, and standardization stages. CPS is more embracing but does not contribute to root cause identification and the development of countermeasures. Finally, 3DP has a more specific use, only relating to the activity of countermeasure development. In this figure, it can be seen that IoT, Cloud, and Sim&Virt are also important for documentation management.

Regarding I4.0's design principles, this framework allows to achieve the principle of interconnection through IoT-enabled data sharing. This way, information availability and exchange are ensured, which is crucial for a data-based CI approach. Information transparency is also guaranteed through systematic Big Data collection, forming the heart of a factory digital twin. This data are then used in conjunction with real-time display systems and several advanced analysis tools, such as data analytics and simulations for the relevant stages of a CI project. Decentralized decisions are also promoted in this framework, as it implies the use of technologies that support the decision-making process, helping to establish both root causes of critical aspects and the estimation of the countermeasure's impact on the production system/product. Thus, this solution allows for the access to information in order to support the different actors of a CI project in decision making. Finally, technical assistance is also assured through technologies that enable the remote guidance of operators during maintenance, production, or quality control activities, displaying relevant information and best practices for those tasks. This is also apparent in the machine's ability to automatically detect failures.

6. Conclusions

Several publications regarding conceptual, empirical, and practical studies on the application of I4.0's technologies towards LM and CI were analyzed, with their findings being organized according to the stages that constitute a typical CI project. It was found that the I4.0′s technologies mentioned in the literature have the potential to enhance a data-driven and more intelligent approach towards CI. It was also found that a holistic and structured framework towards an approach that combines I4.0 and CI was needed, as the vast majority of the authors focus solely on single aspects that are relevant to the subject. Thus, 11 requirements for the proposed approach were established based on functionalities or needs that are identified in the literature, as well as the relevant technologies for their materialization. These requirements were then worked in order to constitute a framework for PDCA 4.0. This framework uses I4.0′s technological concepts to collect production data and constitute a factory digital twin, enabling the identification of critical aspects related to the system's performance through KPI analysis and mapping. These critical aspects are then diagnosed through intelligent data analysis and visualization tools, also enabling the determination of their root causes. Countermeasures can then be tested through simulations and/or through prototypes of the product design. This framework also aids the implementation of the improvements through the prioritization of projects, as well as their actual planning, and their follow-up and standardization activities. Another key component of this framework is related to information management, in which documentation needs to exist in a digital format in order to be stored in a PDCA database, making it accessible to everyone, everywhere. In summary, the application of I4.0 to CI contributes to a faster, more transparent, and efficient data-driven process, allowing to surpass traditional barriers related to such projects.

As future work, the authors are already implementing the PDCA 4.0 approach in industrial companies, allowing to publish and discuss the impact, pros, and cons of the proposed approach. Another direction of research is to understand how aligned PDCA is with the concept of Industry 5.0. A deep and robust understanding of what Industry 5.0 is must be carried out prior to this analysis, but one can say that PDCA 4.0 is highly centered in humans (supporting human-based decisions and enriching humans' activities) and also fosters the sustainability performance (the application of CI has a high impact on resources efficiency, people's motivation, and the improvement of working conditions).

Author Contributions: Conceptualization, P.P. and D.J.; methodology, P.P., D.J., J.E. and M.G.; validation, J.E. and M.G.; formal analysis, M.S., J.E and M.G.; investigation, P.P. and M.S.; resources, P.P.; data curation, J.E. and M.G.; writing—original draft preparation, M.S., J.E. and M.G.; writing—review and editing, P.P.; visualization, J.E. and M.G.; supervision, P.P.; project administration, P.P.; funding acquisition, P.P. All authors have read and agreed to the published version of the manuscript.

Funding: This work was supported by: FCT, through IDMEC, under LAETA, project UIDB/50022/2020; the European Regional Development Fund (FEDER) through a grant of the Operational Programme for Competitivity and Internationalization of Portugal 2020 Partnership Agreement (PRODUTECH4S&C, POCI-01-0247-FEDER-046102) and (PRODUTECHSIF, POCI-01-0247-FEDER-024541).

Institutional Review Board Statement: Not applicable.

Informed Consent Statement: Not applicable.

Data Availability Statement: Not applicable.

Conflicts of Interest: The authors declare no conflict of interest.

Appendix A. Description of the I4.0 Technological Concepts Considered in This Study

Internet of Things [IoT]—The technological concept "IoT" refers to a network characterized by physical devices capable of connecting to the wireless internet. The base components are made up of built-in electronics, such as sensors and transmission hardware. It allows for the rapid generation of data, leveraging the information flow within an organization through physical devices capable of interacting with each other and with control systems, through a network infrastructure [42,43,93].

Cyber-physical Systems [CPS]—According to several authors [36,94,95] "IoT and CPS" should be divided into two different concepts, differing from the categorisation presented by Bibby et al. [44]. The technological concept "CPS" refers to systems capable of combining computational modelling, statistical data and physical data in real-time. The physical devices (machines or production lines) and the physical processes are digitized, creating a digital system identical to the physical system [43].

Big Data [BigData]—Some authors, such as Frank et al. [43], separate Big Data from data analytics. However, both concepts are interdependent, and that is why many authors [40,41,44] choose not to separate them. The technological concept "Big Data" refers to this large data volume [41] and to the technologies of its collection, processing, provisioning, and analysis [44]. This concept is necessary in order to generate factory digital twins [43]. Data analysis refers to data mining, machine learning [43], statistical analysis, and predictive analysis, among others [40]. Regarding cybersecurity, some authors consider it as a separate category [12]. However, there are approaches who do not find it reasonable to approach this concept in an elementary way [40,44]. Following this last choice, and since cybersecurity technologies are associated with data and information, they are now included in the technological concept of Big Data.

Cloud [Cloud]—The technological concept "Cloud" is unanimously adopted by other authors, such as [40], with similar descriptions and associated technologies. This concept refers to any IT services provisioned and accessible from a cloud computing provider [41]. It consists of three IT combinations: internet services, web-based applications, and information management [44].

Sensors and Actuators [Sens&Act]—The logic behind this clustering follows the work by Bai et al. [41], which is entitled "Sensors and Actuators" instead of simply "Sensors". Additionally, Bibby et al. [44] do not address actuators in their article, while other authors [41,43] argue that it is an important topic to be included. This technological concept "Sens&Act" includes basic technologies for the digitalization of objects and physical parameters [43]. It includes all devices that respond to a physical stimulus and transmit a resulting impulse [41]. This concept is also associated with RFID (radio frequency identification) [41].

Autonomous Robotics [AutRob]—This technological concept includes autonomous, collaborative, and intelligent robots and equipment, with embedded sensors, dexterity, artificial intelligence, and machine learning [44]. As there is no common line in the literature about the place of artificial intelligence and machine learning, the present classification follows the same choice of Bibby et al. [44] of having just one technological concept (AutRob) where all of these associated technologies are included.

Simulation and Virtualization [Sim&Virt]—This technological concept includes virtual tools that provide support to the decision-making process. In this scope, assisted reality (AsR), augmented reality (AR), and virtual reality (VR) are emerging technologies that create partial and complete virtual environments, capable of enhancing tasks and speeding up training [43]. Bibby et al. [44] do not individually address the virtualization technologies, namely VR and AR, which is something that happens in several later works [43,45,46]. This indicates the relevance of including a category for these technologies. The literature also suggests that both the simulation and digital twin have an impact on LM practices [8,9]. Ito and Ishida [51] defined the concept of a digital twin as a replica of the real world created in a digital space, bringing together different types of data on site. The replica can be used with artificial intelligence or simulations in order to help to make improvements in the context [51]. Therefore, their inclusion in this technological concept is adequate.

Additive Manufacturing [3DP]—This technological concept is unanimously adopted by other authors, such as [40,41,43], with similar descriptions and associated technologies. Additive manufacturing (AM) or 3D printing (3DP) is a process by which products are produced autonomously, layer by layer [44]. It consists of versatile machines and flexible production systems that transform 3D digital models into physical products [40]. This technology is used especially for rapid prototyping and the creation of custom-made tools [44].

Manufacturing Execution Systems [MES]—MES is a useful tool for organizations that require the accurate traceability of parts, components, and assembly activities to monitor quality, cost, and lead times. The purpose of MES is to initiate, guide, respond, and report shop floor activities as they occur [96]. MES also plays a key role in the central distribution of information [44]. Regarding ERP systems, Bibby et al. [44] include an exclusive classification group: MES is an I4.0 technological concept, and the rest (ERP, SCADA) already existed before I4.0 began. However, other authors define more complete technological categories, integrating MES/SCADA [40] or MES/SCADA/ERP [43]. All things considered, in the present classification approach, SCADA and ERP are directly associated to MES.

e-Value Chains [eVC]—Supported by the digitalization of value chain activities [44], this technological concept consists of collaborative digital platforms [43] together with suppliers, customers, and other parts of the organization. This allows for continuous connectivity, collaboration, and cooperation [44], fostering the synchronization of the production with stakeholders [43]. Slightly different classifications exist in the literature, such as smart supply chain [43]. Furthermore, several other authors mention this technological concept indirectly [4,45]. Despite not being a technological concept, as mentioned in the literature as the base concepts (IoT, Cloud, etc.), eVC includes essential technologies for horizontal integration, and for this reason, this category must be included.

Appendix B. Conceptual Approaches for the Application of I4.0 to LM/CI

Table A1. Conceptual approaches for the application of I4.0 to LM/CI.

Follow-up and Standardization	✓ Enabling data transmission and access [9,47] ✓ Data interchange between machines, systems, and sensors in order to end user software tools [9]	✓ Collect and access real-time data [9,54] ✓ Predict machine failures [52,55]	✓ Data analysis [9,12] ✓ Predictive analysis [11,12,52,53] ✓ Machine learning [11,43,53]	✓ Cloud-based data storage [12,42,83] ✓ Information sharing across organization [9,12] ✓ Cloud computing [35,42,83,85]	✓ Collection of production data [12,82]
Implementation	✓ Enabling data transmission and access [9,47] ✓ Data interchange between machines, systems, and sensors in order to end user software tools [9]	✓ Collect and access real-time data [9,54]	✓ Advanced analytics for planning [8] ✓ Data analysis [9,12]	✓ Cloud-based data storage [12,42,83] ✓ Information sharing across organization [9,12] ✓ Cloud computing [35,42,83,85] ✓ Cloud Kanban [49]	
Countermeasures	✓ Enabling data transmission and access [9,47] ✓ Data interchange between machines, systems, and sensors in order to end user software tools [9]			✓ Cloud-based data storage [12,42,83] ✓ Information sharing across organization [9,12] ✓ Cloud computing [35,42,83,85]	✓ Collection of production data [12,82]

Table A1. *Cont.*

	IoT	CPS	Big Data	Cloud	Sens&Act
Root causes	✓ Enabling data transmission and access [9,47] ✓ Data interchange between machines, systems, and sensors in order to end user software tools [9]		✓ Root cause analysis [11,53] ✓ Correlation analysis [47] ✓ Machine learning [11,43,53]	✓ Cloud-based data storage [12,42,83] ✓ Information sharing across organization [9,12] ✓ Cloud computing [35,42,83,85]	✓ Collection of production data [12,82]
Diagnosis	✓ Enabling data transmission and access [9,47] ✓ Data interchange between machines, systems, and sensors in order to end user software tools [9]	✓ Collect and access real-time data [9,54] ✓ Predict machine failures [52,55]	✓ Data analysis [9,12] ✓ Predictive analysis [11,12,52,53] ✓ Data mining [53] ✓ Correlation analysis [47] ✓ Machine learning [11,43,53]	✓ Cloud-based data storage [12,42,83] ✓ Information sharing across organization [9,12] ✓ Cloud computing [35,42,83,85]	✓ Collection of production data [12,82]
Mapping	✓ Enabling data transmission and access [9,47] ✓ Data interchange between machines, systems, and sensors in order to end user software tools [9]	✓ Collect and access real-time data [9,54] ✓ Predict machine failures [52,55]	✓ Data analysis [9,12] ✓ Predictive analysis [11,12,52,53] ✓ Machine learning [11,43,53]	✓ Cloud-based data storage [12,42,83] ✓ Information sharing across organization [9,12] ✓ Cloud computing [35,42,83,85]	✓ Collection of production data [12,82]
KPI analysis	✓ Enabling data transmission and access [9,47] ✓ Data interchange between machines, systems, and sensors in order to end user software tools [9]	✓ Collect and access real-time data [9,54] ✓ Predict machine failures [52,55]	✓ Data analysis [9,12] ✓ Predictive analysis [11,12,52,53] ✓ Data mining [53] ✓ Correlation analysis [47] ✓ Machine learning [11,43,53]	✓ Cloud-based data storage [12,42,83] ✓ Information sharing across organization [9,12] ✓ Cloud computing [35,42,83,85]	✓ Collection of production data [12,82]
Doc. Management	✓ Enabling data transmission and access [9,47] ✓ Data interchange between machines, systems, and sensors in order to end user software tools [9]			✓ Cloud-based data storage [12,42,83] ✓ Information sharing across organization [9,12]	

Table A2. Conceptual approaches for the application of I4.0 to LM/CI (Continued).

Follow-up and Standardization	✓ Flexible adaptation to changes [9] ✓ Minimization of human errors [8]	✓ AR/ER allow for real-time remote support in manual operations [8,12] ✓ Display of data through digital twin [51] ✓ VR facilitates training [8] ✓ AR/ER to analyze current process condition [11]		✓ Data collection [56,57] ✓ KPIs and charts in real-time [56,57]	✓ Information exchange along supply chain [43,45] ✓ On-demand access to information through digital platforms [43]
Implementation	✓ Flexible adaptation to changes [9] ✓ Automation of routine tasks [11] ✓ Collaboration with operators [9] ✓ Minimization of human errors [8] ✓ Automated logistics [8,9,12]	✓ Simulation of countermeasures and their impacts [12,31,50] ✓ AR/ER allow for real-time remote support in manual operations [8,12] ✓ VR facilitates training [8]		✓ Support planning and resource allocation [56,57]	
Countermeasures		✓ Simulation of countermeasures and their impacts [12,31,50] ✓ Simulation through digital twin [51] ✓ Supporting the creation and planning of standard work procedures [9]	✓ Small batch production (mass customization) [12] ✓ Smart product development processes [9]		
Root causes					
Diagnosis	✓ Automatic detection of machine failures [1,9]	✓ AR/ER to analyze current process condition [11] ✓ Display of data through digital twin [51]		✓ Data collection [56,57] ✓ KPIs and charts in real-time [56,57]	✓ Information exchange along supply chain [43] ✓ On-demand access to information through digital platforms [43]

Table A2. *Cont.*

Mapping	✓ Automatic detection of machine failures [1,9]	✓ AR/ER to analyze current process condition [11]		✓ Data collection [56,57] ✓ KPIs and charts in real-time [56,57]	✓ Information exchange along supply chain [43] ✓ On-demand access to information through digital platforms [43]
KPI analysis	✓ Automatic detection of machine failures [1,9]	✓ AR/ER to analyze current process condition [11] ✓ Display of data through digital twin [51]		✓ Data collection [56,57] ✓ KPIs and charts in real-time [56,57]	✓ Information exchange along supply chain [43] ✓ On-demand access to information through digital platforms [43]
Doc. Management		✓ AR/ER allow real-time remote support in manual operations [8,12]			
	Aut Rob	Sim&Virt	3DP	MES	eVC

Appendix C. Empirical Approaches for the Application of I4.0 to LM/CI

Table A3. Empirical approaches for the application of I4.0 to LM/CI.

Subject	Insights	References
A simultaneous approach is needed	✓ Companies feel that the adoption of I4.0 technologies must be simultaneous with the adoption of LM practices; therefore, a common approach based on LM and I4.0 systems is necessary.	[58,59,61]
Difficulties with I4.0 concepts	✓ Companies find it difficult to deal with some I4.0 technologies and concepts and to foresee the potential of Industry 4.0; thus, they face problems when selecting the relevant and most promising I4.0 technologies for their processes. ✓ High-cost factors and other associated risks, such as cybersecurity, prevent smooth integration. ✓ The effects of Industry 4.0 are still incipient even in manufacturers from developed economies; therefore, research should be carried out in this regard. ✓ Most companies perceive a greater positive impact through LM than I4.0 technologies; this is because they claim high levels of LM implementation, regardless of technologies.	[58–60]

Table A3. *Cont.*

Subject	Insights	References
Benefits of integrating I4.0 into LM	✓ This integration contributes positively to reducing manufacturing waste and operating costs in addition to improving productivity efficiency; the effectiveness of factors including supplier feedback, JIT, and supply development systems can be improved by increasing the use of big data analytics, sensors, cloud computing and other IoT devices; this improved integration will increase the organization's sustainable results, guaranteeing the timely delivery of the product, optimization of the supply network, and logistical reliability; based on the analysis of the data generated by the customer, it is possible to improve the level of customization offered.	[61]
Greater synergies between LM and I4.0	✓ The greatest synergies achieved by companies are between cloud computing/machine learning and waste prevention/increased productivity, Big Data and the principle of zero defects, AR/VR and visual management.	[59,60]
	✓ Some of the greatest correlations with LP principles and I4.0 were identified. Some examples are digital sensors/interfaces and the remote control of production to facilitate the identification of abnormal product/process conditions; IoT as a supporting mechanism to assure interconnectivity among products/processes.	[62]
Frequent interdependencies related to CI	✓ Big Data, RFID/identification, cloud computing, and sensor/actuators have been indicated by many companies as technologies that are interdependent with CI processes.	[59]
Rare interdependencies related to CI	✓ Only a few companies indicated AR/VR, automated guided vehicles, and smart glasses as technologies that are interdependent with CI processes.	[59]
	✓ Companies struggled with the integration of AR/VR, having the lowest mean adoption level	[62]
I4.0 technologies of greatest consideration for implementation	✓ IoT, big data analytics, and cloud computing received greater consideration for implementation, followed by AM, AR, and robotics.	[61]

Appendix D. Practical Approaches for the Application of I4.0 to LM/CI

Table A4. Practical approaches for the application of I4.0 to LM/CI.

References	Use Cases	Doc. Management	KPI Analysis	Mapping	Diagnosis	Root Causes	Countermeasures	Implementation	Follow-Up and Standardization.
[30]	Web-based monitoring tools to collect data, MES to monitor manufacturing processes in real-time, root cause analysis with the aid of digital boards, and 3D printing to test product designs.		X		X	X	X		
[64]	Big Data tool stack that processes a high volume of data, feeding predictive models based on machine learning and a descriptive analysis module that, through graphs, aids in accessing recent problems and their root causes.		X		X	X			
[63]	Data collection system for an actual machine using sensors and a CPS, also allowing for the real-time visualization of KPIs.		X		X				
[65]	RFID-based system that can collect data (e.g., quantity of items in a given place and cycle times). This system has the potential to be adapted for the VSM, allowing a real-time representation of the production system.			X					
[67]	Value chain analysis capable of determining inventory stocks based on a Big Data model that collects product information.			X					
[1,75–77]	Augmented/extended reality applied to manual workstations for the identification of tasks and the display of individualized information. Applicable to maintenance, production, and quality control activities.	X						X	X
[69]	Scheduling solutions based on real-time simulations, allowing to reach on-demand production and JIT delivery.						X		
[67]	Value stream analysis based on a Big Data model that collects information from the products and processes.			X					

Table A4. *Cont.*

References	Use Cases	Doc. Management	KPI Analysis	Mapping	Diagnosis	Root Causes	Countermeasures	Implementation	Follow-Up and Standardization.
[68]	Combination of VSM with simulation models to validate current and future states, allowing for decision makers to perform comprehensive analyses on the system.			X			X		
[70]	Web-based digital Kanban board that allows to visualize and limit WIP for software development projects. Quality of communication and motivation were improved.							X	
[71]	Large screen digital Kanban that can be operated with a smartphone and uses a short-focus projector.							X	
[72]	Implementation of a computer-aided task board with real-time tracking features. Tickets are created online, printed, and pinned to the physical board, which is tracked by a camera. This application can track physical changes in real-time, updating the board online.							X	

Table A5. Practical approaches for the application of I4.0 to LM/CI (continued).

References	Use Cases	Doc. Management	KPI Analysis	Mapping	Diagnosis	Root Causes	Countermeasures	Implementation	Follow-Up and Standardization.
[73]	Measurement-aided welding cell consisting of two handling robots, a welding robot, and an optical measurement system. This system allows the user to quickly change between manufactured products, offering the chance to produce low volume orders.							X	X
[74]	CPS for enabling a safe human–robot collaborative assembly cell, where humans, industrial robots, and moving robots (AGVs) may operate. The system is based on real-time evaluations regarding safety distances and a closed loop control for triggering collision preventive actions.							X	

Table A5. *Cont.*

References	Use Cases	Doc. Management	KPI Analysis	Mapping	Diagnosis	Root Causes	Countermeasures	Implementation	Follow-Up and Standardization.
[80]	*Jidoka*—A *Jidoka* system based on a CPS.		X		X				
[81]	*Jidoka*—Incorporation of a CPS in a milling machine, allowing it to schedule material flow and detect failures automatically.		X		X			X	
[79]	TPM—Transformation of the maintenance model from preventive to predictive through the real-time collection of data. Use of data mining to monitor defect rates and registered breakdowns.		X		X				
[8]	TPM—Machine learning-based condition monitoring along with cloud computing to improve TPM practices.		X		X				
[78]	TPM—Online root cause analysis, tracking maintenance, and repair activities.					X			

References

1. Kolberg, D.; Zühlke, D. Lean Automation enabled by Industry 4.0 Technologies. *IFAC PapersOnLine* **2015**, *28*, 1870–1875. [CrossRef]
2. Bhuiyan, N.; Baghel, A. An overview of continuous improvement: From the past to the present. *Manag. Decis.* **2005**, *43*, 761–771. [CrossRef]
3. Singh, J.; Singh, H. Continuous improvement philosophy—literature review and directions. *Benchmarking* **2015**, *22*, 75–119. [CrossRef]
4. Rosin, F.; Forget, P.; Lamouri, S.; Pellerin, R. Impacts of Industry 4.0 technologies on Lean principles. *Int. J. Prod. Res.* **2020**, *58*, 1644–1661. [CrossRef]
5. Buer, S.V.; Strandhagen, J.O.; Chan, F.T.S. The link between industry 4.0 and lean manufacturing: Mapping current research and establishing a research agenda. *Int. J. Prod. Res.* **2018**, *56*, 2924–2940. [CrossRef]
6. Mayr, A.; Weigelt, M.; Kühl, A.; Grimm, S.; Erll, A.; Potzel, M.; Franke, J. Lean 4.0-A conceptual conjunction of lean management and Industry 4.0. *Procedia CIRP* **2018**, *72*, 622–628. [CrossRef]
7. Miqueo, A.; Torralba, M.; Yagüe-Fabra, J.A. Lean manual assembly 4.0: A systematic review. *Appl. Sci.* **2020**, *10*, 8555. [CrossRef]
8. Moshiri, M.; Charles, A.; Elkaseer, A.; Scholz, S.; Mohanty, S.; Tosello, G. An industry 4.0 framework for tooling production using metal additive manufacturing-based first-time-right smart manufacturing system. *Procedia CIRP* **2020**, *93*, 32–37. [CrossRef]
9. Pereira, A.C.; Dinis-Carvalho, J.; Alves, A.C.; Arezes, P. How Industry 4.0 can enhance lean practices. *FME Trans.* **2019**, *47*, 810–822. [CrossRef]
10. Hambach, J.; Kümmel, K.; Metternich, J. Development of a Digital Continuous Improvement System for Production. *Procedia CIRP* **2017**, *63*, 330–335. [CrossRef]
11. Rittberger, S.; Schneider, M. Continuous Improvement of Lean Processes With Industry 4.0 Technologies. In Proceedings of the 11th International Doctoral Students Workshop on Logistics, Magdeburg, Germany, 9 June 2018.
12. Valamede, L.S.; Akkari, A.C.S. Lean 4.0: A new holistic approach for the integration of lean manufacturing tools and digital technologies. *Int. J. Math. Eng. Manag. Sci.* **2020**, *5*, 854–868. [CrossRef]
13. Liker, J.K. *The Toyota Way: 14 Management Principles from the World's Greatest Manufacturer*; McGraw-Hill: New York, NY, USA, 2004.
14. Ohno, T. *Toyota Production System: Beyond Large-Scale Production*; Productivity Press: New York, NY, USA, 1978.
15. Parmenter, D. *Key Performance Indicators—Developing, Implementing, and Using Winning KPIs*; John Wiley & Sons: Hoboken, NJ, USA, 2007.
16. Kerzner, H. *Project Management Metrics, KPIs, and Dashboards: A Guide to Measuring and Monitoring Project Performance*; John Wiley & Sons: Hoboken, NJ, USA, 2017.

17. Rother, M.; Shook, J. *Learning to See: Value Stream Mapping to Add Value and Eliminate Muda*; Lean Enterprise Institute: Cambridge, MA, USA, 2003.
18. Leite, D.G.; Montesco, R.A.E.; Sakuraba, C.S. Increasing a gas distributor net profit through Lean Six Sigma. *Qual. Eng.* **2018**, *30*, 359–370. [CrossRef]
19. Pietrzak, M.; Paliszkiewicz, J. Framework of Strategic Learning: The PDCA Cycle. *Management* **2015**, *10*, 149–161.
20. Eaton, M. *Lean Practitioner's Handbook*; KoganPage: London, UK, 2013.
21. Liker, J.K.; Meier, D. *The Toyota Way Fieldbook: A Practical Guide for Implementing Toyota's 4Ps*; McGraw-Hill: New York, NY, USA, 2006.
22. Iuga, M.V.; Rosca, L.I. Comparison of problem solving tools in lean organizations. *MATEC Web Conf.* **2017**, *121*, 1–9. [CrossRef]
23. Bassuk, J.A.; Washington, I.M. The a3 problem solving report: A 10-step scientific method to execute performance improvements in an academic research vivarium. *PLoS ONE* **2013**, *8*, e76833. [CrossRef]
24. Sobek, D.K.; Jimmerson, C. A3 reports: Tool for process improvement. In Proceedings of the IIE Annual Conference, Houston, TX, USA, 15–19 May 2004; pp. 1047–1052.
25. Nascimento, D.L.d.; Sotelino, E.D.; Lara, T.P.S.; Caiado, R.G.G.; Ivson, P. Constructability in industrial plants construction: A BIM-Lean approach using the Digital Obeya Room framework. *J. Civ. Eng. Manag.* **2017**, *23*, 1100–1108. [CrossRef]
26. Meister, M.; Beßle, J.; Cviko, A.; Böing, T.; Metternich, J. Manufacturing analytics for problem-solving processes in production. *Procedia CIRP* **2019**, *81*, 1–6. [CrossRef]
27. Tokola, H.; Gröger, C.; Järvenpää, E.; Niemi, E. Designing Manufacturing Dashboards on the Basis of a Key Performance Indicator Survey. *Procedia CIRP* **2016**, *57*, 619–624. [CrossRef]
28. Dogan, O.; Gurcan, O.F. Data perspective of lean six sigma in industry 4.0 era: A guide to improve quality. *Proc. Int. Conf. Ind. Eng. Oper. Manag.* **2018**, *2018*, 943–953.
29. Prause, G.; Atari, S. On sustainable production networks for industry 4.0. *Entrep. Sustain. Issues* **2017**, *4*, 421–431. [CrossRef]
30. Vo, B.; Kongar, E.; Suarez-Barraza, M.F. Root-Cause Problem Solving in an Industry 4.0 Con. *IEEE Eng. Manag. Rev.* **2020**, *48*, 48–56. [CrossRef]
31. Uriarte, A.G.; Ng, A.H.C.; Moris, M.U. Supporting the lean journey with simulation and optimization in the context of Industry 4.0. *Procedia Manuf.* **2006**, *25*, 586–593. [CrossRef]
32. Standridge, C.R.; Marvel, J.H. Why lean needs simulation. In Proceedings of the 2006 Winter Simulation Conference, Monterey, CA, USA, 3–6 December 2006; pp. 1907–1913. [CrossRef]
33. Detty, R.B.; Yingling, J.C. Quantifying benefits of conversion to lean manufacturing with discrete event simulation: A case study. *Int. J. Prod. Res.* **2000**, *38*, 429–445. [CrossRef]
34. Schuh, G.; Reiner, A.; Gausemeier, J.; Hompel, M.t.; Wahlster, W. Industry 4.0 Maturity Index. Available online: https://en.acatech.de/publication/industrie-4-0-maturity-index-managing-the-digital-transformation-of-companies/ (accessed on 15 August 2021).
35. Vinodh, S.; Antony, J.; Agrawal, R.; Douglas, J.A. Integration of continuous improvement strategies with Industry 4.0: A systematic review and agenda for further research. *TQM J.* **2020**, *33*, 441–472. [CrossRef]
36. Hermann, M.; Pentek, T.; Otto, B. Design Principles for Industrie 4.0 Scenarios. In Proceedings of the 49th Hawaii International Conference on System Sciences (HICSS), Koloa, HI, USA, 5–8 January 2016; pp. 3928–3937. [CrossRef]
37. Kagermann, H.; Wahlster, W.; Helbig, J. Recommendations for implementing the strategic initiative INDUSTRIE 4.0. *Final Rep. Ind. 4.0 Work. Group* **2013**, 1–84. Available online: https://en.acatech.de/publication/recommendations-for-implementing-the-strategic-initiative-industrie-4-0-final-report-of-the-industrie-4-0-working-group/download-pdf?lang=en. (accessed on 15 August 2021).
38. Bitkom; VDMA; ZVEI. *Implementation Strategy Industrie 4.0*; Bitkom: Berlin, Germany, 2015.
39. Amaral, A. A Framework for Assessing Manufacturing SMEs Industry 4.0 Maturity. *Appl. Sci.* **2021**, *11*, 6127. [CrossRef]
40. Dalenogare, L.S.; Benitez, G.B.; Ayala, N.F.; Frank, A.G. The expected contribution of Industry 4.0 technologies for industrial performance. *Int. J. Prod. Econ.* **2018**, *204*, 383–394. [CrossRef]
41. Bai, C.; Dallasega, P.; Orzes, G.; Sarkis, J. Industry 4.0 technologies assessment: A sustainability perspective. *Int. J. Prod. Econ.* **2019**, *229*, 107776. [CrossRef]
42. da Costa, M.B.; Santos, L.M.A.L.D.; Schaefer, J.L.; Baierle, I.C.; Nara, E.O.B. Industry 4.0 technologies basic network identification. *Scientometrics* **2019**, *121*, 977–994. [CrossRef]
43. Frank, A.G.; Dalenogare, L.S.; Ayala, N.F. Industry 4.0 technologies: Implementation patterns in manufacturing companies. *Int. J. Prod. Econ.* **2019**, *210*, 15–26. [CrossRef]
44. Bibby, L.; Dehe, B. Defining and assessing industry 4.0 maturity levels–Case of the defence sector. *Prod. Plan. Control* **2018**, *29*, 1030–1043. [CrossRef]
45. Lai, N.Y.G.; Wong, K.H.; Halim, D.; Lu, J.; Kang, H.S. Industry 4.0 Enhanced Lean Manufacturing. In Proceedings of the 8th International Conference on Industrial Technology and Management (ICITM), Cambridge, UK, 2–4 March 2019; pp. 206–211. [CrossRef]
46. Pekarčíková, M.; Trebuňa, P.; Kliment, M. Digitalization effects on the usability of lean tools. *Acta Logist.* **2019**, *6*, 9–13. [CrossRef]

47. Jayaram, A. Lean six sigma approach for global supply chain management using industry 4.0 and IIoT. In Proceedings of the 2nd International Conference on Contemporary Computing and Informatics (IC3I), Greater Noida, India, 14–17 December 2016; pp. 89–94. [CrossRef]
48. Shahin, M.; Chen, F.F.; Bouzary, H.; Krishnaiyer, K. Integration of Lean practices and Industry 4.0 technologies: Smart manufacturing for next-generation enterprises. *Int. J. Adv. Manuf. Technol.* **2020**, *107*, 2927–2936. [CrossRef]
49. Krishnaiyer, K.; Chen, F.F.; Bouzary, H. Cloud Kanban Framework for Service Operations Management. *Procedia Manuf.* **2011**, *17*, 531–538. [CrossRef]
50. Esfandyari, A.; Osman, M.R.; Ismail, N.; Tahriri, F. Application of value stream mapping using simulation to decrease production lead time: A Malaysian manufacturing case. *Int. J. Ind. Syst. Eng.* **2011**, *8*, 230–250. [CrossRef]
51. Ito, D.; Ishida, H. Digital Twin Technology for Continuous Improvement at Manufacturing Sites. *Digit. Solut. Better Futur. Soc.* **2020**, *69*, 66–70.
52. Lee, J.; Jin, C.; Liu, Z. Predictive Big Data Analytics and Cyber Physical Systems for TES Systems. *Proc. Inst. Mech. Eng. Part B J. Eng. Manuf.* **2017**, *231*, 2241. [CrossRef]
53. Karlovits, I. Technologies for using big data in the paper and printing industry. *J. Print Media Technol. Res.* **2017**, *6*, 75–84. [CrossRef]
54. Vaidya, S.; Ambad, P.; Bhosle, S. Industry 4.0—A Glimpse. *Procedia Manuf.* **2018**, *20*, 233–238. [CrossRef]
55. Meesublak, K.; Klinsukont, T. A Cyber-Physical System Approach for Predictive Maintenance. In Proceedings of the IEEE International Conference on Smart Internet of Things (SmartIoT), Beijing, China, 14–16 August 2020; pp. 337–341. [CrossRef]
56. Perico, P.; Arica, E.; Powell, D.J.; Gaiardelli, P. MES as an enabler of lean manufacturing. *IFAC PapersOnLine* **2019**, *52*, 48–53. [CrossRef]
57. Arica, E.; Powell, D.J. Status and future of manufacturing execution systems. In Proceedings of the IEEE International Conference on Industrial Engineering and Engineering Management (IEEM), Singapore, 10–13 December 2017; pp. 2000–2004. [CrossRef]
58. Rossini, M.; Costa, F.; Tortorella, G.L.; Portioli-Staudacher, A. The interrelation between Industry 4.0 and lean production: An empirical study on European manufacturers. *Int. J. Adv. Manuf. Technol.* **2019**, *102*, 3963–3976. [CrossRef]
59. Dombrowski, U.; Richter, T.; Krenkel, P. Interdependencies of Industrie 4.0 & Lean Production Systems: A Use Cases Analysis. *Procedia Manuf.* **2017**, *11*, 1061–1068. [CrossRef]
60. Beifert, A.; Gerlitz, L.; Prause, G. Industry 4.0–For sustainable development of lean manufacturing companies in the shipbuilding sector. *Lect. Notes Netw. Syst.* **2018**, *36*, 563–573. [CrossRef]
61. Kamble, S.; Gunasekaran, A.; Dhone, N.C. Industry 4.0 and lean manufacturing practices for sustainable organisational performance in Indian manufacturing companies. *Int. J. Prod. Res.* **2019**, *58*, 1–19. [CrossRef]
62. Tortorella, G.; Sawhney, R.; Jurburg, D.; de Paula, I.C.; Tlapa, D.; Thurer, M. Towards the proposition of a Lean Automation framework. *J. Manuf. Technol. Manag.* **2020**, *32*, 593–620. [CrossRef]
63. Shafiq, S.I.; Velez, G.; Toro, C.; Sanin, C.; Szczerbicki, E. Designing intelligent factory: Conceptual framework and empirical validation. *Procedia Comput. Sci.* **2016**, *96*, 1801–1808. [CrossRef]
64. Lade, P.; Ghosh, R.; Srinivasan, S. Manufacturing analytics and industrial Internet of Things. *IEEE Intell. Syst.* **2017**, *32*, 74–79. [CrossRef]
65. Aydos, T.F.; Ferreira, J.C.E. RFID-based system for Lean Manufacturing in the context of Internet of Things. In Proceedings of the IEEE International Conference on Automation Science and Engineering (CASE), Fort Worth, TX, USA, 21–25 August 2016; pp. 1140–1145. [CrossRef]
66. Ramadan, M.; Wang, Z.; Noche, B. RFID-Enabled dynamic Value Stream Mapping. In Proceedings of the 2012 IEEE International Conference on Service Operations and Logistics, and Informatics, Suzhou, China, 8–10 July 2012; pp. 117–122. [CrossRef]
67. Wamg, B.; Zhao, J.; Wan, Z.; Ma, J.; Li, H.; Ma, J. Lean Intelligent Production System and Value Stream Practice. *DEStech Trans. Econ. Manag.* **2016**. [CrossRef]
68. Cavdur, F.; Yagmahan, B.; Oguzcan, E.; Arslan, N.; Sahan, N. Lean service system design: A simulation-based VSM case study. *Bus. Process Manag. J.* **2019**, *25*, 1802–1821. [CrossRef]
69. Dallasega, P.; Rojas, R.A.; Rauch, E.; Matt, D.T. Simulation Based Validation of Supply Chain Effects through ICT enabled Real-time-capability in ETO Production Planning. *Procedia Manuf.* **2017**, *11*, 846–853. [CrossRef]
70. Nakazawa, S.; Tanaka, T. Development and application of kanban tool visualizing the work in progress. In Proceedings of the 5th IIAI International Congress on Advanced Applied Informatics (IIAI-AAI), Kumamoto, Japan, 10–14 July 2016; pp. 908–913. [CrossRef]
71. Nakazawa, S.; Komatsu, K.; Tanaka, T.; Matsumoto, K. Development and Evaluation of Large-Screen Digital Kanban with Smartphone Operation. In Proceedings of the 6th IIAI International Congress on Advanced Applied Informatics (IIAI-AAI), Hamamatsu, Japan, 9–13 July 2017; pp. 295–300. [CrossRef]
72. Bacea, I.M.; Ciupe, A.; Meza, S.N. Interactive Kanban—Blending Digital and Physical Resources for Collaborative Project Based Learning. In Proceedings of the IEEE 17th International Conference on Advanced Learning Technologies (ICALT), Timisoara, Romania, 3–7 July 2017; Volume 83, pp. 210–211. [CrossRef]
73. Tuominen, V. The measurement-aided welding cell—Giving sight to the blind. *Int. J. Adv. Manuf. Technol.* **2016**, *86*, 371–386. [CrossRef]

74. Nikolakis, N.; Maratos, V.; Makris, S. A cyber physical system (CPS) approach for safe human-robot collaboration in a shared workplace. *Robot. Comput. Integr. Manuf.* **2019**, *56*, 233–243. [CrossRef]
75. Pejić, P.; Rizov, T.; Krasić, S.; Stajić, B. Augmented reality application in engineering. In Proceedings of the 3rd International Congress, SMAT, Craiova, Romania, 23–25 October 2014; pp. 39–44.
76. Dini, G.; Mura, M.D. Application of Augmented Reality Techniques in Through-life Engineering Services. *Procedia CIRP* **2017**, *38*, 127–134. [CrossRef]
77. de Pace, F.; Manuri, F.; Sanna, A. Augmented Reality in Industry 4.0. *Am. J. Comput. Sci. Inf. Technol.* **2018**, *6*, 17. [CrossRef]
78. Mora, E.; Gaiardelli, P.; Resta, B.; Powell, D. Exploiting lean benefits through smart manufacturing: A comprehensive perspective. *IFIP Adv. Inf. Commun. Technol.* **2017**, *513*, 127–134. [CrossRef]
79. Ghobakhloo, M.; Fathi, M. Corporate survival in Industry 4.0 era: The enabling role of lean-digitized manufacturing. *J. Manuf. Technol. Manag.* **2020**, *31*, 1–30. [CrossRef]
80. Ma, J.; Wang, Q.; Zhao, Z. SLAE–CPS: Smart lean automation engine enabled by cyber-physical systems technologies. *Sensors* **2017**, *17*, 1500. [CrossRef] [PubMed]
81. Stock, T.; Seliger, G. Opportunities of Sustainable Manufacturing in Industry 4.0. *Procedia CIRP* **2016**, *40*, 536–541. [CrossRef]
82. Mrugalska, B.; Wyrwicka, M.K. Towards Lean Production in Industry 4.0. *Procedia Eng.* **2017**, *182*, 466–473. [CrossRef]
83. Wang, S.; Wan, J.; Li, D.; Zhang, C. Implementing Smart Factory of Industrie 4.0: An Outlook. *Int. J. Distrib. Sens. Netw.* **2016**, *12*, 3159805. [CrossRef]
84. Li, Z.; Wang, K.; He, Y. Industry 4.0—Potentials for Predictive Maintenance. *Adv. Econ. Bus. Manag. Res.* **2016**, 42–46. [CrossRef]
85. Taylor, S.; Anagnostou, A.; Kiss, T. High Speed Simulation Analytics. In *Simulation for Industry 4.0*; Springer: Cham, Switzerland, 2019; pp. 167–189. [CrossRef]
86. Aqlan, F.; Al-Fandi, L. Prioritizing process improvement initiatives in manufacturing environments. *Int. J. Prod. Econ.* **2018**, *196*, 261–268. [CrossRef]
87. Maarof, M.G.; Mahmud, F. A Review of Contributing Factors and Challenges in Implementing Kaizen in Small and Medium Enterprises. *Procedia Econ. Financ.* **2016**, *35*, 522–531. [CrossRef]
88. Zakoldaev, D.A.; Gurjanov, A.V.; Shukalov, A.V.; Zharinov, I.O. The life cycle of technical documentation in the smart factory of Industry 4.0. *IOP Conf. Ser. Mater. Sci. Eng.* **2019**, *665*, 5. [CrossRef]
89. Gattullo, M.; Scurati, G.W.; Fiorentino, M.; Uva, A.E.; Ferrise, F.; Bordegoni, M. Towards augmented reality manuals for industry 4.0: A methodology. *Robot. Comput. Integr. Manuf.* **2019**, *56*, 276–286. [CrossRef]
90. White, G.; Cabrera, C.; Palade, A.; Clarke, S. Augmented reality in IoT. *Lect. Notes Comput. Sci.* **2019**, *11434*, 149–160. [CrossRef]
91. Andrade, T.; Bastos, D. Extended reality in iot scenarios: Concepts, applications and future trends. In Proceedings of the 5th Experiment International Conference (exp.at'19), Funchal, Portugal, 12–14 June 2019; pp. 107–112. [CrossRef]
92. Pfeffer, J.; Graube, M.; Reipschlaeger, P.; Arndt, S.; Urbas, L.; Dachselt, R.; Stelzer, R. Towards collaborative plant control using a distributed information and interaction space. In Proceedings of the IEEE 20th Conference on Emerging Technologies & Factory Automation (ETFA), Luxembourg, 8–11 September 2015. [CrossRef]
93. Bairagi, V.K.; Joshi, S.L.; Barshikar, S.H. A Survey on Internet of Things. *Int. J. Comput. Sci. Eng.* **2018**, *6*, 492–496. [CrossRef]
94. Roth, A. *Einführung und Umsetzung von Industrie 4.0*; Springer: Berlin/Heidelberg, Germany, 2016.
95. Lele, A. *Disruptive Technologies for the Militaries and Security*; Springer: Berlin/Heidelberg, Germany, 2019.
96. de Ugarte, B.S.; Artiba, A.; Pellerin, R. Manufacturing execution system—A literature review. *Prod. Plan. Control* **2009**, *20*, 525–539. [CrossRef]

applied sciences

MDPI

Review

From Lean Production to Lean 4.0: A Systematic Literature Review with a Historical Perspective

Francisco Gil-Vilda [1], José A. Yagüe-Fabra [2,*] and Albert Sunyer [3]

1 LEANBOX SL, Distrito 22@. C/Juan de Austria 126, 08018 Barcelona, Spain; francisco.gil.vilda@leanbox.es
2 I3A—Universidad de Zaragoza, C/María de Luna, 3, 50018 Zaragoza, Spain
3 ESEIAAT—Universitat Politècnica de Catalunya, C/Colom, 11, 08222 Barcelona, Spain; albert.sunyer.torrents@upc.edu
* Correspondence: jyague@unizar.es

Abstract: Over recent decades, the increasing competitiveness of markets has propagated the term "lean" to describe the management concept for improving productivity, quality, and lead time in industrial as well as services operations. Its overuse and linkage to different specifiers (surnames) have created confusion and misunderstanding as the term approximates *pragmatic ambiguity*. Through a *systematic literature review*, this study takes a historical perspective to analyze 4962 papers and 20 seminal books in order to clarify the origin, evolution, and diversification of the lean concept. Our main contribution lies in identifying 17 specifiers for the term "lean" and proposing four mechanisms to explain this diversification. Our research results are useful to both academics and practitioners to return to the Lean origins in order to create new research areas and conduct organizational transformations based on solid concepts. We conclude that the use of "lean" as a systemic thinking is likely to be further extended to new research fields.

Keywords: lean manufacturing; lean production systems; lean 4.0; systematic literature review

Citation: Gil-Vilda, F.; Yagüe-Fabra, J.A.; Sunyer, A. From Lean Production to Lean 4.0: A Systematic Literature Review with a Historical Perspective. *Appl. Sci.* **2021**, *11*, 10318. https://doi.org/10.3390/app112110318

Academic Editor: Maurizio Faccio

Received: 6 October 2021
Accepted: 29 October 2021
Published: 3 November 2021

Publisher's Note: MDPI stays neutral with regard to jurisdictional claims in published maps and institutional affiliations.

1. Introduction

Over recent decades, markets have become more and more competitive as they progressively demand customized products and services at lower prices and with shorter delivery times [1]. In the operations field, lean has become a widespread management system that is suitable for achieving these competitiveness targets [2–4] through more efficient processes, shorter lead times, and greater flexibility in supplying a wide variety of products and services in small quantities [5].

As a consequence, the management concept of lean has spread profusely throughout industry and services over the last 40 years [3]. A huge amount of research is now available for scholars and practitioners, with the present work having identified 4962 academic papers with "lean" in the title and "lean manufacturing" generating 8,910,000 results through a Google search.

When a term becomes popular and fashionable, its overuse runs the risk of devaluing its original meaning and may create inconsistencies and ambiguities [6,7]. In addition, the term "lean" leads to more semantic confusion because it is frequently joined with specifiers by way of "surnames" related to a wide variety of fields and uses.

In practice, the term "lean" approximates *pragmatic ambiguity*, as described by Giroux [8] and similarly assessed in terms of the lean culture concept by Dorval et al. [9]. What is even worse, as Schonberger recently warns [7], it may be in risk of disintegration.

Indeed, this misunderstanding is one of the issues that lean practitioners face when implementing an organizational change and they need to align lexicon and terminology with common conceptions [4].

The objective of this study is to provide a historical perspective on the lean concept by clarifying its origins, evolution, and how it became diversified from its original concept up until today. Furthermore, this research aims to help scholars, practitioners, and managers seeking to return to the lean origins in order to better understand the evolution and current state of this field of knowledge.

From a methodological point of view, this research has followed the principles of the systematic literature review (SLR). Templier and Paré [10] have classified literature reviews into four types: narrative (summarizes previous published research); developmental (provides new conceptualizations or methodological approaches); cumulative (compiles empirical evidence and draws conclusions about a topic of interest); and aggregative (tests specific research hypotheses or propositions, with three subtypes: systematic, meta-analysis and umbrella review). The historical approach of this research falls under both cumulative and aggregative literature reviews.

Tranfield et al. [11] propose methodologically adapting SLR from medical science to management science, while Denver and Transfield [12] developed their method even further. SLR has been used in previous studies on lean topics [3,4,9,13–15]. This study follows the SLR methodology as defined by Denver and Tranfield [12], and it uses the PRISMA 2020 checklist [16] to ensure that a rigorous SLR process has been used.

2. Materials and Methods

We conducted an SLR in accordance with the five steps proposed by Denyer et al. [12]: question formulation; locating studies; study selection and evaluation; analysis; synthesis; reporting; and using results.

2.1. Question Formulation

As introduced above, this study aims to answer the following research questions:

- RQ1: What is the historical origin of the term "lean"?
- RQ2: What are the previously used terms (if any) for the lean concept?
- RQ3: How has the term "lean" evolved over time?

2.2. Locating Studies

Three bibliographical materials have been used:LISTORDE

1. English language records at the Web of Science database from 1950 to 2020.
2. English language records at the Scopus database from 1987 to 2020.
3. Books: We analyzed a collection of 20 seminal hardcopy books published between 1977 and 2020.

2.2.1. Locating Database Records

As proposed by Sinha et al. [17] (p. 304), we searched for the keyword "lean" in titles to ensure our focus on both the historical interest and evolution of the topic. A first search on Web of Science conducted on 5 January 2021 provided 13,558 records containing the word "lean" in the title. An initial quick review showed that "lean" is a popular term in other disciplines too. Therefore, our search was refined to some related WoS categories. The final search string is shown in Figure 1.

3.255 (TI="lean") *AND* **IDIOMA:** (English) *AND* **TIPOS DE DOCUMENTOS:** (Article OR Proceedings Paper OR Review)
Refinado por: CATEGORÍAS DE WEB OF SCIENCE: (ENGINEERING INDUSTRIAL OR ECONOMICS OR MANAGEMENT OR INDUSTRIAL RELATIONS LABOR OR ENGINEERING MANUFACTURING OR OPERATIONS RESEARCH MANAGEMENT SCIENCE OR BUSINESS FINANCE OR BUSINESS OR ENGINEERING ENVIRONMENTAL OR ERGONOMICS OR ANTHROPOLOGY OR PSYCHOLOGY APPLIED)
Índices=SCI-EXPANDED, SSCI, CPCI-S, CPCI-SSH, ESCI Período de tiempo=1955-2020

Figure 1. Web of Science search string.

This second bounded search on 5 January 2021 provided a total of 3255 records, which we transferred to a spreadsheet for analysis and classification.

In a first analysis, the most cited studies were analyzed [1,5,6,8,18,19] to uncover any general agreement on the origins of the term "lean" in the field of management, by which we found it was first coined in 1988.

Even though records from 1950 to 1988 were reviewed, only one similar and metaphorical use of "lean" was found [20]: "Managerial Productivity: Who Is Fat and What Is Lean?". However, this instance in management research was not fully associated with its later use in operations management where it was fully developed.

We performed a manual review based on the WoS category and title analysis in order to remove any records of non-related topics (e.g., combustion, food, information, chemicals, etc.). Works published ahead of print (early access) were discarded. Finally, 2932 records were retained for further analysis.

Similarly, a Scopus database search was made on the same date (5 January 2021) for studies published in the English language after 1987, restricting these to items labelled as articles, conference papers, and reviews. These were further limited to the subject areas of business, management, and accounting. The Scopus search string is shown in Figure 2.

TITLE ("lean") AND PUBYEAR > 1987 AND (LIMIT-TO (DOCTYPE , "ar") OR LIMIT-TO (DOCTYPE , "cp") OR LIMIT-TO (DOCTYPE , "re")) AND (LIMIT-TO (SUBJAREA , "BUSI"))   3,607 document results

Figure 2. Scopus search string.

The records were transferred to a spreadsheet, merged with those from the WoS search, and redundancies were removed. Finally, a total of 4962 records were kept for further analysis.

A first manual review based on reading the titles and, if necessary, the abstract allowed us to determine the selection criteria for the records, as described in Section 2.3.

The SLR process is shown in Figure 3.

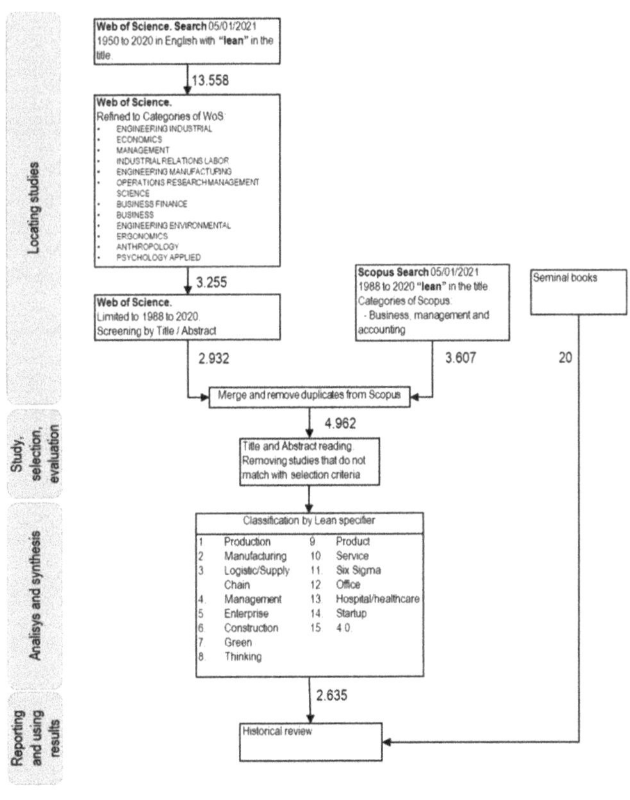

Figure 3. SLR process overview.

2.2.2. Books

Books were collected based on the Holweg's list [19] (p. 434), new titles were added to this list. A total of 20 seminal hardcopy books were analyzed. Books were gathered from private collections or were acquired at https://www.bookfinder.com/ (accessed on 5 September 2021).

2.3. Study Selection and Evaluation

To select the relevant records associated with the research questions, their titles and abstracts were analyzed. The following selection criteria were defined:

- The title allows confirming that the term "lean" is used as a managerial concept.
- The abstract confirms that the term "lean" is related to operations management.

To avoid bias, if neither title nor abstract allowed classifying the record, it was removed.

2.4. Analysis and Synthesis

After selection was done, the database records were analyzed based on the bibliometric measure "papers-per-year" (Figure 4).

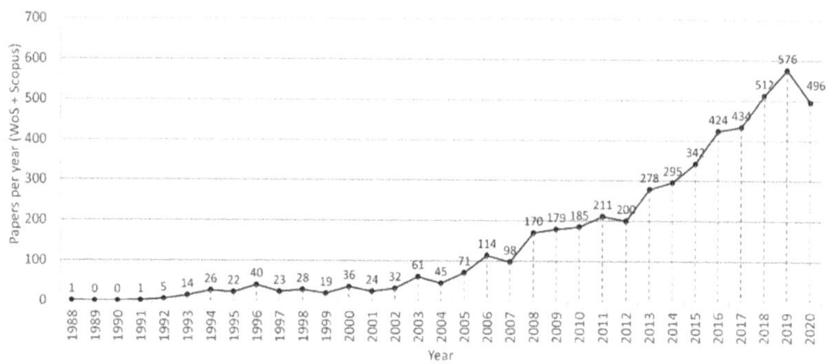

Figure 4. Evolution of papers per year that include "lean" in the title over the study period (1988–2020).

During our analysis of the titles, the main lean specifiers ("surnames") were found and we accordingly classified the records under the following categories (by chronological order of appearance): production, manufacturing, logistics/supply chain, management, enterprise, construction, green, thinking, product, service, six sigma, office, healthcare/hospital, start up, 4.0.

To facilitate the historical analysis, we created a general chart split into categories (Figure 5): the most relevant categories are represented by a line starting at the year of their foundational paper or book; the line thickness is proportional to the paper-per-year bibliometric (see scale in the same Figure 5).

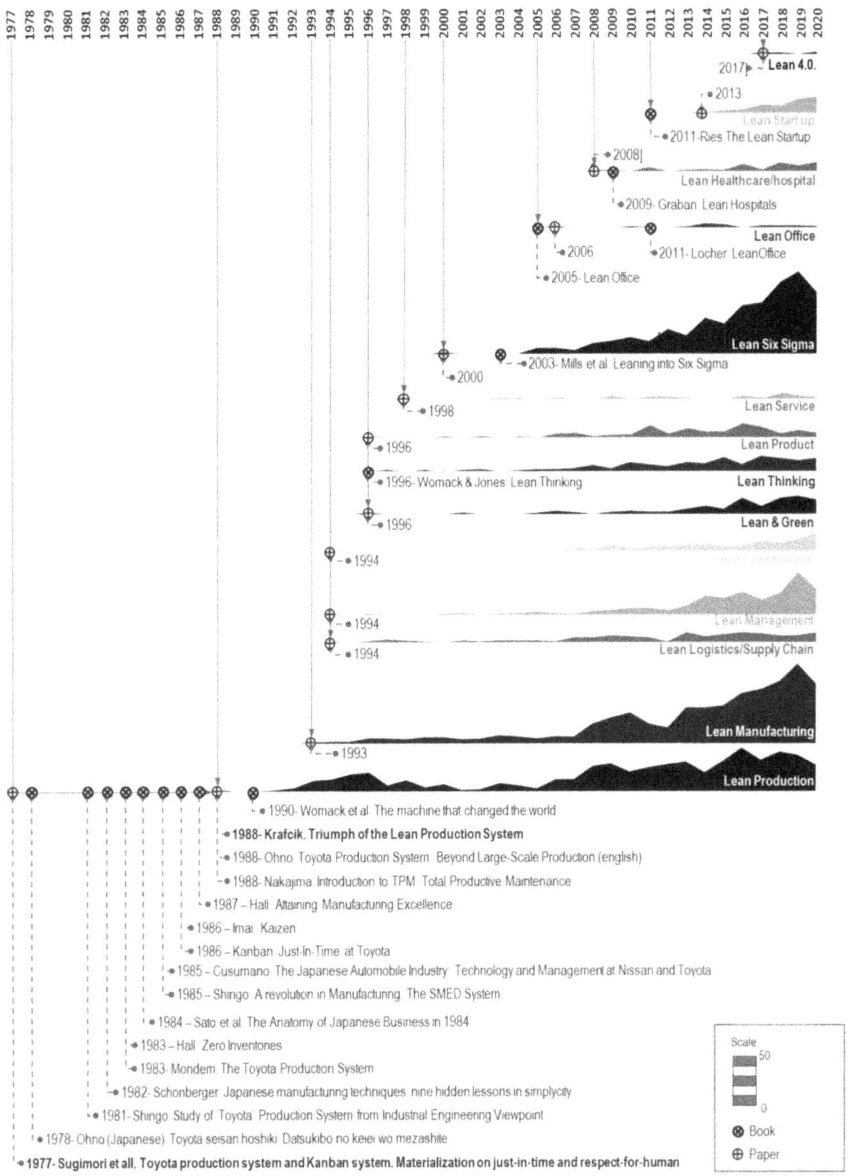

Figure 5. Historical evolution of the main lean categories and their foundational works.

3. Results

3.1. Origin and Previously Used Terms for the Lean Concept in Operations Management

In answering RQ1 (What is the historical origin of the term "lean"?), we found a general consensus [1,2,6,13,19,21,22] that the term "lean" was coined in the International Motor Vehicle Program (IMVP) and published for the first time in 1998 by John F. Krafcik, in the academic paper titled *Triumph of the Lean Production System*, when he stated "lean typology builds on the work of International Motor Vehicle Program researchers Haruo Shimada and John Paul Max Duffie, who use the terms 'robust' and 'fragile' to denote similar concepts" [23] (p. 51). Here, Krafcik is referring to the 1986 working paper *Industrial Relations and "Humanware"* [24].

The word "lean" was chosen for its more positive sense [19] (p. 426) or, as suggested by New, "as an acceptable way of describing Toyota production system without offending the other sponsors of the IMVP" [22] (p. 3547).

Therefore, to answer RQ2 (What are the previously used terms (if any) for the lean concept?), we must return to the foundations of the Toyota Production System. There is wide agreement [5,6,9,22] that the first English language paper introducing the term "Toyota production system" (TPS) was presented in Tokyo by Sugimori et al. in 1977 [25]: *Toyota Production System and Kanban System. Materialization of Just-in-Time and Respect-for-Human System.* This seminal paper based TPS on two pillars: just-in-time and respect-for-humans. The authors acknowledged Taiichi Ohno as having been the promoter and leader of TPS since at least 1957.

Ohno's seminal 1978 book (published only in Japanese) was *Toyota seisan hoshiki: Datsukibo no keiei wo mezashite* [26], and his first paper translated to English dates back to 1982 [27] (*How the Toyota Production System Was Created*, republished in The Anatomy of Japanese Business in 1984 [28]). This early translation offered an alternative and more accurate translation to just-in-time: "right on time" which was not adopted.

The first book in the English language describing TPS was published in 1981 by Shigeo Shingo: *Study of "TOYOTA", Production System from Industrial Engineering Viewpoint* [29]. He acknowledged Ohno as the promoter of TPS (pp. 19–32) and postulated that his own book would provide a more practical explanation. It was a highly influential book in which Shingo insisted repeatedly on an essential systemic view in order to understand TPS. This book was republished with a better English translation in 1989 by Productivity Press [29].

In 1983, Yasuhiro Monden published *Toyota Production System: Practical Approach to Production Management* [30]. The foreword by Ohno highlighted the excellent conceptualization of the TPS. Without losing the holistic vision, tools and methodologies were profusely described.

The first Western researchers interested in the topic were very influenced by these early books, and they published their works in the period between 1983 and 1988, during which different "nicknames" were used to refer to the TPS:

- *Japanese Manufacturing Techniques*, by Schonberger (1982) is the first book written in English and it described "stockless production" and "JIT Production" [31] (p. 17).
- Hall, in his book *Zero Inventory* (1983) [32] (p. 1) adopted the term "stockless production". In another book, *Attaining Manufacturing Excellence* (1987) [33] (p. 23), he summarized the most used terms at the time: "manufacturing excellence", "value-added manufacturing", "continuous improvement manufacturing", and "JIT/TQ".
- Cusumano's book *The Japanese Automobile Industry* (1985) [34] (pp. 262–307) described the foundations and evolution of TPS with a historical perspective, although he did not propose any alternative terms.
- In the context of IMTV, "fragile production" was already being used in 1986 [24].

From 1985 to 1990, different TPS tools were documented in detail, thus providing progressively greater understanding but fragmenting the overall vision by "using the part for the whole", as Shah [6] (p. 786) suggested. Some examples of these TPS tools are: SMED (1985) [35], Kanban (1986) [36], Kaizen (1986) [37], and TPM (1988) [38].

In 1988, the English translation [39] of Ohno's [26] book was published: *Toyota Production System: Beyond Large-Scale Production.*

To summarize the answer to RQ2, the lean production system can be considered a way of naming the Toyota production system without naming Toyota. With the same intention, other terms were proposed prior to 1988 by taking some of the more relevant parts of the system as inspiration: *Japanese manufacturing techniques, stockless production, JIT production, value-added production, continuous improvement manufacturing, non-stock production, fragile production.*

3.2. Answering RQ3. Historical Evolution of the Term "Lean"

As already pointed out, the expression "lean production system" was coined by Krafcik in 1988 [23]. The seminal and best-selling book, *The Machine that Changed the World* [21], popularized the term "lean production"; and earlier similar expressions were completely abandoned (with the only exception being just-in-time, which survived in the supply chain literature until now).

Womack et al. [21] used the term "lean production" in contrast to "mass production" with the intentions of setting a benchmark, although the original systemic vision was lost. This is probably the first symptom of the "lack of distinction between the systems and its components" as Sha et al. [6] suggested.

Thus, 1990 can be considered the year when the term "lean" became popularized as an operations management concept. From 1990 to 1995, the term "lean" was adopted in the literature mainly as "lean production". The first research papers focused on either supporting or questioning lean [40–42] while describing the first lean experiences and the limits of this practice [42].

In 1991, Delbridge et al. [43] introduced "lean manufacturing" as a synonym for "lean production", and this term became more popular after 2000. Nowadays, "lean manufacturing" is the preferred expression for referring to lean in industrial operations (Figure 5).

From 1992 to 1996, some authors intended to upgrade lean to a more conceptual level by introducing the terms "lean management", "lean enterprise" [44], and "lean thinking" [45]. This opened the door to using the term in non-manufacturing contexts, such as the services sector.

In parallel, the period 1994 to 2000 saw the first attempts to apply lean to different production contexts (lean construction) as well as to others outside pure production (lean logistics, lean supply chain) or by combining it with supplementary topics (lean and green, lean product, lean six sigma). The fragmentation of the lean system into its tools continued with books such as *Lean Toolbox* [46]

It was not until 2005 that the lean concept opened its scope to the services sector, mainly under the umbrella of lean six sigma, lean office, lean healthcare/hospital, and, recently, lean startup.

Finally, in 2017, lean 4.0 appeared as a promising way for new developments by fusing lean and Industry 4.0 technologies.

All in all, the term "lean", which was initially conceptualized as a "lean production system", has evolved from 1988 to 2020 as a "living concept". This evolution can be ascribed to the following proposed mechanisms that mostly combined with each other over time:

- Expansion: extending the concept in the operations field.
- Transfer: applying the concept beyond production.
- Targeting: focusing the concept on a particular sector.
- Combination: merging the concept with other concepts.

Through analyses of the scientific literature, this work has identified the most important lean specifiers ("surnames"). They are presented in chronological order of their appearance and describe (a) the first record found in our database analyses; (b) the historical trajectory based on the most relevant publications; (c) the evolution mechanisms; and (d) the present situation in terms of research interests. A chart with the yearly evolution of papers-per-year complements the summary.

3.2.1. Lean Production (1988)

Introduced in 1988 by Krafcik as the "lean production system" [23], this expression was used as an alternative to the "Toyota production system". It was fully adopted after publication of the seminal book *The Machine that Changed the World* [21], which described lean production (LP) as an alternative to mass production [2], whereby the original holistic approach of TPS was partially lost.

Interest in LP grew between 1992 and 1996 (see Figure 6), then diminished from 1997 to 2007. Since 2008, it has once again become of academic interest, along with lean manufacturing.

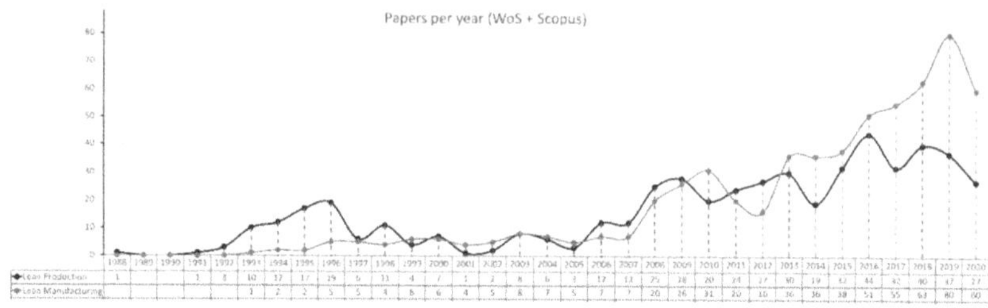

Figure 6. Lean production vs. lean manufacturing evolution.

In 2007, Holweg [19] outlined a detailed historical evolution of the term, which was highly influenced by MIT research. The same year, Sha et al. [6] analyzed the historical context and reported for the first time the semantic confusion surrounding the term "lean production" while also reinforcing the conception of lean "as a system", for which he identified 10 dimensions useful to researchers.

In 2013, Marodin et al. [47] identified six research areas in the field and provided another warning about the system conception becoming fragmented and dissociated. In 2015, Jasti et al. [1] concluded that LP continues to have a high impact on academia, practitioners, and consultants. They further propose a holistic rather than "bits-and-pieces" approach [1] (p. 16).

In 2015, *The Analysis of Industry 4.0 and Lean Production* [48] presented the first comparative study between LP and I4.0. With 17 papers published in the last 5 years, the interrelationship between LP and I4.0 has clearly awakened new interest in this research field [49,50].

3.2.2. Lean Manufacturing (1993)

In 1993, Powell introduced lean manufacturing (LM) in a similar sense to lean production: *Lean Manufacturing Organization, 21st Century* [51].

In 2003, [52], Shah et al. also used this term with the same meaning as lean production and focused the topic on factory management. In 2013, Bhamu et al. [5] presented the evolution of LM definitions.

The most recent main reviews of lean manufacturing that confirm this equivalence with lean production were published between 2019 and 2020 [3–5,53]. These reviews are perhaps targeted more to the fields of industry and, specifically, factory management.

In 2016 [54], the first linkage with Industry 4.0 was made. Until 2020, the relationships between LM and Industry 4.0 were explored. In the last one, published in 2020 [55], Valamede et al. offered a holistic view toward integrating both concepts.

As a conclusion, lean manufacturing can be considered synonymous with lean production, although targeted more at factory operations. After 2000, is the preferred term in the academic literature when referring to lean in the industrial field (Figure 6). That is probably to distinguish the production of goods, since the term "production" is used more and more in the service industries.

3.2.3. Lean Logistics (1994), Lean Supply (1996) and Lean Supply Chain (1999)

The first paper on lean logistics (LL) was authored by Fynes et al. in 1994, *From Lean Production to Lean Logistics: The Case of Microsoft Ireland* [56], which illustrated the expansion from production to supply chain management.

In 1996, the first paper to use using "lean supply" was *Squaring Lean Supply with Supply Chain Management* [57], in which the authors extended the concept to supply chain management.

In 1999, the first paper to use "lean supply chain" (LSC) was *Vertical Integration in a Lean Supply Chain: Brazilian Automobile Component Parts* [58], which related the term "lean" to a broader perspective on supply chain management.

The three surnames can be considered quite equivalent, in that they focus on: the efficiency of material flows inside and outside the factory; the integration and development of suppliers; and the integration of different actors and information across the supply chain [59].

García Buendía et al. (2020) [60] presented a conceptual evolution map of the concepts behind lean supply chain management over the last 22 years.

Relationships between lean supply chain and I4.0 have appeared in recent years in publications on different topics, such as the impact of these on performance improvement [61–64] and their further relationships with information and digital technologies [63].

To conclude, these specifiers appeared as lean expanded to supply chain management: "Lean logistics [. . .] is based around extended TPS right along supply chain from customers right back to raw material extraction" [64] (p. 171).

The three surnames in this field extend lean perspective to supply chain management, with interest having increased moderately since 2007 (Figure 7).

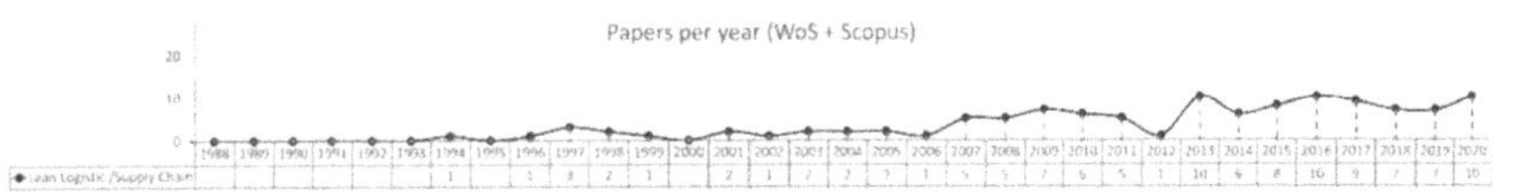

Figure 7. Lean logistics and lean supply chain evolution.

3.2.4. Lean Management (1994)

The origins of the term "lean management" (LMg) are unclear. The first reference in the English language academic literature was introduced by Petrovic et al. in 1994: *Business Process Re-Engineering as an Enabling Factor for Lean Management* [65]. Nevertheless, the German literature has used this term since 1992. It seems to be a first attempt at shifting towards a more managerial concept in a similar way as the later emergence of lean enterprise or lean thinking.

In any case, it was not until 2008 that the literature showed consistent interest in the topic.

In 2014, Martinez-Jurado et al. [66] presented the term in association with organizational sustainability. More recently in 2019, Sinha et al. [17] considered lean management to be an extension "into an inter-disciplinary subject with linkages to operations management, organizational behaviour, and strategic management".

In 2016, the first paper on LMg and I4.0 was published: *Industry 4.0. The End Lean Management?* [67], concluding at the time that the correlations between both concepts were low.

As a conclusion, lean management can be considered as a transfer to a more managerial approach. It refers to adopting lean principles in order to manage an entire organization. Although quite neglected until 2008, it has generated growing interest in the past decade, at least, up until 2019 (Figure 8).

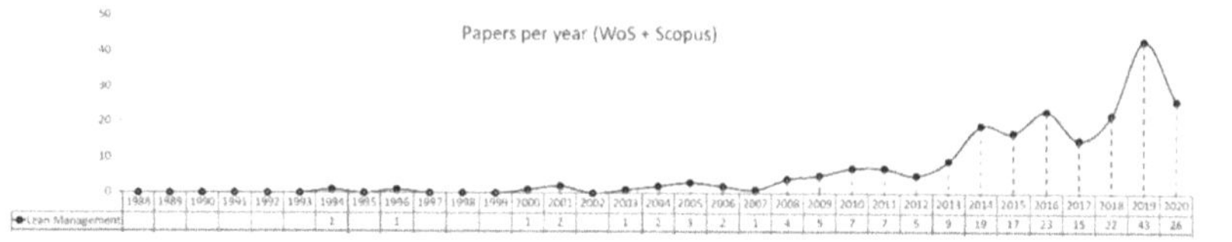

Figure 8. Lean management evolution.

3.2.5. Lean Entreprise (1994)

The expression "lean enterprise" (LE) was coined in 1994 by Womack and Jones in their book *From Lean Production to the Lean Enterprise* [44], in which lean shifts toward a more abstract concept: "the lean enterprise is a group of individuals, functions, and legally separate but operationally synchronized companies that creates, sells, and services a family of product".

As a conclusion, lean enterprise can be considered a transfer to a more abstract concept. The term has not been widely adopted in literature, but it remains alive, as evidenced in the last review published in 2020 [68] and the first proposals linking LE with 4.0 technologies [69] (Figure 9).

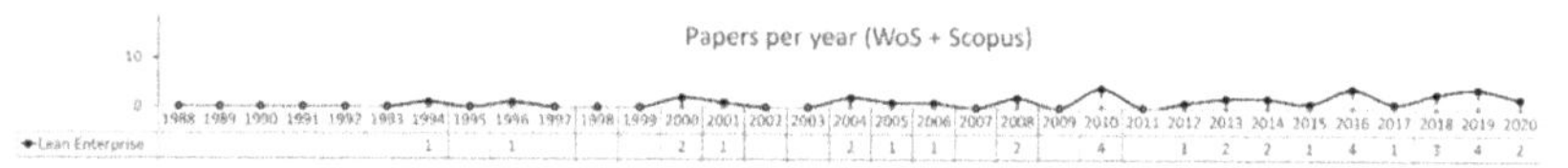

Figure 9. Lean enterprise evolution.

3.2.6. Lean Construction (1994)

The first indexed reference to lean construction (LC) dates back to 1994, when Koskela published the proceedings paper *Lean Construction* [70], which placed lean production in the particular context of a product (a building) that cannot be moved in a continuous flow.

It was not until 2002 when research attention returned to the topic [71], and in 2006 Salem et al. [72] proposed a practical view toward implementing lean tools.

In 2019, Koskela presented an epistemological perspective not only on LC but also on lean and its Japanese origins [73]. In 2020, Lekan et al. [74] proposed "Construction 4.0" as the link between LC and "Industry 4.0" with the aim to go further in construction operations efficiency.

The last review [75] was published in 2020, and it explored the barriers to implementing LC.

As a conclusion, lean construction targeted this specific sector in which the product cannot be moved in a continuous flow. It adapts lean principles and tools to this particular production process. It was quite ignored until 2008, but the topic has generated moderately increasing interest in the past decade, recently linked with I4.0 (Figure 10).

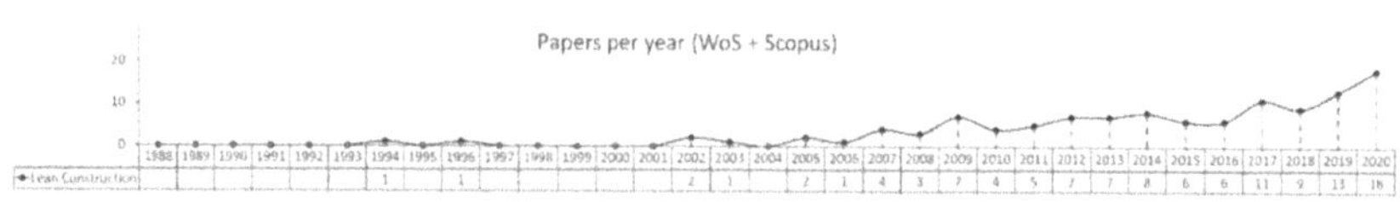

Figure 10. Lean construction evolution.

3.2.7. Lean and Green (1996)

Lean and green (L&G) appeared in a 1996 Florida publication [76]: *Lean and Green: The Move to Environmentally Conscious Manufacturing*. It integrates process improvements with reductions in environmental impact.

The first publications on L&G focused on how to establish a link between lean principles and environmental practices, with an emphasis mainly on manufacturing [77] and supply chain management [78].

Interest in the topic has increased since 2013, and the first literature review (in 2015) [79] proposed it as a specialized research area. In 2019, Farias et al. developed a systemic approach [80,81].

Recently the scope has been extended to products and services [82], as well as combined with Industry 4.0 issues [83,84]

As a conclusion, lean and green (sometimes green lean) emerged as a combination with environmental and sustainability concepts. It refers to the synergy between lean and environmental preservation. More specifically, it focuses on how lean practices can contribute to reducing environmental impact while maintaining profits primarily in operations, but also in services and product design. The topic has generated increasing research interest since 2013 (Figure 11).

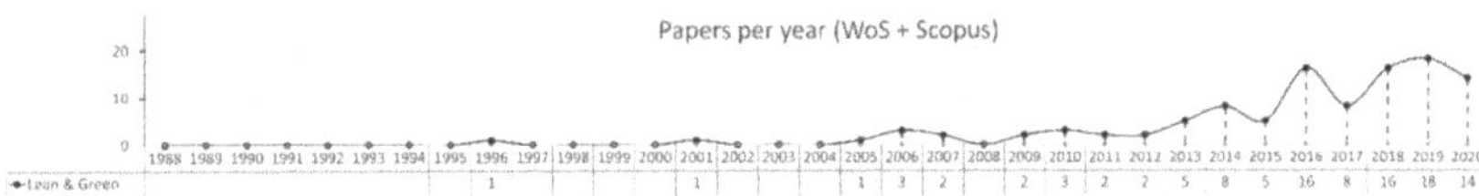

Figure 11. Lean and green evolution.

3.2.8. Lean Thinking (1996)

Lean thinking (LT) was introduced in 1996 by Womack and Jones [45] in their best-selling book *Lean Thinking: Banish Waste and Create Wealth in Your Corporation*. With intentions similar to lean enterprise, the term can be considered a shift toward a philosophy of eliminating waste in organizations. This way of thinking is structured in five steps: specify value; identify the value stream; flow; pull; and pursue perfection.

The first indexed article is from 1997 [85], and it analyzes the impact of LT and LE on the marketing processes. In 2004, Hines et al. [18] published a very detailed study on the topic, beginning with its genesis and moving on to identify both the successes and difficulties of Western companies applying LT. The last review was published in 2020 [86], and it explored the synergies between LT and Industry 4.0 while further suggesting how LT could trigger I4.0 solutions.

As a conclusion, lean thinking is a transfer to a more abstract approach. It refers to adopting a way of thinking in order to make radical improvements in any organization. Research interest in this topic has remained moderate and stable in the past decade (Figure 12).

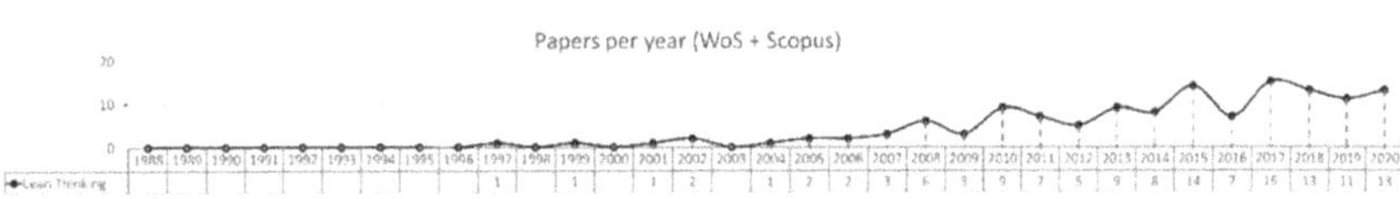

Figure 12. Lean thinking evolution.

3.2.9. Lean Product (1996)

The first paper, *The Difficult Path to Lean Product Development*, by Karlsson et al. [87], introduced the expression "lean product" in 1996 as an extension to product development. It refers to fast, efficient, and low-cost product development [88].

The concept was created for physical goods and generated low interest among researchers until 2006. The same year, Liker et al. [89] proposed the concept in order to go "beyond manufacturing to any technical or service" with a systemic view toward "integrating people, process and tools".

The first review in 2011 [88] showed the historical links between LP and TPS while presenting a list of conceptual principles.

In 2015, Sassanelli et al. [90] introduced a systemic view that focused particularly on services as a lean product service system. This approach was recently analyzed in the latest systematic reviews on the topic [91].

As a conclusion, lean product extends to product development in terms of both goods and services. It has generated moderate and stable interest since 2006 (Figure 13).

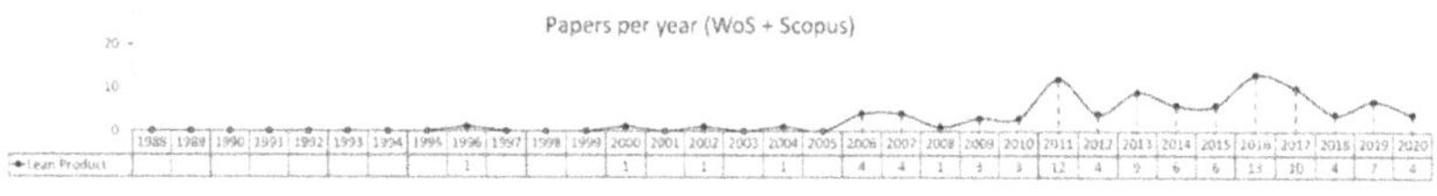

Figure 13. Lean product evolution.

3.2.10. Lean Service (1998)

Lean service (LSe) was proposed in 1998 by Bowen et al. in their article *Lean Service: In defense of a Production-line Approach* [92] in order to extend lean to industrial services. The next indexed paper was published in 2003: *The Lean Service Machine* [93], which adapted lean production to an insurance company (JPF).

LSe refers to applying lean principles and tools toward the improved efficiency of non-manufacturing services [94] such as insurance firms [93], call centers [95], financial services [96], banking, and healthcare services [97].

A systematic review published in 2016 [98] concluded that lean was applicable in services with limitations, and it identified LSe as a nascent research area.

As a conclusion, lean services transfer from manufacturing to service processes. It refers to applying lean manufacturing principles and tools that have been adapted to services production. Researchers have shown little interest in it, probably because this perspective is also explored by lean six sigma scholars (Figure 14).

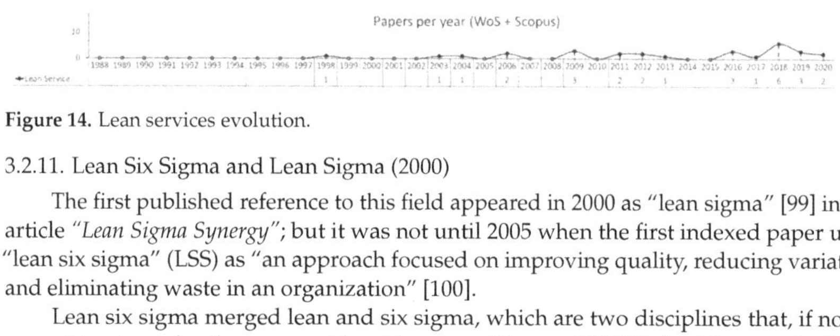

Figure 14. Lean services evolution.

3.2.11. Lean Six Sigma and Lean Sigma (2000)

The first published reference to this field appeared in 2000 as "lean sigma" [99] in the article *"Lean Sigma Synergy"*; but it was not until 2005 when the first indexed paper used "lean six sigma" (LSS) as "an approach focused on improving quality, reducing variation and eliminating waste in an organization" [100].

Lean six sigma merged lean and six sigma, which are two disciplines that, if not in opposition to each other, were at least in competition until 2003. In this year, the book *Leaning into Six Sigma* [101] attempted to join together the best practices from lean and six sigma [102].

After 2005, the concept has generated increasing interest among researchers, as shown in the evolution of published papers (particularly after 2013) in contexts of both manufacturing [103] and service [104]. Recent research has shown interest in the links with Industry 4.0 [105,106].

As a conclusion, lean six sigma and lean sigma is a combination of the principles and tools from lean manufacturing (reducing waste) and six sigma (reducing variability and promoting leadership). Originally created for manufacturing industries, it extended its implementation to services too. Interest in this topic has increased sharply between 2004 and 2019 (Figure 15).

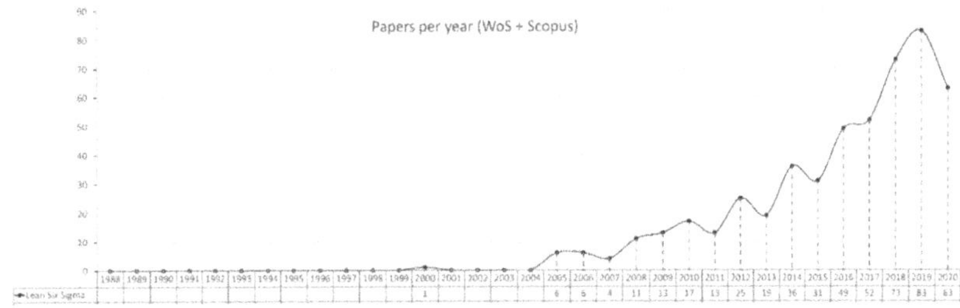

Figure 15. Lean six sigma evolution.

3.2.12. Lean Office (2005)

Lean office (LO) appeared in 2005 with the book *The Lean Office* [107], collecting practical cases from 2000 to 2004 of extending lean to non-manufacturing environments.

In 2011, Locher published *Lean Office* [108] with a methodological and holistic approach to apply lean in services, commercial and administrative environments.

The very limited academic literature available starts on 2006 with Herkommer et al., *Lean Office-System* [109] and focuses on surface optimization, workplace improvements [110], and information flows in administrative processes [111]. A systematic literature review published in 2019 [112] described implementation issues and areas of research.

As a conclusion, lean office is a transfer to non-manufacturing environments with a focus on improving efficiency at the administrative level. The scant academic interest in this topic lies in stark contrast to the term's popularity among practitioners (Figure 16).

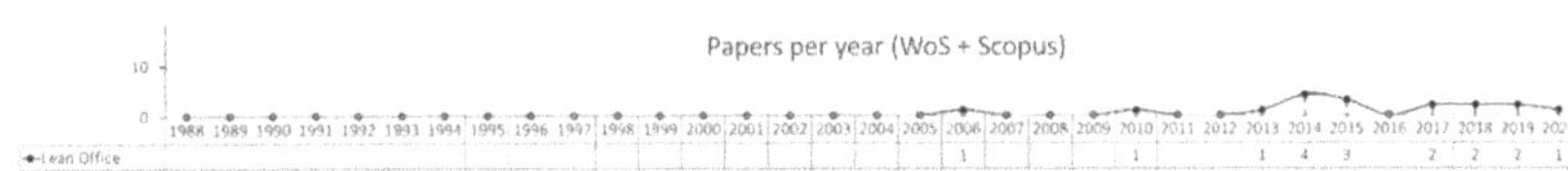

Figure 16. Lean office evolution.

3.2.13. Lean Healthcare/Hospital (2008)

In 2008, Portioli-Staudacher used the term "lean healthcare" for the first time in the paper: *Lean Healthcare. An experience in Italy* [113] published as a lean approach to the healthcare sector. The paper did not focus on service improvement but in how to reduce inventories of drugs and other healthcare supplies by implementing tools from lean logistics.

In 2009, Mark Graban published the book *Lean Hospitals* [114] as a practical guide for adapting lean tools in hospital management.

In 2011 [115], lean healthcare was proposed as a more holistic system for improving healthcare organizations and how to assess them.

In 2016 [116], Costa et al. presented a review based on six parameters: research method, country, healthcare area, implementation, lean tools and methods, and results.

In 2020, Santos et al. [117] highlighted new research areas for the future.

As a conclusion, lean healthcare/hospital is a transfer to services, targeted on healthcare services, and it includes hospital management. It applies lean principles and tools toward improving patient care. Interest in it was very limited until 2015, and it has seen moderate growth in the last 3 years as new research proposals are put forth (Figure 17).

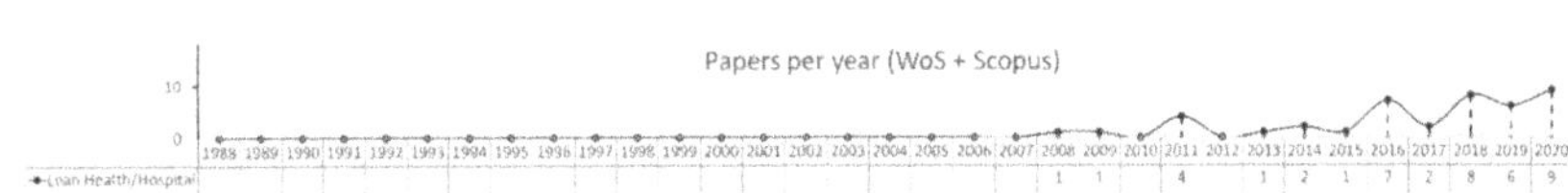

Figure 17. Lean healthcare/hospital evolution.

3.2.14. Lean Startup (2011)

The first reference to the term "lean startup" in the research literature was by Blank in 2013, in *Why the Lean Start-up Changes Everything* [118]. The author expanded on the concept proposed by Ries in his 2011 book *The Lean Startup* [119], proposing it as a new methodology for launching companies faster and cheaper than the methods of a traditional business plan. As a consequence, the term can be considered a variation that applies lean principles to the launching of new businesses.

In 2017, Frederiksen et al. [120] presented evidence from the scientific literature for their in-depth look at the methodological proposals in Ries's book.

In 2018, Bortolini et al. [121] clarified how the foundations of lean startup are linked with lean manufacturing: maximizing customer value while minimizing waste.

In 2020, Silva et al. [122] provided new perspectives on developing a business model and discussed complementary methodologies, such as agile methodologies and customer development.

As a conclusion, lean startup is a transfer to launching a new business. It uses lean principles to launch new business models while reducing time-to-market and minimizing initial investment and risks. Since its appearance in 2011, interest in the topic has seen sustained growth (Figure 18).

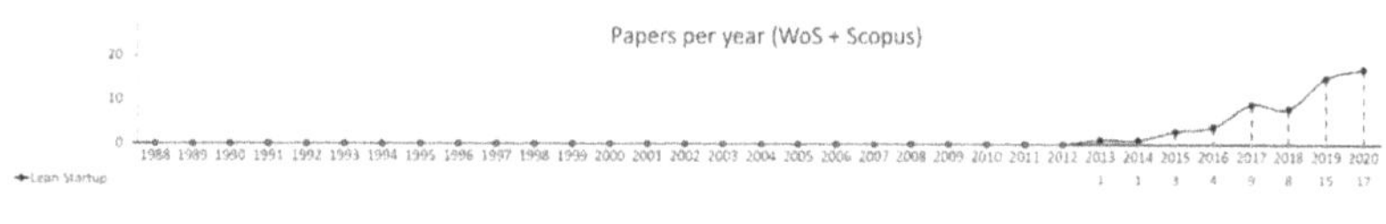

Figure 18. Lean startup evolution.

3.2.15. Lean 4.0 (2017)

Lean 4.0 (L4.0) is the most recent specifier. It was introduced in 2017 by Metternich et al. [123] in the German language paper *Lean 4.0—Between Contradiction and Vision*, which combined lean with Industry 4.0 (I4.0). The authors reflect on the compatibility between lean philosophy and technologies under the umbrella of I4.0, concluding that lean appears to be a prerequisite for digitization.

In 2018, Mayr et al. [124] agreed that lean enabled the successful introduction of I4.0 and concluded that both views complement each other. They present a detailed overview on how the most relevant lean tools can be complemented with I4.0 technologies.

In 2020, Valamede et al. [125] went further by taking a holistic view to identify 25 synergy points between lean tools and 4.0 technologies. Perico et al. [126] proposed new perspectives on how to incorporate artificial intelligence to support human decisions in key lean 4.0 topics (production control, maintaining continuous pull flow, and early prediction of machine failure). Under the denomination "lean Industry 4.0", Ejsmont et al. [127] identified the research trends combining "lean management" and I4.0. to go further in reducing waste to achieve a new level of operational excellence.

At the present moment, only four papers have been found with "lean 4.0" in the title, although increasing interest (see Figure 19) is being generated in the relationships between Industry 4.0 and different lean aspects: lean production [49,128], lean manufacturing [54,129], lean and green [85,130], lean construction [74], lean enterprise [70], lean healthcare [131], lean management [132], lean six sigma [133], lean supply chain [64], and lean thinking [86].

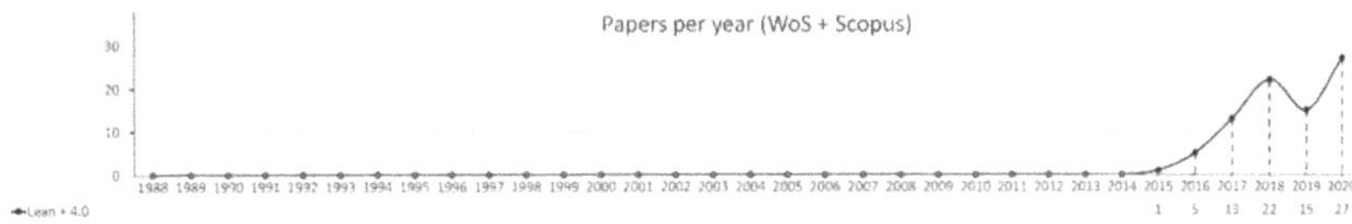

Figure 19. Papers about relationships between lean and Industry 4.0.

In total, 83 papers have been identified linking "lean" and I4.0. An analysis based on the address of the corresponding author shows countries leading the research in this topic: Germany is in the first position as the term "Industry 4.0" was coined in Germany. Nevertheless, an arising interest is shown in different nations, particularly in Brazil, Portugal and Italy (see Figure 20).

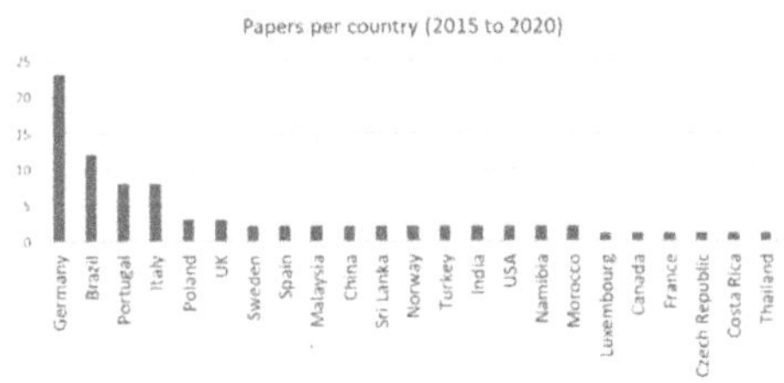

Figure 20. Papers linking lean and Industry 4.0 by corresponding author's country (2015–2020).

As a conclusion, lean 4.0 is a combination of lean manufacturing (or lean production) principles and tools with Industry 4.0 technologies. It deals with the synergies and complementarity of I4.0 with lean with the intention of reducing waste and complexity. It appears to be a promising field of research in the coming years.

4. Conclusions

This article explores the origins and diversification of the term "lean" as a management concept, in both the manufacturing and service sectors. It takes a historical perspective in answering three research questions. To achieve this, 4.962 indexed records and 20 seminal books were analyzed by following a systematic literature review methodology.

Our research questions can be answered as follows:

About the historical origin of the term "lean": it was created in 1988 as "lean production system", a generic denomination for the Toyota production system. The best-selling book, *The Machine that Changed the World* (1990), populated the term "lean production" by absorbing other alternative expressions that existed at that time.

The previously used terms (which had similar intentions of denominating the Toyota production system without naming Toyota) were: *Japanese manufacturing techniques, stockless production, JIT production, value-added production, continuous improvement manufacturing, non-stock production,* and *fragile production.*

Since 1990, the term lean has evolved over time. Its evolution and diversification can be explained through four mechanisms (combined over time): expansion, transfer, targeting, and combination. This resulted in the creation of a confusing puzzle of lean specifier.

This paper has outlined the paths of evolution by using the most cited specifiers in the academic literature:

- Between 1990 and 2000, the term lean remained mainly in its original field of operations management, with the following specifiers: lean production, lean manufacturing, lean logistics, lean supply chain, lean product, lean construction, and lean and green. The first attempt to upgrade the concept to a more conceptual level was greeted with initially limited academic interest: lean management, lean enterprise, and lean thinking.

- In 2000, the combined term "lean six sigma" emerged and up to the present has received much attention in both the manufacturing and service sectors.
- Since 2006, the term "lean" was progressively applied in the service field with new specifiers: lean service, lean hospital, lean healthcare, lean office, lean startup.
- The last specifier, lean 4.0., was created in 2017 as a synergetic combination between lean manufacturing (or lean production) and the Industry 4.0 paradigm. At the moment, it focuses only on the manufacturing field.

The term "lean", as a management concept that allows organizations to remain competitive by removing waste from their processes, has been fully adopted by management researchers. Based on a bibliometric analysis of published papers-per-year, we can say that research interest in this topic has grown exponentially since 1988.

This paper reveals some implications for future research: The use of lean perspective can be further extended beyond its current development, adapting its principles and tools to different sectors or applications. The diversification mechanisms described above can open new research areas in a fast-changing, complex and competitive world. The lean approach combined with the new emerging disruptive technologies (so-called Industry 4.0) open new avenues for future research as intelligent construction, sustainability, smart cities, environmental improvement or public governance.

Author Contributions: Conceptualization, F.G.-V.; methodology, F.G.-V. and A.S.; formal analysis, J.A.Y.-F.; investigation, F.G.-V.; data curation, F.G.-V.; writing—original draft preparation, F.G.-V.; writing—review and editing, A.S.; supervision, J.A.Y.-F. All authors have read and agreed to the published version of the manuscript.

Funding: This paper was funded by the DGA-FSE (Diputación General de Aragón—Fondo Social Europeo) project T56_20R: Grupo de Ingeniería de Fabricación y Metrología Avanzada.

Institutional Review Board Statement: Not applicable.

Informed Consent Statement: Not applicable.

Conflicts of Interest: The authors declare no conflict of interest.

References

1. Jasti, N.V.K.; Kodali, R. Lean production: Literature review and trends. *Int. J. Prod. Res.* **2014**, *53*, 867–885. [CrossRef]
2. Moyano-Fuentes, J.; Sacristán-Díaz, M. Learning on lean: A review of thinking and research. *Int. J. Oper. Prod. Manag.* **2012**, *32*, 551–582. [CrossRef]
3. Psomas, E.; Antony, J. Research gaps in Lean manufacturing: A systematic literature review. *Int. J. Qual. Reliab. Manag.* **2019**, *36*, 815–839. [CrossRef]
4. Antony, J.; Psomas, E.; Garza-Reyes, J.A.; Hines, P. Practical implications and future research agenda of lean manufacturing: A systematic literature review. *Prod. Plan. Control.* **2020**, *32*, 889–925. [CrossRef]
5. Bhamu, J.; Sangwan, K.S. Lean manufacturing: Literature review and research issues. *Int. J. Oper. Prod. Manag.* **2014**, *34*, 876–940. [CrossRef]
6. Shah, R.; Ward, P.T. Defining and developing measures of lean production. *J. Oper. Manag.* **2007**, *25*, 785–805. [CrossRef]
7. Schonberger, R.J.; Schonberger, R.J. The disintegration of lean manufacturing and lean management. *Bus. Horizons* **2019**, *62*, 359–371. [CrossRef]
8. Giroux, H. 'It Was Such a Handy Term': Management Fashions and Pragmatic Ambiguity. *J. Manag. Stud.* **2006**, *43*, 1227–1260. [CrossRef]
9. Dorval, M.; Jobin, M.-H.; Benomar, N. Lean culture: A comprehensive systematic literature review. *Int. J. Prod. Perform. Manag.* **2019**, *68*, 920–937. [CrossRef]
10. Templier, M.; Paré, G. A Framework for Guiding and Evaluating Literature Reviews. *Commun. Assoc. Inf. Syst.* **2015**, *37*, 6. [CrossRef]
11. Tranfield, D.; Denyer, D.; Smart, P. Towards a methodology for developing evidence-informed management knowledge by means of systematic review. *Br. J. Manag.* **2003**, *14*, 207–222. [CrossRef]
12. Denyer, D.; Tranfield, D. Producing a Systematic Review. In *The SAGE Handbook of Organizational Research Methods*; SAGE: Thousand Oaks, CA, USA, 2009.
13. Ciano, M.P.; Pozzi, R.; Rossi, T.; Strozzi, F. How IJPR has addressed 'lean': A literature review using bibliometric tools. *Int. J. Prod. Res.* **2019**, *57*, 5284–5317. [CrossRef]
14. Danese, P.; Manfè, V.; Romano, P. A Systematic Literature Review on Recent Lean Research: State-of-the-art and Future Directions. *Int. J. Manag. Rev.* **2017**, *20*, 579–605. [CrossRef]

15. Soliman, M.; Saurin, T. Lean production in complex socio-technical systems: A systematic literature review. *J. Manuf. Syst.* **2017**, *45*, 135–148. [CrossRef]
16. Page, M.J.; McKenzie, J.E.; Bossuyt, P.M.; Boutron, I.; Hoffmann, T.C.; Mulrow, C.D.; Shamseer, L.; Tetzlaff, J.M.; Akl, E.A.; Brennan, S.E.; et al. The PRISMA 2020 statement: An updated guideline for reporting systematic reviews. *BMJ* **2021**, *372*, n71. [CrossRef]
17. Sinha, N.; Matharu, M. A comprehensive insight into Lean management: Literature review and trends. *J. Ind. Eng. Manag.* **2019**, *12*, 302–317. [CrossRef]
18. Hines, P.; Holweg, M.; Rich, N. Learning to evolve. *Int. J. Oper. Prod. Manag.* **2004**, *24*, 994–1011. [CrossRef]
19. Holweg, M. The genealogy of lean production. *J. Oper. Manag.* **2006**, *25*, 420–437. [CrossRef]
20. Sayles, L.R. Managerial Productivity: Who Is Fat and What Is Lean? *Interfaces* **1985**, *15*, 54–59. [CrossRef]
21. Havn, E.; Womack, J.P.; Jones, D.T.; Ross, D. *The Machine that Changed the World*; Rawson Associates: New York, NY, USA, 1990.
22. New, S.J. Celebrating the enigma: The continuing puzzle of the Toyota Production System. *Int. J. Prod. Res.* **2007**, *45*, 3545–3554. [CrossRef]
23. Krafcik, J.F. Triumph of the lean production system. *Sloan Manag. Rev.* **1988**, *30*, 41–52.
24. Shimada, H.; MacDuffie, J.P. *Industrial Relations and 'Humanware*; Working Paper; Sloan School of Management: Cambridge, MA, USA, 1986.
25. Sugimori, Y.; Kusunoki, K.; Cho, F.; Uchikawa, S. Toyota production system and Kanban system Materialization of just-in-time and respect-for-human system. *Int. J. Prod. Res.* **1977**, *15*, 553–564. [CrossRef]
26. Ohno, T.I. *Toyota Seisan Hoshiki: Datsu-Kibo no Keiei o Mezashite*; DiamondYsha: Tokyo, Japan, 1978.
27. Ohno, T. How the Toyota Production System was Created. *Jpn. Econ. Stud.* **1982**, *10*, 83–101. [CrossRef]
28. Sato, K.; Hoshino, Y. *The Anatomy of Japanase Business*; M.E. Sharpe Inc.: Armonk, NY, USA, 1984.
29. Shingo, S.; Dillon, A.P. *A Study of the Toyota Production System: From an Industrial Engineering Viewpoint (Produce What Is Needed, When It's Needed)*; CRC Press: Boca Raton, FL, USA, 1898.
30. Monden, Y. *Toyota Production System: Practical Approach to Production Management*; Industrial Engineering and Management Press, Institute of Industrial Engineers: Atlanta, GA, USA, 1983.
31. Schonberger, R.J. *Japanese Manufacturing Techniques: Nine Hidden Lessons in Simplicity*; The Free Press: New York, NY, USA, 1982.
32. Hall, R.W. *Zero Inventory*; Dow Jones-Irwin Professional Pub: Homewood, IL, USA, 1983.
33. Hall, R.W.; Hall, J. *Attaining Manufacturing Excellence: Just-in-Time, Total Quality, Total People Involvement*; Dow Jones-Irwin Professional Pub: Homewood, IL, USA, 1987.
34. Cusumano, M.A. *The Japanese Automobile Industry: Technology and Management at Nissan and Toyota. Council on East Asian Studies*; Harvard University Press: Cambridge, MA, USA, 1985.
35. Shingō, S. *A Revolution in Manufacturing: The SMED System*; Productivity Press: New York, NY, USA, 1985.
36. Japan Management Association. *Kanban Just-in Time at Toyota: Management Begins at the Workplace*; Productivity Press: New York, NY, USA, 1986.
37. Imai, M. *Kaizen (Ky'zen): The Key to Japan's Competitive Success*; Random Housre Business Division: New York, NY, USA, 1986.
38. Nakajima, S. *Introduction to TPM: Total Productive Maintenance*; Productivity Press: New York, NY, USA, 1988.
39. Ohno, T. *Toyota Production System: Beyond Large-Scale Production*; Productivity Press: New York, NY, USA, 1988.
40. Williams, K.; Haslam, C.; Williams, J.; Cultler, T.; Adcroft, A.; Johal, S. Against lean production. *Econ. Soc.* **1992**, *21*, 321–354. [CrossRef]
41. Berggren, C. Lean Production—The End of History? *Work. Employ. Soc.* **1993**, *7*, 163–188. [CrossRef]
42. Cusumano, M. The limits of "Lean". *Sloan Manag. Rev.* **1994**, *35*, 4.
43. Delbridge, R.; Oliver, N. Narrowing the gap? Stock turns in the Japanese and Western car industries. *Int. J. Prod. Res.* **1991**, *29*, 2083–2095. [CrossRef]
44. Womack, J.P.; Jones, D.T. From lean production to the lean enterprise. *IEEE Eng. Manag. Rev.* **1996**, *24*, 4.
45. Womack, J.P.; Jones, D.T. *Lean Thinking: Banish Waste and Create Wealth in Your Corporation*; Simon & Schuster: New York, NY, USA, 1996.
46. Bicheno, J. *The Lean Toolbox*; PICSIE Books: Buckingham, UK, 1998.
47. Marodin, G.A.; Saurin, T. Implementing lean production systems: Research areas and opportunities for future studies. *Int. J. Prod. Res.* **2013**, *51*, 6663–6680. [CrossRef]
48. Hai, H.; Wenxing, L. The Analysis of Industry 4.0 and Lean Production. In Proceedings of the Second International Symposium-Management, Innovation and Development, Guangzhou, China, 1 December 2015.
49. Rossini, M.; Costa, F.; Tortorella, G.; Portioli-Staudacher, A. The interrelation between Industry 4.0 and lean production: An empirical study on European manufacturers. *Int. J. Adv. Manuf. Technol.* **2019**, *102*, 3963–3976. [CrossRef]
50. Ciano, M.P.; Dallasega, P.; Orzes, G.; Rossi, T. One-to-one relationships between Industry 4.0 technologies and Lean Production techniques: A multiple case study. *Int. J. Prod. Res.* **2020**, *59*, 1386–1410. [CrossRef]
51. Powell, C. Lean Manufacturing Organization, 21st Century. In Proceedings of the Annual International Conference Proceedings-American Production and Inventory Control Society, 1 November 1993; Available online: https://www.amazon.com/International-Conference-Proceedings-INTERNATIONAL-PROCEEDINGS/dp/1558221042 (accessed on 28 October 2021).

52. Shah, R.; Ward, P.T. Lean manufacturing: Context, practice bundles, and performance. *J. Oper. Manag.* **2002**, *21*, 129–149. [CrossRef]
53. Jasti, N.V.K.; Kodali, R. A literature review of empirical research methodology in lean manufacturing. *Int. J. Oper. Prod. Manag.* **2014**, *34*, 1080–1122. [CrossRef]
54. Sanders, A.; Elangeswaran, C.; Wulfsberg, J. Industry 4.0 implies lean manufacturing: Research activities in industry 4.0 function as enablers for lean manufacturing. *J. Ind. Eng. Manag.* **2016**, *9*, 811–833. [CrossRef]
55. Valamede, L.S.; Akkari, A.C.S. Lean Manufacturing and Industry 4.0: A Holistic Integration Perspective in the Industrial Context. 2020, pp. 63–68. Available online: https://ieeexplore.ieee.org/abstract/document/9080393 (accessed on 28 October 2021).
56. Fynes, B.; Ennis, S. From lean production to lean logistics: The case of microsoft Ireland. *Eur. Manag. J.* **1994**, *12*, 322–331. [CrossRef]
57. Lamming, R. Squaring lean supply with supply chain management. *Int. J. Oper. Prod. Manag.* **1996**, *16*, 183–196. [CrossRef]
58. Huallacháin, B.; Wasserman, D. Vertical Integration in a Lean Supply Chain: Brazilian Automobile Component Parts*. *Econ. Geogr.* **2008**, *75*, 21–42. [CrossRef]
59. Perez, C.; De Castro, R.; Simons, D.; Gimenez, G. Development of lean supply chains: A case study of the Catalan pork sector. *Supply Chain Manag. Int. J.* **2010**, *15*, 55–68. [CrossRef]
60. Garcia-Buendia, N.; Moyano-Fuentes, J.; Maqueira-Marín, J.M.; Cobo, M.J. 22 Years of Lean Supply Chain Management: A science mapping-based bibliometric analysis. *Int. J. Prod. Res.* **2020**, *59*, 1901–1921. [CrossRef]
61. Tortorella, G.L.; Miorando, R.F.; Fries, C.E.; Vergara, A.M.C. On the relationship between Lean Supply Chain Management and performance improvement by adopting Industry 4.0 technologies. In Proceedings of the International Conference on Industrial Engineering and Operations Management, Paris, France, 26–27 July 2018.
62. Tortorella, G. Erratum. *Supply Chain Manag. Int. J.* **2019**, *24*, 301. [CrossRef]
63. Núñez-Merino, M.; Maqueira-Marín, J.M.; Moyano-Fuentes, J.; Martínez-Jurado, P.J. Information and digital technologies of Industry 4.0 and Lean supply chain management: A systematic literature review. *Int. J. Prod. Res.* **2020**, *58*, 5034–5061. [CrossRef]
64. Jones, D.T.; Hines, P.; Rich, N. Lean logistics. *Int. J. Phys. Distrib. Logist. Manag.* **1997**, *27*, 153–173. [CrossRef]
65. Petrovic, O.; Zsifkovits, H.E. Business process re-engineering as an enabling factor for lean management. In Proceedings of the IFIP TC8 Open Conference on Business Process Re-engineering: Information Systems Opportunities and Challenges, Queensland Gold Coast, Australia, 8–11 May 1994.
66. Martínez-Jurado, P.J.; Moyano-Fuentes, J. Lean Management, Supply Chain Management and Sustainability: A Literature Review. *J. Clean. Prod.* **2014**, *85*, 134–150. [CrossRef]
67. Martinez, F.; Jirsak, P.; Lorenc, M. Industry 4.0. the End Lean Management? In *The 10th International Days of Statistics and Economics*; Melandrium: Prague, Czech Republic, 2016; pp. 1189–1197.
68. Jasti, N.V.K.; Kota, S.; Kale, S.R. Development of a framework for lean enterprise. *Meas. Bus. Excel.* **2020**, *24*, 431–459. [CrossRef]
69. Dombrowski, U.; Wullbrandt, J.; Fochler, S. Center of Excellence for Lean Enterprise 4.0. *Procedia Manuf.* **2019**, *31*, 66–71. [CrossRef]
70. Koskela, L. Lean production in lean construction. In Proceedings of the National Construction and Management Conference, Sydney, Australia, 17–18 February 1994.
71. Thomas, H.R.; Horman, M.J.; De Souza, U.E.L.; Završki, I. Reducing Variability to Improve Performance as a Lean Construction Principle. *J. Constr. Eng. Manag.* **2002**, *128*, 144–154. [CrossRef]
72. Salem, O.; Solomon, J.; Genaidy, A.; Minkarah, I. Lean Construction: From Theory to Implementation. *J. Manag. Eng.* **2006**, *22*, 168–175. [CrossRef]
73. Koskela, L.; Ferrantelli, A.; Niiranen, J.; Pikas, E.; Dave, B. Epistemological Explanation of Lean Construction. *J. Constr. Eng. Manag.* **2019**, *145*, 04018131. [CrossRef]
74. Lekan, A.; Clinton, A.; Fayomi, O.S.I.; James, O. Lean Thinking and Industrial 4.0 Approach to Achieving Construction 4.0 for Industrialization and Technological Development. *Buildings* **2020**, *10*, 221. [CrossRef]
75. Mano, A.P.; Da Costa, S.E.G.; De Lima, E.P. Criticality assessment of the barriers to Lean Construction. *Int. J. Prod. Perform. Manag.* **2020**, *70*, 65–86. [CrossRef]
76. Florida, R. Lean and Green: The Move to Environmentally Conscious Manufacturing. *Calif. Manag. Rev.* **1996**, *39*, 80–105. [CrossRef]
77. King, A.A.; Lenox, M.J. Lean and green? An empirical examination of the relationship between lean production and environmental performance. *Prod. Oper. Manag.* **2009**, *10*, 244–256. [CrossRef]
78. Kainuma, Y.; Tawara, N. A multiple attribute utility theory approach to lean and green supply chain management. *Int. J. Prod. Econ.* **2006**, *101*, 99–108. [CrossRef]
79. Garza-Reyes, J.A. Lean and green—A systematic review of the state-of-the-art literature. *J. Clean. Prod.* **2015**, *102*, 18–29. [CrossRef]
80. Farias, L.M.S.; Santos, L.C.; Gohr, C.F.; de Oliveira, L.C.; Amorim, M.H.D.S. Criteria and practices for lean and green performance assessment: Systematic review and conceptual framework. *J. Clean. Prod.* **2019**, *218*, 746–762. [CrossRef]
81. Farias, L.M.S.; Santos, L.C.; Gohr, C.F.; Rocha, L.O. An ANP-based approach for lean and green performance assessment. *Resour. Conserv. Recycl.* **2018**, *143*, 77–89. [CrossRef]
82. Moro, S.R.; Cauchick-Miguel, P.A.; Campos, L.M.S. Product-service systems towards eco-effective production patterns: A Lean-Green design approach from a literature review. *Total. Qual. Manag. Bus. Excel.* **2019**, *32*, 1046–1064. [CrossRef]

83. Duarte, S.; Cabrita, M.D.R.; Cruz-Machado, V. Business Model, Lean and Green Management and Industry 4.0: A Conceptual Relationship. *Adv. Intell. Syst. Comput.* **2019**, *1*, 359–372. [CrossRef]
84. Leong, W.D.; Teng, S.Y.; How, B.S.; Ngan, S.L.; Rahman, A.A.; Tan, C.P.; Ponnambalam, S.; Lam, H.L. Enhancing the adaptability: Lean and green strategy towards the Industry Revolution 4.0. *J. Clean. Prod.* **2020**, *273*, 122870. [CrossRef]
85. Piercy, N.F.; Morgan, N.A. The impact of lean thinking and the lean enterprise on marketing: Threat or synergy? *J. Mark. Manag.* **1997**, *13*, 679–693. [CrossRef]
86. Bittencourt, V.L.; Alves, A.; Leão, C.P. Industry 4.0 triggered by Lean Thinking: Insights from a systematic literature review. *Int. J. Prod. Res.* **2020**, *59*, 1496–1510. [CrossRef]
87. Karlsson, C.; Ahlstrom, P. The difficult path to lean product development. *J. Prod. Innov. Manag.* **1996**, *13*, 283–295. [CrossRef]
88. León, H.C.M.; Farris, J.A. Lean Product Development Research: Current State and Future Directions. *Eng. Manag. J.* **2011**, *23*, 29–51. [CrossRef]
89. Liker, J.K.; Morgan, J.M. The Toyota Way in Services: The Case of Lean Product Development. *Acad. Manag. Perspect.* **2006**, *20*, 5–20. [CrossRef]
90. Sassanelli, C.; Pezzotta, G.; Rossi, M.; Terzi, S.; Cavalieri, S. Towards a Lean Product Service Systems (PSS) Design: State of the Art, Opportunities and Challenges. *Procedia CIRP* **2015**, *30*, 191–196. [CrossRef]
91. Sassanelli, C.; Rossi, M.; Pezzotta, G.; Pacheco, D.; Terzi, S. Defining lean product service systems features and research trends through a systematic literature review. *Int. J. Prod. Lifecycle Manag.* **2019**, *12*. [CrossRef]
92. Bowen, D.E.; Youngdahl, W.E. "Lean" service: In defense of a production-line approach. *Int. J. Serv. Ind. Manag.* **1998**, *9*, 207–225. [CrossRef]
93. Swank, C.K. The lean service machine. *Harv. Bus. Rev.* **2003**, *81*, 123.
94. Suárez-Barraza, M.F.; Smith, T.; Dahlgaard-Park, S.M. Lean Service: A literature analysis and classification. *Total. Qual. Manag. Bus. Excel.* **2012**, *23*, 359–380. [CrossRef]
95. Sprigg, C.A.; Jackson, P.R. Call centers as lean service environments: Job-related strain and the mediating role of work design. *J. Occup. Heal. Psychol.* **2006**, *11*, 197–212. [CrossRef] [PubMed]
96. Piercy, N.; Rich, N. High quality and low cost: The lean service centre. *Eur. J. Mark.* **2009**, *43*, 1477–1497. [CrossRef]
97. LaGanga, L.R. Lean service operations: Reflections and new directions for capacity expansion in outpatient clinics. *J. Oper. Manag.* **2010**, *29*, 422–433. [CrossRef]
98. Gupta, S.; Sharma, M. Lean services: A systematic review. *Int. J. Prod. Perform. Manag.* **2016**, *65*, 1025–1056. [CrossRef]
99. Sheridan, J.H. "Lean Sigma" Synergy. *Ind. Week* **2000**, *249*, 17.
100. Furterer, S.; Elshennawy, A.K. Implementation of TQM and lean Six Sigma tools in local government: A framework and a case study. *Total. Qual. Manag. Bus. Excel.* **2005**, *16*, 1179–1191. [CrossRef]
101. Wheat, B.; Mills, C.; Carnell, M. *Leaning into Six Sigma: A Parable of the Journey to Six Sigma and a Lean Enterprise*; McGraw Hill Professional: New York, NY, USA, 2003.
102. Kumar, M.; Antony, J.; Singh, R.; Tiwari, M.K.; Perry, D. Implementing the Lean Sigma framework in an Indian SME: A case study. *Prod. Plan. Control.* **2006**, *17*, 407–423. [CrossRef]
103. Albliwi, S.A.; Antony, J.; Lim, S.A.H. A systematic review of Lean Six Sigma for the manufacturing industry. *Bus. Process. Manag. J.* **2015**, *21*, 665–691. [CrossRef]
104. Rosa, A.C.M.; Broday, E.E. Comparative analysis between the industrial and service sectors: A literature review of the improvements obtained through the application of lean six sigma. *Int. J. Qual. Res.* **2018**, *12*, 1. [CrossRef]
105. Sony, M. Design of cyber physical system architecture for industry 4.0 through lean six sigma: Conceptual foundations and research issues. *Prod. Manuf. Res.* **2020**, *8*, 158–181. [CrossRef]
106. Chiarini, A.; Kumar, M. Lean Six Sigma and Industry 4.0 integration for Operational Excellence: Evidence from Italian manufacturing companies. *Prod. Plan. Control.* **2020**, *32*, 1084–1101. [CrossRef]
107. Productivity Press Development Team. *The Lean Office, Collected Practices & Cases*; Productivity Press: New York, NY, USA, 2005.
108. Locher, D. *Lean Office and Service Simplified: The Definitive How-To Guide*; Taylor & Francis: Abingdon, UK, 2011.
109. Herkommer, J.; Herkommer, O. Lean Office-System. *Z. Wirtsch. Fabr.* **2006**, *101*, 378–381. [CrossRef]
110. Nieuwenhuis, M.; Knight, C.; Postmes, T.; Haslam, S.A. The relative benefits of green versus lean office space: Three field experiments. *J. Exp. Psychol. Appl.* **2014**, *20*, 199–214. [CrossRef]
111. Freitas, R.D.C.; Freitas, M.D.C.D. Information management in lean office deployment contexts. *Int. J. Lean Six Sigma* **2020**, *11*, 1161–1192. [CrossRef]
112. Yokoyama, T.T.; DE Oliveira, M.A.; Futami, A.H. A Systematic Literature Review on Lean Office. *Ind. Eng. Manag. Syst.* **2019**, *18*, 67–77. [CrossRef]
113. Portioli-Staudacher, A. *Lean Healthcare. An Experience in Italy*; Springer: Berlin/Heidelberg, Germany, 2008; pp. 485–492. [CrossRef]
114. Graban, M. *Lean Hospitals: Improving Quality, Patient Safety, and Employee Engagement*; CCR Press, Taylor & Francis Group: New York, NY, USA, 2009.
115. Dahlgaard, J.J.; Pettersen, J.; Dahlgaard-Park, S.M. Quality and lean health care: A system for assessing and improving the health of healthcare organisations. *Total. Qual. Manag. Bus. Excel.* **2011**, *22*, 673–689. [CrossRef]

116. Costa, L.B.M.; Filho, M.G. Lean healthcare: Review, classification and analysis of literature. *Prod. Plan. Control.* **2016**, *27*, 823–836. [CrossRef]
117. Santos, A.C.D.S.G.D.; Reis, A.; De Souza, C.G.; Dos Santos, I.L.; Ferreira, L.A.F. The first evidence about conceptual vs analytical lean healthcare research studies. *J. Heal. Organ. Manag.* **2020**, *34*, 789–806. [CrossRef]
118. Blank, S. Why the lean start-up changes everything. *Harv. Bus. Rev.* **2013**, *91*, 5.
119. Ries, E. *The Lean Startup: How Today's Entrepreneurs Use Continuous Innovation to Create Radically Successful Businesses*; Crown Business: New York, NY, USA, 2011.
120. Frederiksen, D.L.; Brem, A. How do entrepreneurs think they create value? A scientific reflection of Eric Ries' Lean Startup approach. *Int. Entrep. Manag. J.* **2016**, *13*, 169–189. [CrossRef]
121. Bortolini, R.F.; Cortimiglia, M.N.; Danilevicz, A.D.M.F.; Ghezzi, A. Lean Startup: A comprehensive historical review. *Manag. Decis.* **2018**, *59*, 1765–1783. [CrossRef]
122. Silva, D.S.; Ghezzi, A.; de Aguiar, R.B.; Cortimiglia, M.N.; Caten, C.S.T. Lean Startup, Agile Methodologies and Customer Development for business model innovation. *Int. J. Entrep. Behav. Res.* **2019**, *26*, 595–628. [CrossRef]
123. Metternich, J.; Müller, M.; Meudt, T.; Schaede, C. Lean 4.0—Zwischen Widerspruch und Vision. *ZWF* **2017**, *112*, 346–348. [CrossRef]
124. Mayr, A.; Weigelt, M.; Kühl, A.; Grimm, S.; Erll, A.; Potzel, M.; Franke, J. Lean 4.0—A conceptual conjunction of lean management and Industry 4.0. *Procedia CIRP* **2018**, *72*, 622–628. [CrossRef]
125. Valamede, L.S.; Akkari, A.C.S. Lean 4.0: A New Holistic Approach for the Integration of Lean Manufacturing Tools and Digital Technologies. *Int. J. Math. Eng. Manag. Sci.* **2020**, *5*, 851–868. [CrossRef]
126. Perico, P.; Mattioli, J. Empowering Process and Control in Lean 4.0 with Artificial Intelligence. In Proceedings of the 3rd International Conference on Artificial Intelligence for Industries (AI4I), Irvine, CA, USA, 21–23 September 2020; pp. 6–9. [CrossRef]
127. Ejsmont, K.; Gladysz, B.; Corti, D.; Castaño, F.; Mohammed, W.M.; Lastra, J.L.M. Towards "Lean Industry 4.0"—Current trends and future perspectives. *Cogent Bus. Manag.* **2020**, *7*, 1781995. [CrossRef]
128. Wagner, T.A.; Herrmann, C.; Thiede, S. Industry 4.0 Impacts on Lean Production Systems. *Procedia CIRP* **2017**, *63*, 125–131. [CrossRef]
129. Buer, S.-V.; Strandhagen, J.O.; Chan, F.T.S. The link between Industry 4.0 and lean manufacturing: Mapping current research and establishing a research agenda. *Int. J. Prod. Res.* **2018**, *56*, 2924–2940. [CrossRef]
130. Duarte, S.; Cruz-Machado, V. An investigation of lean and green supply chain in the Industry 4.0. In Proceedings of the 2017 International Symposium on Industrial Engineering and Operations Management (IEOM), Rabat, Morocco, 11–13 April 2017.
131. Ilangakoon, T.; Weerabahu, S.; Wickramarachchi, R. Combining Industry 4.0 with Lean Healthcare to Optimize Operational Performance of Sri Lankan Healthcare Industry. In Proceedings of the International Conference on Production and Operations Management Society (POMS), Peradeniya, Sri Lanka, 14–16 December 2018; pp. 1–8. [CrossRef]
132. Sony, M. Industry 4.0 and lean management: A proposed integration model and research propositions. *Prod. Manuf. Res.* **2017**, *6*, 416–432. [CrossRef]
133. Yadav, N.; Shankar, R.; Singh, S.P. Impact of Industry4.0/ICTs, Lean Six Sigma and quality management systems on organisational performance. *TQM J.* **2020**, *32*, 815–835. [CrossRef]

applied sciences

MDPI

Article

RFID Technology as a Low-Cost and Passive Way to Digitize Industrial Analogic Indicators

Mohammadamin Hosseinifard [1,*], Salam Alzubaidi [1], Andrea Michel [2] and Gualtiero Fantoni [1]

1 Department of Civil and Industrial Engineering, University of Pisa, 56122 Pisa, Italy; salam.al-zubaidi@phd.unipi.it (S.A.); gualtiero.fantoni@unipi.it (G.F.)
2 Department of Information Engineering, University of Pisa, 56122 Pisa, Italy; andrea.michel@unipi.it
* Correspondence: Mohammad.Hosseinifard@phd.unipi.it

Abstract: Simple analog devices like manometers, manual valves, etc., have been ignored in the digitization process that has characterized the transition towards Industry 4.0. The reason behind this is that their substitution with the equivalent digital versions is high cost and needs re-wiring. This study introduces a low-cost wireless and passive model aligned with the Industry 4.0 paradigm to digitize analog indicators. The concept is based on electromagnetic (EM) shielding of the manometer's embedded radio frequency identification (RFID) tag. We designed and tuned a new tiny RFID tag to be embedded into analog devices. Finally, a digitized manometer by RFID electromagnetic shielding concept is simulated in the Ansys HFSS modeling environment.

Keywords: tiny RFID; metallic electromagnetic isolation; analog manometer; Ansys HFSS simulator; passive digitization

Citation: Hosseinifard, M.; Alzubaidi, S.; Michel, A.; Fantoni, G. RFID Technology as a Low-Cost and Passive Way to Digitize Industrial Analogic Indicators. *Appl. Sci.* **2022**, *12*, 1451. https://doi.org/10.3390/app12031451

Academic Editor: Roque Calvo

Received: 31 December 2021
Accepted: 24 January 2022
Published: 29 January 2022

Publisher's Note: MDPI stays neutral with regard to jurisdictional claims in published maps and institutional affiliations.

1. Introduction

The digitization process of raw materials, semifinished and finished products, and machines play a significant role in the Industry 4.0 revolution and rely mainly on the Internet of Things (IoT) and radio frequency identification (RFID) technologies. However, many challenges related to the digitization process include and are not limited to high cost, energy consumption, and information integration. One of the digitization challenges is the design and manufacturing of analogic indicators and manometers, which take into account many industrial and engineering design considerations [1].

Analog devices like manometers, manual valves, handwheels, levers, clamps, etc., are very common on shopfloors. They are present in old machines, but also in the piping system of modern plants, e.g., in fire extinguishers and in security elements such as cage doors, etc. When integrated with old equipment and types of machinery, they are retrofitted using digital devices, switches and IoT cabled or wireless solutions, but alternatively, their monitoring is left to operators with consequent time loss, errors, etc. Sometimes, due to the slow evolution of the phenomena they monitor, they are left unread for a long time with possible consequences on safety (fire extinguishers), performance (wear and corrosion), and so on. Such oblivion is critical for the concept of completely twinning [2] the shop floor in order to be able to digitally manage the production (people and machines) in a cyber-physical way [3].

Data collection based on the passive RFID tags approach is a valuable and competitive solution for the digitization process as a low-cost, battery-less asset and has a long life compared with other techniques [4]. However, a limitation of passive RFID systems is the relatively short read range with respect to other active systems [5]. Larger read ranges can be achieved at the cost of a bigger tag antenna size, which makes the integration process into existing devices more difficult. Therefore, this study aims to simplify the digitization process considering this drawback with low costs, using enabling technologies and embedded systems for data collection in a real-time manner. The challenge is to design and embed small passive RFID tags inside the analog devices, keeping an acceptable reading range. Many researchers have discussed RFID limitations and challenges dealing with the environment and substances and their influences like reading range reductions and wave reflections and suggested different possible solutions [6–8]. The general concept of the methodology is based on the material behavior and its impact on the RFID system. Hence, the radio waves pass through air, plastic, wood, and ceramics efficiently, while there is a considerable reduction and reflection with other materials such as conductors (e.g., metals). The use of metals in industrial environments creates blind spots for RFID readers and disrupts their performance [9,10]. Covering the RFID tag with metals causes electromagnetic (EM) shielding, but a movement of the metal cover removes the isolation and turns the RFID chip on. The adverse effects of metals on RFID performance have also been investigated by Arora [11].

This work aims to digitize analog devices by integrating new tiny passive RFID tags [9]. We performed experiments in large dimensions with commercial RFID tags to evaluate the technical feasibility of the concept. A tag antenna was designed, virtually localized in an analog manometer, and numerically simulated using Ansys HFSS V2021 R2 software.

Then, the EM shielding cover was optimized, and the final digitized analog manometer performance was simulated. The following sections illustrate the state of the art, then the methodology will be explained in detail. Later, in the experiments, Ansys HFSS simulation practices and results are described, and finally, conclusions are drawn.

2. The State of the Art

2.1. Potential of Wireless Networks in the Industry

Wireless network protocols as an IoT technology are used to enable physical objects to collect and exchange data nowadays of Industry 4.0 era [12]. Wireless networks are promising for current and future industrial aspects in terms of the digital transformation of companies, especially in automating the process of logistics, asset tracking digitization, monitoring, and controls [13]. The growing influence of IoT has created great potential for increasing the share of wireless communication networks in the industry. Compiling Hardware Meets Software (HMS) team annual reports for the last seven years (Figure 1) show that the majority of industrial networks' share (approximately 94%) is related to industrial networks with cable platforms such as Fieldbus and industrial Ethernet [14]. However, the wireless network graph has constantly been rising, from about zero percent in 2015 to 7% in 2021. This trend illustrates a clear picture of the future position of wireless networks in the industry.

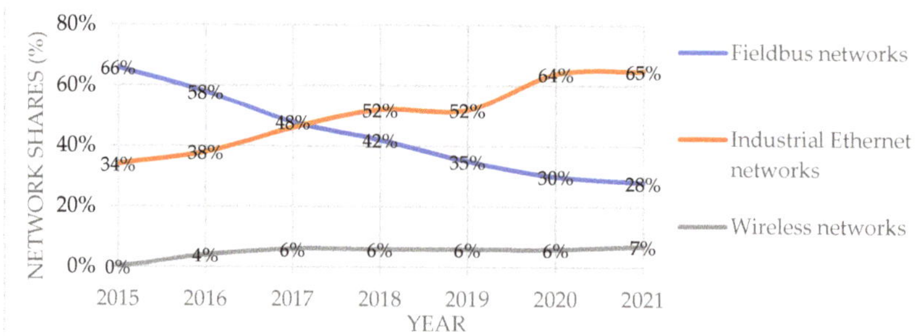

Figure 1. Industrial networks' market shares.

In [15], the authors merged the wireless network and RFID technology to manage the production line in a traditional manufacturing laboratory to avoid errors due to the manual recording, remotely monitoring the manufacturing process via the internet, and evaluating the workers' progress. In [16], authors proved that the transmission speed, low energy consumption, and a higher number of devices could be simultaneously operated using a wireless network. Furthermore, when discussing IoT applications in industry, we found the acronym IIoT that stands for Industrial IoT and that it is building a new domain with respect to IoT in everyday products. All the considerations mentioned above demonstrate that passive RFID could be a promising solution for digitizing analogical devices.

2.2. Embedded Passive RFID Tags

Passive RFID tags are increasingly used to automate systems, identify products, trace assets, monitor, measure the process, and control their physical parameter [17]. But embedding passive RFID tags requires many considerations related to the RFID itself, the object or systems, and/or the space that contains the RFID [18]. In [19], the authors explained the importance of the RFID tag design parameters in terms of size, shape, configuration, the material of the antenna and chip, and the objects that RFID tags install on or are embedded. The RFID tag design parameters are essential, so embedding passive RFID tags in the analog indicators and manometers is a practical challenge and potential research area. Digital transmitters could replace these manometers, but safety standards emphasize the need to keep them as redundancies for analogical ones. These restrictions ensure that operators/supervisors can monitor critical parameters such as boiler temperature and pressure in both methods, on-site/physically and in an automation system/digitally [20]. In this regard, the purpose of our practical application is to digitize hotspots and trigger points in an analog manometer passively and wirelessly.

Moreover, many industrial environments present "ATmosphere EXplosive" areas where digital devices have to follow particular restrictions and their ATEX certified versions have a higher cost than the standard ones.

2.3. Electromagnetic Isolation

The EM isolation between transmitter (RFID reader) and receiver (RFID tag) is a critical problem in a pulse-modulated RFID system, where transmitter and receiver use identical frequencies [21]. Metals can create blind spots in RFID reading zones by completely reflecting high-frequency EM waves [22]. These blind spots cause signal interruption and receiving signal strength indicator (RSSI) variations [23]. RSSI assesses the attenuation of a received signal with respect to the source. The value of this parameter is always negative, and numbers closer to zero indicate better power and quality of the network [24]. Therefore, embedding the passive RFID tags in the analogic devices will be affected by the composed material of the devices, especially when the cover is made of metallic components. The isolation of passive RFID tags from EM waves transmitted from the RFID reader is not always a disadvantage, especially when the physical parameters required are not continuously varied. So, the metallic interaction could play an essential role in RFID isolation from the reader whenever needed in an automatic mode, and the passive RFID tags could be detected whenever the metallic cover is (re)moved. This dynamic scenario supported by the facts mentioned in the paragraphs above shall guide us to the main research questions:

- Can RFID technology digitize the measurement process of analogic devices?
- Are passive RFID tags appropriate to be embedded in the analogic systems collecting the data wirelessly?

The following sections will clearly answer and explain how RFID technology can support the digitization of the measurement process of analogic devices.

3. Materials and Methods

Passive RFID tags, due to the lack of internal power supply, are dependent on the energy absorption sent from the RFID reader antenna, making the tag's performance highly linked to the environment and the installation location. For this reason, the RFID must be tuned precisely for its installation substrate (analog manometer) then simulation must be performed for all RFID tag, substrate, and metallic cover parameters [25,26]. For this reason, the research methodology is divided into three sections.

- Experiment and conceptual test;
- Design and tune a new RFID tag;
- Electromagnetic isolation cover simulation and optimization.

The study methodology started with modeling EM isolation effects on commercial RFID tags [26]. Due to the large dimension of general purpose RFID tags (a few centimeters), a test bench was designed on a large scale. Modeling was done to test the initial concept and determine the influential variables. In the second part, an analog manometer is introduced as one of the possible applications to be digitized with this method. A tiny RFID tag was designed and tuned to be implanted into the analog manometer. Finally, we execute a compound EM simulation to validate the analog manometer digitization concept for the embedded RFIDs, manometer, and metallic cover.

3.1. Experiment and Conceptual Test

The general idea was to design a bench test based on covering the RFID tags with different metal plates. Circular openings with various radiuses were manufactured in metal plates that allow electromagnetic waves to pass through. The RFID tag was slowly moved towards or backward of the isolation area by rotating the metallic sheet, thus examining the effect of changing the amount of EM isolation in RFID tags. The operating frequency of UHF RFID was in the range of 860 to 930 Mhz. The size of the smallest RFID tag was 44 × 44 mm, therefore, the smallest size of the circular window was set to 40 mm, while the largest size of the circular window to 80 mm, which corresponds to about $\lambda/4$ at 900 MHz. The EM waves emitted from the RFID reader antenna charge the tag antenna, and the absorbed energy turn on the tag chip by moving the cover. A stepper motor drives the shielding cover with an accuracy of 3° degree of rotation per step. The stepper motor was driven in half-step mode by a 4ZeroBox (a microcontroller unit suitable for industrial applications) and controls the metallic cover [27]. The measurements were performed by a CAEN UHF RFID reader model R4300, and the data was sent to the computer through the serial port (Figure 2a). The reader was connected to a 9 dB linearly polarized antenna by SMA adapters and a 20 foot RG-58 cable with 1db per 10 feet loss in the 900 MHz frequency range. According to Figure 2b, the metal cover moved from position 1 in steps of 3° to position 2, so that half of the tag was exposed to EM waves, then moving from position 2 to 3 caused the entire surface of the tag antenna to be uncovered. The same pattern was repeated in positions 4 and 5, and the whole surface of the tag antenna was to pass from covered to uncovered and vice versa.

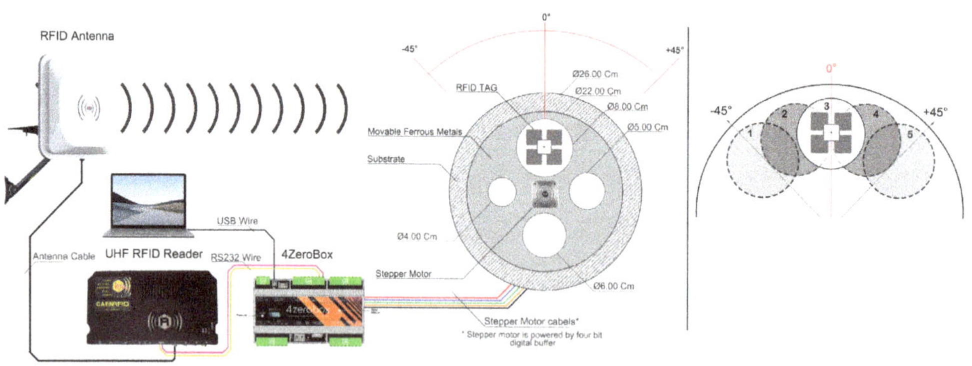

(a) (b)

Figure 2. Test bench design and covering process.

RSSI measurements were performed in 1080 samples for three different RFID tags and three different metal covers. The passive tags specifications are given in Table 1. The chips used in the RFID tags belong to the same family, but the structure and design of their antennas were entirely different and designed for various purposes.

Table 1. Passive tags specification.

RFID Passive Tag	Protocol	Chip Type	EPC Memory	User Memory	Antenna Size	Read Range
True 3D QTTM Long Reading	ISO/IEC 18000-6C (class 1 Gen2)	Impinj Monza-4QT QTTM	128 bits	512 bits	72 × 72 mm	10–12 m
Impinj h47	ISO/IEC 18000-6C (class 1 Gen2)	Impinj Monza 4E	496 bits	128 bits	44 × 44 mm	6–7 m
Impinj Monza 4D	ISO 18000-6C, EPC Class 1 Gen 2	Impinj Monza 4D	128-bit	32-bit	86 × 24 mm	8–10 m

3.2. *Design and Tuning of RFID Tags*

The commercial tags' large dimension increased the modeling test bench size and made it impossible to implement the concept directly in practical applications. Therefore, an RFID tag was designed and simulated with the priority of keeping the antenna dimensions so limited that they could be embedded in an analog manometer. The proposed solution, shown in Figure 3, is based on creating an EM shield for the RFID tags located between the dial layer of the manometer and a metallic cover. At least two RFID tags need to be embedded under the shield layer to determine the measuring area of the manometer (green or red zone).

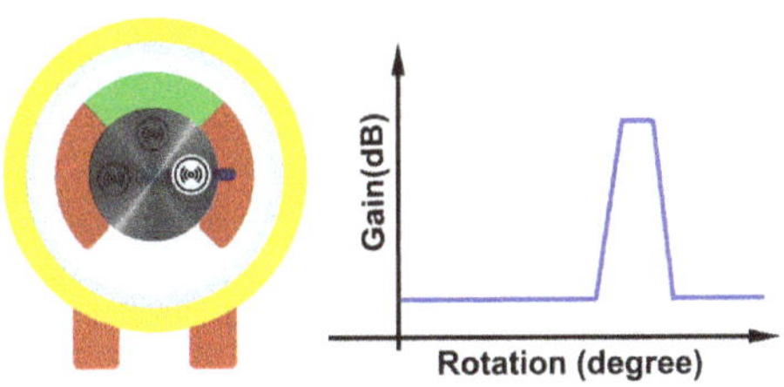

Figure 3. Analog manometer conceptual solution design.

3.2.1. Analog Manometer Digitization Concept

Analog manometers are widely used in industry to measure various physical parameters. Despite the diverse applications for which these manometers are designed, they all have similar mechanical mechanisms. The basic concept of this study was to implant small RFID tags on a single-side printed circuit board (PCB) FR4 substrate in analog manometers. The FR4 layer is placed on the manometer dial layer and covered with a metal cover to rotate with the manometer's pointer. The impact of cover addition weight on the manometer performance was neutralized by two offset adjustment screws inside the manometer. A circular opening in line with the pointer direction in metallic cover let EM waves pass to turn on the red or green zone RFID tag. A typical commercial RFID reader determines the manometer zone (green or red) by reading the tag's electronic product code (EPC) numbers.

3.2.2. RFID Tag Design

Commercial RFIDs in the 900 MHz frequency band are usually designed in the size of a few centimeters. Many restrictions, including (i) lambda wavelength in this frequency band; (ii) limited sensitivity of RFID microchips; (iii) chip and antenna impedance matching, have made it impossible to reduce RFID tag dimensions smaller than a specific size. Engineers designing RFID tags for general use consider the optimal point between antenna size and antenna gain (which determines the reading range) [28]. For this reason, their dimensions are too large to be embedded in a manometer with a diameter of 100 mm. Therefore, designing and simulating a new RFID tag with a tiny size preference was necessary. Despite their variety and diversity of uses, RFID tags all have fixed components that must be designed in perfect harmony with each other. The steps for creating an RFID tag are shown in Scheme 1 [29].

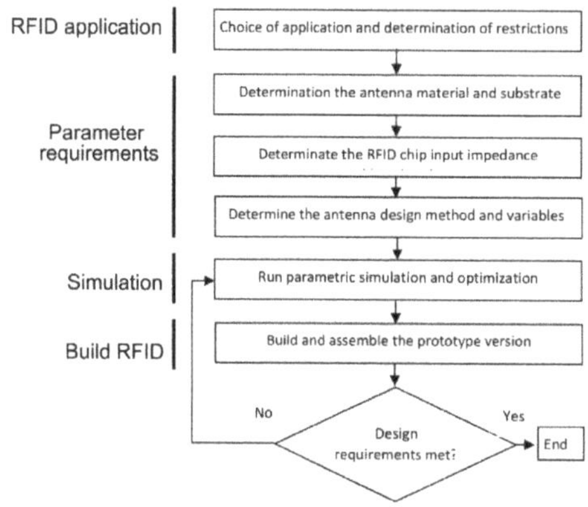

Scheme 1. RFID tag antenna design process.

Application Requirements and Restrictions

Frequency Band

RFID technology covers a wide bandwidth from 125 kHz to 5.4 GHz. In the low-frequency band, the antenna size increases due to the increase in wavelength, and the data transmission speed and reading range decrease. Despite the advantages of the high-frequency band, such as reducing the antenna size and increasing the speed and reading range, it is not an appropriate option for industrial environments due to reducing the penetration of EM waves in objects. Therefore, we have selected the 866 MHz (mid-frequency band), the most common frequency band in industry applications.

Antenna Pattern and Dimension Limitation

Bench test outputs showed that their RSSI chart pattern changed in two ways whenever the cover rotated and the tags exited the EM shield. Rectangular tags create an M-shaped pattern with two peaks, and square and circular tags create an A-shaped pattern with one maximum point. Therefore, due to the manometer shapes, the best option for our application was a circular dipole RFID tag. In terms of dimensions, manometers are designed and manufactured in less than one centimeter to several tens of centimeters. We chose the WIKA pressure gauge PGS21.100 as a case study for digitization. Detailed specification material and dimensions of the manometer are given in [30]. Due to the manometer dial plate limitation (94 mm diameter), we had to keep the RFID tag dimension under about 20 mm. The green and red area tags were located at a distance of $\lambda/(8)$ from each other.

Antenna Material and Substrates

Choosing the suitable material and substrate is one of the most important parts of designing an RFID tag. As shown in Figure 4, the entire body of the manometer, except for the central part, is made of stainless steel. The RFID tags are all implanted on a standard single-layer FR4 PCB substrate with 1.6 mm of thickness coated with a 35 µm thick copper layer as the RFID tag antenna. This layer is located directly on the dial plate.

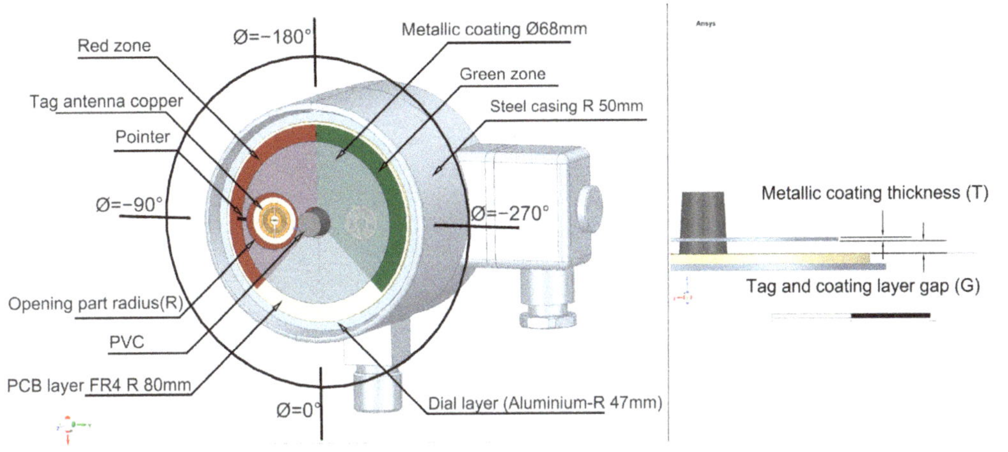

Figure 4. Digitized manometer components and materials.

RFID Chip Impedance Matching and Resistant

Numerous chips are designed and manufactured in the RFID frequency bands, but they are mostly not solderable and not available for commercial use in mass-production assembly lines. NXP UCODE G2iM is a series of passive identification transponder chips in the 900 Mhz frequency band. UCODE G2iM SL3S1003 is a six-pin chip in the solderable SOT886 package [31], which provides −17.5 dBm read sensitivity. As shown in Figure 5, pins 1 and 3 are the RFP and RFN pins connected to the tag antenna, and pins 4 and 6 are the chip external battery pins (active mode configuration).

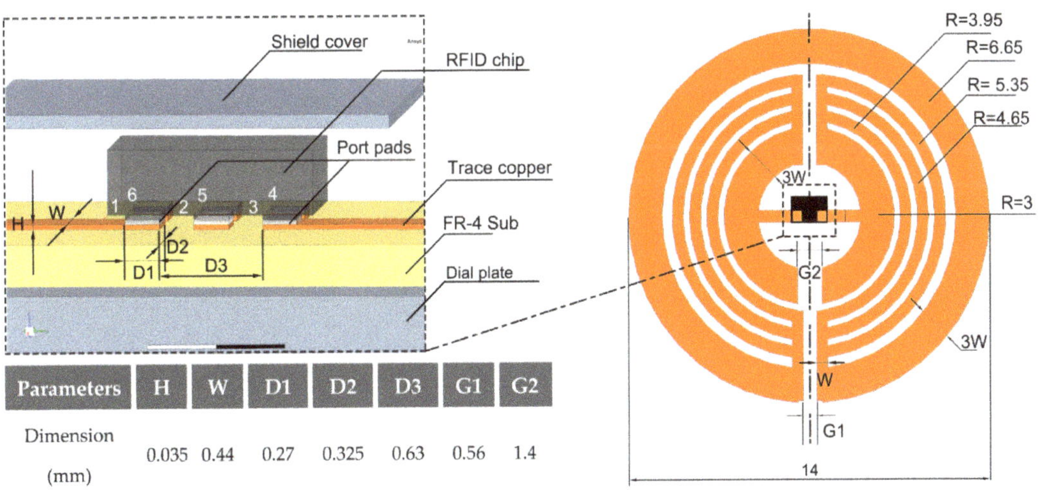

Parameters	H	W	D1	D2	D3	G1	G2
Dimension (mm)	0.035	0.44	0.27	0.325	0.63	0.56	1.4

Figure 5. RFID tag Antenna and chip Geometry.

To transfer the maximum power absorbed by the tag antenna to the chip, the internal impedance of the designed antenna must be precisely equal to the conjugate input impedance of the chip. The higher the sensitivity and operating frequency of the tag, the more it allows the designer to design tags in smaller sizes. The internal impedance of the chips has a complex parameter that can be expressed as follows:

$$Z_c = R_c + jX_c \cdot Z_a = R_a + jX_a \tag{1}$$

The chip impedance (Z_c) and antenna impedance (Z_a) are both frequency-dependent and contain a real impedance part (R) and complex impedance part (X). The internal resistance table of the SL3S1003 chip in the frequency band is given in Table 2 [31]. Due to the European frequency band standard, the impedance of 866 MHz in designing and tuning the tag antenna has been considered.

Table 2. SL3S1003 chip Input impedance.

Symbols	Parameter	Frequency	Real Part (R)	Complex Part (X)	Units
Z	Impedance	866 MHz	27	−234	Ω
		915 MHz	24	−222	
		953 MHz	23	−213	

3.3. Electromagnetic Shielding Cover Simulation and Optimization

EM isolation in the form of frequency shift and reduced transmission power can affect the proper functioning of the tags. For this reason, after adjusting all the parameters at the best point in the last section, we examine the effect of the metal plane in S11 and gain diagrams for a complete rotation of the metal cover. S11 is the most quoted parameter regarding antennas which represents how much power is reflected from the antenna and hence is known as the reflection coefficient. RFID antennas are two-way antennas (transmitter and receiver), and the gain diagram shows how well the waves are absorbed or propagated by the antenna. All simulation in this study was carried out with Ansys HFSS 2021 R2 software based on Finite Element Method (FEM) solver.

4. Results

This section presents the results of practical experiments and simulations in the following three sections:

- Isolation modeling results;
- RFID tag and isolator simulation results;
- Digitization simulation results.

The first part presents the modeling results of isolation and the effect of different metals and openings in EM isolation rate and its impact on RSSI parameters. The second part presents the simulation and tuning results of the designed RFID antenna. We sweep the W parameter (Figure 5) to adjust the resonant frequency of the embedded tag inside the manometer without an isolator cover. Then, we tune the whole manometer with the metallic cover at the optimum points. In the last section, we examined the performance of RFID tags in digitizing manometers (green and red zone detection) by simulating their gain changes.

4.1. Experimental Test Bench Results

The RSSI parameter changes were measured for the commercial tags against EM isolation. EM waves can reach the RFID antenna through circular openings in the metallic coating layer. 4Zerobox rotates the metallic layer to change the EM isolation rate and measure the RFID RSSI response via an RFID reader. The tests were performed in an environment without EM reflection to prevent the effects of unwanted noise on the outputs. The RSSI changes of each tag are examined against the EM isolation created by different metals, including stainless steel, galvanized steel, and iron (actually, it was AISI 304, a carbon steel used in sheet metalworking, but we use the label "Iron" to easily distinguish it from stainless steel). The isolation covers and circular opening made a different percentage of EM isolation for the tags. RSSI is shown in the separate curves in Figure 6, and the average change of all curves is given as a black line in charts. Metals cause different effects in the EM field, depending on their shape, thickness, density, and EM properties. The QTTM and H47 tags were both dual-dipole designs so that they could be read in any polarization without limiting the installation angle. The Impinj tag was a simple dipole with a rectangular antenna shape, leading to a different RSSI pattern with two peaks (M shape) against magnetic shielding. The maximum RSSI occurred when half of the tag was covered by metals (steps 2 and 4 in Figure 2b). The True 3D QTTM Long Reading tag had the highest amount of RSSI against various metals in an A-shaped pattern with a peak. The difference between the maximum and minimum RSSI in iron and stainless-steel diagrams was more significant than galvanized Steel.

RFID Tag	No Shield RSSI	Cover Materials		
		Stainless Steel	**Galvanized Steel**	**Iron**
True 3D QTTM Long Reading 	−23.2 dBi	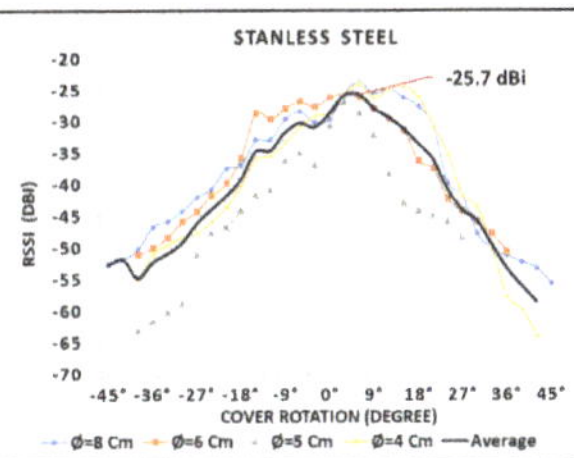	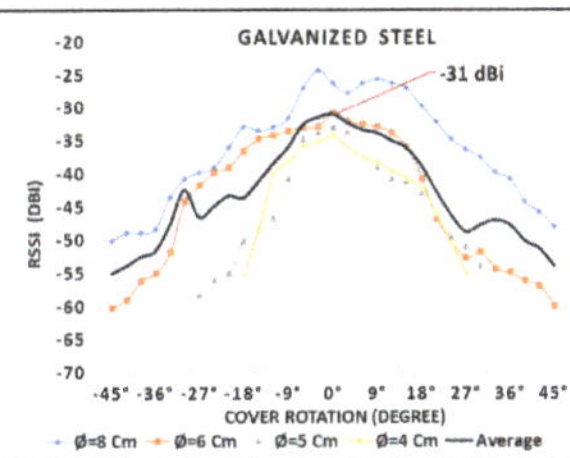	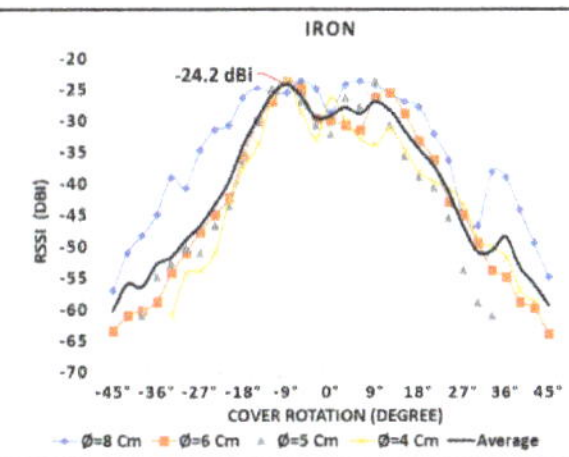
Impinj h47 	−35.2 dBi	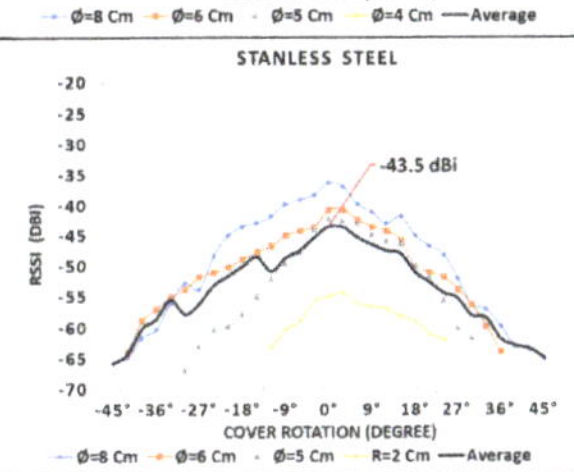	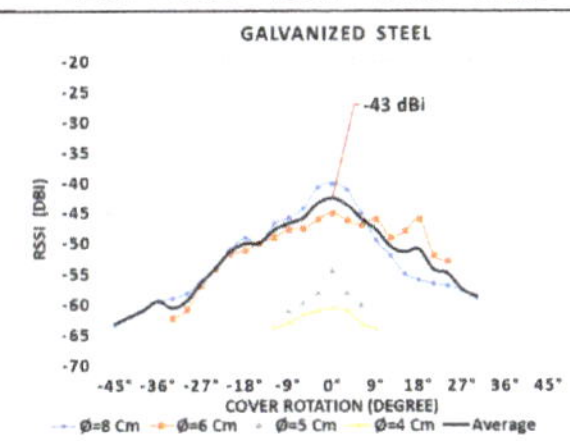	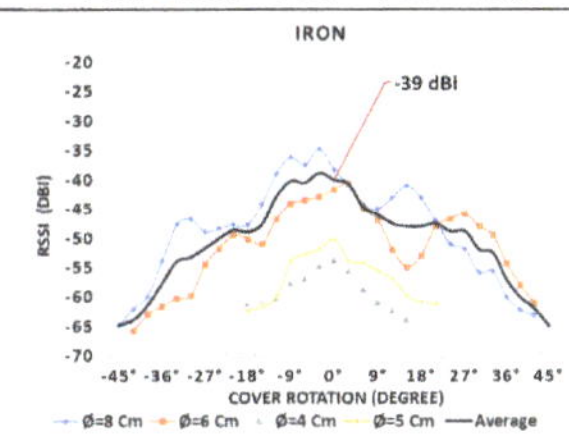
Impinj Monza 	−33.4 dBi	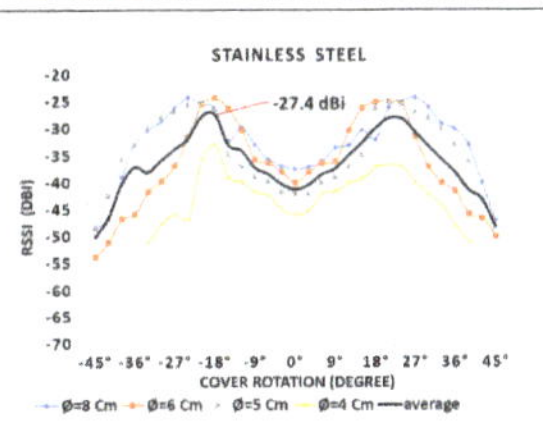	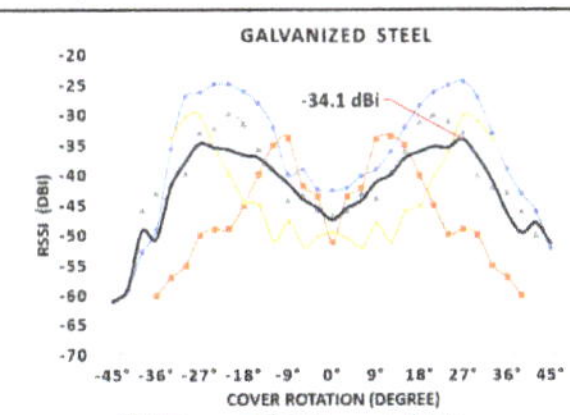	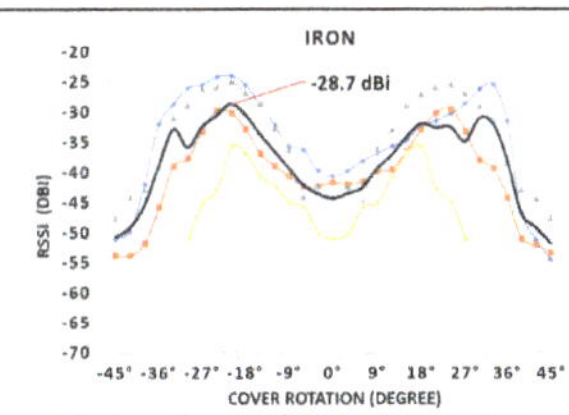

Figure 6. RSSI measured for RFID tags with different covering materials.

4.2. RFID Tag and Isolator Simulation Results

In Section 3.2, after explaining the reasons for designing a new RFID for embedding in the manometer, we described the designing steps and antenna geometry. In this step, to determine the dimensions of the antenna and adjust the parameters, we must first determine the parameters of the RFID tag by the EM wave simulator and then choose the metal cover parameters by embedding them in the manometer. Finally, by rotating the metal cover in 360 degrees cycles, we simulate the performance of the entire assembly against the EM field to digitize the manometer.

4.2.1. RFID Parametric Simulation

In the next step, after determining and selecting the material and the initial geometry of the antenna to perform impedance matching, we should examine the parameters affecting the input resistance of the antenna and the S11 diagram. The optimal point of each variable was determined separately and in combination by EM simulation software and parametric simulation. Parameters D1- to D3 were static, and their dimensions were defined according to the size of the antenna port and RFID chip. The H parameter was also fixed due to the same thickness of the copper coating on the FR4 layer (about 35 μm for FR4 1 oz). Due to the limited size of the designed RFID tag described in Section Antenna Pattern and dimension limitation, R parameters were specified in some way to keep the tag dimension less than 20 mm. Therefore, we tried to create the impedance matching by performing parametric analysis in only variables parameter W. The cross-section area and distance of conductors are two critical components in determining the size of capacitors, inductors, and resistors. The sweep in the W parameter simultaneously leads to change in both these components. As a result, the impedance matching can be tuned accurately to the primary operating frequency (900 MHz) by creating a frequency shift. The W parameter was swept in the range of 0.31–0.37 mm, and its effect in the S11 plot was investigated in Figure 7a. As shown in Figure 7b, the simulated input impedance at the UHF RFID central frequency is Zin = 10.3 + j234. The electric field or "E" plane shows the polarization or direction of the radio wave. The magnetic field or "H" plane lies at a right angle to the "E" plane. Co and cross-polarization in the H and E plan are shown in Figure 7c,d.

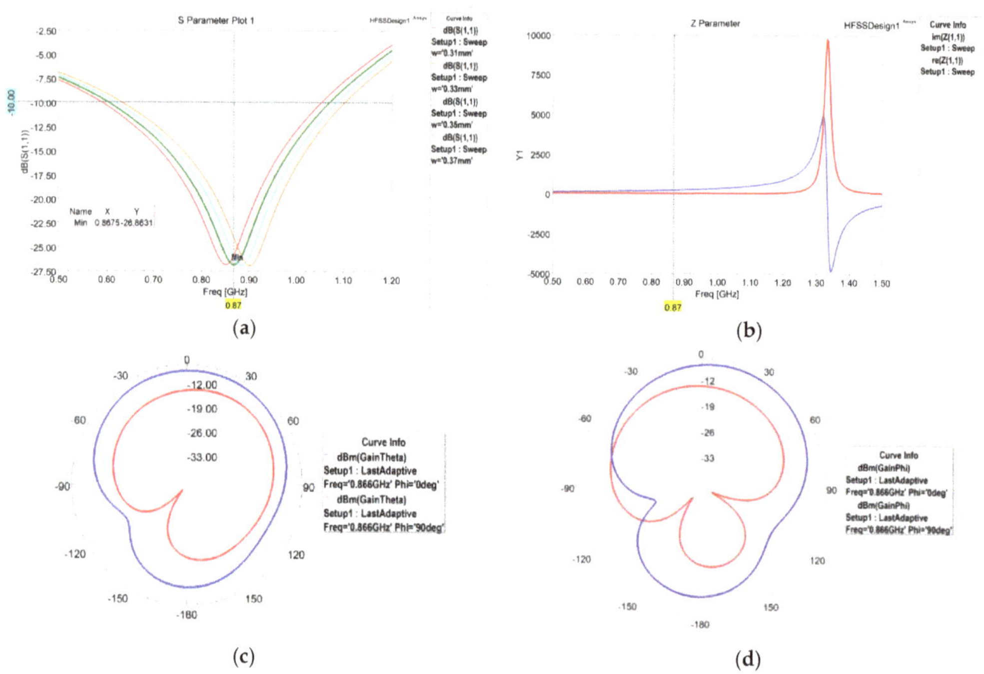

Figure 7. (**a**) Frequency tuning by sweeping W parameter; (**b**) antenna input impedance; (**c**) co-and cross-polarization of E plane (XZ plane); (**d**) co-and cross-polarization of H plane (YZ plane).

4.2.2. Isolation Cover Parametric Simulation

We tried to optimize the tag at the best working point in the previous step. In this step, we seek to optimize the metal cover to create the most damaging effect in the gain diagram and maximize the difference in both the gain graphs of shielded and non-shielded modes. Eventually, this difference in gain will cause the tags to be turned on or off by moving the manometer's pointer. According to the measurements made in the bench test, we expected that the EM shield coating would eliminate the frequency matching by creating a frequency shift. It would be led to reducing the power transfer received by the tag antenna to the chip and eventually causing the chip to turn off. Therefore, after adding the metallic cover step by step, each parameter is checked by rotating the cover in two modes: Ø = 0° (the tag is covered) and Ø = −90° (the tag is not covered). The ultimate goal is to create the highest significant possible difference in the gain of the RFID tag in 2 cases when the cover angle is 0° and uncovered at −90° (Figure 4).

RFID Tag and Cover Gap

The first parameter that was examined after adding the cover was the distance between the RFID tag and the EM cover, which was simulated as a hybrid parametric simulation at the same time as the cover was rotated. According to the results shown in Figure 8a, the smaller the gap between the tag and the metallic cover led to the more significant the difference in RFID gain between covered and without cover conditions (the best case is the enormous difference), which creates a difference of about 19 dBm in RFID gain. Therefore, according to the dimensions of the RFID tag chip, a distance of 0.15 mm was determined as the optimal distance.

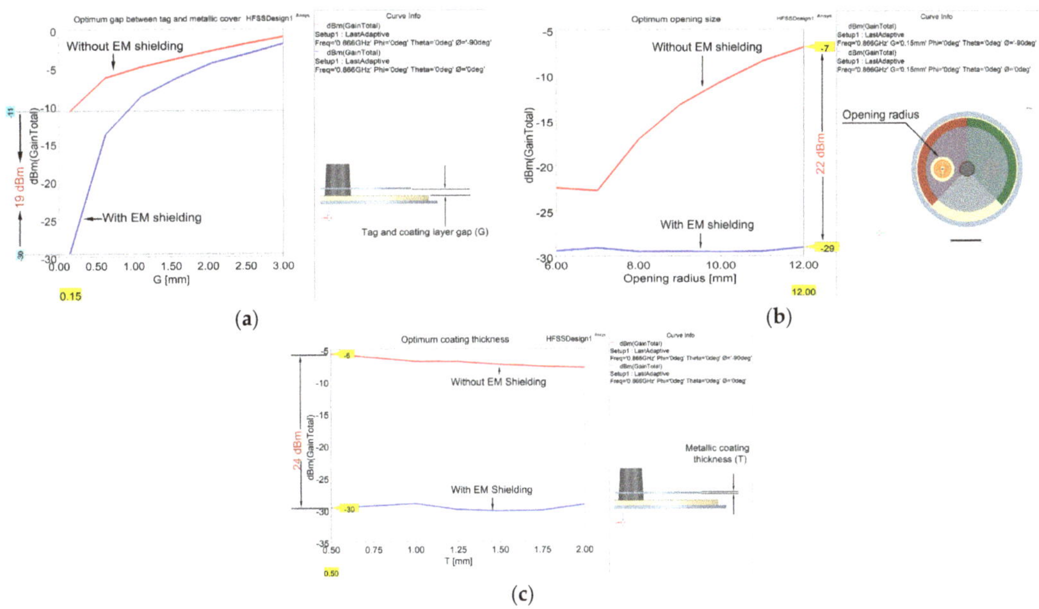

Figure 8. Metallic EM shield cover tuning parameters. (**a**) The optimal point of tag and cover gap; (**b**) the optimal point of the cover opening part radius; (**c**) the Optimal point of electromagnetic cover thickness.

Cover Opening Part Size

Based on the bench test results, it emerged that the RSSI rate improved with increasing the diameter of the cover opening part, the dimensions of the opening part would have a substantial effect on the amount of tag received gain. For this reason, we simulated this parameter in combination with the rotation of the EM cover. The results of the cover opening radius sweeping in the range of 6 to 12 mm are shown in Figure 8b. As expected, the larger opening size increased the gain of the tag when the tag was not covered (Ø = −90°). Due to the size limitation, we considered the maximum possible size of opening radius (12 mm) for the opening part. This opening size increases the difference gain (with and without cover) from 19 dBm to 22 dBm.

Isolation Cover Thickness

In the last step, we investigated the thickness of the EM shield cover and the effects of this parameter on the output gain by parametric simulation. Figure 8c shows the sweep results of this parameter from 0.5 to 2 mm. The results clearly show that the thickness of the protective layer does not have a significant effect on the gain of the RFID tag, so we consider its minimum value of 0.5 mm as the optimal point.

4.3. *Digitization Simulation Results*

4.3.1. Metallic Isolation Effect on the Reflection Coefficient

After optimizing all the variables of the RFID tag and the magnetic cover separately and determining all the parameters, we can simulate the whole set with a combined simulation and examine the result of adding an EM shield. For this purpose, we must examine the reflection coefficient (or S11 parameter) in the total bandwidth at two angles Ø = 0° (EM isolated) and Ø = −90° (without electromagnetic isolation). As shown in Figure 9, rotation of the electromagnetic cover creates a frequency shift of about 300 MHz in the S11 diagram. The bandwidth was reduced from 0.46 MHz to 0.26 MHz after creating the magnetic shield so that there is no overlap bandwidth between the two graphs.

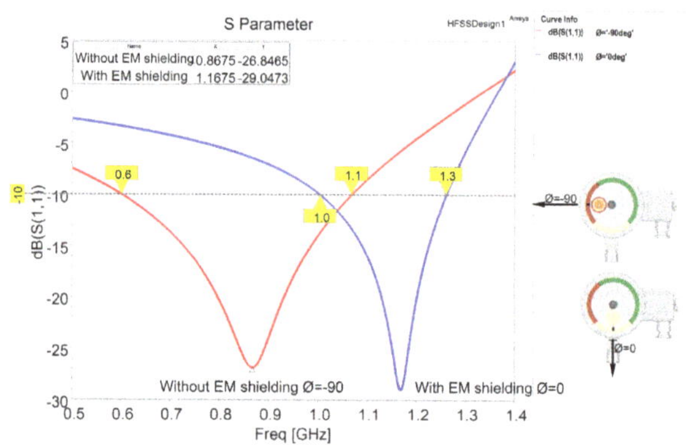

Figure 9. Electromagnetic isolation impact on S11 grape.

4.3.2. RFIDs Total Gain and Digitization

After tuning all the effective parameters in the RFID tag and the insulation cover layer, we can now check the performance of the tags embedded in the manometer. For this purpose, we measure the total gain changes of both tags in a complete rotation of the metallic cover. As shown in Figure 4, the red zone tag was mounted at Ø = −90°, and the green zone tag was mounted at Ø = −270° manometers. By rotating the metal cover, the EM waves reached the tags from the opening part and caused them to be read by the RFID reader. This gain increased in 2 intervals of 100° degrees to the center of angles −90° and −270° (Figure 10).

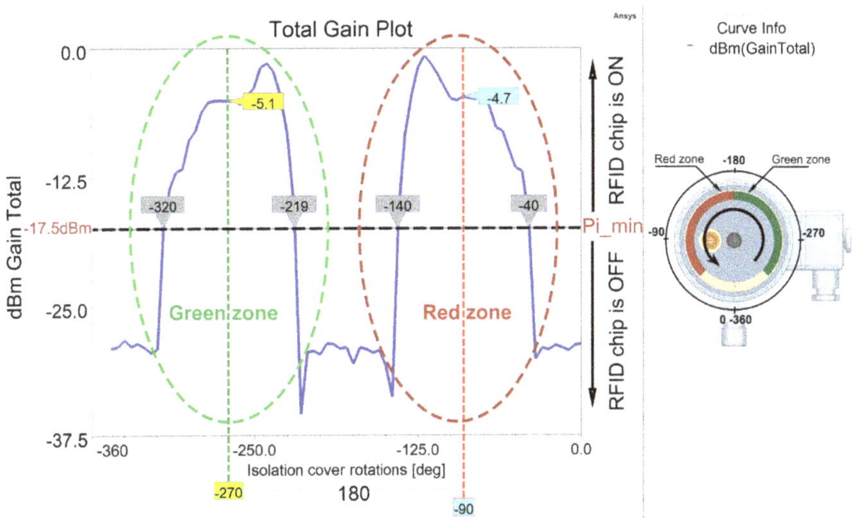

Figure 10. RFID tag total gain and switching angles.

5. Discussion

The metallic covers and parasitic elements cause EM isolation for the RFID tags. Near-field metal covers create an inductive and capacitive coupling that changes the antenna's input impedance. Chip and antenna impedance mismatch causes only a tiny portion of the absorbed energy to be transferred to the chip. If the energy transferred to the chip is less than the P_{imin}, it leads to turning off the chip. P_{imin} or reading sensitivity means the minimum input power of the chip to turn it on. NXP states that this parameter is −17.5 dBm in SL3s1003, which means that the chip is off as long as the power gain chart is below this value [31]. The intersection of this line (−17.5 dBm) with the gain diagram in Figure 10 shows that the red tag from Ø = −40° to −140° and the green tag from Ø = −219° to −320° are clear readable for the RFID reader. To better understand the performance of the tags and switching points, Figure 11 shows the trigger angles during a 360-degree rotation of the manometer pointer. The locating of the manometer pointer in each zone (red or green) leads to only activating the same zone tag, and the RFID reader easily determines the zone through the EPC code of the chips. In this way, thanks to RFID technology and the EM isolation effect as an advantage, we introduced a method to digitize manometer passively and wirelessly.

Such an integrated passive RFID tag and the metallic cover allow the manometer to be easily read by already installed RFID readers or by installing an antenna on movable objects (e.g., forklifts, bicycles often used in large industrial plants, or even in the industrial cleaning trolleys), thus reducing the necessary number of readers. The study demonstrated the feasibility of a Passive RFID tag and EM isolation to retrofit industrial analogical devices such as manometers, handwheels, etc., in an effective and low-cost way.

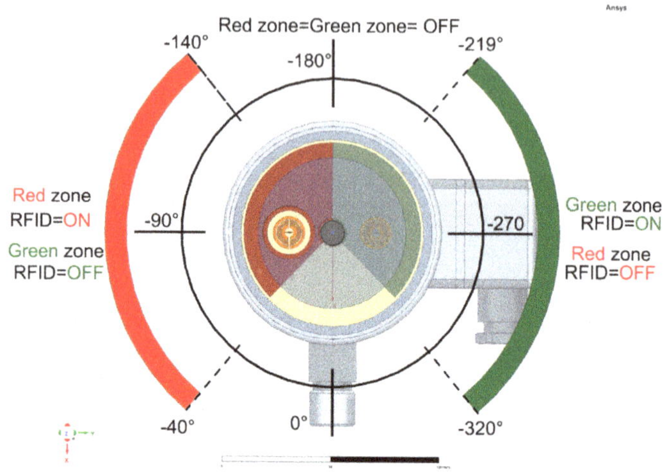

Figure 11. Digitized Analog manometer.

Author Contributions: Conceptualization, G.F.; methodology, G.F. and S.A.; software, M.H.; validation, G.F. and A.M.; formal analysis, M.H.; investigation, M.H. and S.A.; resources, A.M.; data curation, G.F.; writing—original draft preparation, M.H. and S.A.; writing—review and editing, A.M.; visualization, A.M.; supervision, A.M.; project administration, G.F.; funding acquisition, G.F. All authors have read and agreed to the published version of the manuscript.

Funding: This research work was undertaken in the context of DIGIMAN4.0 project ("DIGItal MANufacturing Technologies for Zero-defect Industry 4.0 Production", http://www.digiman4-0.mek.dtu.dk/, accessed on 21 July 2021). DIGIMAN4.0 is a European Training Network supported by Horizon 2020, the EU Framework Program for Research and Innovation (Project ID: 814225).

Institutional Review Board Statement: Not applicable.

Informed Consent Statement: Not applicable.

Data Availability Statement: Not applicable.

Acknowledgments: The authors express their gratitude to the anonymous reviewers for their valuable comments that helped us to improve the paper significantly.

Conflicts of Interest: The authors declare no conflict of interest.

References

1. Varriale, V.; Cammarano, A.; Michelino, F.; Caputo, M. Sustainable Supply Chains with Blockchain, IoT and RFID: A Simulation on Order Management. *Sustainability* **2021**, *13*, 6372. [CrossRef]
2. Stark, R.; Damerau, T.; Chatti, S. Digital twin. *CIRP Encycl. Prod. Eng.* **2019**, *66*, 1–8.
3. *DIN SPEC 91345: Reference Architecture Model Industrie 4.0 (RAMI4. 0)*; DIN: Berlin, Germany, 2016; Volume 4, p. 2016.
4. Cook, B.S.; Vyas, R.; Kim, S.; Thai, T.; Le, T.; Traille, A.; Aubert, H.; Tentzeris, M. RFID-based sensors for zero-power autonomous wireless sensor networks. *IEEE Sens. J.* **2014**, *14*, 2419–2431. [CrossRef]
5. Dimitriou, A.G.; Bletsas, A.; Sahalos, J.N. Room-coverage improvements in UHF RFID with commodity hardware [wireless corner]. *IEEE Antennas Propag. Mag.* **2011**, *53*, 175–194. [CrossRef]
6. Zhong, R.Y.; Dai, Q.Y.; Qu, T.; Hu, G.J.; Huang, G.Q. RFID-enabled real-time manufacturing execution system for mass-customization production. *Robot. Comput.-Integr. Manuf.* **2013**, *29*, 283–292. [CrossRef]
7. Ramirez, R.A.; Rojas-Nastrucci, E.A.; Weller, T.M. UHF RFID tags for on-/off-metal applications fabricated using additive manufacturing. *IEEE Antennas Wirel. Propag. Lett.* **2017**, *16*, 1635–1638. [CrossRef]
8. Ghahremani, A.; Rezaei, V.D.; Bakhtiar, M.S. A UHF-RFID transceiver with a blocker-canceller feedback and+ 30 dBm output power. *IEEE Trans. Circuits Syst. I* **2013**, *60*, 3043–3054. [CrossRef]
9. Michel, A.; Franchina, V.; Nepa, P.; Salvatore, A. A UHF RFID Tag Embeddable in Small Metal Cavities. *IEEE Trans. Antennas Propag.* **2019**, *67*, 1374–1379. [CrossRef]

10. Pino, M.R.; Michel, A.; Narciandi, G.A.; Nepa, P. Read-Range Estimation of UHF-RFID Tags Placed Inside Metal Pipes. In Proceedings of the 2019 13th European Conference on Antennas and Propagation (EuCAP), Krakow, Poland, 31 March–5 April 2019.
11. Arora, K.; Mallinson, H.; Kulkarni, A.; Brusey, J.; McFarlane, D. The practical feasibility of using RFID in a metal environment. In Proceedings of the 2007 IEEE Wireless Communications and Networking Conference, Shanghai, China, 21–25 September 2007; IEEE: Shanghai, China, 2007.
12. Tuyishime, C.R.; Nzanywayingoma, F.; Gatera, O. IoT-based Intelligent Energy Efficiency Management System for Smart Industries (IoT-IEEMS). In Proceedings of the 2021 IEEE International Conference on Industry 4.0, Artificial Intelligence, and Communications Technology (IAICT), Bali, Indonesia, 27–28 July 2021; IEEE: Bali, Indonesia, 2021.
13. Hänel, T.; Brüggemann, L.; Loske, F.; Aschenbruck, N. Long-Term Wireless Sensor Network Deployments in Industry and Office Scenarios. In Proceedings of the 2021 IEEE 22nd International Symposium on a World of Wireless, Mobile and Multimedia Networks (WoWMoM), Pisa, Italy, 7–11 June 2021; IEEE: Pisa, Italy, 2021.
14. Nideborn, J. *Industrial Network Market Shares 2015–2021*; Hardware Meets Software (HMS) Group: Halmstad, Sweden.
15. Cheng, Y.J.; Chen, M.H.; Cheng, F.C.; Cheng, Y.C.; Lin, Y.S.; Yang, C.J. Developing a decision support system (DSS) for a dental manufacturing production line based on data mining. *Appl. Syst. Innov.* **2018**, *1*, 17. [CrossRef]
16. Michalski, A.; Watral, Z. Problems of Powering End Devices in Wireless Networks of the Internet of Things. *Energies* **2021**, *14*, 2417. [CrossRef]
17. Neal, A.D.; Sharpe, R.G.; van Lopik, K.; Tribe, J.; Goodall, P.; Lugo, H.; Segura-Velandia, D.; Conway, P.; Jackson, L.M.; Jackson, T.W.; et al. The potential of industry 4.0 Cyber Physical System to improve quality assurance: An automotive case study for wash monitoring of returnable transit items. *CIRP J. Manuf. Sci. Technol.* **2021**, *32*, 461–475. [CrossRef]
18. Vagaš, M.; Galajdová, A.; Šimšík, D.; Onofrejová, D. Wireless data acquisition from automated workplaces based on RFID technology. *IFAC-PapersOnLine* **2019**, *52*, 299–304. [CrossRef]
19. Chen, S.-L.; Kuo, S.-K.; Lin, C.-T. A metallic RFID tag design for steel-bar and wire-rod management application in the steel industry. *Prog. Electromagn. Res.* **2009**, *91*, 195–212. [CrossRef]
20. CEN, EN 837-1:1996/AC:1998 Pressure Gauges—Part 1: Bourdon Tube Pressure Gauges—Dimensions, Metrology, Requirements and Testing. 11 June 1998, p. 24. Available online: https://standards.iteh.ai/catalog/standards/cen/812a8aab-49be-4466-9113-428d36d4543e/en-837-1-1996-ac-1998 (accessed on 20 October 2021).
21. Kim, J.; Yoon, H.; Park, J.; Burm, J. A method to improve isolation for RFID applications. In Proceedings of the 2005 European Microwave Conference, Paris, France, 4–6 October 2005; IEEE: Paris, France, 2005.
22. Loo, C.H.; Elsherbeni, A.Z.; Yang, F.; Kajfez, D. Experimental and simulation investigation of RFID blind spots. *J. Electromagn. Waves Appl.* **2009**, *23*, 747–760. [CrossRef]
23. Buffi, A.; Michel, A.; Nepa, P.; Tellini, B. RSSI measurements for RFID tag classification in smart storage systems. *IEEE Trans. Instrum. Measur.* **2018**, *67*, 894–904. [CrossRef]
24. Shen, L.; Zhang, Q.; Pang, J.; Xu, H.; Li, P.; Xue, D. ANTspin: Efficient Absolute Localization Method of RFID Tags via Spinning Antenna. *Sensors* **2019**, *19*, 2194. [CrossRef] [PubMed]
25. Babar, A.A.; Bjorninen, T.; Bhagavati, V.A.; Sydanheimo, L.; Kallio, P.; Ukkonen, L. Small and Flexible Metal Mountable Passive UHF RFID Tag on High-Dielectric Polymer-Ceramic Composite Substrate. *IEEE Antennas Wirel. Propag. Lett.* **2012**, *11*, 1319–1322. [CrossRef]
26. Chen, S.-L.; Lin, K.-H. A slim RFID tag antenna design for metallic object applications. *IEEE Antennas Wirel. Propag. Lett.* **2008**, *7*, 729–732. [CrossRef]
27. Mazzei, D.; Montelisciani, G.; Baldi, G.; Baù, A.; Cipriani, M.; Fantoni, G. Improving the Efficiency of Industrial Processes with a Plug and Play Iot Data Acquisition Platform. In *Enterprise Interoperability: Smart Services and Business Impact of Enterprise Interoperability*; Wiley Online Library: Hoboken, NJ, USA, 2018; pp. 315–321.
28. Marrocco, G. The art of UHF RFID antenna design: Impedance-matching and size-reduction techniques. *IEEE Antennas Propag. Mag.* **2008**, *50*, 66–79. [CrossRef]
29. Rao, K.S.; Nikitin, P.V.; Lam, S.F. Antenna design for UHF RFID tags: A review and a practical application. *IEEE Trans. Antennas Propag.* **2005**, *53*, 3870–3876. [CrossRef]
30. Alexander Wiegand WIKA. Bourdon Tube Pressure Gauge with Switch Contacts Stainless Steel Case, NS 100 and 160 Models PGS21.100 and PGS21.160. 2020, p. 12. Available online: https://www.wika.com/en-en/pgs21_100_pgs21_160.WIKA (accessed on 28 December 2021).
31. NXP Semiconductors. SL3S1003_1013 UCODE G2iM and G2iM+, NXP, Editor. 5 February 2019, p. 47. Available online: https://www.nxp.com/docs/en/data-sheet/SL3S1003_1013.pdf (accessed on 28 December 2021).

applied sciences

MDPI

Article

Low-Cost Digitalization Solution through Scalable IIoT Prototypes

Marko Vuković *, Oliver Jorg, Mohammadamin Hosseinifard and Gualtiero Fantoni

Department of Civil and Industrial Engineering, University of Pisa, Largo Lucio Lazzarino, 56122 Pisa, Italy
* Correspondence: marko.vukovic@phd.unipi.it

Abstract: Industry 4.0 is fast becoming a mainstream goal, and many companies are lining up to join the Fourth Industrial Revolution. Small and medium-sized enterprises, especially in the manufacturing industry, are the most heavily challenged in adopting new technology. One of the reasons why these enterprises are lagging behind is the motivation of the key personnel, the decision-makers. The factories in question often do not have a pressing need for advancing to Industry 4.0 and are wary of the risk in doing so. The authors present a rapid, low-cost prototyping solution for the manufacturing companies with legacy machinery intending to adopt the Industry 4.0 paradigm with a low-risk initial step. The legacy machines are retrofitted through the Industrial Internet of Things, making these machines both connectable and capable of providing data, thus enabling process monitoring. The machine chosen as the digitization target was not connectable, and the retrofit was extensive. The choice was made to present the benefits of digitization to the stakeholders quickly and effectively. Indeed, the solution provides immediate results within manufacturing industrial settings, with the ultimate goal being the digital transformation of the entire factory. This work presents an implementation cycle for digitizing an industrial broaching machine, supported by state-of-the-art literature analysis. The methodology utilized in this work is based on the well-known DMAIC strategy customized for the specifics of this case study.

Keywords: industrial IoT; Industry 4.0; prototyping; retrofitting solutions; embedded solutions; low-cost

Citation: Vuković, M.; Jorg, O.; Hosseinifard, M.; Fantoni, G. Low-Cost Digitalization Solution through Scalable IIoT Prototypes. *Appl. Sci.* **2022**, *12*, 8571. https://doi.org/10.3390/app12178571

Academic Editors: Abílio Manuel Pinho de Jesus, Roque Calvo, José A. Yagüe-Fabra and Guido Tosello

Received: 19 July 2022
Accepted: 22 August 2022
Published: 27 August 2022

1. Introduction

Many industries are embracing Industry 4.0 advancements since the term was first introduced in 2011. The term Industry 4.0 resulted from a German government initiative to increase the global competitiveness of the country's manufacturing industry [1]. In the following years, many countries adopted and started researching and refining the principles set in the initial report. A set of guidelines and principles was shaped and presented as the nine pillars of Industry 4.0 [2]. Major industrial stakeholders took notice and started investing in the research, advancement followed, and the growth has been steady ever since. Leading adopter industries in 2020 were automotive, computer, electronic and electric, and metals and mining, as well as process industries, with adoption rates ranging up to 36% [3]. It is also worth noting that major advancements have been observed in North America, Europe, and Asia. While this is expected, due to the level of financial investment in those regions, there are still a lot of industries in other parts of the world that have not started the process of adoption. One of the main culprits for this is that the starting position of those industries is further back than the industries in the aforementioned regions. Other reasons include a low level of education, and the unawareness of the potential for improvement through Industry 4.0 adoption, as well as the implementation cost. Therefore, there is still room for improvement and more research that has the goal of stimulating the adoption process.

According to Eurostat [4], in 2018, small and medium enterprises (SME) made up the vast majority of enterprises in the EU, ranging from 97% in Germany and Luxembourg, to 98% and above in the remaining EU Member States. Therefore, developing solutions to help SMEs in the process of digital transformation is a topic worth investigating and investing in. Digital transformation is the integration of new digital technologies into all business areas, leading to a fundamental change in the way the organization works [5]. Marushchak et al. [6] deem that the digital transformed business is approximately 25% more successful than other business forms. Nevertheless, leaping from the current level of industrialization to Industry 4.0 is sometimes deemed too risky, especially for SMEs. These companies do not have big development teams and are not able to dedicate a substantial amount of investment to something that is not certain to bring immediate results and profit. Survey results conducted in [7,8] show the responsibility for the decision-making in SMEs is mostly in the hands of a small group, if not only one person, the entrepreneur. These decision-makers are interested in advancing the company and its operations, but it is difficult to motivate them into big changes or investments, especially if their current operation is successful. Considering Industry 4.0 as a luxury rather than a necessity diminishes their motivation for change. Kumar et al. [9] have observed that motivation and the support of top management, among other factors, are essential for the adoption of smart technologies. Therefore, the focus lies on overcoming the lack of motivation in adopting Industry 4.0 paradigms [10]. The lack of motivation among decision-makers has been singled out as one of the biggest problems for the Industry 4.0 adoption rate, and the authors desire to further investigate it in this article.

The decision-makers need to have a clear benefit from Industry 4.0 to be truly convinced that it is a smart business move. Thorough knowledge of the factory state, including insights into ongoing operations, is the underlying basis for the entrepreneurial improvement of a manufacturing company's business. In [11], the author investigated the aspects that can enable a company to identify areas of improvement, as well as unessential or replaceable tasks. Knowledge is gained by understanding information. Information, in turn, can be created from data that first of all have to be acquired. The acquisition of all the necessary data play a fundamental role, and it is especially challenging when dealing with originally disconnected machines.

Most machines of the latest generation possess new sensor technology and can easily be connected in order to provide production data. In contrast, SMEs, especially in the manufacturing industry, mostly have machines that belong to Industry 3.0. Industry 3.0 is defined by the rise of Information and Communication Technology, and the use of robots, actuators, and servomotors became a major part of the manufacturing industry [12]. These machines are considered disconnected, and thus, cannot seamlessly integrate into Industry 4.0. According to Vuković et al. [13] machines can be classified as talking or mute with regard to their level of connectivity. It is thus necessary to teach machines to "talk", to enable them to provide the data. As mentioned before, the data provide insight into the entire factory operation. This will, in turn, provide tools to make informed and smarter data-driven decisions. The process of getting the data from analog machines is the process of machine digitization. Digitization is one of the first steps in transforming Industry 3.0 to Industry 4.0. The authors would like here to differentiate between the terms digitization and digitalization. According to Gartner, digitalization is "the use of digital technologies to change a business model and provide new revenue and value-producing opportunities" [14].

Machine digitization enables the creation of the digital twin, a virtual representation of a physical system. It is a closed-loop system with the information being exchanged in both directions between the virtual and physical counterparts. Once the digital twin is created, it presents a way to simulate the system operations and to orchestrate the production system in an optimal way [15]. Digital twins of manufacturing systems show great promise in diverse applications, according to Mourtzis et al. [16]. However, Brauner et al. [17] does not consider a complete digital twin to be feasible, due to the massive amount of data that the virtual replica would require. They propose the use of a digital shadow, which, in their view, represents reality in a more compact fashion and with better performance than a fully integrated digital twin. The authors, therefore, have chosen the digital shadow as a preferable option to the digital twin, especially considering the requirements for the solution. The "downgrade" from the digital twin additionally makes sense from a business perspective, especially in cases where even a limited number of information can create enough value for the different internal or external stakeholders.

The authors extrapolated a set of requirements the digitization solution for SME shop floors machine should fulfill. The solution should enable the creation of the digital shadow and therefore provide the necessary data, and in extension, the insight that will benefit the decision-makers. These requirements are:

1. **Cost effectiveness** —A low initial cost of investment for the digitization of a single machine or manufacturing unit decreases the entrepreneurial risk, and hence the hesitation of a company [18–20]. After seeing immediate improvements, the company can gradually upgrade the entire machinery. Furthermore, the benefits grow progressively with the degree of digitization.
2. **Implementation simplicity**—A lower complexity increases the speed of development and implementation. The increase in speed provides the results quickly, thus increasing the confidence of the decision-makers that the entire process can be finished quickly. Many authors have identified a high complexity and a lack of qualified workforce as big barrier towards the Industry 4.0 adoption process [21,22].
3. **Security**—Security is the focus of any IoT development, the industrial setting only emphsizes this. It is one of the main challenges faced by companies in the context of the Industrial Internet of Things (IIoT) [23].
4. **Scalability**—There are two aspects of scaling to consider; the first is scaling the solution from one machine to the entire factory. Digitizing the entire factory floor at once is exposing the business to the high initial cost before they are sufficiently convinced of the digital transformation benefits. The second aspect is the "maturation" of the solution, developing it from the prototype phase to the fully operational product.
5. **Unobtrusiveness**—There is less motivation to change a process that already works. It is beneficial if the processes and operations already present in the factory are not overly modified. Wee et al. [24] state that due to the high cost of downtime in production, the manufacturing companies will carefully consider introducing new technologies due to possible risks to process reliability.

IIoT solutions fit the aforementioned requirements and can make machines connectable, e.g., "teaching them how to talk". The current IIoT landscape provides several such solutions. However, as observed by Martikkala et al. [18], SMEs do not possess financial capabilities, or they lack a skilled workforce to operate and maintain IIoT systems. One more drawback is that those commercial solutions often lack the required level of adaptability. SMEs normally have very specific use case tailored solutions that need to be developed and tested. This means developing a prototype that can quickly demonstrate the capabilities and benefits of digitization. This is also in line with the "test before invest" principle that has been touted as a strategic means for driving the digital transformation of European SMEs [25] by European Commission Digital Innovation Hubs. Moreover, the culture of digital processes is not sufficiently introduced, and prototyping is the key to raising awareness of the benefits.

This work focuses on the Industry 4.0 adoption problem in the manufacturing industry, especially considering the motivation factor of the decision-makers. The authors selected an actual case study, and the chosen company had not adopted the Industry 4.0 paradigm yet. It is the same situation that many companies in the manufacturing industry find themselves in. Pirola et al. [26] have conducted an assessment of Italian SMEs to determine Digital Readiness Level, and they have found that 40% of manufacturing SMEs have only partially digitized their operations. The machines were made before the introduction of Industry 4.0 technologies and are thus not equipped for digitization. It is important to note that adopting the Industry 4.0 paradigm is not perceived as a necessity from the stakeholder point of view; nevertheless, the study aims to demonstrate its benefits to them and ultimately motivate them to upgrade.

In the scope of this paper, this principle was applied as a means to introduce and test the solution to one machine in one factory. The final goal involves scaling up to a complete solution ready for Industry 4.0 adoption in the entire production line, in one or more factories. In particular, the approach wants to digitize the machine and then turn it into the final industrial-grade system in a seamless manner and at almost zero additional cost. This will cheaply and quickly show the benefits of digitalization, and it will lead to further adoption of Industry 4.0 paradigms.

Section 2 further investigates the motivation effect in SMEs, the current state of the digital transformation, and finally, the state of the art in retrofitting and automation solutions. In Section 3, the authors present the methodology that was used in the development of the IIoT solution, while Section 4 will focus on the specific case study that demonstrates the solution development, implementation, and installation. Section 5 deals with the explanation of the prototyping process, as well as the solution selection process. Finally, the conclusion and a look toward future development are given in Section 6.

2. State of the Art

The authors wanted to find support in the literature for their assumption that one of the main culprits of the low adoption rate is the lack of motivation. Müller and Voigt [19] have been investigating low levels of Industry 4.0 adoption in SMEs, especially with regards to IIoT in the scope of "Industrie 4.0" and "Made in China 2025" [27] programs. They interviewed a large number of Chinese and German SME stakeholders (predominately CEOs) and they observed their confidence levels in their respective programs. The stakeholders felt that these programs are more suited to large enterprises rather than SMEs. Furthermore, Veile et al. [28] noted that companies remain unconvinced about the benefits of the new technologies, having in mind they are untested, and that old technology still has better cost–benefit ratios. Masood and Sonntag [8] also identified through a survey that most SMEs struggle with the implementation of Industry 4.0 technologies, both from the financial point of view as well as from the lack of expertise and knowledge of those technologies. Müller et al. [29] also noted that for SMEs to not only exploit but to explore new business opportunities through Industry 4.0, they require implementation support and financial security to experiment, as they have limited resources and have to avert risks because of their limited size. The authors, therefore, presume that there is a motivation issue and that there is a need to develop a solution that takes this into account.

In the search for the appropriate digitization techniques, the authors focused on the various approaches toward digitization in the literature. In their work, Sorger et al. [30] state that to utilize the full potential of Industry 4.0, all entities in the supply chain must be able to communicate with each other. They suggest that different layers must have standardized technical communication that is also supported by the Reference Architectural Model for Industry 4.0 (RAMI 4.0) [31]. There are different approaches to the digital transformation of the manufacturing SME. In most cases, old machines do not provide a level of connectivity and digitization necessary for the ensuing process. There are two choices, either the acquisition of new machines or the retrofitting of the old ones. The analysis [32] showed that the retrofitting solutions were more cost-effective than the purchasing solutions, and that strong retrofitting in particular was the best in terms of sustainability. Contreras Pérez et al. [33] demonstrated that the adoption of the Industry 4.0 in SMEs can be achieved through retrofitting existing equipment using mostly open hardware and software, and that it does not imply a high investment in new equipment and technologies.

Retrofitting old industrial machines to adhere to Industry 4.0 paradigms has been subject to intensive scientific research in recent years [34–38]. Industrial IoT is seen as an ideal tool for machine digitization. Liu et al. [39] conducted a literature review that focused on the digitalization of the machine tools in the Industry 4.0. They observed that the IIoT topics have seen a rise in the research articles. Actually, the only Industry 4.0 technology in manufacturing that had more published articles is Artificial Intelligence/Machine Learning. Lins et al. [34] focus on retrofitting in their work, where they noted that for automated industrial equipment to reach Industry 4.0 level of modernization, they need integration with the IoT sensors.

The leading companies in the automation field have recognized the benefits of retrofitting old machines, and several studies have utilized those solutions. Lima et al. [40] on the other hand, have proposed a retrofit solution using the Siemens IoT-2040 gateway and the data transmission to the Mindsphere Cloud from Siemens and GRV Software to monitor the energy data in real-time. Hesser and Markert [41] have performed a retrofit of a CNC machine with an accelerometer using the Bosch XDK sensor prototyping platform.

On the other end of the spectrum, many more works have explored small microcontroller-based boards as a means for digitizing machines with IoT. In contrast with the solutions provided by the leading automation companies, these boards provide a low-cost way to digitize, although they sacrifice many features that mature solutions deliver. However, they are very interesting from the point of view of this research, as they mitigate one of the biggest obstacles and fears the decision makers have when facing Industry 4.0 adoption. Various works use Arduino boards, and the applications are covering diverse topics and areas. This type of approach is researched at length for home automation, such as [42–45]. There are also several works coping with the use of Arduino in an industrial setting. The applications are numerous, monitoring the humidity and the temperature in industrial storage rooms [46], a remote bio-gas monitoring system [47], a monitoring system for renewable energy generation facilities [48], and real-time monitoring of photovoltaic systems [49]. All these works have been proposing the utilization of Arduino boards as a low-cost solution for sensor control in data gathering. In the work of Arjoni et al. [50], three different retrofitting applications have been presented in the manufacturing sector. Two of the machines were retrofitted with Arduino boards while in the third one, a Raspberry Pi was used. Additionally, pairing Arduino as a data-gathering system with Raspberry Pi was considered by several authors. Their presented architectures use the Raspberry Pi as a communication link and the data transmission hub [51–54]. The reason behind the inclusion of the Raspberry Pi board is that the microcontrollers often lack the power necessary for applications that require higher processing power.

3. Materials and Methods

The development approach used in this work is based on the Six Sigma set of techniques for process improvement in industrial environments [55]. This five-step methodology is called DMAIC, the acronym of Define-Measure-Analyze-Improve-Control [56]. In the following subsections, we will explain how these steps correspond to the development of the IIoT solution covered in this article. It is important to note that there are other proceedings available for SMEs to conduct Industry 4.0 projects. Schmitt et al. [57] performed a literary review and presented an evaluation of those proceedings in their work.

In the previous section, we have established the issues that SMEs have in the Industry 4.0 adoption. There is a need for solutions that address these issues and that provide SMEs with a way to embrace these paradigms The ultimate goal of this process is to raise the level of digitalization of the factory. The reluctance of factories can be reduced by creating a digitization solution where they can clearly see the improvement it brings. It is preferable if this solution is, according to the requirements set in Section 1, low-cost, secure, non-intrusive to the processes of the factory and, integrally, provides clear value-adding quickly. The authors have determined that this goal would be reached by the digitalization of the machines and their transformation.

In the scope of this work, the Measure phase of DMAIC determines the current state of the factory from an Industry 4.0 readiness point of view. This phase ascertains the necessary steps needed to reach the desired digitization of the company. A specific target is chosen, and the criteria show the benefits of digitization in the most effective way. To assess the current state, it is helpful to start from the five-layer automation pyramid. The automation pyramid was used before as the reference architecture by Martinez et al. [58] in their work on deploying a digital twin in the manufacturing system. The automation pyramid is an effective means of visualizing the current state, and therefore, the improvement that each layer needs to achieve the goal of digitalization.

After determining the state that should be reached, the next phase focuses on the means to reach it. There are various means to improve a process and to enable machine digitization. Many of the solutions available in the market provide the elements needed to perform the digitization process. The general requirements for an IIOT solution from Section 1 are taken into account. However, specific needs of the factory should also be considered and the solution must adhere to these as well. The situation in SMEs is more specific because of the narrow circle of decision-makers, and more often than not, the decision-making process is limited to entrepreneurs alone. Therefore, the solution evaluation process is affected by the desires of the entrepreneur.

The improvement process is performed by implementing the solution that was selected in the previous phase as the most suitable for the purpose. The solution needs to be developed and adjusted for the actual state of the machine in the factory. Having in mind that the solutions consist of hardware and software parts, both of them need to be developed and tested before installation. Installation is performed with the least disturbance to the regular factory operation. It is preferable to keep the possibility of quick updates or algorithm modification, even after the installation completes. This would allow for quicker installation with only the essential features implemented and with minimal disturbance of the factory. The speed of implementation is paramount, as it allows for quick control and improvement. Quick implementation cycles allow for a high customization and higher factory decision-makers engagement. This is very important as they should be the primary drivers in the advancement process, and will improve their motivation.

After prototype implementation, the installation, and the verification in the factory, the data are processed and analyzed. The analysis provides information about the machine operation as well as the operation of the installed solution. The solution is improved to provide more information, either by improving its software algorithm or by adding new sensors. By having many analyze-improve-control cycles, the solution becomes more and more beneficial for the factory process, and useful data are kept and analyzed. The machine operation is quantified by the data, and the data analysis gives the decision-makers a clear correlation with the actual factory operation. It is important to visualize data properly to allow for the extraction of valuable information. This process leads to solution improvement, but the goal is the improvement of the machine and, ultimately the entire manufacturing process.

The entire flow of both process improvement and the specific machine digitization is presented in Figure 1.

3.1. Hardware

Numerous solutions can fulfill the requirements and provide a means of performing the machine digitization. The solutions vary in their complexity, scale, robustness, and readiness for implementation in the industrial setting etc. The chosen hardware must enable speeding up the measurement phase of the development. The requirements these solutions must fulfill are given in Section 1:

- Cost effectiveness—Depending on an application, some solutions do not provide enough benefits to convince decision-makers of their usefulness.
- Implementation simplicity—The solutions should aim to provide quick results because long development and complex implementation demand a level of expertise not present in the SMEs.
- Unobtrusiveness—The normal operation of the factory must not be negatively affected by the solution, so a wireless connection and an installation that does not require extensive changes to the machine are necessary.

- Scalability—The goal of the solution is to provide the digitalization of the entire factory process; therefore, the solution must provide easy vertical and horizontal scalability.
- Security—The security of data is always very important, especially in an industrial context. This requirement is not as important in the prototyping phase of the development, but is paramount in the final version of the solution.

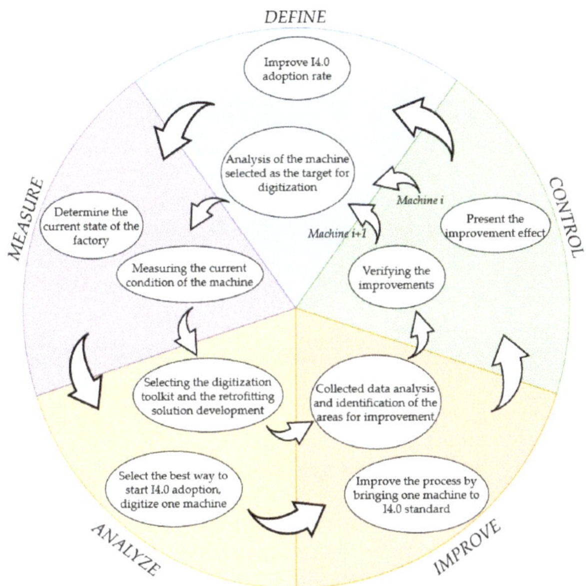

Figure 1. Industry 4.0 adoption process with the specific machine digitization flow.

3.2. Software

The second part of the solution is the software. Software solutions can be divided into firmware and cloud application software. Various possibilities exist on the market for both elements of the software part of the solution. The selected software must enable a quick improve-control cycle so that the actions aimed at the process improvement can be tested quickly, and the corresponding data collected and visualized. A similar set of requirements to the ones used for the hardware part also apply to the software part of the solution.

- Cost effectiveness—While most of the firmware solutions are open-source or available with the purchase of the hardware part, some are acquired separately. The cloud application software and device management is the element that needs to be purchased separately, and the possibilities are numerous.
- Implementation simplicity—Having in mind various elements of the solution and differing levels of developer's expertise, the solution benefits greatly if the technologies used are the ones that are popular in the developer community.
- Scalability—The firmware should provide a modular structure so that it is easy to expand and to add new functionalities. Both the firmware and cloud application software should provide this, and the technology used for development can provide greater flexibility.
- Development speed—Technologies that are already tested and integrated with the hardware part of the solution would provide an easier and quicker setup phase. The popularity and simplicity of the chosen technology are considered when the solution is selected.
- Security—The development of both software elements should provide the necessary security elements. The cloud application is more vulnerable than the firmware ele-

ment. Also worth noting is that most cloud IoT platforms take particular measures in ensuring the safety and security of customers' data.

Several solutions are considered with regard to the requirements set in Section 1 and the two previous subsections, as well as the state of the art (Section 2):

The first solution concept considered is the industry automation standard; a ready-made industrial-grade IIoT solution with a high level of robustness, safety, and reliability. It offers wireless connections and proprietary software that easily connect and interface the machine with an already existing connection. These solutions are not as easily obtained, and often require specific training with regard to software. These solutions are provided by Siemens, ABB, Bosch, etc. The drawback of this solution is that they are expensive and the implementation is not as quick as the other solutions. With the involvement of the big automation companies, the fear of a vendor "lock-in" is a well-known phenomenon [59,60] that is perceived to be even more critical in the case of SMEs. This is an even bigger issue when the high cost of implementation is considered.

The second solution concept is a low-end prototyping solution. These solutions are very cheap and the devices are readily available in the market. A typical example is Arduino boards, microcontroller-based boards that offer very high customization possibilities. Having in mind the lack of processing power and the limited connectivity abilities of these boards, they would need to be paired with a more powerful board such as Raspberry Pi. However, the solutions need to be heavily adapted for them to scale up to the fully matured solution. The industrial setting demands a certain level of safety features that these boards do not provide as-is.

The authors have chosen the conceptual solution that provides industrial-grade equipment and can be used for a relatively inexpensive development of the prototype. The real value of these solutions lies in the possibility to use the same equipment in the final implementation after digitalization to the scope of the entire factory. The chosen solution is easy to install and maintain, and is low cost when comparing it to the industry standard. For prototyping purposes that are very dependent on time, it is preferable to use a more "user-friendly" [61,62] language, such as Python. Actually, Python is a high-level interpretative programming language that is gaining popularity in embedded development [63]. It has extensive support libraries and clean object-oriented designs that can increase the productivity of the programmer up to 10-fold [64]. The drawback is the lower execution speed. The most optimal solution is to have a combination of both. It is preferable if the solution can support data storage and visualization natively, because the implementation can be lengthy and the goal is to provide quick results. Having in mind that this solution must provide the platform for both the prototyping and production phase, the safety and security concerns are very important.

Having in mind all of the above, the solution chosen in this work is the Zerynth platform [65]. The hardware element is the 4ZeroBox, while the Zerynth Cloud provides device management and data visualization, as well as storage. It enables wireless connectivity and support for the most common industrial communication protocols. Zerynth OS is the base for firmware development, and it can be programmed in Python and C. It is easily connected to the cloud and provides a remote firmware update feature. The security requirement is fulfilled by the secure crypto element [66], and on the software side, the "hardened" TLS v1.2 and v1.3 protocol [67]. The storage is large enough so that data can be buffered and uploaded to the cloud in an appropriate format. The data are collected and sent to the cloud where they are available for export or visualization on the dashboards. The cloud solution has an option to visualize the data through an integrated dashboarding system. This system is based on Grafana [68], an open-source analytics and monitoring solution. The visualization support is native, and the time for setting it up is almost zero.

Another very important benefit of a prototyping solution is its ability to be updated remotely. This capability is present in a chosen solution through the Firmware over the Air (FOTA) feature. FOTA allows for a new version of firmware to be installed remotely through the wireless connection. The operators need to be notified and FOTA can be scheduled during the downtime of the broaching machine operation, thus diminishing the intrusiveness of the update. Of course, for changes and improvements concerning the hardware and the broaching machine itself, these disturbances are necessary. Remote firmware update capability is important, because going to the site means disturbing the factory operation, as well as another cycle of installation and testing. Finally, the solution can be upgraded from a prototyping one to a full production level solution with almost zero effort. Both vertical and horizontal solutions are greatly improved, compared to the Arduino-based solutions.

4. Case Study

The case study presented in this paper is the digitization of a broaching machine in Toscana Spazzole Industriali S.r.l (TSI). TSI is an SME with 40 employees and an annual turnover of €5 million. It is based in Tuscany, Italy, and manufactures industrial brushes, which are mainly used in the textile industry. The installation and implementation of an IIoT system to collect and analyze relevant machine data have been performed on a hydraulic broaching machine on the company's shop floor. Before the retrofitting of the broaching machine, performance indicating data, if at all, has solely been displayed analogically and not registered.

4.1. Defining the Goals of the Process Improvement

The goal of the case study was to demonstrate the feasibility of the digital transformation of the plant and its resulting benefits to the entrepreneurs, on the basis of the digitalization of a key machining step in their manufacturing process. The broaching operation of the polypropylene tubes was chosen as it is an essential step of the entire production process. The company's products cannot be manufactured without it, and the occurrence of problems affect all consecutive steps, and interrupts or delays the entire production. The digitalization of the broaching machine enables predictive maintenance, and thereby avoids tool breakage and machine failure. Furthermore, a comprehensive knowledge of the machining process allows for design and process improvements, which eventually decrease energy consumption and increase production efficiency, as well as product quality. The improvement of Industry 4.0 readiness level and the rise of decision-makers' motivation towards the adoption is the most desired outcome of the entire process.

4.2. Introduction of the Case Study

Broaching belongs to the subtractive manufacturing processes and is especially used when high accuracy is needed. A special tool, called a broaching tool, is employed to progressively remove material from a workpiece. The broaching tool moves linearly and the cutting mechanism is mostly orthogonal. The tool possesses a longitudinal series of teeth, which are arranged on a shaft. Consecutive teeth rise in height so that each tooth cuts material in the form of chips from the surface of the workpiece.

In the investigated relevant industrial environment, internal broaching is used to machine the inner surface of long polypropylene tubes to create a precise inner diameter and a high accuracy in roundness. Figure 2a shows the broaching machine. A 2.5 m polypropylene tube is clamped with four double-V brackets on the workbench of the machine where it touches the axial stop with its right end.

4.3. Digitization Target Analysis

To start the broaching process the piston of the hydraulic cylinder is extracted and the broaching tool is mounted on its tip, as shown in Figure 2b. The pressure difference $\Delta p = p - p_0$ between the cylinder chambers creates a force that retracts the piston and hence pulls the broaching tool through the tube. The teeth on the outer surface of the broaching tool cut material from the inner surface of the tube during this process. The most relevant performance indicator is the cutting force of the broaching tool. There are four main motives for the measurement of this force:

- Recording of the cutting force progression during single cuts;
- Distinction between tools of different sizes used for tubes of different diameters;
- Comparison of a new and an old tool design used for tubes of the same diameter;
- Detection of cutting force increase for specific tools over time, due to edge wear.

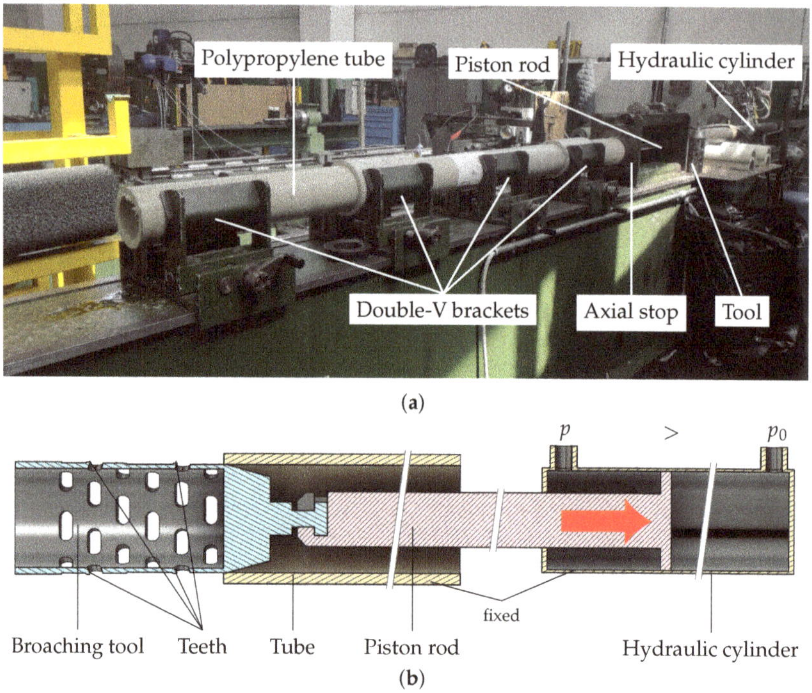

Figure 2. Hydraulic broaching: (**a**) Machine on the shop floor with mounted polypropylene tube to be broached; (**b**) Scheme of the working principle of the hydraulic broaching machine.

The measurement has been performed indirectly via the measurement of the cylinder pressure using a digital pressure transmitter.

4.4. Measuring the Current State of the Factory

The factory was not equipped with any Industry 4.0 technology on the shop floor, and extensive retrofitting was needed. As mentioned in Section 3, the automation pyramid was used to evaluate the current state of the factory to provide the functionality of lower layers. Implementation of the full stack would require extensive changes in factory operation and structure, and this was not the goal of this research. The solution that was implemented provided the possibility to include the upper layers at a later date when the decision-makers decide to make the step up. Therefore, the first three layers of the automation pyramid were implemented to digitize and visualize the broaching machine data. Figure 3a shows the broaching machine automation pyramid before and after IIoT enabling. The blue sections of the automation pyramid represent the IIoT-enabled layers.

The broaching machine is represented by the following automation pyramid layers:

- Field layer—This layer contains devices, actuators, and sensors in the field or production floor. The broaching machine was equipped with an analog pressure indicator (manometer), and the process was actuated locally via remote control with a couple of relays. It was impossible to use the manometer data because the manometer did not have the capability to transmit said data. Therefore, the analog manometer was replaced with a digital pressure transmitter to gather the data and describe the behavior of the broaching machine, and thus, also the broaching tools.

- Control layer—This layer includes all programmable controllers in the field layer assets, such as Programmable Logic Controller (PLC) or Distributed Control System (DCS). The broaching machine did not benefit from any programmable elements; therefore, an IIoT edge device was embedded in the electrical cabinet to collect the data from the sensors in the Field layer and transmit the data to the upper layer.
- Supervision layer—This layer focuses on visualizing data collected from the control layer in various ways, such as control room screens, Human Machine Interface (HMI), and online dashboards. The ability of data visualization was provided thanks to the Zerynth cloud and the local internet network of the shop floor.
- Planning layer—or Manufacturing Execution System (MES) layer, has not yet been implemented in the factory.
- Management layer—or Enterprise Resource Planning (ERP) layer, has not yet been implemented in the company, aside from basic business software.

The solution presented in this work did not entail direct changes to the Planning or Management layers, although the development of the layers was enabled by providing the data from the lower pyramid layers. The broaching machine control became possible because of the device management system that is part of the Cloud contained within the solution.

Having in mind that the automation pyramid architecture is not suitable for the IIoT development, the solution in this work was based on a presented set of Technical Specifications (TS) presented by Mazzei et al. [69], as shown in Figure 3b.

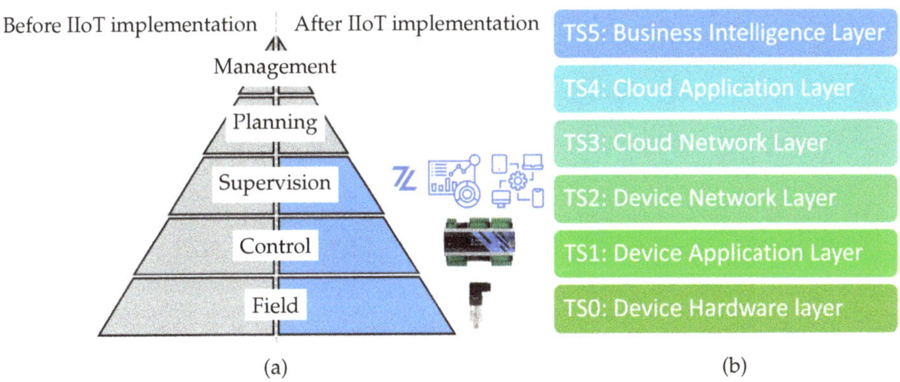

Figure 3. (**a**) Automation pyramid architecture used for ascertaining the current state of the factory and (**b**) IIoT stack used to present solution architecture.

4.5. Selecting the Digitization Toolkit

The displayed tool in Figure 2 is a special hollow broaching tool with 36 teeth arranged in a circular helical pattern around its outer cylindrical surface. It has been designed in previous research work by Jorg and Fantoni [70]. Its unique chip evacuation concept eliminates the timely cleaning step required in traditional broaching, and thus reduces the lead time significantly. Furthermore, it decreases the cutting forces due to minimized friction. The cutting force is the sum of the axial forces of all teeth of the broach that are intruding in and removing material from the tube. It is equivalent to the longitudinal piston force required to pull the broaching tool through the tube. The following two different measurement principles according to the instruments appear generally suitable for achieving the required task:

1. Direct measurement of the force through a dynamometer mounted between the broaching tool and the piston;
2. Indirect measurement of the force through the measurement of the cylinder pressure p using a digital pressure transmitter.

The relationship between the hydraulic force and the pressure p is described by Jorg and Fantoni [70]. A safety valve at 170 bar protects the hydraulic system against overload. Cylinder pressures up to 150 bar are common during broaching operations. These result in cutting forces of up to approximately 65 kN. Both measurement methods have their pros and cons. Therefore, several aspects had to be weighed for the specific application case. The direct measurement of the force via a load cell delivers the absolute force value accurately. A digital pressure transmitter does not measure the force, but the pressure. The force can be calculated [70]. There remains uncertainty regarding the correct absolute force value due to the potential losses of the hydraulic system. However, the prior formulated four main goals of monitoring the progression, and thus, the relative changes of the performance indicator, can still be satisfied. Additionally, to investigate the behavior adequately, especially during the entering and exiting phase of the broaching tool, a resolution of $d = 1$ mm is required. At an average forward rate of the broach of $v = 3$ m/min, this results in a minimum measurement frequency of $f_{\text{min}} = v/d = 50$ Hz or a sample time of $t_{\text{max}} = 20$ ms, respectively. This requirement can easily be achieved by both systems. A standard 100 kN dynamometer for tension and compression costs about 2000 EUR. However, these standard instruments are too big and do not fit inside the tube. Using option one would hence require specially manufactured measurement equipment, which is even more expensive and has long delivery times. Furthermore, the installation and implementation of a dynamometer inside the tube (regarding also the cabling) are complex. A digital pressure transmitter, on the other hand, can easily be swapped with or added to the existing analog manometer, and it can immediately be bought off-the-shelf for under 100 EUR. Lastly, a solution with the least possible intrusiveness to the existing system is desirable. The most critical factors that finally led the authors to choose the second option of the indirect force measurement via a digital pressure transmitter, were its instant availability and its immense cost advantage. These factors play a central role in IIoT prototyping.

4.6. The Retrofitting Solution Development

The broaching machine was controlled using a simple electrical circuit, including non-intelligent components such as relays and contactors. The electrical cabinet was updated by adding and embedding the Zerynth 4ZeroBox [65] as an IIoT edge device. The 4ZeroBox was mounted on the DIN-35 rail and powered by a 24 VDC power supply unit that had already been present in the electrical cabinet. The IIoT-enabled electrical cabinet of the broaching machine is shown in Figure 4. Then, in order to evaluate the performance and the efficiency of the broaching tool, the existing analog manometer was replaced with a digital pressure transmitter, WIKA S-20 (measuring range 0–250 bar). The piston movement signals are transmitted to the digital input–output pins of the 4ZeroBox using the reserve contacts of the R1 and R2 relays. The 4ZeroBox was wirelessly connected to the internet through an internal Wi-Fi network in the factory.

A software algorithm was developed, and the 4ZeroBox was programmed, using Python programming language. The software utilized for programming is Visual Studio Code; more precisely, the Zerynth expansion. The Zerynth Visual Studio Code expansion uses the Zerynth Software Development Kit (SDK) to program and manage the 4ZeroBox. The connection to the hardware was established, and the 4ZeroBox was connected to the Zerynth Cloud through Zerynth Device Manager (ZDM) [71]. The communication between the 4ZeroBox and the Zerynth Cloud is implemented through an MQTT protocol.

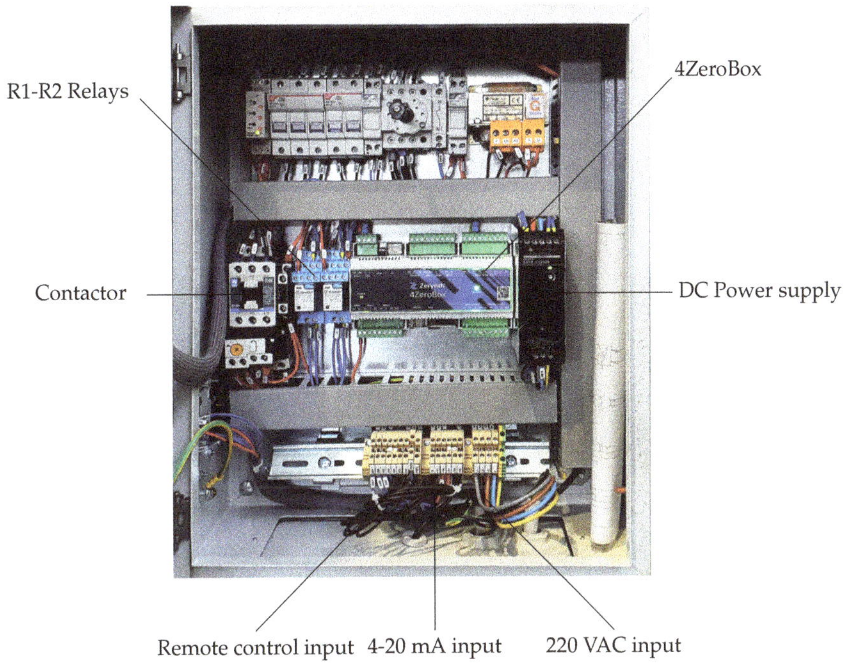

Figure 4. IIoT-enabled electrical cabinet layout.

The firmware was programmed in Python programming language with the libraries from Zerynth SDK. The multi-threaded structure was utilized with one main thread and one thread used for data gathering, the acquisition thread. The acquisition thread has only one task, sampling the digital pressure transmitter every 20 ms and storing the data in the queue. A data frame is created by storing every sample with the UNIX timestamp. An accurate timestamp was obtained by the process of synchronization of the 4ZeroBox to the ZDM. The acquisition thread starts asynchronously after detecting the press of any of the buttons on the remote control. A button press is detected as the change of the logical level on the digital input pins of the 4ZeroBox. Furthermore, any press of the button is registered as an event and stored in the queue, along with the UNIX timestamp.

The main thread is tasked with initialization, establishing the connection and synchronization to the ZDM, and data storage and publishing to the ZDM, as well as the control of the acquisition thread. After detecting data in the queue, the main thread takes the data frame and stores it on the local SD card. All the data collected during one active period of the acquisition thread is stored on the SD card and treated as one cut of the tube. The cut signifies one piston extension and one retraction, as explained in Section 4.2. The main thread detects the block of the acquisition thread and starts sending the data from the SD card storage to the Zerynth Cloud storage.

The data are sent to the Zerynth Cloud, where it is stored and handled by the ZDM. The communication between the device and the cloud is protected by the TLS v1.2 and v1.3 protocols based on private and public keys [67]. Every data frame is assigned the timestamp the data were received. The data are in the form of a JSON file. After the data are stored, it is possible to download it in a couple of ways: scheduling the periodic download or using REST API to retrieve the data from the Zerynth Cloud. After the data are downloaded, they can be analyzed. The entire system can be seen in Figure 5, while Figure 6 shows the overview of the entire process.

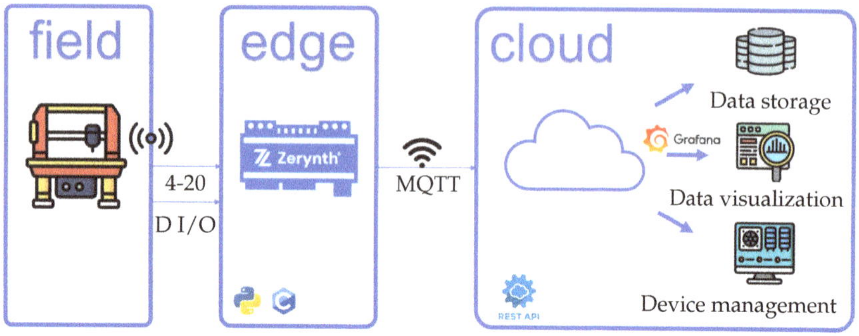

Figure 5. Final system overview.

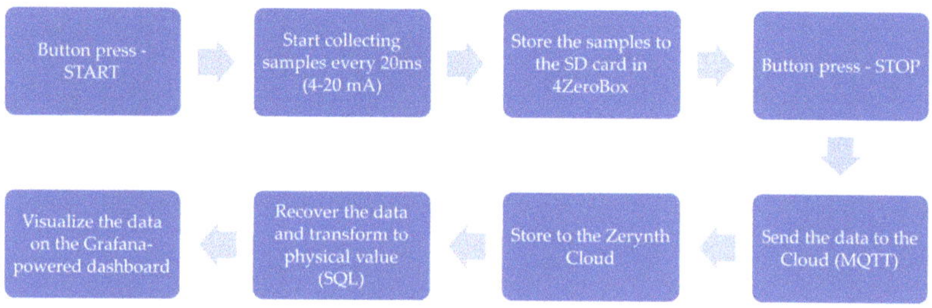

Figure 6. Flowchart of the broaching process digitization solution.

4.7. Analysis of the Collected Data

The analysis of the broaching process was performed on the pressure. Figure 7 shows the characteristic pressure progression during one single cut in the Zerynth Dashboard.

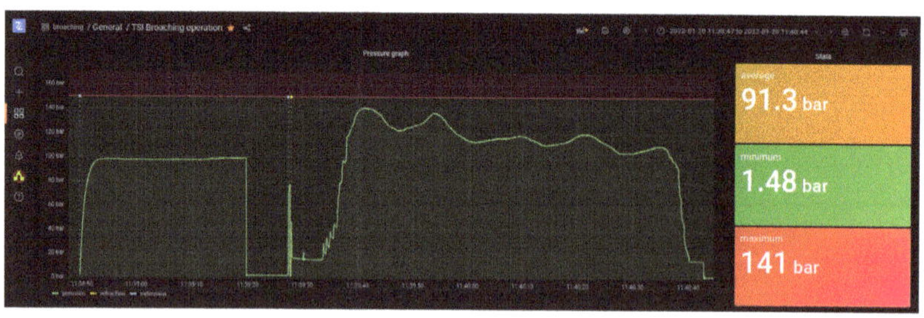

Figure 7. Characteristic pressure progression during one single cut shown in the Zerynth Dashboard.

The pressure in bar is plotted in light green over time in the format HH:MM:SS. The user can zoom and/or slide along the time axis to see the data of the entire day or to search for a cut conducted at a known time. However, any operation of the broach is detected, and cuts are identified automatically by the system. Any day can be selected through a calendar. The dashboards also display the timestamps for every time a control button has been pressed. The pressing of the extension button is indicated with a blue dotted vertical line, and the retraction button with an orange dotted vertical line. The system provides a basic statistical analysis of the data shown on the screen by displaying the minimum, the average, and the maximum pressure values. The initial plateau after the blue dotted vertical line resembles the extraction phase of the piston. This comparably high value without an external load can be explained through the design of the control valve. During the forward movement of the piston, the valve connects both cylinder chambers equally to the oil pump. In both chambers, the same pressure prevails. Nevertheless, the piston moves to the left due to the different effective surface areas of its piston head. When the piston is fully extracted, the valve is closed and the pressure drops. This characteristic plateau can be observed before every cut. Hence, it can be used to identify cuts from a large number of data points automatically. The entering phase of the broaching tool can be well observed in the plot too. The pressure increases step-wise whenever one circular array of teeth enters the tube and starts cutting the material. The opposite, less pronounced behavior occurs at the end of the pipe.

During the cut, the digital pressure transmitter detected fluctuations in the pressure. The plateau is not as flat as during the extraction, but shows several peaks instead. The delivered polypropylene tubes have deviations in circularity, concentricity, and diameter, so that the broaching tool has to cut more material in some areas and less material in others. Additionally, a misalignment of the four bracket pairs and too-tight clamping of the tubes could deform them. The implementation of the measurement system eventually allows the authors to investigate these effects.

Figure 8 shows the pressure progressions while broaching tubes of three different diameters with three different tools (tool diameters of 80, 102, and 110 mm). Every tool has its characteristic cutting pressure and can therefore be identified easily. The cutting force, and hence, the hydraulic pressure, are proportional to the removed cross-sectional area inside the tube. The cut area increases with the tube, and respectively, the tool diameter, which explains the difference between the 80 and 110 mm tool diameters (yellow and orange graphs). The graph with the highest pressure belongs to the new tool presented in Figure 2, with a diameter of 102 mm. This tool cuts in two stages instead of one stage, as in previous tools. Therefore, the material removal is about twice that of a previous tool with the same diameter [70]. This explains the higher pressure. The step-wise increase and decrease do not occur in the previous tools, due to their different design. Additionally, in the chosen representation in Figure 8, the distance traveled by the piston is plotted on the x-axis instead of the time. This, for instance, allows a pressure peak to be related to a position, and also to understand the length of the different tubes.

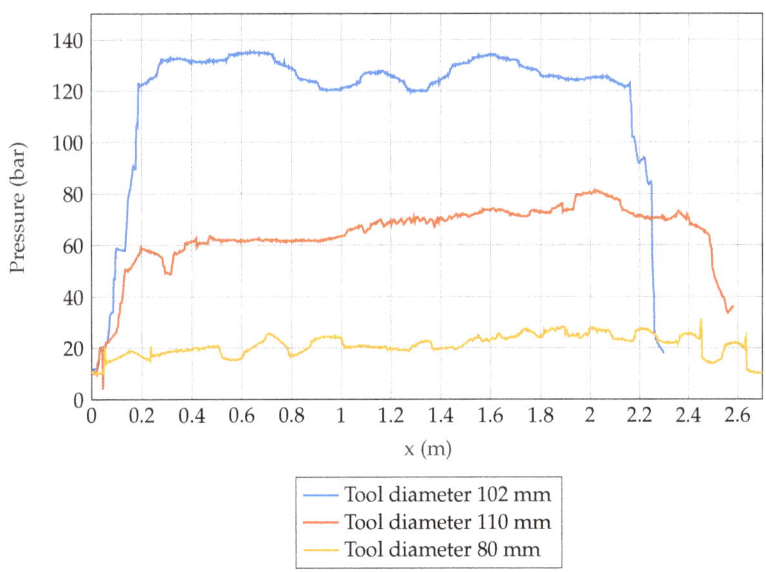

Figure 8. Pressure progressions when broaching tubes of three different diameters with three different tools.

4.8. Determining the Areas for Improvement

The observation of pressure fluctuations in the plateau corresponding to the distance between clamps led the authors to rethink the clamping mechanism of the tubes. They modified the standard clamping system and were able to reduce the pressure deviation significantly. Figure 9 shows the deviation of the plateau pressure from its average value in percent for two cuts with the 102 mm tool before and after the modification of the clamping system.

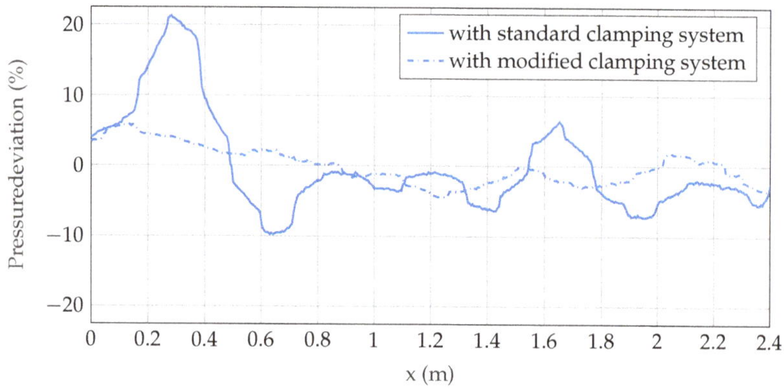

Figure 9. Reduction in the pressure deviation during the cut-through modification of the clamping system.

While the pressure deviated by more than 20% with the standard system, the value can be kept under 5% through the undertaken modifications. A more regular cut results in a better surface finish and increases the lifetime of the tool. Modifications like these, which were only made possible through the IIoT system, enable the factory to produce products of higher quality at lower manufacturing costs. In addition to the purchase of a broaching machine itself, the biggest investment is the manufacturing of new broaching tools. As these wear out, it is inevitable. Enabling predictive maintenance through data collection and analysis allows for the determination of the ideal point in time for the resharpening of a tool, and thereby avoids overstressing and increases the tool's lifetime. The tool presented in Figure 2 already outperforms previous tools in terms of efficiency by far. However, it is still manufactured from one solid piece of steel which requires many machining hours on a very expensive five-axis CNC mill. Therefore, the design had been further upgraded by using a modular architecture with a hollow body carrying ring-shaped blades. The body only has to be manufactured once, and the blades can easily be resharpened and cost-effectively reproduced when completely worn out. The tool has already been successfully tested.

4.9. Verifying the Improvements

The installed system enabled the collection of the data and quick data presentation to the operators. The benefits to the factory were that they can collect the data during the broaching process. In this way, they can track the number of tubes being cut, and which tubes had an issue with cutting. The detection of a blockage was present before, but it was not recorded in the data. The factory wants to continue digitizing by including environmental sensors that can measure air temperature, pressure, and humidity. One of the reasons for this is that in the last few months, it was noticed that the pressure drops in consecutive cuts during the extension process. The extension process is performed without load, so the expected behavior is that this pressure does not change significantly. However, as seen in Figure 10, the pressure gradually drops from cca 90 bar to the 70 bar. The research team suggested installing an analog temperature sensor on the broaching machine to ascertain whether the temperature of the machine is correlated with this variation in pressure.

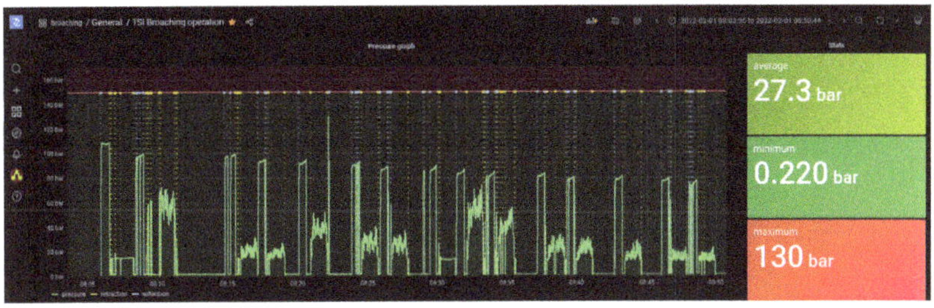

Figure 10. Gradual decline in the pressure during the first few hours of operation.

The operators in the factory wanted to include environmental measurements in the next iteration of the solution. This can be achieved easily by installing an appropriate Click board from MikroElektronika, and creating another thread in the firmware that periodically reads the environmental data.

The factory production manager could observe other anomalies in pressure data. Those anomalies were heightened pressure during several broaching operations. Finally, it was determined that operators were not adequately lubricating the broaching tool before the broaching operation. This resulted in the production manager creating an operating procedure that removed this problem in the future. An increased level of cooperation between factory workers and a research team was also observed with the additional features requested in the future. Better collaboration also provided the research team with the possibility for conducting further experiments remotely, with the operator's assistance. The ability to remotely update the firmware and control the installed system has provided this option to the research team.

5. Discussion

The digitalization process of small or medium enterprises, especially those with a traditional approach to their work, is complex since it involves not only the technical, but organizational and cultural aspects as well.

The described application tries to engage the entrepreneur primarily, giving him/her a solid base to calculate the return value of the investment. Then, the target is the production manager deploying the digitalization as a means to supervise the machines and to provide better communication with the operators. The operators, in turn, promote and execute numerous improvements to the manufacturing process in collaboration with the production manager. A new culture of experiments and data analysis seems to be emerging in the company. Digital data recording is replacing manual monitoring for the first time, on the shop floor of the company. The engagement of key stakeholders in the industrial process improvement was the primary goal of this research, and there is evidence of a long-term commitment to the digital transformation of the factory floor.

It is important to note that various experiments were performed by the factory production managers and operators, without an explicit request from the research team. The company changed the design of the broach, as previously mentioned, and has asked the research team to digitize another machine on the shop floor. The machine in question is the next in the manufacturing process chain, a hydraulic press that inserts a steel roll into the broached polypropylene tube. As the reader can understand, the new machine can be digitalized using the same hardware, sensors, and firmware as the previous one, thus reducing the cost of the twinning. This indicates that the Industry 4.0 principles are being integrated into the factory processes, as well as demonstrating the development of a data-driven approach, which was one of the goals of the solution.

The low cost of the solution is best reflected through the reduced installation and development cost, and especially in the easy transition between the prototype and the production-ready solution. This is the main benefit of this solution concept, as was shown in Section 3. The approximate cost of the entire hardware solution is cca USD 600, which is higher than that of the Arduino-based solution, but with the added benefit of using the same hardware for the production version of the solution.

From a technical perspective, the broaching machine digitalization was executed in a fast and smooth way, thanks to the Zerynth environment. The researchers were able to quickly install the hardware and acquire the data. The entire development, testing, and installation process was completed in 6 days with the installation, and therefore, the break of operation lasted only one day. It is important to note here that the open libraries and standards used in the digitization process greatly reduced its complexity, and therefore improved the development speed. The entrepreneur especially appreciated the rapid setup and the fact that the prototype immediately started providing machine data. Data visualization was paramount in helping the entrepreneur, as well as the production manager, to gain valuable insights from the acquired data. It also enabled an immediate analysis and an improvement of the broaching tool, as well as the broaching process in general. These results lead to new requests for the digitalization of other machines in the production line, thus demonstrating the positive impact of the entire process. Of course, the retrofitting of the machine was made easy by the simple architecture of the broaching machine, and a more complex system could require additional efforts.

Furthermore, since the requirements in Section 1 have been elaborated to address the needs collected in manufacturing SMEs, the described solution could fulfill other SMEs with the same goal of digitalization and similar electromechanical machines. As shown, the solution is not dependent on the type of the collected data; therefore, almost any SME with similar problems with old and non-connected machines can benefit from it.

The main drawback of the study was not having access to the costs before and after our implementation. Due to this lack of information, we were unable to calculate the return on investment nor its payback period. Another aspect that is difficult to quantify is related to the additional quality documentation that TSI can provide to its customers. These detailed records would properly document each broaching process, thus providing a better guarantee of the final performance of the brushes.

6. Conclusions

The paper presents a rapid, low-cost prototyping solution for retrofitting legacy machinery in manufacturing SMEs intending to adopt the Industry 4.0 paradigm, reducing the perceived risk of the unknown digitalization process. The digital transformation has been implemented through a well-known improvement process (Six Sigma) that runs in parallel with machine retrofitting. This allowed the company to have immediate benefits and positively influence the motivation of decision-makers to continue the scale-up of the solution. A set of requirements for the solution was elaborated, and the solution was developed with an eventual scale-up in mind. The factory decision makers started the process of digitizing another machine, and they want to incorporate the data from both machines to further improve the entire process. The authors have observed the factory appreciating the benefits of the digitalization, and are confident that more Industry 4.0 technologies will be adopted with time.

In fact, additional improvements of the broaching process are planned, and will be investigated in further work. Having in mind that the broaching tool is expensive and time-consuming to manufacture, its breakage or performance deterioration would require a change, and therefore, a lengthy cease of operation. Because the broaching machine is just one of several steps in the manufacturing chain of the factory, this cease can affect the entire factory. The Zerynth solution enabled the collection, processing, and analysis of the data, which provided insight into machine operation. Currently, the data are used for the monitoring of the process as it happens. Further insight can be obtained as the database grows. Future work regarding the broaching machine involves developing artificial intelligence with the ultimate goal of enabling predictive maintenance.

The good results let us foresee a possible digitalization of the entire shop floor, adopting the same approach and technologies.

Author Contributions: Conceptualization, G.F. and M.V.; methodology, G.F.; software, M.V.; validation, M.V., O.J. and M.H.; formal analysis, G.F. and O.J.; writing—original draft preparation, M.V.; writing—review and editing, O.J.; visualization, M.H., O.J. and M.V.; supervision, G.F.; project administration, M.V.; funding acquisition, G.F. All authors have read and agreed to the published version of the manuscript.

Funding: This research work was undertaken in the context of DIGIMAN4.0 project ("DIGItal MANufacturing Technologies for Zero-defect Industry 4.0 Production", http://www.digiman4-0.mek.dtu.dk/, accessed on 1 January 2020). DIGIMAN4.0 is a European Training Network supported by Horizon 2020, the EU Framework Program for Research and Innovation (Project ID: 814225).

Institutional Review Board Statement: Not applicable.

Informed Consent Statement: Not applicable.

Data Availability Statement: Restrictions apply to the availability of these data. Data was obtained from TSI and are available from the authors with the permission of TSI.

Acknowledgments: The authors would like to give special thanks to Salam Qaddoori Dawood Al-Zubaidi from the University of Pisa; Lorenzo Biagini, Alessio Costanzi, and Dario Vanzi from Toscana Spazzole Industriali; and Ugo Scarpellini, Davide Neri, and Elia Guglielmin from Zerynth.

Conflicts of Interest: The authors declare no conflict of interest.

References

1. Kagermann, H.; Wahlster, W.; Helbig, J. *Recommendations For implementing the Strategic Initiative Industrie 4.0: Final rEport of the Industrie 4.0 Working Group*; Forschungsunion: Berlin, Germany, 2013.
2. Erboz, G. How to define industry 4.0: Main pillars of industry 4.0. *Manag. Trends Dev. Enterp. Glob. Era* **2017**, *761*, 761–767.
3. Wopata, M. Industry 4.0 Adoption 2020–Who is Ahead? 2021. Available online: https://iot-analytics.com/industry-4-0-adoption-2020-who-is-ahead/ (accessed on 1 January 2020).
4. Entrepreneurship—Statistical Indicators. Available online: https://ec.europa.eu/eurostat/statistics-explained/index.php?title=Entrepreneurship_-_statistical_indicators (accessed on 16 March 2021).
5. Schwertner, K. Digital transformation of business. *Trakia J. Sci.* **2017**, *15*, 388–393. [CrossRef]
6. Marushchak, L.; Pavlykivska, O.; Khrapunova, Y.; Kostiuk, V.; Berezovska, L. The economy of digitalization and digital transformation: Necessity and payback. In Proceedings of the 2021 11th International Conference on Advanced Computer Information Technologies (ACIT), Deggendorf, Germany, 15–17 September 2021; IEEE: Manhattan, NY, USA, 2021; pp. 305–308.
7. Stentoft, J.; Jensen, K.W.; Philipsen, K.; Haug, A. Drivers and barriers for Industry 4.0 readiness and practice: A SME perspective with empirical evidence. In Proceedings of the 52nd Hawaii International Conference on System Sciences, Maui, HI, USA, 8–11 January 2019.
8. Masood, T.; Sonntag, P. Industry 4.0: Adoption challenges and benefits for SMEs. *Comput. Ind.* **2020**, *121*, 103261. [CrossRef]
9. Kumar, R.; Singh, R.K.; Dwivedi, Y.K. Application of industry 4.0 technologies in SMEs for ethical and sustainable operations: Analysis of challenges. *J. Clean. Prod.* **2020**, *275*, 124063. [CrossRef] [PubMed]
10. Vieru, D.; Bourdeau, S.; Bernier, A.; Yapo, S. Digital competence: A multi-dimensional conceptualization and a typology in an SME context. In Proceedings of the 2015 48th Hawaii International Conference on System Sciences, Kauai, HI, USA, 5–8 January; IEEE: Manhattan, NY, USA, 2015; pp. 4681–4690.
11. Moeuf, A.; Pellerin, R.; Lamouri, S.; Tamayo-Giraldo, S.; Barbaray, R. The industrial management of SMEs in the era of Industry 4.0. *Int. J. Prod. Res.* **2018**, *56*, 1118–1136. [CrossRef]
12. Popkova, E.G.; Ragulina, Y.V.; Bogoviz, A.V. *Industry 4.0: Industrial Revolution of the 21st Century*; Springer: Berlin/Heidelberg, Germany, 2019; Volume 169.
13. Vuković, M.; Mazzei, D.; Chessa, S.; Fantoni, G. Digital twins in industrial IoT: A survey of the state of the art and of relevant standards. In Proceedings of the IEEE International Conference on Communications Workshops (ICC Workshops), Virtual, 14–23 June 2021; IEEE: Manhattan, NY, USA, 2021; pp. 1–6.
14. Inc, G. Definition of Digitalization—Gartner Information Technology Glossary. Available online: https://www.gartner.com/en/information-technology/glossary/digitalization (accessed on 12 August 2022).
15. Kritzinger, W.; Karner, M.; Traar, G.; Henjes, J.; Sihn, W. Digital Twin in manufacturing: A categorical literature review and classification. *IFAC-PapersOnLine* **2018**, *51*, 1016–1022. [CrossRef]

16. Mourtzis, D.; Angelopoulos, J.; Panopoulos, N. Digital Manufacturing: The evolution of traditional manufacturing toward an automated and interoperable Smart Manufacturing Ecosystem. In *The Digital Supply Chain*; Elsevier: Amsterdam, The Netherlands, 2022; pp. 27–45.
17. Brauner, P.; Dalibor, M.; Jarke, M.; Kunze, I.; Koren, I.; Lakemeyer, G.; Liebenberg, M.; Michael, J.; Pennekamp, J.; Quix, C.; et al. A computer science perspective on digital transformation in production. *Acm Trans. Internet Things* **2022**, *3*, 1–32. [CrossRef]
18. Martikkala, A.; David, J.; Lobov, A.; Lanz, M.; Ituarte, I.F. Trends for low-cost and open-source iot solutions development for Industry 4.0. *Procedia Manuf.* **2021**, *55*, 298–305. [CrossRef]
19. Müller, J.M.; Voigt, K.I. Sustainable industrial value creation in SMEs: A comparison between industry 4.0 and made in China 2025. *Int. J. Precis. Eng.-Manuf.-Green Technol.* **2018**, *5*, 659–670. [CrossRef]
20. Horváth, D.; Szabó, R.Z. Driving forces and barriers of Industry 4.0: Do multinational and small and medium-sized companies have equal opportunities? *Technol. Forecast. Soc. Chang.* **2019**, *146*, 119–132. [CrossRef]
21. Müller, J.; Voigt, K. Industry 4.0—Integration strategies for small and medium-sized enterprises. In Proceedings of the 26th International Association for Management of Technology (IAMOT) Conference, Vienna, Austria, 14–18 May 2017; pp. 14–18.
22. Shamim, S.; Cang, S.; Yu, H.; Li, Y. Management approaches for Industry 4.0: A human resource management perspective. In Proceedings of the 2016 IEEE Congress on Evolutionary Computation (CEC), Vancouver, BC, Canada, 24–29 July 2016; IEEE: Manhattan, NY, USA, 2016; pp. 5309–5316.
23. Corallo, A.; Lazoi, M.; Lezzi, M.; Luperto, A. Cybersecurity awareness in the context of the Industrial Internet of Things: A systematic literature review. *Comput. Ind.* **2022**, *137*, 103614. [CrossRef]
24. Wee, D.; Kelly, R.; Cattel, J.; Breunig, M. Industry 4.0-how to navigate digitization of the manufacturing sector. *Mckinsey Co.* **2015**, *58*, 7–11.
25. Asplund, F.; Macedo, H.D.; Sassanelli, C. Problematizing the Service Portfolio of Digital Innovation Hubs. In *Proceedings of the Working Conference on Virtual Enterprises*; Springer: Berlin/Heidelberg, Germany, 2021; pp. 433–440.
26. Pirola, F.; Cimini, C.; Pinto, R. Digital readiness assessment of Italian SMEs: A case-study research. *J. Manuf. Technol. Manag.* **2019**, *31*, 1045–1083. [CrossRef]
27. Wübbeke, J.; Meissner, M.; Zenglein, M.J.; Ives, J.; Conrad, B. Made in China 2025. *Mercat. Inst. China Stud. Pap. China* **2016**, *2*, 74.
28. Veile, J.W.; Kiel, D.; Müller, J.M.; Voigt, K.I. Lessons learned from Industry 4.0 implementation in the German manufacturing industry. *J. Manuf. Technol. Manag.* **2019**, *31*, 977–997. [CrossRef]
29. Müller, J.M.; Buliga, O.; Voigt, K.I. The role of absorptive capacity and innovation strategy in the design of industry 4.0 business Models-A comparison between SMEs and large enterprises. *Eur. Manag. J.* **2021**, *39*, 333–343. [CrossRef]
30. Sorger, M.; Ralph, B.J.; Hartl, K.; Woschank, M.; Stockinger, M. Big data in the metal processing value chain: A Systematic digitalization approach under special consideration of standardization and SMEs. *Appl. Sci.* **2021**, *11*, 9021. [CrossRef]
31. Adolphs, P.; Berlik, S.; Dorst, W.; Friedrich, J.; Gericke, C.; Hankel, M.; Heidel, R.; Hoffmeister, M.; Mosch, C.; Pichler, R.; et al. *DIN SPEC 91345:2016-04*; Reference Architecture Model Industrie 4.0. Beuth Verlag GmbH: Berlin, Germany 2016.
32. Ilari, S.; Carlo, F.D.D.; Ciarapica, F.E.E.; Bevilacqua, M. Machine tool transition from Industry 3.0 to 4.0: A comparison between old machine retrofitting and the purchase of new machines from a triple bottom line perspective. *Sustainability* **2021**, *13*, 10441. [CrossRef]
33. Contreras Pérez, J.D.; Cano Buitrón, R.E.; García Melo, J.I. Methodology for the retrofitting of manufacturing resources for migration of SME towards industry 4.0. In Proceedings of the International Conference on Applied Informatics, Bogota, Colombia, 1–3 November 2018; Springer: Berlin/Heidelberg, Germany, 2018; pp. 337–351.
34. Lins, T.; Oliveira, R.A.R.; Correia, L.H.; Silva, J.S. Industry 4.0 Retrofitting. In Proceedings of the 2018 VIII Brazilian Symposium on Computing Systems Engineering (SBESC), Salvador, Brazil, 6–9 November 2018; IEEE: Manhattan, NY, USA, 2018; pp. 8–15.
35. Lins, T.; Oliveira, R.A.R. Cyber-physical production systems retrofitting in context of industry 4.0. *Comput. Ind. Eng.* **2020**, *139*, 106193. [CrossRef]
36. Haskamp, H.; Orth, F.; Wermann, J.; Colombo, A.W. Implementing an OPC UA interface for legacy PLC-based automation systems using the Azure cloud: An ICPS-architecture with a retrofitted RFID system. In Proceedings of the 2018 IEEE Industrial Cyber-Physical Systems (ICPS), Saint Petersburg, Russia, 15–18 May 2012; IEEE: Manhattan, NY, USA, 2018; pp. 115–121.
37. Waters, M.; Waszczuk, P.; Ayre, R.; Dreze, A.; McGlinchey, D.; Alkali, B.; Morison, G. Open source IIoT solution for gas waste monitoring in smart factory. *Sensors* **2022**, *22*, 2972. [CrossRef]
38. Mourtzis, D.; Angelopoulos, J.; Panopoulos, N. Recycling and retrofitting for industrial equipment based on augmented reality. *Procedia CIRP* **2020**, *90*, 606–610. [CrossRef]
39. Liu, C.; Zheng, P.; Xu, X. Digitalisation and servitisation of machine tools in the era of Industry 4.0: A review. *Int. J. Prod. Res.* **2021**. [CrossRef]
40. Lima, F.; Massote, A.A.; Maia, R.F. IoT energy retrofit and the connection of legacy machines inside the Industry 4.0 concept. In Proceedings of the IECON 2019–45th Annual Conference of the IEEE Industrial Electronics Society, Lisbon, Portugal, 14–17 October 2019; IEEE: Manhattan, NY, USA, 2019; Volume 1, pp. 5499–5504.
41. Hesser, D.F.; Markert, B. Tool wear monitoring of a retrofitted CNC milling machine using artificial neural networks. *Manuf. Lett.* **2019**, *19*, 1–4. [CrossRef]

42. Mandula, K.; Parupalli, R.; Murty, C.A.; Magesh, E.; Lunagariya, R. Mobile based home automation using Internet of Things (IoT). In Proceedings of the International Conference on Control, Instrumentation, Communication and Computational Technologies (ICCICCT), Kumaracoil, India, 18–19 December 2015; IEEE: Manhattan, NY, USA, 2015; pp. 340–343.
43. Baraka, K.; Ghobril, M.; Malek, S.; Kanj, R.; Kayssi, A. Low cost arduino/android-based energy-efficient home automation system with smart task scheduling. In Proceedings of the Fifth International Conference on Computational Intelligence, Communication Systems and Networks, Hong Kong, China, 5–7 June 2013; IEEE: Manhattan, NY, USA, 2013; pp. 296–301.
44. Piyare, R.; Tazil, M. Bluetooth based home automation system using cell phone. In Proceedings of the 2011 IEEE 15th International Symposium on Consumer Electronics (ISCE), Singapore, 14–17 June 2011; IEEE: Manhattan, NY, USA, 2011; pp. 192–195.
45. Viciana, E.; Alcayde, A.; Montoya, F.G.; Baños, R.; Arrabal-Campos, F.M.; Zapata-Sierra, A.; Manzano-Agugliaro, F. OpenZmeter: An efficient low-cost energy smart meter and power quality analyzer. *Sustainability* **2018**, *10*, 4038. [CrossRef]
46. Roy, A.; Das, P.; Das, R. Temperature and humidity monitoring system for storage rooms of industries. In Proceedings of the 2017 International Conference on Computing and Communication Technologies for Smart Nation (IC3TSN), Gurgaon, India, 12–14 October 2017; IEEE: Manhattan, NY, USA, 2017; pp. 99–103.
47. Acharya, V.; Hegde, V.V.; Anjan, K.; Kumar, M. IoT (Internet of Things) based efficiency monitoring system for bio-gas plants. In Proceedings of the 2017 2nd International Conference on Computational Systems and Information Technology for Sustainable Solution (CSITSS), Bengaluru, India, 21–23 December 2017; IEEE: Manhattan, NY, USA, 2017; pp. 1–5.
48. Choi, C.S.; Jeong, J.D.; Lee, I.W.; Park, W.K. LoRa based renewable energy monitoring system with open IoT platform. In Proceedings of the 2018 International Conference on Electronics, Information, and Communication (ICEIC), Honolulu, HI, USA, 24–27 January 2018; IEEE: Manhattan, NY, USA, 2018; pp. 1–2.
49. Khan, M.S.; Sharma, H.; Haque, A. IoT enabled real-time energy monitoring for photovoltaic systems. In Proceedings of the 2019 International Conference on Machine Learning, Big Data, Cloud and Parallel Computing (COMITCon), Faridabad, India, 14–16 February 2019; IEEE: Manhattan, NY, USA, 2019; pp. 323–327.
50. Arjoni, D.H.; Madani, F.S.; Ikeda, G.; Carvalho, G.d.M.; Cobianchi, L.B.; Ferreira, L.F.; Villani, E. Manufacture equipment retrofit to allow usage in the industry 4.0. In Proceedings of the 2017 2nd international conference on Cybernetics, Robotics and Control (CRC), Chengdu, China, 21–23 July 2017; IEEE: Manhattan, NY, USA, 2017; pp. 155–161.
51. Kumar, S.; Jasuja, A. Air quality monitoring system based on IoT using Raspberry Pi. In Proceedings of the 2017 International Conference on Computing, Communication and Automation (ICCCA), Greater Noida, India, 5–6 May 2017; IEEE: Manhattan, NY, USA, 2017; pp. 1341–1346.
52. Al-Shahethi, M.G.; Muneer, A.; Ghaleb, E.A.; Darshi, S.; Almohammedi, A.A. Real life monitoring of conveyor line speed using IoT and raspberry Pi. In Proceedings of the 2021 International Conference on Intelligent Technology, System and Service for Internet of Everything (ITSS-IoE), Sana'a, Yemen, 1–2 November 2021; IEEE: Manhattan, NY, USA, 2021; pp. 1–5.
53. Agrawal, N.; Singhal, S. Smart drip irrigation system using raspberry pi and arduino. In Proceedings of the International Conference on Computing, Communication & Automation, Greater Noida, India, 15–16 May 2015; IEEE: Manhattan, NY, USA, 2015; pp. 928–932.
54. Minchala, L.I.; Peralta, J.; Mata-Quevedo, P.; Rojas, J. An approach to industrial automation based on low-cost embedded platforms and open software. *Appl. Sci.* **2020**, *10*, 4696. [CrossRef]
55. De Mast, J.; Lokkerbol, J. An analysis of the Six Sigma DMAIC method from the perspective of problem solving. *Int. J. Prod. Econ.* **2012**, *139*, 604–614. [CrossRef]
56. Shankar, R. *Process Improvement Using Six Sigma: A DMAIC Guide*; Quality Press: Seattle, WA, USA, 2009.
57. Schmitt, P.; Schmitt, J.; Engelmann, B. Evaluation of proceedings for SMEs to conduct I4. 0 projects. *Procedia Cirp* **2019**, *86*, 257–263. [CrossRef]
58. Martinez, E.M.; Ponce, P.; Macias, I.; Molina, A. Automation pyramid as constructor for a complete digital twin, case study: A didactic manufacturing system. *Sensors* **2021**, *21*, 4656. [CrossRef]
59. Opara-Martins, J.; Sahandi, R.; Tian, F. Critical analysis of vendor lock-in and its impact on cloud computing migration: A business perspective. *J. Cloud Comput.* **2016**, *5*, 4. [CrossRef]
60. What is Vendor Lock-In and How can You Avoid It? Available online: https://www.ionos.com/digitalguide/hosting/technical-matters/vendor-lock-in/ (accessed on 14 July 2022).
61. Sotomayor-Beltran, C.; Segura, G.W.Z.; Roman-Gonzalez, A. Why should python be a compulsory introductory programming course in Lima (Peru) universities? In Proceedings of the 2018 IEEE International Conference on Automation/XXIII Congress of the Chilean Association of Automatic Control (ICA-ACCA), Concepcion, Chile, 17–19 October 2018; IEEE: Manhattan, NY, USA, 2018; pp. 1–4.
62. Perkel, J.M. Programming: Pick up Python. *Nature* **2015**, *518*, 125–126. [CrossRef]
63. Python for Microcontrollers. Available online: https://micropython.org/ (accessed on 14 March 2022).
64. Advantages and disadvantages of python programming language. Available online: https://medium.com/@mindfiresolutions.usa/advantages-and-disadvantages-of-python-programming-language-fd0b394f2121 (accessed on 28 July 2022).
65. 4ZeroBox—Zerynth Industrial IOT Device. 2022. Available online: https://www.zerynth.com/products/hardware/4zerobox/ (accessed on 7 March 2021).
66. The Veil of IOT Security—How Zerynth Plugs Security Breaches in IOT. 2022. Available online: https://www.zerynth.com/resources/white-papers/iot-security/ (accessed on 1 January 2020).

67. Gigante, M. How Zerynth Ensures Data Security and Data Integrity. 2022. Available online: https://www.zerynth.com/blog/how-zerynth-ensures-data-security-and-data-integrity/ (accessed on 12 March 2021).
68. Grafana: The Open Observability Platform. Available online: https://grafana.com/ (accessed on 16 March 2021).
69. Mazzei, D.; Baldi, G.; Montelisciani, G.; Fantoni, G. A full stack for quick prototyping of IoT solutions. *Ann. Telecommun.* **2018**, *73*, 439–449. [CrossRef]
70. Jorg, O.; Fantoni, G. Design, development and test of a novel broach for long polypropylene tubes. *Precis. Eng.* **2022**, *74*, 224–231. [CrossRef]
71. Zerynth Device Manager. 2022. Available online: https://www.zerynth.com/products/software/zdm/ (accessed on 7 March 2021).

applied sciences

MDPI

Review

Study of Augmented Reality Based Manufacturing for Further Integration of Quality Control 4.0: A Systematic Literature Review

Phuong Thao Ho *, José Antonio Albajez, Jorge Santolaria and José A. Yagüe-Fabra

I3A, Universidad de Zaragoza, 50018 Zaragoza, Spain; jalbajez@unizar.es (J.A.A.); jsmazo@unizar.es (J.S.); jyague@unizar.es (J.A.Y.-F.)
* Correspondence: thao.ho@unizar.es

Abstract: Augmented Reality (AR) has gradually become a mainstream technology enabling Industry 4.0 and its maturity has also grown over time. AR has been applied to support different processes on the shop-floor level, such as assembly, maintenance, etc. As various processes in manufacturing require high quality and near-zero error rates to ensure the demands and safety of end-users, AR can also equip operators with immersive interfaces to enhance productivity, accuracy and autonomy in the quality sector. However, there is currently no systematic review paper about AR technology enhancing the quality sector. The purpose of this paper is to conduct a systematic literature review (SLR) to conclude about the emerging interest in using AR as an assisting technology for the quality sector in an industry 4.0 context. Five research questions (RQs), with a set of selection criteria, are predefined to support the objectives of this SLR. In addition, different research databases are used for the paper identification phase following the Preferred Reporting Items for Systematic Reviews and Meta-Analysis (PRISMA) methodology to find the answers for the predefined RQs. It is found that, in spite of staying behind the assembly and maintenance sector in terms of AR-based solutions, there is a tendency towards interest in developing and implementing AR-assisted quality applications. There are three main categories of current AR-based solutions for quality sector, which are AR-based apps as a virtual Lean tool, AR-assisted metrology and AR-based solutions for in-line quality control. In this SLR, an AR architecture layer framework has been improved to classify articles into different layers which are finally integrated into a systematic design and development methodology for the development of long-term AR-based solutions for the quality sector in the future.

Keywords: augmented reality; industry 4.0; quality 4.0; metrology; assembly

Citation: Ho, P.T.; Albajez, J.A.; Santolaria, J.; Yagüe-Fabra, J.A. Study of Augmented Reality Based Manufacturing for Further Integration of Quality Control 4.0: A Systematic Literature Review. *Appl. Sci.* **2022**, *12*, 1961. https://doi.org/10.3390/app12041961

Academic Editor: Emanuele Carpanzano

Received: 19 January 2022
Accepted: 9 February 2022
Published: 13 February 2022

1. Introduction

The industry 4.0 revolution has enabled many improvements and benefits for manufacturing as well as service systems (see Figure 1). However, the rapid and remarkable changes that appeared in manufacturing also led to higher requirements in technological knowledge, increasing the degree of task complexity or variability of tasks on the shop-floor level for the operators [1–3]. This leads to the demands of systems that intensively adopt the enabling technologies of industry 4.0 to reduce those burdens for the operators.

The latest key facilitating technologies of Industry 4.0 are Advanced Simulation, Advanced robotics, Industrial "Internet of Things" (IoT), Cloud computing, Additive manufacturing, Horizontal and vertical system integration, Cybersecurity, Big Data and analytics, Digital-twin, Blockchain, Knowledge Graph and Augmented Reality (AR) [4].

Besides these key technologies, there are some fundamental technologies such as sensors and actuators, Radio Frequency Identification (RFID) and Real-Time Locating Solution (RTLS) technologies, etc. to support them. For the long-term adaptation progress of Industry 4.0, seven design principles need to be considered when designing and developing a solution in general and for manufacturing specifically [5]. These principles are real-time data management, interoperability, virtualization, decentralization, agility, service orientation and integrated business processes. An important aspect of Industry 4.0 is the synthesis of the physical environment and the virtual elements [6], which can be achieved by the advantages of AR together with the Cyber-Physical System (CPS).

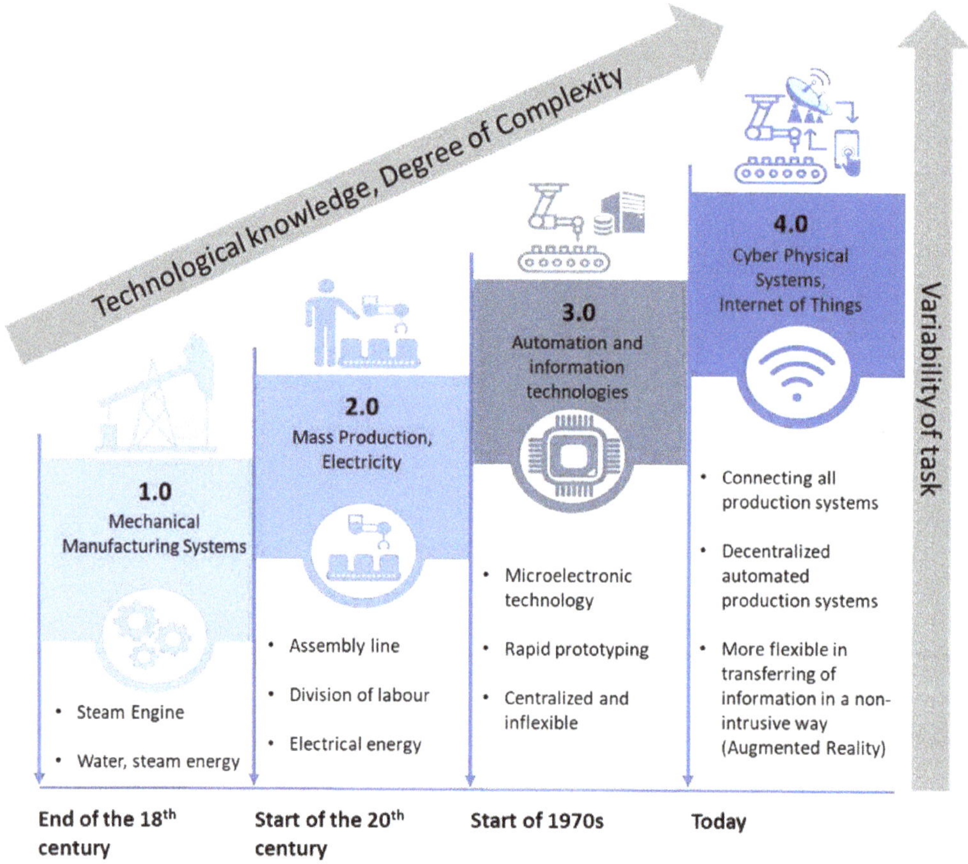

Figure 1. Industrial revolutions and their characteristics based on [1].

In the last few years, Augmented Reality (AR) has been constantly adapted by key companies on industrial innovation, such as General Electrics, Airbus [7] and Boeing. It has been employed for productivity improvement, product and process quality advancement (reducing error rates) [8] or higher ergonomics in diverse manufacturing phases to boost the transformation of Industry 4.0. AR studies in the quality sector have emerged, and have shown potential results in enhancing human performance in technical quality control tasks, supporting the Total quality management (TQM) and autonomizing operators' decision making. Despite the mentioned advantages, there are still limited examples of AR's concrete implementation in manufacturing, especially for the quality sector.

For this reason, the main objective of this paper is to conduct a state-of-the-art review of AR systematically in terms of technology used, applications and limitations, focusing on the quality context. This is to prepare for the digital transformation of Industry 4.0, which also leads to the change in the quality sector, known as Quality 4.0, and the adaptation of AR for quality control. However, not only the studies focusing on quality context but also the other relevant studies in the manufacturing context are considered to build a long-term roadmap for AR-based applications supporting Quality 4.0. To achieve this, a Systematic Literature Review (SLR) was performed to assure the reproducibility and scalability of the study, together with the objectivity of the results [9]. An investigation of the status of AR-based manufacturing applications on the shop-floor level in the context of Industry 4.0 was carried out to give a holistic view about future challenges and to propose roadmaps to implement AR technology for the quality control sector in the short term and Quality 4.0 in the long term.

The paper is structured in four sections. Section 1 introduces the project, AR technology and Quality 4.0. Section 2 describes the methodology applied for the SLR. Section 3 reports on the results and answers the research questions (RQs) to provide a holistic view about the current AR-based manufacturing in general and AR-based quality control in particular. The final Section 4 concludes and proposes future works.

1.1. Augmented Reality (AR)

From a technical point of view, AR is a technology superimposing digital, computer-generated information onto the physical world to enrich humans' perspectives about the surrounding environment. This innovates regarding the interaction of humans with digital information and the real world. There are different types of augmented information, which are visual augmentation [8], audio [10], haptic feedback [11] and multimodal feedback [12]. AR applications based on visual augmentation are currently dominant in the manufacturing context. However, there is an emerging interest in multimodal AR applications, which mainly implement visual augmentation with another sensing feedback.

Although the research interest in AR technology has rapidly evolved and been intensively investigated over the past 20 years, the first immersive reality prototype can be dated back to 1968. In that year, the first head-mounted display (HMD) device connecting to the computer, which provided the earliest of humankind's experience into augmented reality named "Sword of Damocles", was invented by Ivan Sutherland [13]. The way humans interact with industrial AR today is influenced by this invention. However, the term Augmented Reality was first officially formulated years later in 1992 by Thomas Caudell, who was a Boeing researcher. He implemented a heads-up display (HUD) application to demonstrate his idea of designing and prototyping an application to support the manual manufacturing process [14]. In 1996, immersive reality was classified by different levels of immersive experience, which depends on the type of dominant content—reality information or virtual information—and was introduced into the Reality-Virtuality Continuum (RV continuum) by Milgram et al. [15] as shown in Figure 2.

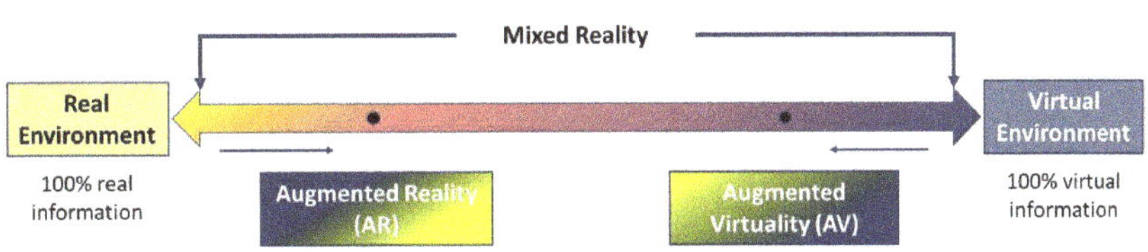

Figure 2. Reality-Virtuality Continuum adapted from [15].

One year later, Azuma defined the three main technical characteristics of AR based on its technology, which are combining real and virtual objects, interacting with real/virtual objects in real-time and registering (aligning) virtual objects with real objects [16].

Technically, a general AR system is constructed of software built on a selection of four fundamental hardware components: a processing unit, a tracking device, a display device and an input device. The processing unit creates augmentation models, controls devices' connections and adjusts the position of superimposed information into the real world with respect to the pose and position of the user by employing the information coming from a tracking device. The tracking device is used to track the exact position and orientation of the user to align/register the augmentations accurately to the desired positions. This device usually consists of at least one element of image capture (a Charge-Coupled Device CCD, stereo, or depth-sensing camera Kinect) [17]. Regarding the tracking technology of AR, depending on the selected tracking devices and tracking methods, it can be classified into three groups: computer vision-based tracking (CV-based tracking), sensor-based tracking and hybrid tracking. The input device is used to obtain the stimulation of the environment or users to trigger the augmentation functionalities. However, the input device is optional because there are some built-in input methods integrated with display devices, especially HMD and HHD. In some cases, the activating elements (images, GPS positions, sensor values, markers, etc.) are pre-defined, thus an input device is not essential in those cases. The current input techniques for HMD are hand-tracking, head/eye-gaze and voice. The processing data are visualized onto the display device via a user interface (UI) enhancing two-way communication between the user and the system. The current display devices can be classified into two groups: in situ display (desktop monitor, projection-based augmentation, spatial augmentation, etc.) and mobile display (hand-held device HHD, head-mounted device HMD).

Depending on the selection of devices, the overlaying augmentations technique onto the user's scene can be different. Currently, there are three superimposing techniques. With the first technique, the augmentation can be directly projected to the field of view (FoV) of the user. This one is called optical combination and is implemented with an optical see-through HMD (OST-HMD). The second technique is known as video mixing. The user's scene is taken by the camera and processed by a computer. After inserting the augmentations on the processed scene, the result is displayed on a display device, on which the user views the real scene indirectly. The last technique is image projection, which directly projects the augmentations onto the physical objects.

Tracking and registration are the crucial and challenging aspects of AR applications. The accuracy in tracking and registration determines the alignment quality of augmentations. According to [18], the tracking and registration algorithms are divided into three groups: (1) Marker-based algorithms, (2) Markerless (or Natural feature-based) algorithms and (3) Model-based algorithms. For marker-based tracking, the 2D markers having unique shapes/patterns are placed on the real objects, where the digital information is planned to be overlayed. A digital augmentation is programmatically assigned for each marker in the workplace. When the camera recognizes the markers, the pre-assigned augmentation is displayed onto the marker. In some situations, the markers are occluded and it is not efficient. Thus, natural feature-based tracking (NFT) is more commonly used in computer vision-based tracking. Some well-known natural feature-based tracking algorithms are Speeded Up Robust Features (SURF), Scale Invariant Feature Transform (SIFT) and Binary Robust Independent Elementary Features (BRIEF). This technique extracts the characteristic points in images to train the AR system's real-time detection of those points. Despite providing the seamless integration of augmentations into the real world, natural feature tracking (NFT) intensively depends on computational power and is slower and less effective at long distances. Therefore, small artificial markers knowns as fiducial markers are used to mitigate the disadvantages of NFT by accelerating the initial recognition, decreasing the computational requirements and improving the system's performance. Model-based tracking algorithms utilize a predefined list of models, which are then compared with the real-time extracted features.

In general, a basic pipeline of AR system/application consists of image capturing, digital image processing, tracking, interaction handling, information management, rendering and displaying [17]. It starts by capturing a frame with the device's camera. Then comes the Digital Image Processing step of AR software to process the captured image in order to estimate the camera position in relation to a reference point/object (a marker, an optical target, etc.). This estimation can also utilize the internal sensors, which help in tracking the reference object. The camera positioning accuracy is crucial for displaying AR content because it needs to be scaled and rotated according to the scenarios. After that, the processed image is rendered for the relevant perspective and is shown to the user on a display device. In some cases, when certain remote or local information is required, the Information Management module is responsible for accessing it. The interaction handling module is to enable the users' interaction with the image.

1.2. Quality 4.0

Besides cost, time and flexibility, quality is one crucial dimension of manufacturing attributes in terms of products and processes [19]. Its objective is to assure that the service or final product meets the specifications and satisfies the customers' requirements.

Total quality management (TQM) is the current highest level of quality in an organization context, which holistically considers internal and external customers' needs, cost of quality and system development to organize and assist quality improvement. Quality control (QC) is a part of TQM, playing an essential role in fulfilling technical specifications with inspection applying techniques such as statistical process control (SPS), which is statistical sampling to manage the in-line quality on the shop-floor manufacturing level [20]. Contrastingly, Quality assurance (QA) concentrates more on the pre-manufacturing phases, such as planning, design, prototyping, etc., to ensure the achievement of quality requirements for manufacturing products. The international standard ISO 9001:2015 describes the specific standards of the quality management system [21]. Many organizations have employed various methods and approaches to improve quality performance such as TQM, Lean Six-Sigma, Failure mode and effect analysis (FMEA), quality function deployment (QFD) and benchmarking [22]. Furthermore, certain behaviors in the factories, such as process management, customer focus, involvement in the quality of supply and small group activity, are required for the successful application of quality management [23].

Industry 4.0 is a new industrial digitization paradigm that may be seen at all levels of modern industry. Quality 4.0 can be considered an integral part of Industry 4.0 when the status of quality and industry 4.0 are combined. It is the digitalization of TQM or the application of Industry 4.0 technology to improve quality. The value propositions for Quality 4.0 include the augmentation or improvement of human intelligence; the enhancement of productivity and quality for decision-making; the improvement of transparency and traceability; human-centered learning; change prediction and management [24–26].

In a holistic view of a smart factory, Big Data, in conjunction with CPS, can be applied to manage the data understanding. Big data analytics play a critical role in supporting early failure detection during the manufacturing process, providing valuable insight into factory management such as productivity enhancement [27]. IoT provides a superior global vision for the industrial network (including intelligent sensors and humans) as well as the ability to take real-time actions based on data comprehension [28]. Then it comes to AI, which is currently used to perform visual inspections of products towards quality control evaluation. One of the most critical issues in manufacturing is the ability to visually assess product quality [29]. AI methods (Machine learning techniques) proved their advancement in assisting inspections based on data analysis. This latter is frequently taken as images collected from sensors/cameras inside manufacturing environments. Finally, AR technologies can be applied to facilitate the inspection process with an immersive experience by superimposing digital information onto the working environment [30].

At the time of conducting this study, most enabling technologies of Industry 4.0, especially AR technology, have reached a mature point that could enhance the transformation of quality 4.0. This means that a systematic literature review about AR applied in the quality sector is essential and crucial not only for the digital transformation in quality 4.0 but also for the long-term integration of AR technology in quality sector. All relevant AR-assisted quality control solutions in the manufacturing context are considered for this SLR to observe how the cutting edge AR technology has been applied and evolved in quality sector. The findings of this SLR then can be used as references for further improvement and implementation of AR in quality 4.0. to save costs and resources, as well as to improve productivity, accuracy and autonomy

2. Research Methodology

The literature methodology is demonstrated in this section. Two successive searches were carried out following the Preferred Reporting Items for Systematic Reviews and Meta-analyses (PRISMA) [31], which is a straightforward reporting framework for systematic reviews that supports authors to develop their reviews and meta-analysis reporting. The primary search was done on 15 September 2020 and the extended search on 13 August 2021.

Then, the predefined inclusion and exclusion criteria were allocated into relevant stages of the PRISMA flowchart to support the paper selection process (see Section 2.2 Paper selection).

2.1. Planning

The initial step was to identify exactly which areas the study should cover and which are excluded. To fill an essential gap in the AR-based quality control sector, as well as to provide a road map for the further implementation of AR technology to support Quality 4.0 in the future, this study focuses on AR systems and their applications in manufacturing, especially shop-floor processes that require intensive involvement of operators' activities such as assembly, maintenance and quality control. Hence, the following five research questions are defined (see Table 1).

Table 1. Research questions.

Code	Research Question (RQ)	Knowledge Extraction from the RQ
RQ1	What is the current state of AR-based applications in manufacturing?	The current gaps in adopting AR-based applications into the industry context
RQ2	How does AR-based quality control benefit manufacturing in the context of Industry 4.0?	Problems that AR-based quality control applications are providing supporting for
RQ3	What are the available tools to develop AR-based applications for the quality sector?	Focus on the current display devices, tracking methods and software development platforms
RQ4	How can AR-based applications for the quality sector be evaluated?	Methods and metrics to analyze and evaluate results and effectiveness of AR applications
RQ5	How to develop an AR-based solution for long-term benefits of quality in manufacturing?	A conceptual framework for AR-based solutions in quality and manufacturing

RQ1: What is the current state of AR-based applications in manufacturing?

The motivation of this question is to understand the current industry adoption status of AR-based applications, and to determine the gap between applications that were tested in the industry through field experiments and the ones that were still in the novel stage as pilot projects or only tested in a laboratory context.

RQ2: How does AR-based quality control benefit manufacturing in the context of Industry 4.0?

The objective is to observe the evolution of AR-based quality control applications in the manufacturing context. In addition, it is to understand how AR technology is currently applied to support each specific case in the quality sector.

Thus, a holistic identification of application areas for AR-based quality control in industrial manufacturing based on technology suitability can be carried out in the future.

RQ3: What are available tools to develop AR-based applications for quality sector?

The objective is to systemize the current development tools and frameworks supporting in the AR-assisted manufacturing process development. Thus, when it comes to developing AR-based quality control applications in the future, a useful set of tools and frameworks would be available to consider.

RQ4: How can AR-based applications for the quality sector be evaluated?

The motivation is to know which metrics, indicators and methods are utilized to evaluate effectiveness and improvement when applying AR technology to support a quality-related activity.

RQ5: How to develop an AR-based solution for long-term benefits of quality in manufacturing?

Based on the results concerning previous RQs (RQ1, RQ2, RQ3, RQ4), a concept of development framework for AR-assisted quality can be generalized and used to answer this question.

The next step was choosing the databases for document identification. Four well-known technology research databases, which are Scopus, Web of Science (WoS), SpringerLink and ScienceDirect, were used for finding high-quality literature resources. Those databases were selected due to their broad coverage of journals and disciplines. Mendeley was used as a reference manager software. This program is chosen due to its user-friendly aspects such as fast processing of large numbers of references, word citation add-in, integrated pdf-viewer and teamwork collaboration. Microsoft Excel was used for data extraction and evaluation.

2.2. Paper Selection

For the systematic search of documents, a set of search strings was determined to search the databases mentioned in the planning phase. The Search strings and search syntax for each database are listed in Table 2.

Table 2. Database query strings.

No.	Database Name	Search Strings [Search: by Title, Abstract, and Keywords]
1	Scopus	TITLE-ABS-KEY (("augmented reality" OR "mixed reality") AND ("industry 4.0" OR "manufacturing" OR "Production" OR "factory" OR "industrial application" OR "quality" OR "assembly" OR "maintenance")) AND (LIMIT-TO (DOCTYPE, "ar")) AND (LIMIT-TO (LANGUAGE, "English"))
2	Web of Science	(TS = (("augmented reality" OR "mixed reality") AND ("industry 4.0" OR "manufacturing" OR "Production" OR "factory" OR "industrial application" OR "quality" OR "assembly" OR "maintenance"))) AND LANGUAGE: (English) AND DOCUMENT TYPES: (Article)
3	Springerlink	("augmented reality" OR "mixed reality") AND ("industry 4.0" OR "manufacturing" OR "Production" OR "factory" OR "industrial application" OR "quality" OR "assembly" OR "maintenance")
4	ScienceDirect	1st search: ("augmented reality" OR "mixed reality") AND ("industry 4.0" OR "manufacturing" OR "Production" OR "factory") 2nd search: ("augmented reality" OR "mixed reality") AND ("industrial application" OR "quality" OR "assembly" OR "maintenance")

Comparing to assembly and maintenance, the AR-based applications supporting the quality sector have not been comprehensively investigated in the last few years. Thus, assembly and maintenance sectors can be considered as good references for the development of AR-based quality control applications. In addition, assembly, maintenance and quality control are normally carried out in similar working conditions and all require intensive involvement of the operators. Therefore, keywords such as "assembly" and "maintenance" are included in the search strings besides keywords such as "manufacturing", "industrial application", etc.

After systematically searching with the above search strings on respective databases, there were 1248 documents found (see Table 3). On the left side of Figure 3, a chart referring to the numbers of publications is illustrated, which were systematically found on databases. The large number of articles found using databases were from Scopus (38%) and WoS (36%). Regarding duplication, 78% of found articles only belong to one single database, 18% to two databases and 4% to three databases shown on the right side of Figure 3. In addition, the manual search resulted in 48 more articles by scanning cited references of influential review papers.

Table 3. Identified papers by database.

Database	Identified Papers	Duplicate Papers	Non-Duplicate Papers
Scopus	476	0	476
Web of Science	446	223	223
Springerlink	73	0	73
ScienceDirect	253	95	158
Total	1248	318	930

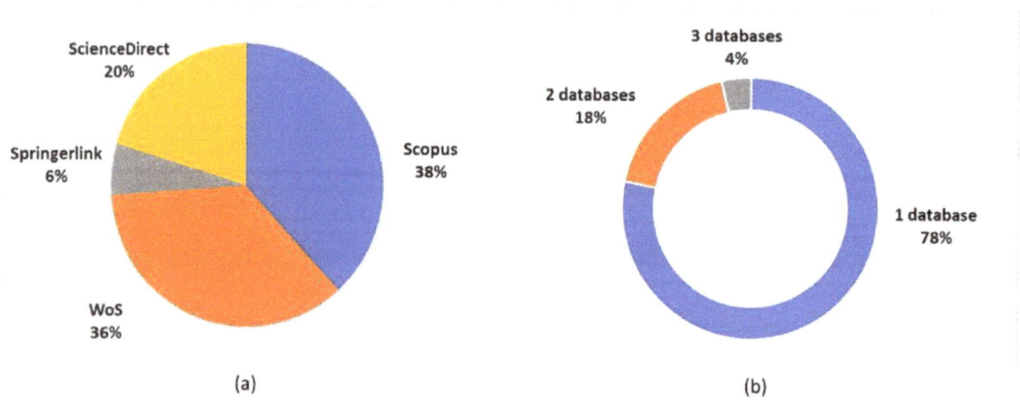

Figure 3. Search by database: (**a**) Database distribution; (**b**) Database duplication.

The AR articles were selected and approved based on the following criteria described in Table 4: Inclusion and exclusion criteria. The number of publications excluded referring to specific criteria were also included in the table. The relevant exclusion criteria were applied for each stage of paper selection flowing PRISMA flowchart as in Figure 4.

The main strategy for paper selection following the adapted PRISMA flowchart is intensively applying exclusion criteria for the first screening and full-text eligibility evaluation. If a paper meets any exclusion criteria, it would be immediately excluded from the search results. The inclusion criteria were used for the second screening to check the quality of articles.

Table 4. Selection criteria.

Inclusion/ Exclusion	Criteria	Code	Description	Identified Papers
Exclusion	Duplication	D	Duplicated articles	318
	Not relevant	NR1	The screened content demonstrates that the article is completely irrelevant to AR or applies AR outside the context of Manufacturing	555
		NR2	VR is mainly applied instead of AR	38
	Loosely relevant	LR	AR in manufacturing is only mentioned as an example	23
	Other exclusion	OE1	Publication year: older than 2010	56
		OE2	Not a peer-reviewed article from a conference or journal	6
		OE3	Publication language: not English	3
		OE4	Full text is not available	5
		OE5	Excluded by the quality check	92
Inclusion	Quality check	HQ1	The full text of the article provides a clear methodology	
		HQ2	The full text of the article provides results	
		HQ3	The article is relevant to the research questions	
Total identified articles			1296	
Total excluded articles			1096	
Total included articles			200	

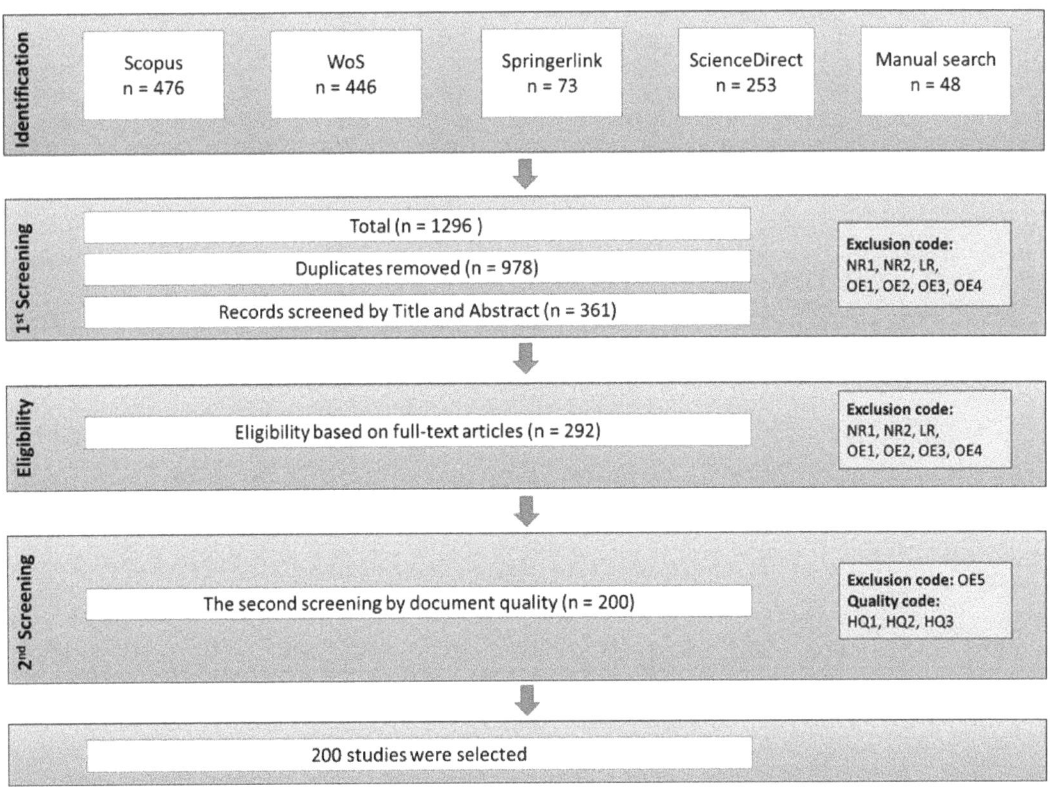

Figure 4. Search results following PRISMA flow chart adapted from [31].

In the document identification step, there were a total number of 1296 papers found from both systematic searches on the aforementioned databases and a manual search. After removing duplicated documents (318 papers), 978 publications were analyzed by Title and Abstract, referring to exclusion criteria to identify the relevant papers supporting the study's objective. After the first screening, 361 publications were remaining, which were carefully considered following the exclusion code (NR1, NR2, LR, OE1, OE2, OE3, OE4). A total of 69 publications were rejected, including five papers that could not be accessed as a full text. There were 292 articles qualified for the next screening. The quality assessment at the second screening phase was achieved by evaluating each document through binary decision compliance with a set of criteria HQ1, HQ2, HQ3. If a paper did not satisfy the quality check, it was listed as an exclusion result regarding the OE5 code. The quality check criteria were:

HQ1: The full text of the article provides a clear methodology

HQ2: The full text of the article provides results

HQ3: The article is relevant to the research questions

However, there were some exceptions, namely that a paper was not required to fulfill all the quality check criteria. If a document did not provide methodology or results but exhibited an interesting concept or potential development, it could be accepted. For example, paper [8] provides a clear methodology and implementation of in-line quality assessment of polished surfaces in a real manufacturing context, but there is no test or evaluation to validate the results. However, the prototype in the paper was built on a robust development approach, which can be further improved to adopt for long-term implementation. Besides that, most of the remote collaboration articles focusing more on the Human-Computer Interaction (HCI) and human cognition fields, which do not support the RQs of this study, rather than the AR context were also excluded at this step. As a result of paper selection, a total of 200 studies were selected to conduct the systematic review. Figures and tables summing up these papers and their research are provided in the following sections.

2.3. Data Extraction and Analysis

2.3.1. Classification Framework

The classification framework used to analyze AR-based publications in manufacturing and extracting relevant information for answering RQs consists of three parts:

1. Application area in manufacturing mixed categories of papers:

At the beginning of the 200 selected publications, each can be allocated into five solution groups according to their application field in an industry 4.0 context: (1) Maintenance, (2) Assembly, (3) Quality, (4) Others and (5) General manufacturing context. Next, they were classified into 4 different categories of papers following the benchmark in [32]: review papers, technical papers, conceptual papers and application papers (see Table 5).

Table 5. Literature retrieved and organized based on the classification framework.

Paper type/ Applied Sector	Maintenance	Assembly	Quality	Others	General Manufacturing Context
Review paper	2 articles	3 articles	0	0	16 articles
	[33,34]	[35–37]			[2,17,18,38–50]
Technical paper	1 article	18	2 articles	0	5 articles
	[51]	[52–69]	[70,71]		[72–76]
Conceptual paper	15 articles	22 articles	5 articles	3 articles	16 articles
	[77–91]	[92–113]	[114–118]	[119–121]	[122–137]
Application paper	20 articles	32 articles	25 articles	9 articles	6 articles
	[138–157]	[7,11,158–187]	[8,30,188–210]	[211–219]	[220–225]

The correlation between these two classifications formed a matrix giving an overview of the current interest in AR-based solutions in the industry. In more detail, review papers are the ones summing up the current literature on a specific topic to provide the state of the art of that area. Technical papers are mainly about solutions and algorithms for the development of hardware/software and AR systems. Conceptual papers consider specific characteristics of AR solutions to propose advanced concepts for their further practical adoption. Finally, application papers provide works that develop and test AR solutions in a case study or real environment.

With this classifying approach, the results in Figure 5 show that there is currently no systematic review paper about AR technology enhancing quality sector.

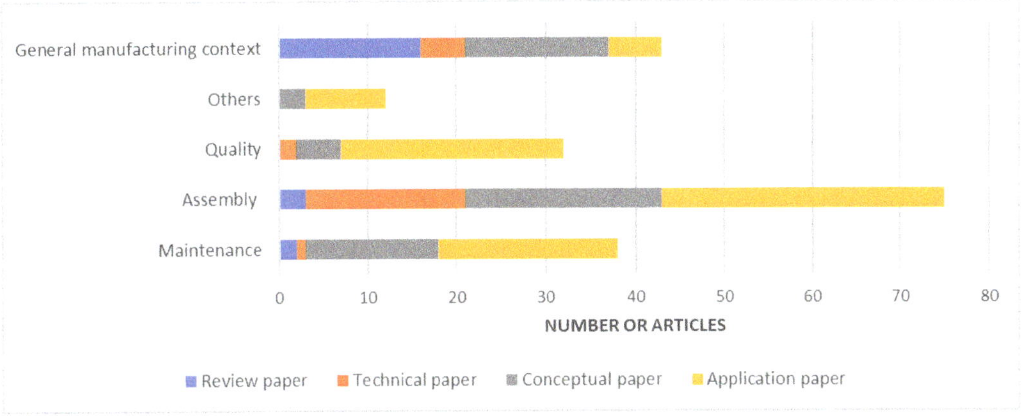

Figure 5. Type of papers and application field.

The general manufacturing context has the highest number of review papers, while Assembly has the highest number of application papers. Although the total number of publications of AR-based quality is less than the total number of articles in AR-based maintenance, AR-based application papers in quality are slightly higher than in maintenance. Considering that the investigation for AR-based quality solution was behind maintenance in the past, this proves that the interest for implementing AR technology in quality sector has significantly grown in recent years.

2. The architecture layer framework of AR systems in manufacturing

After the first classification, each paper's content was analyzed following the architecture layer framework of the AR system adapted from [226], as in Figure 6, to extract relevant data for answering the RQs.

This architecture layer framework of AR systems was adapted and improved from a study in the built environment sector. The framework was chosen for the analysis step because its architecture was constructed in accordance with the standard architecture layer criteria for developing information technology concepts and tools. Besides that, it could cover all essential aspects of an AR application based on system point of view (layer 1, 2), industry point of view (layer 4, 5) and user point of view (layer 3: usability, layer 2: interaction design, content design).

In more detail, the framework in the study consists of five layers covering most of the important characteristics of AR-based solutions, from fundamental aspects to advanced intelligent solutions, as in the following:

- Concept & Theory
- Implementation
- Evaluation

- Industry adoption
- Intelligent AR solution

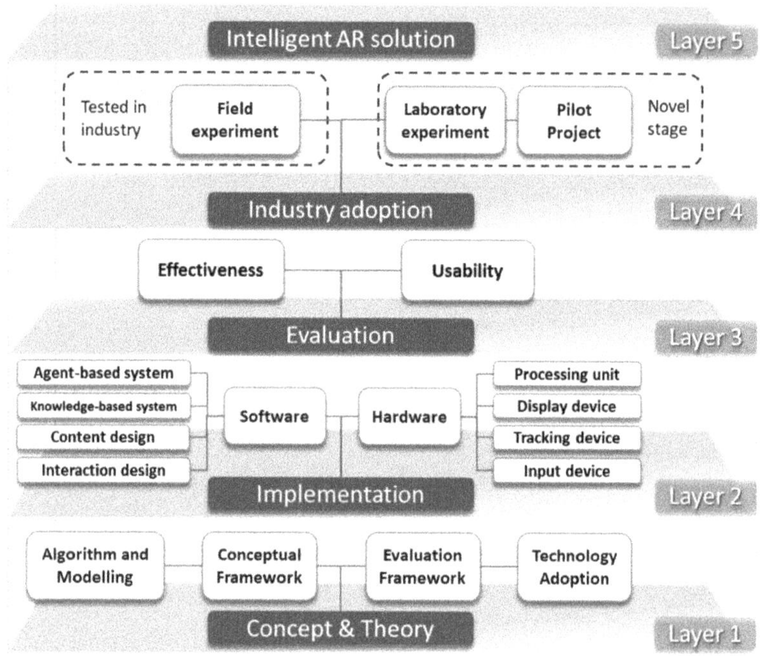

Figure 6. The architecture layer framework of an AR system, adopted from [226].

Layer 1: Concept & Theory

This layer includes Algorithm, Conceptual Framework, Evaluation Framework and Technology Adoption. Algorithm relates to technical aspects of AR/Registration/Tracking methodology. Conceptual Framework supports the development or proposal of AR solutions for proof-of-concept cases. Evaluation Framework assists in grading and selecting the right enabling elements for an AR concept or AR systems. Finally, Technology adoption is relevant to the papers that point out the current challenges, limitations and gaps which needs to be solved to facilitate a wide adoption of AR-based solutions.

Layer 2: Implementation

This layer consists of two sublayers, which are Software and Hardware layers.

Hardware sublayer includes the fundamental elements of an AR system, which are a Processing Unit, an Input device, a Tracking device and a Display device.

Due to the fact that the Processing Unit can be flexibly selected depending on the computing workloads of the desired tracking methods and the chosen display techniques, this paper does not consider extracting this information. Besides that, the input device is an optional element of the system because it depends on the system design and specific use case. The stimuli to trigger the AR modules could be automatically included (sensors data, camera, tracking algorithms) in the approach itself. Therefore, the paper aims to concentrate more on extracting the data regarding the Display device and Tracking methods to support for RQs.

In this paper, the display devices are classified into 2 groups: In-situ display and Mobile display. An In-situ display involves a Spatial display/Projector and a Monitor/Large screen. Mobile display involves HHD and HMD. Tracking methods are categorized into 3 groups: Computer vision-based tracking (CV-based), including marker-based; markerless (NFT), model-based tracking; Sensor-based tracking and Hybrid tracking.

Software sublayer consists of Interaction design and Content design, as well as Agent-based and Knowledge-based elements.

Content design is relevant to those papers that focus more on demonstrating how the AR information is constructed and used. In this, there is no interaction between the user and the virtual information; no external database is required.

Interaction design focuses more on developing and enhancing the interaction between the user and virtual objects/contents.

An Agent-based system (ABS) applies an agent or multi-agent system, which originates from Artificial Intelligent (AI), enabling the autonomous, adaptive/learning, intelligent characteristics of a system. Agent-based software is a higher evolution of object-oriented software [227–229].

A Knowledge-based system (KBS) is a type of AI targeting that captures human experts' knowledge to support the autonomy of decision-making. The typical architecture of a KBS consists of a knowledge base, which contains a collection of information in each field, and an inference engine, which deduces insights from the information captured/encoded in the knowledge base. Depending on the KBS problem-solving method/approach, it can be referred as a rule-based reasoning (RBS) system that encodes expert knowledge as rules, or a case-based reasoning (CBS) system that substitutes cases for rules [230–232].

Layer 3: Evaluation

This layer consists of Effectiveness and/or Usability categories that involve a user study. There is a close relationship between these two categories. The more usable a system is, the more effective it could become.

Effectiveness evaluation is designed to measure the system's capability of getting the desired result for a specific task or activity. For example: reducing assembly time, enhancing productivity, etc. [30].

Usability evaluation utilizes expert evaluations, needs analysis, behaviors measures, user interviews, surveys, etc. to measure the ease of adaption of AR-based systems. Thus, the system flaws can be identified at the early stages of development [194].

Layer 4 Industry adoption

This layer considers whether an AR prototype/application is tested in industry or not. A prototype/application can be classified into two classes depending on its industry adoption status, which are "Tested in the industry" and "Novel stage". If the field experiment is carried out for a prototype, it is classified into the "Tested in industry" class. The "novel stage" is relevant to applications, which focus on solving specific issues of AR technology such as tracking, calibration, etc., rather than finding holistic solutions for real industrial case studies, or are only tested in a laboratory environment. A pilot project solves real case studies and has the potential to be applied in the manufacturing environment, but there were no in-depth experiments carried out for it to verify/validate the results. Thus, pilot projects are also classified into the "novel stage" category.

Layer 5 Intelligent AR solution

To support this layer, an article should satisfy at least one of the following questions.

- Does the prototype or application integrate with another industry 4.0 technology such as AI, IoT, CPS, Digital Twin, etc.?
- Does the solution/concept potentially establish the fundamental base for the further integration of AI, IoT, etc. in AR environment to support manufacturing?
- Do the algorithms try to solve a limitation in AI-supporting AR systems?

3. Categories of current AR assisted quality sector

All the AR-based solutions for the quality sector can be classified into 3 groups:

- AR supporting quality as a virtual Lean tool for error prevention (virtual Poka Yoke)
- AR-based applications for metrology
- AR-based solutions for in-line quality control (process, product, machine, human)

Metrology: Applied metrology is a subset of metrology and a measuring science created to ensure the appropriateness of measurement devices, as well as their calibration and quality control, in manufacturing and other operations. Nowadays, measurement technologies are utilized not only for assuring the completed product, but also for proactive management of the entire production process. With AR's superimposition advantage and metrology's power, metrology integrated AR might be a promising research area for the long-term success of quality 4.0.

2.3.2. Analysis

Based on the proposed classification framework, a pilot datasheet was designed using Excel to extract relevant data for RQs (see Table 6). All the selected publications were systematically scanned and extracted by the main author. Two main reviewers were used, as well as a third to resolve any disagreements. Mendeley was used to keep track of references. The final decision to modify, keep or remove any defined categories was made by cross-checking each step of the reviewers, who also verified the extracted information.

Table 6. Example of data extraction from selected papers for the SLR.

Article	[30]	[188]
Type of paper	Application paper	Application paper
Application field	Quality	Quality
Objectives	Supporting operator Improving process Preventing human errors Reducing movements between workstation and display system	Reducing human mental workload to reduce errors in performing a task
Layer 1	n/a	n/a
Layer 2	Interaction design	Interaction design
Layer 3	Usability + Effectiveness	Usability + Effectiveness
Layer 4	Field experiment	Field experiment
Layer 5	n/a	n/a
Display devices	Monitor/Large screen	Android tablets
Tracking method	Markerless	Marker-based
Development platform	OpenCV, PCL, ROS	ARCore framework and SDK, Unity

3. Results and Discussion

In this section, the results of the SLR are reported and the analyzed papers are synthesized. The objective of the SLR is to answer the defined RQs. In order to guarantee the requirement of the PRISMA method in terms of transparency, there is a table providing all relevant articles of specific classification criteria at the end of each subsection.

These RQs are discussed, analyzed and answered in the following subsections. While the RQ1 and RQ2 utilized all selected papers to provide a holistic picture about current AR-based applications in manufacturing and their benefits to the quality sector, the RQ3 to RQ5 focus more comprehensively on finding the practical answers to support AR solutions development for the quality field.

3.1. Answering RQ1 and RQ2

RQ1: What is the current state of AR-based applications in manufacturing?

RQ2: How does AR-based quality control benefit manufacturing in the context of Industry 4.0?

The distribution of AR-based solutions in the Maintenance, Assembly, Quality, Other and General manufacturing contexts are 19%, 38%, 16%, 6% and 21%, respectively, as depicted in Figure 7. In more detail, the number of AR articles in each application field considered within this paper's objectives from the year 2010 to 2021 is illustrated in Figure 8, which provides a longitudinal viewpoint for analyzing patterns, themes and trends concerning the application field in the quantity of publication. The timeframe from 2010 to 2021 is extensive enough to determine the evolution of literature in each field.

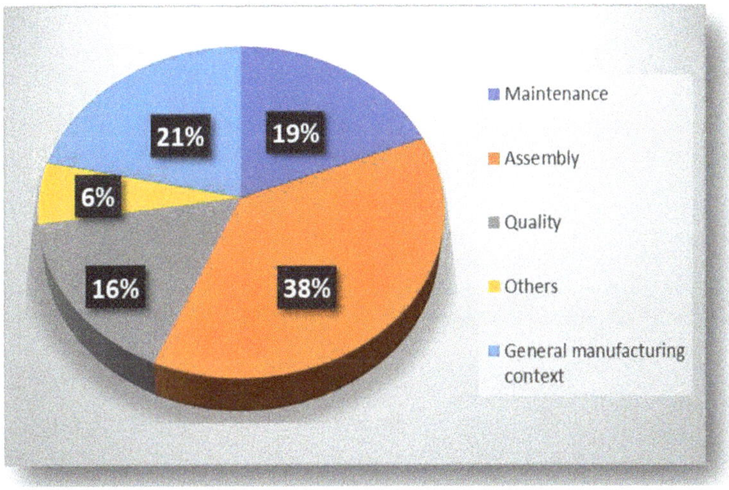

Figure 7. Distribution of application field in manufacturing.

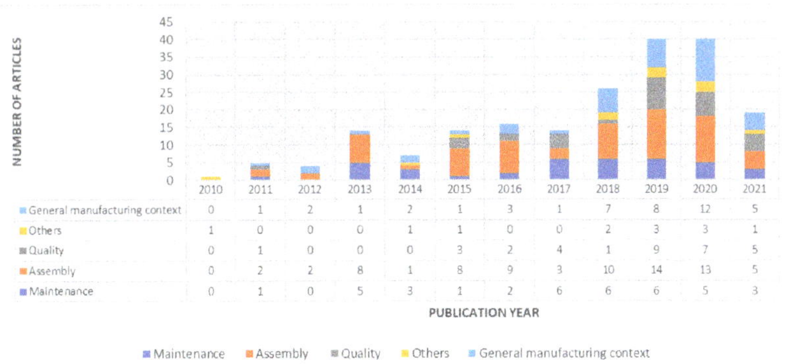

	2010	2011	2012	2013	2014	2015	2016	2017	2018	2019	2020	2021
General manufacturing context	0	1	2	1	2	1	3	1	7	8	12	5
Others	1	0	0	0	1	1	0	0	2	3	3	1
Quality	0	1	0	0	0	3	2	4	1	9	7	5
Assembly	0	2	2	8	1	8	9	3	10	14	13	5
Maintenance	0	1	0	5	3	1	2	6	6	6	5	3

Figure 8. Distribution of AR solutions in different fields over years.

It is not surprising that assembly is the leading adopter with 75 articles, or 38% of the total. This demonstrates a sustained interest in AR-assisted assembly, which peaked in 2019. Undoubtedly, assembly is the dominant sector in manufacturing to embrace AR technology. This is due to the nature of manual and semi-manual assembly activities that required the intensive involvement of operators, whose work is visual-oriented and who are in need of visual aid supporting. Next, when it comes to AR-based industrial applications in a specific field, maintenance is the second dominant sector, with 38 articles, or 19% of the total. Although the amount of AR-based maintenance applications fluctuates over time, they get the consistent consideration in 3 consecutive years from 2017 to 2019. Despite the investigation of AR solutions for the quality sector, this area is still far behind with 32 articles, or 16% of the total. Recently, this area has significantly emerged, reaching its peak in 2019, catching up with the AR articles for maintenance sector. Other sectors consisting of AR-assisted robot programming [211,214], machine tool setup [213], real-time manufacturing modeling and simulation [121] have slowly been considered. General manufacturing context solutions are relevant to those articles investigating generic AR-based solutions that can be customized and adopted for any particular field later to support the objectives of that field [123,221].

Virtual and real context fusion, as well as the intuitive display, is the main advantage of implementing AR-based solutions for maintenance and assembly instructions. Thus, media representation in the forms of text, symbols, indicators, 2D symbols, 3D models, etc. could be directly projected on the relevant objects [100,104,144,147,165]. A comparative study for AR-based assisted maintenance was conducted to compare maintenance efficiency in using different assisting tools such as video instructions, AR instructions and paper manuals. The results showed that AR technology could help in productivity enhancement, maintenance time reduction and quality assurance of maintenance tasks compared to other traditional tools [233]. Similarly, Fiorentino et al. [147] and Uva et al. [142] conducted a series of studies comparing AR-based instructions to 2D documents for assembly, finding that AR-based instructions dramatically increased assembly efficiency [147]. Nevertheless, AR-assisted instructions also enhanced the assembly order memorization of operators [142].

Considering the quality field, the AR-supported quality process has evolved from a basic indicating tool of projecting 2D information onto processed parts to support in situ quality inspection of welding spots using Spatial AR (SAR) [209] to a higher level that combines real-time 3D metrology data and the MR glasses HoloLens for in-line assessing of the quality of parts' polished surfaces [8]. In another scenario, SAR is also applied to improve the repeatability of manual spot-welding in the automotive industry to assure the precision and accuracy of the process [201]. Several types of cues visualized with different sizes and colors (red, green, white, yellow and blue) are defined and superimposed on the welding area to support operators in focusing the weld guns onto the correct welding spot. In a real case at the Igamo company in Spain, AR technology was adopted to work as an innovative Poka-Yoke tool. In the packaging sector, setting up the die cutters is crucial to ensuring the final quality of the cardboard. However, this process is error-prone, causing defects and low-quality products. Thus, correction templates, which are made of paper marked with tapes using different colors, are applied to balance the press differences of die cutters. These correction templates are made based on the traditional Poka-Yoke method for error prevention. The templates are then digitalized and directly projected onto the die cutter, resulting in warehouse cost reduction, which comes from storing correction templates, and data loss prevention, which is caused by damaged templates [198]. Additionally, 3D models or CAD are implemented into AR tools for design discrepancies [206] and design variations inspection [195]. In a quality assurance of sheet metal parts in the automotive industry, an interactive SAR system integrating point cloud data is implemented and validated [234].

In recent studies [30,210], an AR-based solution for improving the original quality control procedure used on the shop floor to check error deviation in several key points of an automotive part has been investigated and developed to automatically generate virtual guidance content for operators during measuring tasks. The main problem of the original procedure is that quality control consists of repetitive and precise tasks, which are frequently complex, requiring a high mental workload for the operator. Although quality control tests are facilitated with documents of static media such as video recording, photos or diagrams to support the operators, they still need to divide the attention between the task and the documents, which also lack in-time feedback. This leads to a slowing of the processes as well as movement waste due to the operator's need to move between a workstation and a computer to validate measuring results after a certain number of tests. In detail, the original quality control is to measure deviation errors of an automotive part at specific positions in accordance with the essential specification of clients. A wireless measurement device (a comparator) is manually positioned by operators at specific locations for evaluation. During the nine measures, the operators need to move back and forth between the working cell and a display device to verify the measurements. For the AR-based solution, a camera is mounted on a tripod, pointing downwards at the gauge where the test takes place. The correct position for the comparator in each step indicated by green boxes is augmented onto the RGB-D live video stream using the same screen with other methods. In this method, whenever the comparator is detected in the correct position, the measure is taken automatically. The validation of the correct comparator positioning is also used to trigger the transition to the next assembly stage. With this approach, an AR-based quality control system provides automatic in-process instructions for the next steps of measuring and accurate guidance to speed up the workers' efficiency. A test is carried out with seven operators: four inexperienced users and three experienced users. As a result, the experienced participants performed faster in both non-AR and AR-based methods, but the difference was smaller with the AR-based method. After implementation in an industrial setup by operators working on the shop floor in the metal industry, it was shown that AR-based systems help to reduce by 36% the execution time of a complex quality control procedure, allowing an increase of 57% in the number of tests performed in a certain period of time. It is also concluded that the AR system can prevent users from making costly errors during peak production times, though this has not been tested yet. Besides that, the risk of human errors is also reduced. In another scenario, an AR inspection tool is developed based on a user-centered design approach, following the standard ISO 9241-210:2019 to support workers during assembly error detection in an industry 4.0 context [188]. Once again, it is mentioned that the inspection activities naturally require high mental concentration and time when using traditional paper-based detection methods. Besides that, when the geometric complexity of the product grows, the probability that an operator makes mistakes also increases. In order to solve this, the research proposed and developed a novel AR tool to assist operators during inspection activities by overlaying 3D models onto real prototypes. When errors are detected, the users can add an annotation by using the virtual 3D models. The AR tool is then tested in a case study of assembly inspection of an auxiliary baseplate system (14 m long and 6 m wide) used for providing oil lubrication of turbine bearings and managing oil pressure and temperature. 16 engineers and factory workers of the Baker Hughes plant, skilled in the use of smartphone and tablet devices but novices to AR technology, were selected for the test. Five markers (rigid plastic QR code size of 150mm x 150mm x 1mm) were placed 1.5 m apart, along with the system for the tracking method. The users went through a demo, performed training steps and indicated a set of six tasks: framing a marker and visualizing the AR scene; detecting a design discrepancy and adding the relative 3D annotation; taking a picture of design discrepancies detected during the task; changing the size of the 3D annotations added during task 2; framing marker 4 and hiding the 3D model of the filter component; sending a picture and 3D annotations to the technical office. By adopting multiple markers to minimize tracking errors, freedom of movement for the user when inspecting large-size products is ensured. Analysis of Variance (ANOVA) is used to evaluate the number of errors and completion time, while System Usability Scale (SUS) and NASA Task Load Index (NASA-TLX) are applied to evaluate user acceptance. The ANOVA and SUS results showed that a low number of errors occurred during the interaction of the user with the proposed tool, which means that the AR tool is easy and intuitive to use. Thus, the AR tool could be efficiently adopted to support workers during the inspection activities for detecting design discrepancies. Nevertheless, the NASA-TLX test proved that the developed AR tool minimizes the cognitive load of divided attention induced to both the physical prototype and the related design data.

Another interesting AR-assisted quality study relevant to the automotive industry is investigated for car body fitting, correcting alignment errors [189]. Alignment car panels of exterior bodywork to satisfy the specific tolerances is a challenging task in automotive assembly. The workers need to be guided during the panel fitting operations to reduce errors and performance time. In addition, correcting the positioning of body work components is a key operation in automotive assembly, which is time-consuming and characterized by a strong dependence between the achievable results and the skill level of the worker performing the operation. To solve this, an AR prototype system is developed for supporting the operator during complex operations relating to the dedicated phase of panel fitting for car body assembly by providing gap and flushness information to correct the alignment errors. This system also provides the feature of converting the information on gap and flushness between car panels measured by sensors into AR instructions to support the workers for correcting alignment errors. The main elements of the solution consist of measuring sensors positioned on the wrist of a 6-axis articulated robot for gap and flushness data acquisitions and an AR system utilized for providing instructions and visual aids to the worker through a Head-Mounted Device (HMD). Gap and flush measurements of the component are first acquired for each control point (CP) and analyzed by comparing the extracted features with reference values to decide whether the component position needs correcting with further manual adjustments. Thus, the AR system starts guiding the operator by showing proper assembly instructions. During adjustment operations, gap and flushness are continuously measured and checked, creating, if necessary, further instructions until the assembly phase is completed. With this approach, the system has some outstanding features: immediate detection of alignment errors, in-process selection of the recovery procedure, accurate guidance for reduced time and procedural errors in task execution, real-time information without diverting the worker from the assembly process, fast feedback after adjusting and easy use, thanks to the integration of the real environment and the AR instructions in the user's field of view. A verified step and a test is carried out for the developed system. The results show a potential for further integration and industry adoption. With the advantage of immediate detection of alignment errors, the same assembly procedure has been able to be completed almost 4 times faster with the AR tool. The data collected from 10 tests are also less dispersed, indicating the robustness of the procedure conducted with the support of the AR system. The gap and flushness are reduced from 12.77 mm and 3.05 mm to 7.17 mm and 0.33mm, respectively. Besides that, the AR system also helped in increasing assembly effectiveness and efficiency as well as reducing errors. The correct positioning of bodywork components no longer depends on the experience and dexterity of the operator. For further improvement, system setup time needs to be minimized, implementing an Artificial Neutral Network (ANN) to support the measurement for gap and flushness error detection as well as reducing the collected data.

At this point, it is found in this SLR that although the AR-based application proved its strength in assisting quality activities, there are still challenges and limitations. In general, the current AR-assist quality applications can be classified into three groups depending on the features and objectives of their approach: AR as a virtual Lean tool, AR-assisted metrology and AR-based solutions for in-line quality control. The details are included in the following Table 7:

Table 7. Categories of current AR-assisted quality.

Ref.	Year	Quality Category			Application	Technology	Sample	Results
		AR as a Virtual Lean Tool	AR Assisted Metrol-ogy	AR-Based Solution for In-Line Quality Control				
[30]	2021		x	x	Shop floor procedures	Monitor, Markerless tracking	7 operators: 4 inexperienced users and 3 experienced users	36% reduction of execution time besides reducing the risk of human errors
[188]	2021			x	Inspection activities	HHD, Hybrid tracking (marker-based tracking, markerless tracking, sensor-based tracking)	16 engineers and factory workers of Baker Hughes plant	High satisfaction from the selected user
[189]	2021		x		Car body alignment	HHD, HMD, Markerless tracking	N/A	Immediate detection of alignment errors Reducing the gap and flushness from 12.77mm and 3.05mm to 7.17 mm and 0.33mm, respectively
[190]	2021			x	Machining process monitoring	HMD, Hybrid tracking (markerless tracking, sensor-based tracking)	N/A	Robust registration method User-friendly AR interface to support the integration of operators with digital twin data
[191]	2021			x	Aviation system inspection	HMD, Model-based tracking	N/A	Achieved accuracy of pin-detection up to 99%
[192]	2020			x	PCBA inspection	HHD, HMD, Hybrid tracking (marker-based tracking, markerless tracking)	31 users from university	High usability of the system while ensuring the quality of inspection of PCBAs
[193]	2020		x		Casting process	Monitor, Marker-based tracking	N/A	AR marker-tracking method can be used to measure the workers' pouring motion Revealing the relations between workers' motion and the casting defect

Table 7. *Cont.*

Ref.	Year	Quality Category			Application	Technology	Sample	Results
		AR as a Virtual Lean Tool	AR Assisted Metrol-ogy	AR-Based Solution for In-Line Quality Control				
[194]	2020	x			Car body quality control	HMD, Model-based tracking	Pre-test: 20 participants with non-MR experience Test: 7 experts from Alfatec Sistemas company	The MR-based interface was weighted 80.25/100 for the usability test => high potential of industrial adoption but still needs improvement
[210]	2020		x	x	Shop floor procedures	Monitor, Markerless tracking	N/A	36% reduction of execution time besides reducing the risk of human errors
[235]	2020		x		Aviation system inspection	HHD, Hybrid tracking (markerless tracking, edge-based tracking)	N/A	The Edge-based tracking algorithms are developed and tested, which shows potential in a new way of tracking for the AR system
[8]	2019		x	x	Polished surfaces quality assessment	HMD, Model-based tracking	N/A	The metrology data is successfully shown on the real parts
[195]	2019		x		Design variations detection	HHD, Marker-based tracking	20 participants (8 factory workers, 12 engineers)	Provides medium-to-high levels of usability
[196]	2019			x	Car body quality control	HMD, Marker-based tracking	41 users without experience	The MR solution get high results in usability test comparing to another method
[197]	2019			x	IGBT Wafer Condition Monitoring	Markerless tracking	N/A	A prototype was shown for proof-of-concept
[198]	2019	x			Packaging	Spatial display, Marker-based tracking	4 operators	Preventing data loss, reducing costs Less error-prone Potential functionality using data analytics for decision making
[204]	2019			x	Bottleneck identification	HHD, HMD, Markerless tracking	20 participants	Performance in bottlenecks detection with AR App outperformed the traditional lean observers

Table 7. *Cont.*

Ref.	Year	Quality Category			Application	Technology	Sample	Results
		AR as a Virtual Lean Tool	AR Assisted Metrology	AR-Based Solution for In-Line Quality Control				
[206]	2019		x		Design discrepancies detection	HHD, Hybrid tracking (marker-based tracking, markerless tracking)	34 volunteers	Providing similar results to the other instruments in terms of effectiveness
[199]	2017		x		Shop floor procedures	Monitor, HHD, Marker-based tracking	43 students	Reducing the measuring time High usability
[200]	2017			x	Machining process monitoring	Monitor, CNC feedback-based tracking	N/A	Validated the concept with an implementation
[201]	2017	x			Spot-welding	Spatial display, Markerless tracking	8 trained operators	Reduction of 52% of the standard deviation of manual spot-weld placement with AR visual cues
[71]	2016		x		Design discrepancies detection	Monitor, Hybrid tracking (marker-based tracking, markerless tracking)	N/A	Able to detect discrepancies in the range of approximately 0.01 m
[202]	2016			x	Sampling acceptance	HHD, Marker-based tracking	N/A	Showed a good performance when operators used the proposed tool for AS
[205]	2015	x			Spot-welding	HHD, Marker-based tracking	N/A	Improving the repeatability and precision of the manual spot-welding process
[207]	2015		x		Lean process	HHD, N/A	N/A	Savings 27.36% in lathe process, 26.54% in milling and 45.16% in dimensional validation
[209]	2011	x			Spot-welding	Spatial display, Marker-based tracking	N/A	Successfully developed a prototype

To continue answering RQ1, all the collected data reported are shown in Table 8:

Table 8. Number of articles classified by the framework.

Classification Criteria		Number or Articles	Global Percentage (%)
Application field		*200*	*100*
Maintenance		38	19.0
Assembly		75	37.5
Quality		32	16.0
Others		12	6.0
General manufacturing context		43	21.5
Layer 1	*Concept & Theory*	*127*	*63.5*
1.1	Algorithm & Modelling	47	23.5
1.2	Conceptual framework	48	24
1.3	Evaluation framework	9	4.5
1.4	Tech adoption	57	28.5
Layer 2	*Implementation*	*150*	*75*
2.1	*Software*	*150*	*75*
2.1.1	Content design	40	20
2.1.2	Interaction design	95	47.5
2.1.3	Agent-based AR	4	2
2.1.4	Knowledge-based AR	12	6
2.2	*Hardware*	*148*	*74*
In-situ	Spatial display/Projector	19	9.5
	Monitor/Large screen	34	17
Mobile	HHD	42	21
	HMD	50	25
Multimodal		16	8
Others		1	0.5
Layer 3	*Evaluation*	*62*	*31*
3.1	Usability	13	6.5
3.2	Effectiveness	24	12
3.3	Effectiveness + Usability	27	13.5
Layer 4	*Industry adoption*	*153*	*76.5*
Tested in industry	Field experiment	25	12.5
Novel stage	Laboratory experiment	69	34.5
	Pilot project	59	29.5
Layer 5	*Intelligent AR*	*20*	*10*

In terms of Layer 1, 63.5% (127 articles) of the selected publications provide important concepts and theory in their research (Layer 1). The largest percentage of AR articles is in Layer 2 Implementation (150 articles, 75%). This indicates that AR technology has matured to the point where it can be implemented using off-the-shelf commercial packages or self-development using less expensive software infrastructure. Significant assessment (Layer 3) of the Effectiveness and Usability studies using scientific and formal methods is found in 31% (62 articles) of the publications. 25 works, or 12.5% of the total, perform field experiments and have significant industry adoption context (Layer 4). An interesting point is that 20 articles, or 10% of the total, have contributed proof-of-concept or a conceptual framework supporting the current stage and further integration of intelligent elements for AR solutions (Layer 5). These five layers' percentage distribution have already illustrated a holistic view of the ongoing stage and the trend of AR-based application for the manufacturing context. They also depicted a general view of what AR-based solutions for manufacturing context have accomplished (see Figure 9). The AR technology has rapidly evolved and reached its mature point to be integrated into manufacturing, equipping operators with immersive interaction tools on the shop-floor level and providing essential manufacturing information for decision making in a short time.

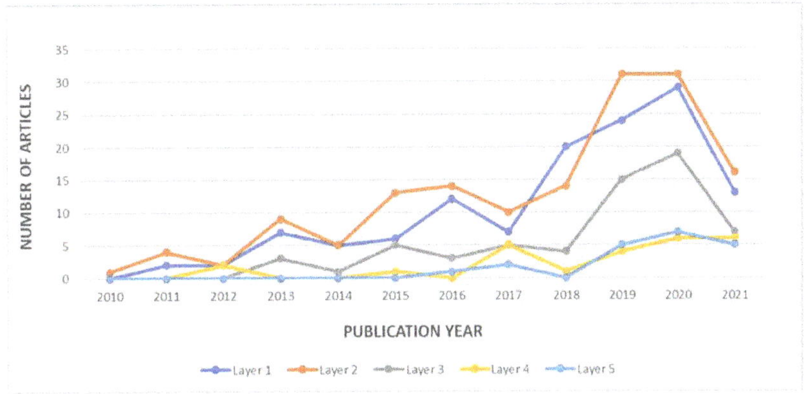

Figure 9. Number of articles in each layer from 2010 to 2021.

Regarding Layer 1, "AR Concept and Theory", the publications can be categorized into four subjects (see Table 9 and Figure 10). This layer is dedicated to the concept of how AR adoption benefits in solving problems in one specific field of manufacturing: the new theories and fundamentals to build and utilize AR for manufacturing contexts. The algorithm is a crucial element in developing an AR system. It consists of studies relevant to Artificial Intelligent methodology, establishing the base for AR to grow into intelligent systems [191]. A conceptual framework provides a general view of what the AR systems are and how they can be implemented. It may be relevant to the systems' capabilities, the system functions of the AR user interface, the system data flow or system management [190]. The evaluation framework forms the fundamental of heuristic guidelines either for the evaluating and selecting of AR elements for implementation or for analyzing and evaluating the usability of AR solutions in the context of manufacturing. For example, Quality function deployment mixed with an Analytic hierarchy process (QFD-AHP) methodology was applied for the selection of the appropriate AR visual technology in creating an implementation for the aviation industry in [123] or to support the decision-makers with quantitative information for a more efficient selection of single AR devices (or combinations) in manufacturing [87]. Technology transfer and adoption in the industry is relevant to articles that provide a holistic view about the current challenges, limitations and potential improvements that could support the adoption of AR technology in an industry context while satisfying the business requirements of companies. Literature review articles about AR in manufacturing usually contribute to the Technology Adoption category [77,190,235].

Table 9. Articles on Layer 1 Concept and Theory Layer.

Classification Criteria		References
Concept and Theory	Algorithm and Modelling	[11,40,51,54–76] [95,121,126,139,143,159] [161,163–165,167,168,170,190] [191,193,197,202,213–215,235]
	Conceptual Framework	[45,49,67,73,77,79,81,83,86,88–93] [95,96,102,109,111,112] [118,120,122,123,125,127,130–132] [135,137,140,141,143,145,149,150,154] [171,183,190,217,219,221–223,225]
	Evaluation Framework	[45,63,82,83,107,123,128,133]
	Technology Adoption	[2,17,18,33–53] [73,77,78,84,85,87,90–93,101] [102,106,114,116,117,122,124,130–132] [134–137,141,162,166,190] [219,221–223,235]

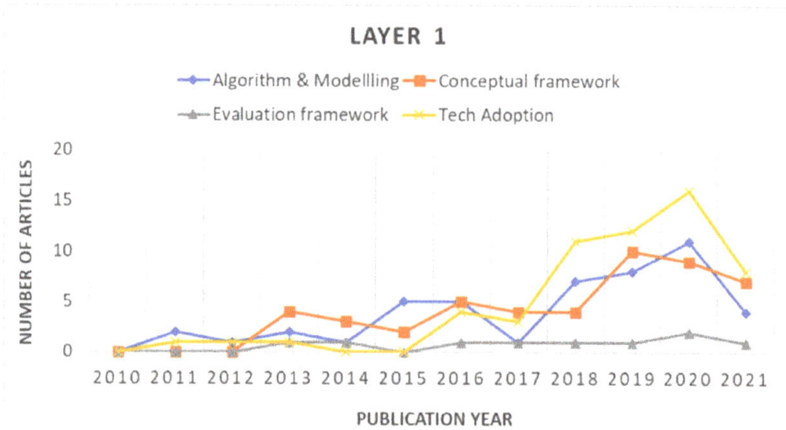

Figure 10. Number of articles in each category of layer 1 from 2010 to 2021.

After analyzing the data, it showed that there is a nearly balanced investigation into "Conceptual Framework" (48 articles, or 24% of the total) and "Algorithm and Modelling" (47 articles, or 23.5% of the total). Only 9 articles, or 4.5% of the total, contribute to the "Evaluation framework", while there is a high interest at the moment in "Technology transfer/adoption", with 57 articles, or 28.5% of the total. Considering this, all selected AR literature review articles can be considered to reuse for AR technology transfer and adoption in the long-term in a manufacturing context.

In layer 4, the relevant articles are analyzed and divided into two categories based on their industry adoption stages: "tested in the industry" and "novel stage." When an application has been tested in a real manufacturing context or field experiments are carried out, it is characterized as "tested in the industry." The "novel stage" is more relevant to applications or implementations that focus on solving specific issues of AR technology, such as tracking, calibration, etc., and is only tested in the laboratory environment. Besides that, pilot projects are relevant to those that implemented AR in an industrial environment but with no comprehensive tests carried out. Those proposing and developing a comprehensive AR-based solution that has high potential to integrate further in a real manufacturing context are also considered as pilot projects. The results reveal that 84% of applications are still in the novel stage, while the remaining 16% are tested in the industry and can be improved further for industry adoption (see Figure 11).

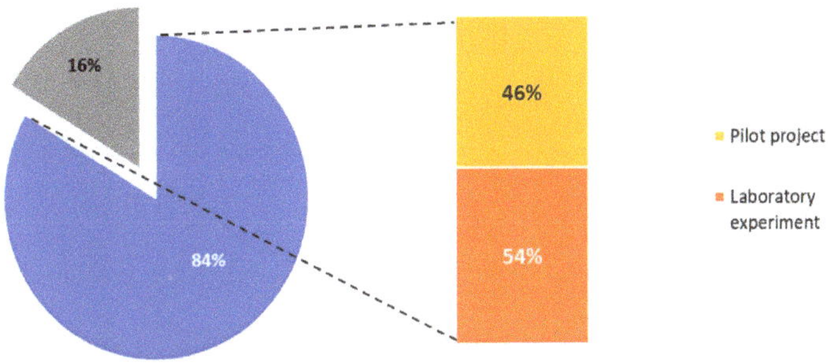

Figure 11. Industry adoption distribution of AR solutions 2010–2021.

Although the novel stage projects achieve potential results, user acceptance, human-centric issues, seamless user interaction and user interface are still challenges that need investigating for long-term industry adoption in manufacturing [2]. Because AR is a technology enhancing human perspectives by virtual and real context fusion, a universal human-centered model for AR-based solutions development can help in closing the gap between academia and industry implementations [236,237]. Following the international human-centered design standards ISO 9241-210, 2019 [188,195], a human-centered model can be developed by combining a simplified AR pipeline [17] and AR system elements [238] with a value-sensitive design approach for smart 4.0 operators [239]. All the AR-based implementations and their industry adoption status are shown in Table 10.

Table 10. Layer 4 AR solutions and their industry adoption status 2010–2021.

Classification Criteria			References
Concept and Theory	Tested in industry	Field experiment	[7,30,122,138,143,146,151,152,161,162] [165,167,181,188,193–196,198,199] [201,220,222,223,225]
	Novel stage	Laboratory experiment	[11,51–57,63,64] [66,68,71–76,79,81] [86,90,92,98,113,119,130,140–142] [145,147–150,154,156,159] [157,163,164,166,168–180] [182,184,186,187,189,192,201] [204,206,211,214–216,218,222]
		Pilot project	[8,62,65,67,69,77,84,87–89] [93–97,99,100,102,103,110] [115,116,121,123,125–129,131] [132,139,144,153,155,158,160,177,183,185] [190,191,197,200–203,205,207–210] [212,213,217,219,221,224,235]

Layer 5 provides an overview of the emerging trend in integrating AR with AI, industrial IoT, Digital twin and other comprehensive industry 4.0 technologies that result in the development of layer 5 Intelligent AR (IAR) solutions. This layer provides the holistic approach for implementing intelligent industry 4.0 elements with AR to enhance the robust and smart features of AR systems/solutions for long-term adoption in industry. This layer considers all studies that propose or involve interesting, significant concepts, algorithms and implementations that show high potential in the further development of the IAR system in the future (see Table 11).

Table 11. Layer 5 Intelligent AR relevant articles 2010–2021.

Classification Criteria	References	Intelligent Elements
Intelligent AR solution	[37,155,190]	Digital twins
	[72,74,116,164] [166,172,191,219]	Deep learning, CNN, AI
	[219,225]	Industrial IoT
	[49,102,127,132] [138,199,217]	Big Data, Cloud computing, Cloud architecture, Expert system for decision making, Ubiquitous system

Artificial Neural Networks (ANN) and Convolutional Neural Networks (CNN) are recently applied or proposed for further improvement of the registration methods by enhancing the CV-based tracking algorithms [74,191], while Industrial IoTs and Digital Twin are frequently considered in recent studies [37,190] to utilize Big data information and the advantages of AR in visualization, fusing digital data with a real working context.

3.2. Answering RQ3 to RQ5

This section considers all selected articles to establish a broad view about current AR development tools in manufacturing, thus making conclusions about how those tools could be utilized for developing AR solutions for the quality sector.

RQ3: What are available tools to develop AR-based applications for quality sector?

- Software design

Regarding Layer 2 Implementation, the number of articles dealing with the software side (150 articles, 75%) and hardware side is nearly balanced (148 articles, 74%). These numbers once more emphasize that AR technology has reached its mature point, where the improvement in either the AR software side or the AR hardware side would boost the technology adoption speed for AR solutions in manufacturing. There are a dominant number of Interaction design articles dealing with high functional user interfaces (95 articles, or 47.5% of the total), which is understandable due to high demands in interactive activities on the shop-floor level in manufacturing. Content design, with 40 articles, or 20% of the total, is the second dominant interest when considering software design. There are 4 articles, or 2% of the total, and 12 articles, or 6% of the total, dedicated to Agent-based AR and Knowledge-based AR systems, respectively. Although these two percentages of Agent-based AR and Knowledge-based AR are not significant, they are essential for the further integration of AI elements into AR systems supporting manufacturing in long term. The articles of each category relating to this sublayer Software design are listed in Table 12 and the number of articles of each category over the period 2010–2021 is depicted in Figure 12.

There has been a steady interest in Interaction design for AR-based in a manufacturing context over the years, which reached its peak in 2019. This is a positive trend for the long-term adoption of AR solutions in manufacturing be-cause manufacturing at the shop-floor level consists of lots of interactive activities between operators, especially interaction of operators with working spaces as well as and operators that need essential manufacturing information/data in the right manner of time [8,63,195].

Content design is the second dominant category in software design for AR solutions in manufacturing. In 2018–2020, AR content design especially focusing on visual elements and the conversion of manufacturing actions into standard symbols for AR content are key points [104,112,129,186].

Knowledge-based AR applications are designed to incorporate the domain knowledge of experts during the authoring phase to create a knowledge-based system (KBS) built on technical documents, manuals and other relevant documentation of the authoring domain (assembly, maintenance, quality control, etc.) [30,138,191].

Table 12. Articles on Layer 2 Implementation-sublayer Software from 2010–2021.

Classification Criteria		References
Software	Content design	[7,51,60,62,65,66,68,71,73,75] [84,92,98,100,104,108,109,112,122,129] [139,141,142,146,150,182,184–186,193] [197,201,202,205,207,210,212–214,223]
	Interaction design 95	[8,11,52,53,55,57,63,64,67,69] [58,59,61,72,77,79,81,86,87,90] [93–95,97,99,103,105,113,119,125–127] [130,131,140,144,145,147–149] [151–163,165–181] [183,187–189,192,194,195,198,200,204] [206,208,209,211,215–220] [221,222,224,225,235]
	Agent-based AR	[88,102,190,203]
	Knowledge-based AR	[30,56,116,121,132,138,143,164] [191,196,199,210]

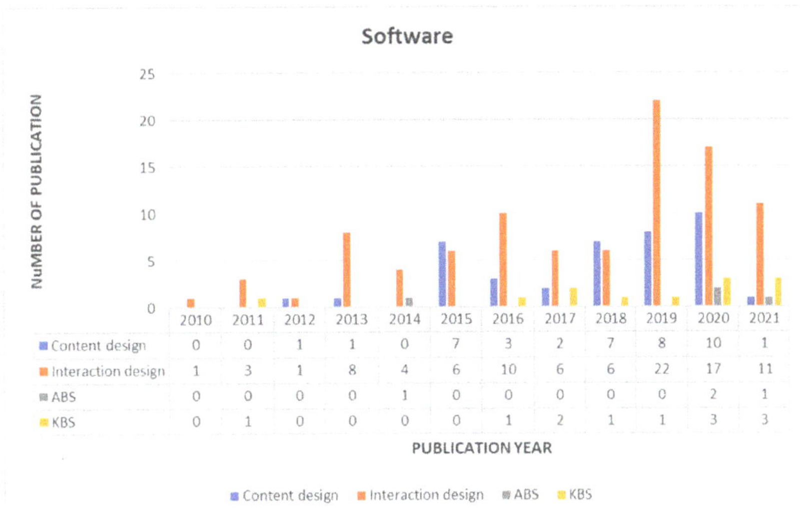

Figure 12. Number of articles about AR software design in manufacturing over years 2010–2021.

Agent-based AR utilizes the available entities of a system and their attributes to integrate into the AR solutions, supporting autonomy decision making [102,190].

- Display devices

This subsection presents an overview of the most popular display devices used in the development of AR solutions in manufacturing, which provide good references for the development of AR-assisted quality activities later. Table 8 and Figure 13 depict in detail the main display devices mentioned and applied in the selected and analyzed articles for this SLR. The implementation of one device rather than another depends on the purpose justification of the AR application. The evolution of display technology is also considered for the analysis of display devices.

Starting from the most dominant display device, HMD is mentioned in 50 articles, which is 25% of the selected articles, all of which comprehensively included HMD into their content. This is thanks to the advantages of HMD, which are portability, hands-free interaction and user experience enhancement through direct overlaying of computer-generated information onto users' views. The HMDs mentioned in the selected articles are usually commercial devices that are available on the market. Hololens is the one that is utilized the most among commercial optical see-through HMDs (OST-HMD) [8,52,53,220]. According to [240,241], another type of HMD is video see-through HMD (VST-HMD). In this category of HMD, Samsung Gear VR or Oculus are remarkable candidates. A customized VST-HMD based on the use of Z800 3D visor by Emagin combined with a USB camera—Microsoft LifeCam VX6000—was made to create an interactive AR implementation for virtual assembly in [167]. In terms of technology, the current HMD devices' ergonomics (weight, resolution, field of view (FOV)) have improved compared to the past, and have moved closer to the industrial requirements for long-term implementation. The technology of OST-HMD allows users to observe the real context through a transparent panel while at the same time seeing the computer-generated information projected onto it. The VST-HMD has cameras affixed to the front of the HMD which captures the real-world images, superimposes the digital information onto those images and then displays AR content through a small display area in front of the user's eyes [242]. Due to the amount of information that needs processing, VST-HMDs usually have higher latency (the time gap between the real world's occurring events and the ones perceived by users' eyes). The current challenges of both types of HMD technology are system latency, FoV, costs, ergonomics and view distortion [2].

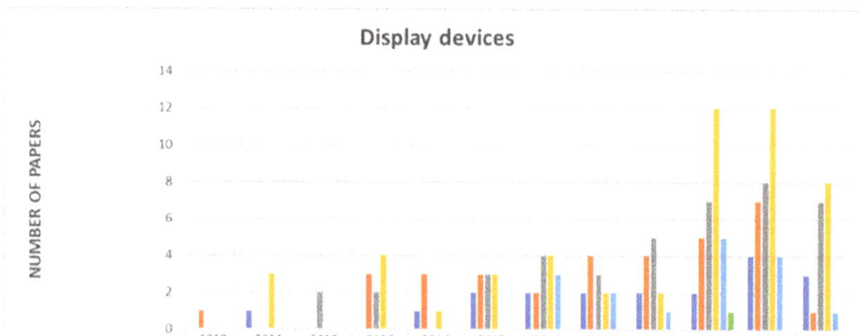

	2010	2011	2012	2013	2014	2015	2016	2017	2018	2019	2020	2021
Spatial display/Projector	0	1	0	0	1	2	2	2	2	2	4	3
Monitor/Large screen	1	0	0	3	3	3	2	4	4	5	7	1
HHD	0	0	2	2	0	3	4	3	5	7	8	7
HMD	0	3	0	4	1	3	4	2	2	12	12	8
Multimodel	0	0	0	0	0	0	3	2	1	5	4	1
Others	0	0	0	0	0	0	0	0	0	1	0	0

Figure 13. Number of articles about AR display devices in manufacturing over years 2010–2021.

The second dominant display device is HHD, with 42 articles, or 21% of selected articles for this SLR. HHDs utilized in the articles are mainly commercial devices such as tablets and mobiles. The greatest advantage of using these HHDs is that the users are more familiar with the technology because mobiles and tablets are also used in daily activities such as work, entertainment, etc. In addition, their portability, cross-platform development, cost and capabilities also make them promising alternatives to HMDs [77,195]. However, in tasks or activities requiring both free hands and intensive manual interaction on the shopfloor, such as assembly [160], quality control [30,194], etc., HHDs were not an appropriate selection in some cases.

The third and fourth trending display devices providing in-situ, hands-free AR displaying content are Monitors/large screens (34 articles, or 7% of the total) and Project for spatial display (19 articles, or 9.5% of the total). Monitors and large screens are also commonly selected for developing a human-centered smart system to support assembly tasks or quality assurance activities [30,147,164], provide assembly training assistance tools [166] or real-time data for cyber-physical machine tools monitoring [200], etc. Monitors and large screens are popular devices that are available in any manufacturing or shop-floor context. By utilizing them for developing AR solutions, the cost aspect can be satisfied. However, these systems usually require an external camera system or webcam for capturing real-world images to support the tracking and registration modules of AR applications. Nevertheless, portability is a limitation of using these display devices. Regarding Spatial AR (SAR) display with a projector, this is a favorite display method applied in spot-welding by utilizing the advantages of direct display digital information onto the work piece to enhance workers' concentration, thus reducing the process errors [201]. The system is considered as SAR when the projection is directly displayed onto the physical object. In another scenario of spot-welding inspection, SAR was applied to directly indicate the welding spots for the operators to check during the quality control process of the welding spots, which helped in reducing the inspection process time [209]. For these applications that support spot-welding relevant processes, one important note is the correct rendering for the readability of the indicator text. Besides that, text legibility is also essential, which was comprehensively investigated in a study to enhance the quality and effectiveness of applying SAR in industrial applications [75]. The projector is also popularly used in assisting the assembly process. A projection-based AR system was proposed to monitor and provide instruction for the operator during the assembly process in [161], or to provide the picking information together with assembly data for the operation as in [185]. Then, in a higher-level conceptual system, real-time instructions for assembly using SAR were comprehensively studied and demonstrated in [63]. Finally, during implementation at a real working environment in a factory, projected AR was utilized together with data digitalization to support the setting up of a die cutters process, which resulted in effective cost saving and processing error reduction [198].

It is crucial to note that the capabilities of the above display hardware are changing rapidly, but they both have their advantages and drawbacks, which are detailed in Table 13.

There is a small percentage of articles—16 articles, or 8% of the total—applying the multimodal displaying technique, and only 1 article, or 0.5% of the total, completely used another technique, which was haptic AR to support assembly process [11]. Multimodal display provides the immersive experience for more than one human sense, which can be mixed between visual displaying and audio to enrich the capabilities of AR applications in industry 4.0 [122] or effectively support and attract the awareness of the worker during the mechanical assembly process [164]. A combination of haptic feedback and visual information is commonly utilized in self-aware worker-centered intelligent manufacturing systems [166] or in applications that require lots of bare-hand interactions with physical objects such as assembly or maintenance [90,127,171].

By having a general view about how displaying devices are utilized in manufacturing through the above analysis, a conclusion regarding display device selection for AR assisting in the quality field can be made. By considering the specific working conditions/requirements such as human working environment, movements, flexibility, etc., as well as the advantages/disadvantages of each display technology, the appropriate display devices can be evaluated and selected for each specific AR-assisted quality application. The QFD-AHP methodology mentioned in Layer 1 could be utilized to systematically evaluate all these elements.

- Tracking methods

By having a general view about how displaying devices are utilized in manufacturing through the above analysis, a conclusion of display devices selection for AR assisting in the quality field can be made. By considering the specific working conditions/requirements such as human working environment, movements, flexibility, etc., as well as the advantages/disadvantages of each display technology, the appropriate display devices can be evaluated and selected for each specific AR-assisted quality application. The QFD-

AHP methodology mentioned in Layer 1 could be utilized to systematically evaluate all these elements.

Table 13. Articles on Layer 2 Implementation-sublayer Hardware from 2010–2021.

Display Devices	Representative Works	Advantages	Disadvantages
HMD	[8,52,53,55,56,64,72,79,88,97,102] [104,108,113,125,129–131,143,145] [146,149–151,155–158,160] [163,165,167,173–175] [178,189–192,196,203] [204,211,212,219–222,225]	Portability Hands-free	Ergonomics FoV Resolution
HHD	[7,51,52,65,67,73,77,84,86] [100,131,132,138,139,141,144] [153,157,159,162,169,177,181] [182,187–189,192,195,199] [202,204–209] [214,217,220,221,225,235]	Portability Mobile	Hand-occupied FoV Resolution
Projector/ SAR	[60,63,75,92,105,113,116] [131,142,161,163,170,176] [184,185,198,201,209,224]	Hands-free Directly project onto the object User tracking is not essential User movements do not affect the visualization	Low light-intensity Objects displaying in mid-air
Monitor/ Large screen	[30,57,59,61,62,66,68,71,81] [93,94,98,99,109,112,119,121,144,147] [148,150,154,159,172,180,186] [193,199,200,210,213,215,216,218]	Hands-free Low-cost Common devices in working environment	Portability
Multimodal	[58,90,103,122,126,127,140,152] [164,166,168,171,179,183,194,223]	Enriching immersive information Compensating the issues of each displaying techniques	User distraction could be a problem if information in different senses is provided at the same time
Others	[11]	Low-cost	Not intuitive Limitation in transmitted information

This subsection provides an insight into the current tracking methods utilized in developing AR solutions in manufacturing, thus giving useful references for working on AR assistance in the quality field in the future. Tables 14 and 15 list in detail the tracking methods used in articles on this SLR. In Table 14, besides Global Percentage, which is calculated for 200 selected articles, Relative Percentage demonstrates the composition of each tracking method that contributes to 133 tracking relevant articles. Then, Figures 14 and 15 illustrate the holistic view of the distribution and evolution of tracking methods over the period from 2010 to 2021.

Table 14. Number of articles classified by tracking methods.

Classification Criteria		Number of Articles	Relative Percentage (%)	Global Percentage (%)
Tracking methods		133	100	66.5
CV-based tracking	Marker-based tracking	63	47	31.5
	Markerless tracking	33	25	16.5
	Model-based tracking	13	10	6.5
Sensor-based tracking		3	2	1.5
Hybrid tracking		21	16	10.5

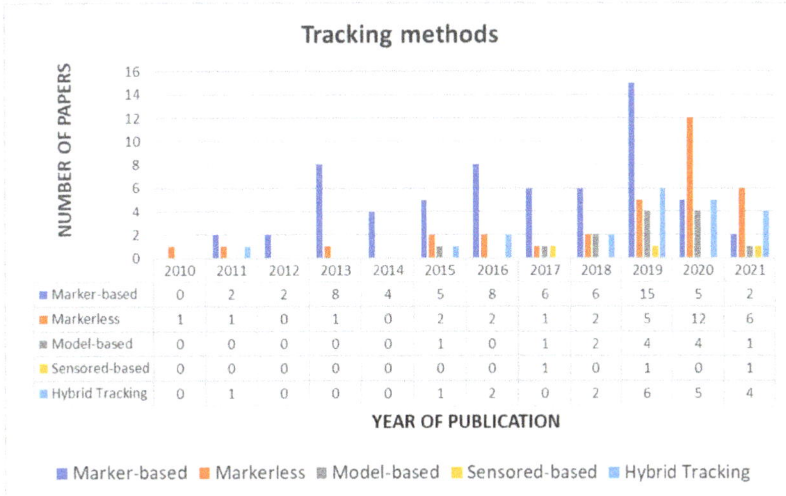

Figure 14. Number of articles about AR tracking methods in manufacturing over years 2010–2021.

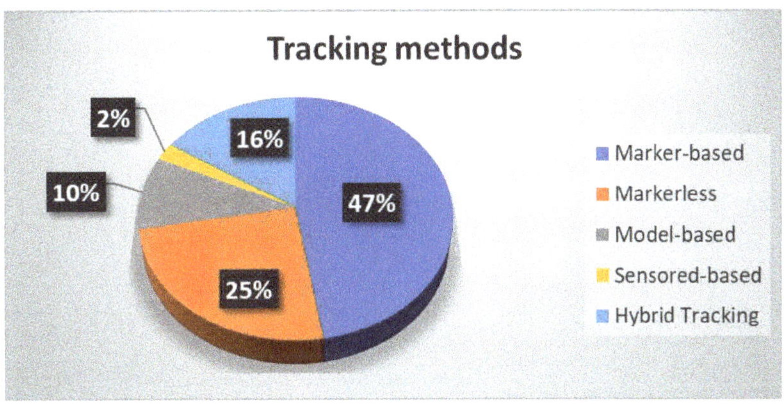

Figure 15. Distribution of tracking methods in terms of tracking methods from 2010–2021.

Tracking plays an important role in the real-time AR assisting in manufacturing application. It calculates the pose and position of the physical components as well as the relative pose of the camera to those components in real-time. The orientation (6DoF) and the position of an object form a pose. High accuracy in tracking this provides the users' location and their movements in reference to the surrounding environments, which is an important requirement for AR-based manufacturing applications [243], except in some applications using SAR. Tracking technology is one of the main challenges affecting AR application in support of intelligent manufacturing [2]. Robustness and low latency at an acceptable computation cost also need considering in terms of AR tracking. It is essential to distinguish between recognition and tracking. Recognition seeks to estimate the camera posture without relying on any previous information provided by the camera. When the AR system is initialized or at any time when there is a tracking failure, recognition is made. In contrast, tracking aims to analyze the camera pose based on the camera's previous frame [244].

Currently, there are three main categories of tracking techniques, known as computer vision-based (CV-based), sensor-based and hybrid-tracking, the latter of which utilizes both

CV-based and sensor-based tracking techniques at the same time [188]. CV-based tracking techniques are usually utilized for indoor environments and can be classified into three categories in terms of "a priori" methods, which are: marker-based tracking, markerless tracking (feature-based or natural feature tracking NFT) and model-based tracking. "A priori" is predefined knowledge about the object, which would be tracked. It could be a marker, a feature map or a model regarding marker-based, markerless and model-based tracking techniques, respectively. In order to initialize "a-priori" knowledge to support the CV-based tracking methods, "ad-hoc" methods can be applied to create the information that establishes "a priori" knowledge. In addition, "ad-hoc" could provide marker tracking methods or feature tracking methods based on Optical flow, Parallel Tracking and Mapping (PTAM) and Simultaneous Localization and Mapping (SLAM) [34]. A sensor-based method tracks the location of the sensor, which could be Radio Frequency Identification (RFID), a magnetic sensor, an ultrasonic sensor, a depth camera, an inertial sensor, infra-red (IR), GPS, etc.

Table 15. Articles on Tracking methods from 2010–2021.

Classification Criteria			References
Tracking method	CV-based tracking	Marker-based tracking	[7,55,64,77,79,81,84,88,93,94,97–100] [103,104,109,112,121,122,125,130,132] [142–145,147–149,152,155,157,158,160] [164,165,168,169,171,174,177,178] [180–184,186,193,198,199,202,205] [206,209,211,212,214,215,218,221,223]
		Markerless tracking	[30,51,54,57,61,62,65,66,69,72,74] [119,138,150,156,159,162] [172,175,187,189,192,197] [201,203,204,210,211] [216,217,219,220,225]
		Model based tracking	[8,67,68,73,86,139,141,153] [154,191,194,222,235]
	Sensor-based tracking		[11,161,200]
	Hybrid tracking		[52,53,56,58,59,71,90,102,127] [140,163,166,167,173,179,188,190] [195,196,208,213]

Marker-based tracking is the most utilized method for AR-based solutions in the manufacturing context, with 63 articles—31.5% of 200 selected articles or 47% of 133 articles that have a contribution to tracking content. This shows a steady trend in applying marker-based tracking methods for AR-based manufacturing solutions over the period from 2010 to 2021, which reached a peak in 2019 and then slowly reduced in 2020 and 2021. In these two years 2020 and 2021, markerless tracking has grown as the dominant method for AR-based manufacturing solutions. The markerless tracking method also takes the second place in dominant tracking methods for AR-based solutions in manufacturing, with 33 articles, equivalent to 16.5% of 200 selected articles or 25% of 133 tracking relevant articles. Next comes the hybrid tracking method that is applied in 21 articles, equaling 10.5% of 200 selected articles or 16% of 133 articles containing the tracking method. As in Figure 15, it also shows that there is a tendency towards using the hybrid tracking method in manufacturing in the last three years. Similar to hybrid tracking, model-based tracking has significantly increased in recent years, which is due to the built-in model tracking packet of popular AR software development platforms such as Unity and Vuforia. This method is implemented in 13 articles—6.5% of 200 selected articles or 10% of 133 articles contributing to tracking content. The sensor-based tracking method is the least favorite tracking technique in implementing AR solutions for manufacturing, with only 3 articles. This can be explained by the high cost and complex hardware that is usually required for indoor tracking, utilizing ultrasound, magnetic sensors, etc. In addition, the indoor environments of factory plants, production lines or laboratories usually consist of several types of equipment, machines, objects and so on, which could block sensor signals, thus reducing this tracking method's effectiveness [245]. This is why it only appeared in 1.5% of 200 selected articles or 2% of 133 articles regarding tracking methods.

In order to gain a more comprehensive understanding of how these tracking methods have been applied in AR-based solutions in manufacturing, some representative works relevant to each type of tracking method are depicted ij more detail in the following, while the rest are listed in Table 15.

The marker-based tracking method is fast, simple and robust, thus it is broadly utilized for various scenarios in the manufacturing context. For instance, it was implemented to facilitate a human-centered intelligent manufacturing workplace with an AR instructional system for assembly processes regarding highly customized products in some cases [64,93,142,164,165]. On the other hand, markers could be added to machines as a priori for tracking to assist maintenance processes [77,143,144] or to provide useful instructions to newcomer shop-floor operators [199]. In an innovative case, the AR marker-based method was applied for measuring and evaluating casting quality through hand-pouring motion tracking [193]. Furthermore, this tracking method is widely used for applications that provide guidance for operators during process preparation, which is time-consuming, error-prone and costly for small batch production, such as the setup of die cutters in the packaging industry [198], the setup of machine tools in the smart factory [213] or programming the trajectory of touching probes for aligning the raw material with the machine reference frame [214]. Finally, maker-based tracking is the prominent choice when it comes to AR-based solutions that support the welding process in the automotive industry. It was used in an AR-based quality control solution to enhance in-situ inspection of spot welding in the automotive industry [209], to guide the manual spot-welding process in order to ensure the in-line quality [205] or to assist intuitively in welding robot programming [218].

Despite its advantage, the effectiveness of the marker-based tracking method is not reliable in a working environment that may cause occlusion or damage the marker tag, such as workshops, production lines, etc. Therefore, several AR-assisted manufacturing applications also considered and employed other tracking methods. In a series of works [30,210], a markerless tracking methodology is developed to optimize the quality control procedure for automotive parts in terms of measuring deviation errors. An algorithm for extracting and comparing two consecutive 3D point clouds of the workstation, which are captured by the camera, is developed for this specific industrial case. In [189], the same approach, with control points instead of 3D point clouds, was used in enhancing the panel alignment process in car body assembly. In another scenario, some applications integrate the cloud database with the asset determining a priori in different locations of the plant to support fault diagnosis of the aseptic production line [220], enhance event-driven for AR-assisted maintenance and scheduling for a remote machine [138]. Other relevant applications using markerless tracking can be seen in Table 15. It is crucial to note that, the markerless or NFT methods often require lots of computing algorithms [65,72,74,197], as well as a powerful computing system [30,189,192,210] and a robust information system architecture or cloud platform [138,220].

The AR-hybrid tracking method in manufacturing usually employs data from additional sensors to increase the tracking speeds, reduce the latency and lighten the computing burden of markerless tracking algorithms. In addition, using sensory data also boosts the tracking performance of marker-based tracking or the model-based tracking method. In more detail, a hybrid tracking method using the marker and sensor data was applied in a series of studies to support evaluating the impact of using Hololens 1 for assembly work instructions [52,53]. A mock wing assembly task following paper instructions, traditional digital instructions including tablet model-based instructions (MBIs), desktop MBIs and AR-based instructions using different AR hardware such as a tablet and Hololens 1 was implemented to compare the data in terms of completion time, net promoter score (NPS) and error count. The results of these papers provide good references for manufacturing stakeholders regarding the benefits of diverse AR technologies that could be used for manual assembly tasks as well as to address some limitations of using a Hololens for larger-scale applications. In other scenarios, additional sensors are integrated to reduce the latency of markerless tracking systems to support design discrepancy inspections and annotations for flange systems in the Baker Hughes company [188,195]. Although the sensor-based technique is usually integrated with other tracking techniques to form hybrid tracking situations, as mentioned in the above part, it is rarely applied alone for AR-based solutions in manufacturing. This is because sensor-based tracking technically requires intensive tracking algorithms [11] and advanced deep learning methods such as CNN [161] to boost the tracking performance and reduce latency, thus achieving the real-time feature. Finally, a model-based tracking method is applied when a 3D CAD model of the tracked object's parts are available to extract, analyze and determine the pose and position for later recognition and tracking [8,73,139].

To conclude this subsection, similar to the selection of tracking method for a manufacturing context, when considering AR-based solutions for quality sector, different use cases and conditions of the working environment could be taken into account and evaluated to choose the most appropriate tracking method. This subsection provides solid references in terms of tracking methods for further utilization to develop AR-based solutions for the quality sector.

- Software development platform

Currently, there are lots of available Software Development Kits (SDKs) supporting AR application development in a fast and robust approach. The most well-known SDKs are Vuforia, ARToolKit, Vuforia, Metaio and Hololens. These SDKs provide detailed documentation and functionalities to develop AR applications while not requiring high coding skills and experience. One or more software development platforms can be utilized to develop an AR system. Table 16 lists in detail the different software development platforms mentioned across the selected articles in this SLR, which provides good references for the development of AR-based solutions in the quality sector later.

The development process of software is not always specified in articles. Thus, it is difficult to demonstrate this information about which programming language is mostly used by data collection. Not considering all articles, the most mentioned and utilized languages are C# and C++, shown in Table 16. In addition, libraries of functions such as OpenCV (Open-Source Computer Vision), Matlab and OpenGL (for 2D/3D graphics rendering) are also applied for the development of AR solutions. Mid/low programming languages and libraries of functions allow an AR application's development from scratch, providing high flexibility. However, to be highly skilled in programming is required for developing those systems. The use of SDK was mentioned across selected articles. SDKs' utilization has increased recently because new devices on the market usually provide them (Hololens, HHD). However, in terms of developing a high functionality AR application, SDKs are not sufficient. More extensive software built with a game engine or mid/low-level programming language must be integrated to achieve a full functionality AR application. Unity3D and Unreal are the two most popular game engines utilized for AR application development. A game engine is technically a user-friendly platform allowing the building of AR applications with minimum programming knowledge, but skilled AR developers are required to use them. In order to create the 3D AR contents, other supporting development platforms are mentioned, such as Blender, Autodesk Inventor, Rhinoceros, SolidWorks, Catia and 3ds-Max.

Unity 3D engine, developed by Unity Technologies in 2005, is a commercial cross-platform game creation system, which supports C# scripting together with 3D/2D graphics and animations. This development platform consists of a graphical user interface (GUI) and five fundamental windows, which are ProjectWindow, HierarchyWindow, Inspection-Window, SceneView and Toolbar. It is one of the most effective platforms applied for AR applications to assist human-machine interaction (HCI). Unity is compatible with the Vuforia SDK plug-in which enhances the 3D object tracking and detection in AR applications [8]. In addition, Unity provides pre-defined functionalities to develop interactive 3D content for real scenarios. Nevertheless, it offers the flexibility to export the designed application in different formats of executable files which are compatible with various building platforms such as Windows, Mac, iOS, Android and Universal Windows Platform (UWP).

Vuforia is the most common SDK for AR application development, launched by Qualcomm for a wide variety of devices. In addition, Vuforia provides some distinguishing features supporting different tracking protocols such as image target, object target, model target, feature tracking, cloud recognition and video playback. CV algorithms are used for Vuforia to assist object recognition in the image frame and 3D model presentation or data visualization in a real-time interface. In summary, Vuforia provides high-speed partly covered recognition, robust tracking of objects and consistent efficient tracking in low-light situations.

The OpenGL graphics library is usually integrated to render the scene, while the ARToolkit, which is a marker-based tracking library, is used to track and place virtual elements. To detect markers in a captured scene, the ARToolkit employs image processing functions. OpenSceneGraph is a free and open-source framework for computer graphics applications that could be used for rendering 3D graphics. OpenCV is an open-source framework for computer vision programming.

Table 16. Software development elements mentioned in selected articles 2010–2021.

Year	Ref.	Programming Language	Functionalities Library/SDK	3D Content Creation
2021	[30]	C++	ROS, OpenCV, PCL	Based on 3D point cloud extraction and algorithm
2021	[90]	C#	Vuforia, Unity3D	N/A
2021	[138]	C#	Unity3D, Vuforia Engine	N/A
2021	[173]	C#	Unity, Vuforia	N/A
2021	[188]	N/A	ARCore framework, ARCore SDK, Unity	N/A
2021	[189]	C#	Unity3D, Vuforia Engine	N/A
2021	[219]	C#	Unity3D, Mixed Reality Toolkit (MRTK 2), Microsoft Visual Studio	N/A
2021	[220]	Java: for Android mobile devices C#: for Holelens	Unity 3D game engine	Blender
2021	[221]	C#	ROS, ARToolkit, OpenCV, Vuforia	N/A
2021	[225]	N/A	Holo Toolkit	N/A
2020	[52]	C#	Unity3D, HoloToolkit, Vuforia, MicrosoftMixedRealityToolkit	N/A
2020	[53]	C#	Unity3D, Vuforia	N/A
2020	[62]	C++	OpenGL, OpenCV	N/A
2020	[59]	N/A	Unity, Kinect SDK, OpenCV, Visual Studio	N/A
2020	[72]	C#	Hololens, Unity3D engine	N/A
2020	[102]	N/A	Unity3D	N/A
2020	[119]	C#	Vuforia, Unity3D, Matlab	N/A
2020	[122]	N/A	Unity3D, Hololens, MicrosoftMixedRealityToolkit	N/A

Table 16. Cont.

Year	Ref.	Programming Language	Functionalities Library/SDK	3D Content Creation
2020	[126]	C++	N/A	N/A
2020	[139]	C#	YOLO, Unity3D, Vuforia Engine	Autodesk Inventor
2020	[150]	C#	Unity3D	N/A
2020	[151]	N/A	Holo Toolkit	N/A
2020	[162]	C#	Unity3D, ARCore SDK	Catia
2020	[163]	N/A	ROS, Hololens	N/A
2020	[164]	N/A	Unity3D	N/A
2020	[182]	C#	Unity3D, Vuforia	N/A
2020	[192]	C#	ARToolkit, Vuforia, Wikitude, EasyAR	N/A
2020	[194]	N/A	Hololens	Siemens NX, Blender
2020	[203]	N/A	Unity 3D, MicrosoftMixed RealityToolkit	N/A
2020	[210]	C++	ROS, OpenCV, PCL	Based on 3D point cloud extraction and algorithm
2020	[211]	N/A	Optitrack, Oculus Rift DK2	SolidWorks
2020	[222]	C#	Hololens, Unity3D, Holotoolkit	N/A
2020	[235]	C++, C#	OpenCV, Unity3D	N/A
2019	[8]	N/A	Unity3D, Hololens, MicrosoftMixedRealityToolkit, Vuforia	N/A
2019	[64]	C++	Visual Studio, OpenCV, Eigen 3.2.8 for OST, VST calibration	N/A
2019	[73]	C++	OpenCV	N/A
2019	[103]	C#	Unity3D, Vuforia	N/A
2019	[104]	N/A	Vuforia, Catia, PiXYZ software, Unity3D	N/A
2019	[112]	C++	Microsoft Visual Studio, OpenCV	Coin3Ds
2019	[130]	C#	Unity3D	N/A
2019	[140]	PHP	Matlab, Unity, Vuforia, Optitrack	N/A
2019	[141]	C#	Hololens, Unity3D, Vuforia	3Ds Max or Blender, Catia
2019	[154]	C++	ViSP library (Visual Servoing Platform), OpenCV	Ogre3D
2019	[155]	C#	Unity3D, MixedRealityToolkit	N/A
2019	[157]	C#	Unity3D, Vuforia	N/A
2019	[165]	Visual Basic	Unity3D, Vuforia	N/A
2019	[174]	N/A	ROS, Unity3D, Microsoft Mixed Reality Toolkit	N/A
2019	[175]	C#	Unity3D, HoloToolKit, Vuforia	N/A
2019	[183]	C#	Unity3D, Vuforia	N/A
2019	[187]	C#	Unity, Vuforia	3ds Max
2019	[195]	N/A	Google Project Tango SDK (instead of Vuforia), Unity3D	N/A
2019	[196]	N/A	Hololens, Unity, Visual Studio, Vuforia Engine, MixedRealityToolkit	Blender

Table 16. *Cont.*

Year	Ref.	Programming Language	Functionalities Library/SDK	3D Content Creation
2019	[197]	N/A	Mathlab	N/A
2019	[204]	N/A	Apple ARKit	N/A
2019	[206]	N/A	Google Tango Project SDK	N/A
2019	[212]	C#	ARKit™ API tool, Unity3D	3ds Max™ software, Solidworks
2019	[213]	C#	Unity3D, Vuforia	Solidworks, 3DsMax, Rhino3D
2019	[218]	N/A	MoCap Studio, ARToolkit	N/A
2018	[65]	C++	Unity3D	N/A
2018	[66]	C++	Unity3D, ARToolKit	3D Studio Max
2018	[67]	C/C++	ARToolKit, OpenGL, VRML toolkit	N/A
2018	[84]	C#	Unity3D, Vuforia	N/A
2018	[94]	N/A	AR Toolkit, Optical Flow Lib	Solid Edge
2018	[108]	C#	Unity3D, HoloToolKit	N/A
2018	[109]	N/A	Microsoft Visual Studio, OpenCV	Coin3D
2018	[121]	C++	OpenCV, OpenGL, Matlab	N/A
2018	[153]	C#	Unity3D, Vuforia	N/A
2018	[167]	N/A	Leap motion control, VST AR architecture, Visual Studio, ARToolkit libraries	N/A
2018	[176]	C#	Unity, Vuforia	Catia
2018	[214]	N/A	OptiTrack, ARUCO library	N/A
2017	[86]	C#	Unity3D, Vuforia	N/A
2017	[99]	C#	Unity3D, Vuforia	N/A
2017	[143]	C#	Unity3D, NET libraries, Vuforia	Catia
2017	[144]	C#	Unity3D, Vuforia	N/A
2017	[152]	C#	Unity3D, Vuforia	N/A
2017	[199]	C#	Unity 3D, Vuforia	N/A
2017	[200]	C#	Visual Studio, OpenCV	N/A
2017	[223]	C#	Unity3D, Vuforia	N/A
2016	[125]	N/A	COLLADA, ARToolkit	SolidWorks
2016	[131]	C#	Unity, Vuforia	N/A
2016	[132]	C#	Unity, Vuforia	N/A
2016	[145]	N/A	OpenNI (cross-platform framework dedicated to Natural Interaction), OpenCV, ARToolKit library, OpenGL, OpenSceneGraph (OSG)	N/A
2016	[168]	N/A	OpenGLj, ViSP tracking platform (Visual Servoing Platform)	N/A
2016	[158]	C#	Unity3D, Vuforia	N/A
2016	[171]	C++, Java	Visual Studio, OWL API	N/A
2016	[177]	C#	Unity3D, Vuforia, ROS, JSON library	N/A
2016	[179]	N/A	Unifeye SDK, Metaio platform	N/A

Table 16. *Cont.*

Year	Ref.	Programming Language	Functionalities Library/SDK	3D Content Creation
2016	[185]	C++	Metaio SDK	N/A
2016	[202]	C++	ARToolkit, OpenGL libraries	N/A
2015	[68]	N/A	ARToolkit	Solidworks
2015	[69]	C++	OpenCV	Coin3D
2015	[75]	C++	Qt	N/A
2015	[95]	C++	OpenCV	Coin3D
2015	[98]	N/A	OpenSceneGraph, OpenCV, ARToolkit	N/A
2015	[105]	Java	N/A	3Ds Max
2015	[146]	Java, C++	Microsoft foundation class (MFC), OpenSceneGraph (OSG), ARToolKit, OpenGL	N/A
2015	[159]	N/A	OpenGL, ARToolkit	N/A
2015	[178]	N/A	Unity, Metaio for Oculus Rift	N/A
2015	[186]	N/A	Unifeye, Metaio SDK	N/A
2014	[88]	C++	OWL API, ARToolkit	N/A
2014	[81]	N/A	OpenCV, OpenInventor	N/A
2014	[215]	C++	ARToolKit	N/A
2013	[55]	N/A	ARToolKitPlus, OpenGL	SolidWorks
2013	[79]	N/A	ARToolkit	N/A
2013	[100]	C#. NET	Unity3D, Vuforia, ROS, JSON library	N/A
2013	[148]	C++	OWL API, ARToolKit	Solidworks
2013	[149]	C++	ARToolkit, OpenGL	N/A
2013	[169]	C++	OpenSceneGraph, OpenNI library, Windows Forms	N/A
2013	[180]	N/A	Unity3D, Zigfu plugin	N/A
2011	[56]	N/A	OpenCV, OpenGL	N/A
2011	[156]	N/A	OpenGL	N/A

RQ4: How can AR-based applications for quality sector be evaluated?

Layer 3 consists of two main categories, which are effectiveness evaluation and usability evaluation. While effectiveness evaluation technically considers evaluation metrics such as completion time, the number of errors, productivity performance and other quantitative key performance indicators (KPIs), usability evaluation concentrates on the study of the user experience via interviews, questionnaires and field evaluations, as well as utilizing expert evaluations of the AR systems. As the complexity of tasks and AR systems has evolved over time, a hybrid evaluation of effectiveness and usability should be applied to holistically consider all impact factors that could help for the improvement of the investigated AR system. In terms of effectiveness evaluation, quantitative methods such as descriptive statistics, t-test and analysis of variance (ANOVA) were applied in several studies included in this SLR [92,188,195]. Regarding usability evaluation, two standard questionnaires, which are NASA Task Load Index (NASA-TLX) and the System Usability Scale (SUS), are usually utilized [194,220]. The NASA-TLX questionnaire is widely used to evaluate physical and digital experiences in working environments. The SUS questionnaire is short, concise and widely used too. However, other similar tests are also available for relevant use cases such as the Subjective Workload Assessment Technique (SWAT) [246], the Unified Theory on Acceptance and Use of Technology (UTAUT) [247], a Likert Scale questionnaire [142], a Computer System Usability Questionnaire (CSUQ) [141], etc. Nevertheless, the standards ANSI 2001 and ISO 9241 [195,206] are essential when considering metrics to evaluate the usefulness of a developed AR tool via analyzing human performance towards some target acquisition tasks.

There are 62 articles (31% of 200 articles in this SLR) that provided rigorous evaluation work, with 13 articles (6.5%) relating to usability evaluation, which is slightly more than half of the number of articles relevant to effectiveness evaluation (24 articles, 12%), and 27 articles (13.5%) conducting both effectiveness and usability evaluation. Considering the nature of AR technology, which mainly enhances user perspectives, usability evaluation should be more frequently employed to heuristically address all potential impacts of the AR system. Thus, more holistic and comprehensive evaluating results can be achieved for further development of the AR system and AR technology in the long term.

In summary, this subsection provides a useful evaluation methodology as well as different standards that could be reused to analyze the AR-based solutions for the quality sector later. Table 17 includes and classifies articles that are relevant to the evaluation of AR systems in manufacturing.

Table 17. Articles contributing for Layer 3 Evaluation from 2010–2021.

Classification Criteria		References
Evaluation	Usability	[53,57,92,99,139,141,145,151,157] [167,187,194,196]
	Effectiveness	[11,52,62,64,72,73,98,107,129,138] [142,147,159,162,164,166,169,183] [186,189,192,201,204,206]
	Usability + Effectiveness	[30,63,110,113,122,140,146,148] [150,163,165,170,171,173,175,176] [178,182,184,188,195,198,199,211] [220,222,223]

RQ5: How to develop an AR-based solution for long-term benefits of quality in manufacturing?

Quality control procedures or activities relating to the quality aspect in manufacturing frequently include intensive repetitive and precise tasks which are regularly complex and require human involvement. In some findings, it was mentioned that in terms of human error controlling, there is a current bottleneck in AR-assisted manufacturing systems. This relates to the artificial intelligence capabilities of AR systems [114,189]. Thus, the first factor that should be considered for the long-term benefits of using AR technology for the quality sector is solving the challenge of integrating intelligent agents into each AR-based solution at the beginning. In order to achieve that, a methodology adopted from [196], combined with the findings in this SLR based on AR architecture layer framework, is created to systematically consider all elements and factors that contribute to the development of AR-based assisted applications. Thus, the further enhancement and integration of AI elements to improve the intelligent capabilities of the AR systems in the long term are also benefited. This methodology is depicted in the following Figure 16. On the left side of the model is the systematic flow for the design and development of AR-based applications. Through each stage (Design, Development, Implementation, Evaluation, Improvement) and each step (Mockup design, Client validation, etc.) of the development flow, several valuable findings in this SLR (reference tables) and tools provided on the right-hand side of the model could be adapted as well as utilized to systematically create an AR-based application with a human-centered approach following several standards.

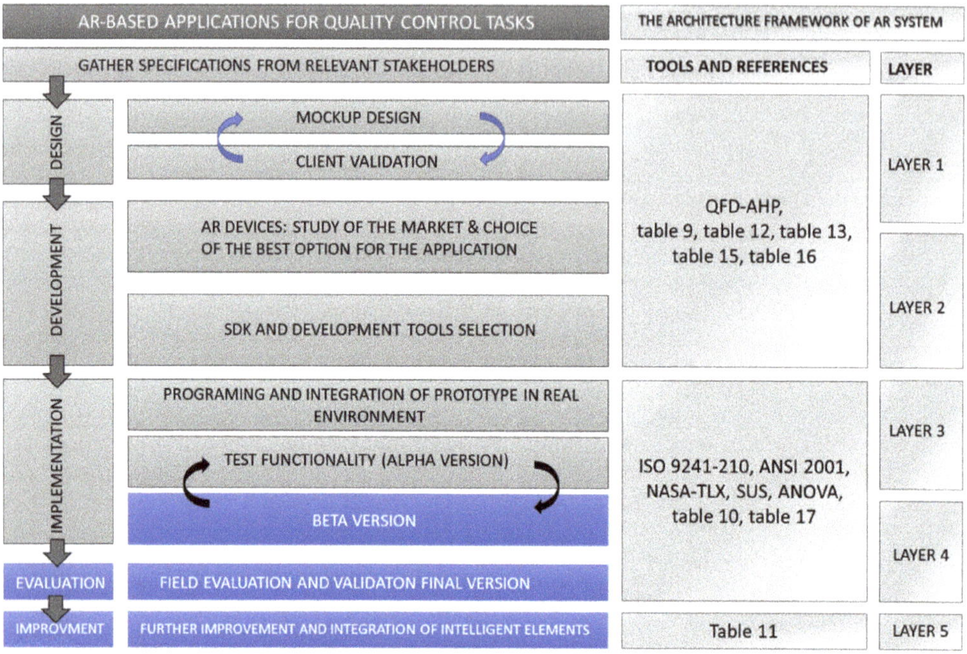

Figure 16. Methodology to design and develop long-term AR-based solution for quality sector adapted from [196].

The second crucial factor for the long-term implementation of AR-based solutions in the quality sector is to employ the ubiquitous computing system or boost the fusion of manufacturing information systems with AR systems [37,221], not only to save the information resource and boost the performance speed of AR applications but also to slowly transform to data-driven AR solutions and achieve the real-time feature for AR systems [102].

4. Conclusions and Outlook

The main objective of this study is to contribute to the current research by providing a holistic view of AR systems and AR-based applications in manufacturing, especially focusing on shop-floor processes that require the intensive involvement of operators' activities, such as assembly, maintenance and quality control from the year 2010 to 2021. Thus, this review fills an essential gap in the quality sector and provides a systematic model for the further implementation of AR technology to enhance and support Quality 4.0 in the future.

The main contributions of the study are based on the systematic literature review, which has answered the five research questions relating to AR-based applications in the quality sector within a manufacturing context. The conclusions are drawn as follows:

Firstly, quality control and quality-relevant activities themselves are important to ensure customer specification. However, quality is a special sector that belongs to non-value-added activities for the functionalities of the product. Thus, the fewer errors and the less process time for these quality control and management activities, the more resources (cost, time, human, etc.) can be utilized. In this regard, the advantages of AR technology could be applied to supporting the reduction of human errors and process time, as well as in-line error-prone process controlling. Although AR-based applications benefit the quality sector in several scenarios, there are also some drawbacks. Therefore, before the implementation of AR technology as a solution for quality sector, pre-implementation evaluation is essential to gain insights into which specific cases AR should be utilized in, how the technology could be integrated, whether employing AR is a long-term solution, etc.

Secondly, AR technology in manufacturing comes to a point that implementation of software and hardware has improved over time and has gradually reached the essential maturity for long-term industrial AR solutions. However, the current barrier to shop-floor level implementation is user acceptance, which has a long-term impact on the efficiency of the integration of AR solutions in manufacturing. Thus, when it comes to creating a long-term AR-based solution for the quality sector, all relevant elements of AR systems must be systematically evaluated at each step of the design and development process. The model in Figure 16 provides a comprehensive approach to address all impact factors, ensuring that a robust and practical AR-based solution is established step by step.

Finally, there are several available software development toolkits and hardware devices that have been improved over time to support the development of AR applications in manufacturing. In order to know which ones are appropriate for a specific AR-based solution to support quality enhancement activities, the working environment conditions need to be considered as well as the requirements in terms of cost, time and effectiveness. By considering all these factors using QFD-AHD, the selection of suitable hardware devices, SDKs and tracking methods could be made in a holistic way.

As the result of the SLR and the analysis of current AR technology development, the potential research areas on the subject of AR applied in the quality sector in industry 4.0 context could be one of the following topics:

- Transformation of a traditional quality Lean tool to a virtual quality tool by identifying and implementing AR technology when it is feasible.
- Integration of AR to assist manual and automatic metrology activities to prevent human errors, reduce setup time, ensure the accuracy of metrology data, etc.
- Standardization of quality-relevant knowledge representation and quality data formats to make AR systems and manufacturing information systems compatible.
- Development of a universal human-centered model for the adaption of AR-based solutions in the quality sector following the international human-centered design standards ISO 9241-210, 2019, to close the gap between industrial and academic implementations. In addition, this human-centered model could also boost AR technology adoption not only for the quality sector but also for manufacturing in general.
- Integration of AR solutions with other enabling technologies of Industry 4.0 such as industrial IoT, AI, Digital twin, etc., to improve the effectiveness, intelligence and real-time performance of the AR-assisted quality sector. Thus, the concept of ubiquitous AR applied in the quality sector in an industry 4.0 context could be achieved in the long term
- Enhancing AR registration and tracking methods by applying ANN and CNN to improve the accuracy of superimposing an AR model onto the real object in a shorter time, which is important for AR-assisted quality control for large-volume size parts in the automotive and aerospace industries.

Besides these key topics, the usability and effectiveness of innovative AR quality systems also depend on how quality knowledge is implemented and fused with AR technology. This matter relates to the familiarity of developers and users with the scientific principles and experiences underlying quality tasks that the new AR applications support. This insight is also crucial to the design and development of suitable system features for AR quality applications, thus ensuring the success of the implementation of AR for the quality sector in the long term.

Author Contributions: Conceptualization, P.T.H.; methodology, P.T.H.; formal analysis, P.T.H.; investigation, P.T.H.; data curation, P.T.H.; writing—original draft preparation, P.T.H.; writing—review and editing, J.A.A., J.A.Y.-F. and J.S.; supervision, J.A.A., J.A.Y.-F. and J.S. All authors have read and agreed to the published version of the manuscript.

Funding: This research work was undertaken in the context of DIGIMAN4.0 project ("DIGItal MANufac-turing Technologies for Zero-defect Industry 4.0 Production", http://www.digiman4-0.mek.dtu.dk/, accessed on 02 January 2022). DIGIMAN4.0 is a European Training Network supported by Horizon 2020, the EU Framework Program for Research and Innovation (Project ID: 814225).

Institutional Review Board Statement: Not applicable.

Informed Consent Statement: Not applicable.

Data Availability Statement: Not applicable.

Acknowledgments: The authors express their gratitude to the anonymous reviewers for their valuable comments that helped us to improve the paper significantly.

Conflicts of Interest: The authors declare no conflict of interest.

Abbreviations

AR	Augmented Reality
VR	Virtual Reality
MR	Mixed Reality
HMD	Head Mounted Display
HHD	Hand-Held Display
SD	Spatial Display
HUD	Heads-up Display
RV	Reality- Virtuality
AV	Augmented Virtuality
CV	Computer Vision
OST	Optical see-through
VST	Video see-through
CPS	Cyber Physical System
RFID	Radio Frequency Identification
RTLS	Real-time locating system
SDK	Software Development Kit
NASA-TLX	NASA Task Load index
SUS	System Usability Scale
MBI	Model-based instructions
ANOVA	Analysis of Variance
SURF	Speeded Up Robust Features
SIFT	Scale Invariant Feature Transform
BRIEF	Binary Robust Independent Elementary Features
NFT	Natural Feature Tracking
FoV	Field of View
ANN	Artificial Neural Networks
CNN	Convolutional Neural Networks
HCI	Human-Computer Interaction
IoT	Internet of Things

References

1. Popkova, E.G.; Ragulina, Y.V.; Bogoviz, A.V. *Industry 4.0: Industrial Revolution of the 21st Century*; Studies in Systems, Decision and Control; Springer International Publishing: Berlin/Heidelberg, Germany, 2019; Volume 169.
2. Egger, J.; Masood, T. Augmented Reality in Support of Intelligent Manufacturing—A Systematic Literature Review. *Comput. Ind. Eng.* **2020**, *140*, 106195. [CrossRef]
3. Lall, M.; Torvatn, H.; Seim, E.A. Towards Industry 4.0: Increased Need for Situational Awareness on the Shop Floor. In *IFIP International Conference on Advances in Production Management Systems (APMS)*; APMS: Hamburg, Germany, 2017; pp. 322–329.
4. Mubarok, K. Redefining Industry 4.0 and Its Enabling Technologies. *J. Phys. Conf. Ser.* **2020**, *1569*, 032025. [CrossRef]
5. Lidong, W.; Guanghui, W. Big Data in Cyber-Physical Systems, Digital Manufacturing and Industry 4.0. *Int. J. Eng. Manuf.* **2016**, *6*, 1–8. [CrossRef]
6. Hermann, M.; Pentek, T.; Otto, B. Design Principles for Industrie 4.0 Scenarios. In *2016 49th Hawaii International Conference on System Sciences (HICSS)*; IEEE: Piscataway, NJ, USA, 2016; pp. 3928–3937.

7. Serván, J.; Mas, F.; Menéndez, J.L.; Ríos, J. Using Augmented Reality in AIRBUS A400M Shop Floor Assembly Work Instructions. *AIP* **2012**, *1431*, 633–640.
8. Ferraguti, F.; Pini, F.; Gale, T.; Messmer, F.; Storchi, C.; Leali, F.; Fantuzzi, C. Augmented Reality Based Approach for On-Line Quality Assessment of Polished Surfaces. *Robot. Comput. Integr. Manuf.* **2019**, *59*, 158–167. [CrossRef]
9. Santos, A.C.C.; Delamaro, M.E.; Nunes, F.L.S. The Relationship between Requirements Engineering and Virtual Reality Systems: A Systematic Literature Review. In *2013 XV Symposium on Virtual and Augmented Reality*; IEEE: Piscataway, NJ, USA, 2013; pp. 53–62.
10. Sundareswaran, V.; Wang, K.; Chen, S.; Behringer, R.; McGee, J.; Tam, C.; Zahorik, P. 3D Audio Augmented Reality: Implementation and Experiments. In *The Second IEEE and ACM International Symposium on Mixed and Augmented Reality, 2003. Proceedings*; IEEE Computer Society: Washington, DC, USA, 2003; pp. 296–297.
11. Arbeláez, J.C.; Viganò, R.; Osorio-Gómez, G. Haptic Augmented Reality (HapticAR) for Assembly Guidance. *Int. J. Interact. Des. Manuf.* **2019**, *13*, 673–687. [CrossRef]
12. Gang, P.; Hui, J.; Stirenko, S.; Gordienko, Y.; Shemsedinov, T.; Alienin, O.; Kochura, Y.; Gordienko, N.; Rojbi, A.; López Benito, J.R.; et al. *User-Driven Intelligent Interface on the Basis of Multimodal Augmented Reality and Brain-Computer Interaction for People with Functional Disabilities*; Springer: Berlin/Heidelberg, Germany, 2019; Volume 886.
13. Sutherland, I.E. A Head-Mounted Three Dimensional Display. In *Proceedings of the December 9–11, Fall Joint Computer Conference, Part I on—AFIPS '68 (Fall, Part I)*; ACM Press: New York, NY, USA, 1968; p. 757.
14. Caudell, T.P.; Mizell, D.W. Augmented Reality: An Application of Heads-up Display Technology to Manual Manufacturing Processes. In Proceedings of the Twenty-Fifth Hawaii International Conference on System Sciences, Kauai, HI, USA, 7–10 January 1992; IEEE: Piscataway, NJ, USA, 1992; Volume 2, pp. 659–669.
15. Milgram, P.; Takemura, H.; Utsumi, A.; Kishino, F. Augmented Reality: A Class of Displays on the Reality Virtuality Continuum. In Proceedings of the SPIE Volume 2351, Telemanipulator and Telepresence Technologies (SPIE), Boston, MA, USA, 31 October–4 November 1994; SPIE: Boston, MA, USA, 1995; pp. 282–292.
16. Azuma, R.T. A Survey of Augmented Reality. *Presence Teleoperators Virtual Environ.* **1997**, *6*, 355–385. [CrossRef]
17. Fraga-Lamas, P.; Fernandez-Carames, T.M.; Blanco-Novoa, O.; Vilar-Montesinos, M.A. A Review on Industrial Augmented Reality Systems for the Industry 4.0 Shipyard. *IEEE Access* **2018**, *6*, 13358–13375. [CrossRef]
18. Nee, A.Y.C.; Ong, S.K. Virtual and Augmented Reality Applications in Manufacturing. *IFAC Proc. Vol.* **2013**, *46*, 15–26. [CrossRef]
19. Chryssolouris, G. *Manufacturing Systems: Theory and Practice*; Mechanical Engineering Series; Springer: Berlin/Heidelberg, Germany, 2006.
20. Tomic, B.; Spasojevic Brkic, V.K. Customer Satisfaction and ISO 9001 Improvement Requirements in the Supply Chain. *TQM J.* **2019**, *31*, 222–238. [CrossRef]
21. Sanchez-Marquez, R.; Albarracín Guillem, J.M.; Vicens-Salort, E.; Jabaloyes Vivas, J. Diagnosis of Quality Management Systems Using Data Analytics—A Case Study in the Manufacturing Sector. *Comput. Ind.* **2020**, *115*, 103183. [CrossRef]
22. Yamada, T.T.; Poltronieri, C.F.; do Gambi, L.N.; Gerolamo, M.C. Why Does the Implementation of Quality Management Practices Fail? A Qualitative Study of Barriers in Brazilian Companies. *Procedia Soc. Behav. Sci.* **2013**, *81*, 366–370. [CrossRef]
23. Phan, A.C.; Abdallah, A.B.; Matsui, Y. Quality Management Practices and Competitive Performance: Empirical Evidence from Japanese Manufacturing Companies. *Int. J. Prod. Econ.* **2011**, *133*, 518–529. [CrossRef]
24. Thoben, K.-D.; Wiesner, S.; Wuest, T. "Industrie 4.0" and Smart Manufacturing—A Review of Research Issues and Application Examples. *Int. J. Autom. Technol.* **2017**, *11*, 4–16. [CrossRef]
25. Milunovic Koprivica, S.; Maric, A.; Ristic, O.; Arsovski, S. Social Oriented Quality: From Quality 4.0 towards Quality 5.0. *Proc. Eng. Sci.* **2019**, *1*, 405–410. [CrossRef]
26. Radziwill, N.M. Quality 4.0: Let's Get Digital-The Many Ways the Fourth Industrial Revolution Is Reshaping the Way We Think about Quality. *arXiv Prepr.* **2018**, arXiv:1810.07829.
27. Kumar, A.; Shankar, R.; Thakur, L.S. A Big Data Driven Sustainable Manufacturing Framework for Condition-Based Maintenance Prediction. *J. Comput. Sci.* **2018**, *27*, 428–439. [CrossRef]
28. Kumar, S.; Tiwari, P.; Zymbler, M. Internet of Things Is a Revolutionary Approach for Future Technology Enhancement: A Review. *J. Big Data* **2019**, *6*, 111. [CrossRef]
29. García-Alcaraz, J.L.; Maldonado-Macías, A.A.; Cortes-Robles, G. (Eds.) *Lean Manufacturing in the Developing World*; Springer International Publishing: Berlin/Heidelberg, Germany, 2014.
30. Alves, J.B.; Marques, B.; Dias, P.; Santos, B.S. Using Augmented Reality for Industrial Quality Assurance: A Shop Floor User Study. *Int. J. Adv. Manuf. Technol.* **2021**, *115*, 105–116. [CrossRef]
31. Liberati, A. The PRISMA Statement for Reporting Systematic Reviews and Meta-Analyses of Studies That Evaluate Health Care Interventions: Explanation and Elaboration. *Ann. Intern. Med.* **2009**, *151*, 65–94. [CrossRef]
32. Cohen, Y.; Faccio, M.; Pilati, F.; Yao, X. Design and Management of Digital Manufacturing and Assembly Systems in the Industry 4.0 Era. *Int. J. Adv. Manuf. Technol.* **2019**, *105*, 3565–3577. [CrossRef]
33. Fernández del Amo, I.; Erkoyuncu, J.A.; Roy, R.; Palmarini, R.; Onoufriou, D. A Systematic Review of Augmented Reality Content-Related Techniques for Knowledge Transfer in Maintenance Applications. *Comput. Ind.* **2018**, *103*, 47–71. [CrossRef]
34. Palmarini, R.; Erkoyuncu, J.A.; Roy, R.; Torabmostaedi, H. A Systematic Review of Augmented Reality Applications in Maintenance. *Robot. Comput. Integr. Manuf.* **2018**, *49*, 215–228. [CrossRef]

35. Wang, X.; Ong, S.K.; Nee, A.Y.C. A Comprehensive Survey of Augmented Reality Assembly Research. *Adv. Manuf.* **2016**, *4*, 1–22. [CrossRef]
36. Danielsson, O.; Holm, M.; Syberfeldt, A. Augmented Reality Smart Glasses for Operators in Production: Survey of Relevant Categories for Supporting Operators. *Procedia CIRP* **2020**, *93*, 1298–1303. [CrossRef]
37. Qiu, C.; Zhou, S.; Liu, Z.; Gao, Q.; Tan, J. Digital Assembly Technology Based on Augmented Reality and Digital Twins: A Review. *Virtual Real. Intell. Hardw.* **2019**, *1*, 597–610. [CrossRef]
38. Dey, A.; Billinghurst, M.; Lindeman, R.W.; Swan, J.E. A Systematic Review of 10 Years of Augmented Reality Usability Studies: 2005 to 2014. *Front. Robot. AI* **2018**, *5*, 37. [CrossRef]
39. Damiani, L.; Demartini, M.; Guizzi, G.; Revetria, R.; Tonelli, F. Augmented and Virtual Reality Applications in Industrial Systems: A Qualitative Review towards the Industry 4.0 Era. *IFAC-PapersOnLine* **2018**, *51*, 624–630. [CrossRef]
40. Carmigniani, J.; Furht, B.; Anisetti, M.; Ceravolo, P.; Damiani, E.; Ivkovic, M. Augmented Reality Technologies, Systems and Applications. *Multimed. Tools Appl.* **2011**, *51*, 341–377. [CrossRef]
41. Kim, K.; Billinghurst, M.; Bruder, G.; Duh, H.B.-L.; Welch, G.F. Revisiting Trends in Augmented Reality Research: A Review of the 2nd Decade of ISMAR (2008–2017). *IEEE Trans. Vis. Comput. Graph.* **2018**, *24*, 2947–2962. [CrossRef]
42. Gallala, A.; Hichri, B.; Plapper, P. Survey: The Evolution of the Usage of Augmented Reality in Industry 4.0. *IOP Conf. Ser. Mater. Sci. Eng.* **2019**, *521*, 012017. [CrossRef]
43. Boboc, R.G.; Gîrbacia, F.; Butilă, E.V. The Application of Augmented Reality in the Automotive Industry: A Systematic Literature Review. *Appl. Sci.* **2020**, *10*, 4259. [CrossRef]
44. de Souza Cardoso, L.F.; Mariano, F.C.M.Q.; Zorzal, E.R. A Survey of Industrial Augmented Reality. *Comput. Ind. Eng.* **2020**, *139*, 106159. [CrossRef]
45. Masood, T.; Egger, J. Augmented Reality in Support of Industry 4.0—Implementation Challenges and Success Factors. *Robot. Comput. Integr. Manuf.* **2019**, *58*, 181–195. [CrossRef]
46. Bottani, E.; Vignali, G. Augmented Reality Technology in the Manufacturing Industry: A Review of the Last Decade. *IISE Trans.* **2019**, *51*, 284–310. [CrossRef]
47. Nee, A.Y.C.; Ong, S.K.; Chryssolouris, G.; Mourtzis, D. Augmented Reality Applications in Design and Manufacturing. *CIRP Ann.* **2012**, *61*, 657–679. [CrossRef]
48. Santi, G.M.; Ceruti, A.; Liverani, A.; Osti, F. Augmented Reality in Industry 4.0 and Future Innovation Programs. *Technologies* **2021**, *9*, 33. [CrossRef]
49. Baroroh, D.K.; Chu, C.-H.; Wang, L. Systematic Literature Review on Augmented Reality in Smart Manufacturing: Collaboration between Human and Computational Intelligence. *J. Manuf. Syst.* **2021**, *61*, 696–711. [CrossRef]
50. Evangelista, A.; Ardito, L.; Boccaccio, A.; Fiorentino, M.; Messeni Petruzzelli, A.; Uva, A.E. Unveiling the Technological Trends of Augmented Reality: A Patent Analysis. *Comput. Ind.* **2020**, *118*, 103221. [CrossRef]
51. Lee, W.-H.; Lee, K.-H.; Lee, J.-M.; Nam, B.-W. Registration Method for Maintenance-Work Support Based on Augmented-Reality-Model Generation from Drawing Data. *J. Comput. Des. Eng.* **2020**, *7*, 775–787. [CrossRef]
52. Hoover, M.; Miller, J.; Gilbert, S.; Winer, E. Measuring the Performance Impact of Using the Microsoft HoloLens 1 to Provide Guided Assembly Work Instructions. *J. Comput. Inf. Sci. Eng.* **2020**, *20*, 061001. [CrossRef]
53. Miller, J.; Hoover, M.; Winer, E. Mitigation of the Microsoft HoloLens' Hardware Limitations for a Controlled Product Assembly Process. *Int. J. Adv. Manuf. Technol.* **2020**, *109*, 1741–1754. [CrossRef]
54. Radkowski, R. Object Tracking With a Range Camera for Augmented Reality Assembly Assistance. *J. Comput. Inf. Sci. Eng.* **2016**, *16*, 011004. [CrossRef]
55. Wang, Z.B.; Ng, L.X.; Ong, S.K.; Nee, A.Y.C. Assembly Planning and Evaluation in an Augmented Reality Environment. *Int. J. Prod. Res.* **2013**, *51*, 7388–7404. [CrossRef]
56. Zhang, J.; Ong, S.K.; Nee, A.Y.C. RFID-Assisted Assembly Guidance System in an Augmented Reality Environment. *Int. J. Prod. Res.* **2011**, *49*, 3919–3938. [CrossRef]
57. Wu, L.-C.; Lin, I.-C.; Tsai, M.-H. Augmented Reality Instruction for Object Assembly Based on Markerless Tracking. In Proceedings of the 20th ACM SIGGRAPH Symposium on Interactive 3D Graphics and Games—I3D '16, Redmond, WA, USA, 26–28 February 2016; ACM Press: New York, NY, USA, 2016; pp. 95–102.
58. Radkowski, R.; Kanunganti, S. Augmented Reality System Calibration for Assembly Support With the Microsoft HoloLens. In *Volume 3: Manufacturing Equipment and Systems*; ASME: Houston, TX, USA, 2018.
59. Liang, J.; He, H.; Wu, Y. Bare-Hand Depth Perception Used in Augmented Reality Assembly Supporting. *IEEE Access* **2020**, *8*, 1534–1541. [CrossRef]
60. Costa, C.M.; Veiga, G.; Sousa, A.; Rocha, L.; Sousa, A.A.; Rodrigues, R.; Thomas, U. Modeling of Video Projectors in OpenGL for Implementing a Spatial Augmented Reality Teaching System for Assembly Operations. In Proceedings of the 2019 IEEE International Conference on Autonomous Robot Systems and Competitions (ICARSC), Porto, Portugal, 24–26 April 2019.
61. Radkowski, R.; Oliver, J. *Natural Feature Tracking Augmented Reality for On-Site Assembly Assistance Systems*; Springer: Berlin/Heidelberg, Germany, 2013; pp. 281–290.
62. Wang, K.; Liu, D.; Liu, Z.; Duan, G.; Hu, L.; Tan, J. A Fast Object Registration Method for Augmented Reality Assembly with Simultaneous Determination of Multiple 2D-3D Correspondences. *Robot. Comput. Integr. Manuf.* **2020**, *63*, 101890. [CrossRef]

63. Wang, Z.; Bai, X.; Zhang, S.; He, W.; Zhang, X.; Yan, Y.; Han, D. Information-Level Real-Time AR Instruction: A Novel Dynamic Assembly Guidance Information Representation Assisting Human Cognition. *Int. J. Adv. Manuf. Technol.* **2020**, *107*, 1463–1481. [CrossRef]
64. Yin, X.; Fan, X.; Zhu, W.; Liu, R. Synchronous AR Assembly Assistance and Monitoring System Based on Ego-Centric Vision. *Assem. Autom.* **2019**, *39*, 1–16. [CrossRef]
65. Wang, Y.; Zhang, S.; Wan, B.; He, W.; Bai, X. Point Cloud and Visual Feature-Based Tracking Method for an Augmented Reality-Aided Mechanical Assembly System. *Int. J. Adv. Manuf. Technol.* **2018**, *99*, 2341–2352. [CrossRef]
66. Wang, Y.; Zhang, S.; Yang, S.; He, W.; Bai, X. Mechanical Assembly Assistance Using Marker-Less Augmented Reality System. *Assem. Autom.* **2018**, *38*, 77–87. [CrossRef]
67. Xiao, H.; Duan, Y.; Zhang, Z. Mobile 3D Assembly Process Information Construction and Transfer to the Assembly Station of Complex Products. *Int. J. Comput. Integr. Manuf.* **2018**, *31*, 11–26. [CrossRef]
68. Chen, C.J.; Hong, J.; Wang, S.F. Automated Positioning of 3D Virtual Scene in AR-Based Assembly and Disassembly Guiding System. *Int. J. Adv. Manuf. Technol.* **2015**, *76*, 753–764. [CrossRef]
69. Liu, Y.; Li, S.; Wang, J.; Zeng, H.; Lu, J. A Computer Vision-Based Assistant System for the Assembly of Narrow Cabin Products. *Int. J. Adv. Manuf. Technol.* **2015**, *76*, 281–293. [CrossRef]
70. Qiu, S.; Yang, X.; Shu, Y.; Fan, X.; Wang, J. Edge-Feature-Based Aircraft Cover Recognition and Pose Estimation for AR-Aided Inner Components Inspection. In Proceedings of the 2019 IEEE 8th Joint International Information Technology and Artificial Intelligence Conference (ITAIC), Chongqing, China, 24–26 May 2019.
71. Wasenmuller, O.; Meyer, M.; Stricker, D. Augmented Reality 3D Discrepancy Check in Industrial Applications. In Proceedings of the 2016 IEEE International Symposium on Mixed and Augmented Reality (ISMAR), Merida, Mexico, 19–23 September 2016.
72. Park, K.-B.; Kim, M.; Choi, S.H.; Lee, J.Y. Deep Learning-Based Smart Task Assistance in Wearable Augmented Reality. *Robot. Comput. Integr. Manuf.* **2020**, *63*, 101887. [CrossRef]
73. Zubizarreta, J.; Aguinaga, I.; Amundarain, A. A Framework for Augmented Reality Guidance in Industry. *Int. J. Adv. Manuf. Technol.* **2019**, *102*, 4095–4108. [CrossRef]
74. KIM, Y.H.; LEE, K.H. Pose Initialization Method of Mixed Reality System for Inspection Using Convolutional Neural Network. *J. Adv. Mech. Des. Syst. Manuf.* **2019**, *13*, JAMDSM0093. [CrossRef]
75. Di Donato, M.; Fiorentino, M.; Uva, A.E.; Gattullo, M.; Monno, G. Text Legibility for Projected Augmented Reality on Industrial Workbenches. *Comput. Ind.* **2015**, *70*, 70–78. [CrossRef]
76. Neubert, J.; Pretlove, J.; Drummond, T. Rapidly Constructed Appearance Models for Tracking in Augmented Reality Applications. *Mach. Vis. Appl.* **2012**, *23*, 843–856. [CrossRef]
77. Kollatsch, C.; Klimant, P. Efficient Integration Process of Production Data into Augmented Reality Based Maintenance of Machine Tools. *Prod. Eng.* **2021**, *15*, 311–319. [CrossRef]
78. Ceruti, A.; Marzocca, P.; Liverani, A.; Bil, C. Maintenance in Aeronautics in an Industry 4.0 Context: The Role of Augmented Reality and Additive Manufacturing. *J. Comput. Des. Eng.* **2019**, *6*, 516–526. [CrossRef]
79. Zenati, N.; Benbelkacem, S.; Belhocine, M.; Bellarbi, A. A New AR Interaction for Collaborative E-Maintenance System. *IFAC Proc. Vol.* **2013**, *46*, 619–624. [CrossRef]
80. Engelke, T.; Keil, J.; Rojtberg, P.; Wientapper, F.; Webel, S.; Bockholt, U. Content First—A Concept for Industrial Augmented Reality Maintenance Applications Using Mobile Devices. In Proceedings of the 6th ACM Multimedia Systems Conference, Portland, OR, USA, 18–20 March 2015.
81. Wang, J.; Feng, Y.; Zeng, C.; Li, S. An Augmented Reality Based System for Remote Collaborative Maintenance Instruction of Complex Products. In Proceedings of the 2014 IEEE International Conference on Automation Science and Engineering (CASE), New Taipei, Taiwan, 18–22 August 2014.
82. Palmarini, R.; Erkoyuncu, J.A.; Roy, R. An Innovative Process to Select Augmented Reality (AR) Technology for Maintenance. *Procedia CIRP* **2017**, *59*, 23–28. [CrossRef]
83. del Amo, I.F.; Galeotti, E.; Palmarini, R.; Dini, G.; Erkoyuncu, J.; Roy, R. An Innovative User-Centred Support Tool for Augmented Reality Maintenance Systems Design: A Preliminary Study. *Procedia CIRP* **2018**, *70*, 362–367. [CrossRef]
84. Scurati, G.W.; Gattullo, M.; Fiorentino, M.; Ferrise, F.; Bordegoni, M.; Uva, A.E. Converting Maintenance Actions into Standard Symbols for Augmented Reality Applications in Industry 4.0. *Comput. Ind.* **2018**, *98*, 68–79. [CrossRef]
85. Quint, F.; Loch, F.; Bertram, P. The Challenge of Introducing AR in Industry—Results of a Participative Process Involving Maintenance Engineers. *Procedia Manuf.* **2017**, *11*, 1319–1323. [CrossRef]
86. Erkoyuncu, J.A.; del Amo, I.F.; Dalle Mura, M.; Roy, R.; Dini, G. Improving Efficiency of Industrial Maintenance with Context Aware Adaptive Authoring in Augmented Reality. *CIRP Ann.* **2017**, *66*, 465–468. [CrossRef]
87. Elia, V.; Gnoni, M.G.; Lanzilotto, A. Evaluating the Application of Augmented Reality Devices in Manufacturing from a Process Point of View: An AHP Based Model. *Expert Syst. Appl.* **2016**, *63*, 187–197. [CrossRef]
88. Zhu, J.; Ong, S.K.; Nee, A.Y.C. A Context-Aware Augmented Reality System to Assist the Maintenance Operators. *Int. J. Interact. Des. Manuf.* **2014**, *8*, 293–304. [CrossRef]
89. Ong, S.K.; Zhu, J. A Novel Maintenance System for Equipment Serviceability Improvement. *CIRP Ann.* **2013**, *62*, 39–42. [CrossRef]
90. Siew, C.Y.; Ong, S.K.; Nee, A.Y.C. Improving Maintenance Efficiency and Safety through a Human-Centric Approach. *Adv. Manuf.* **2021**, *9*, 104–114. [CrossRef]

91. Kunnen, S.; Adamenko, D.; Pluhnau, R.; Loibl, A.; Nagarajah, A. System-Based Concept for a Mixed Reality Supported Maintenance Phase of an Industrial Plant. *Procedia CIRP* **2020**, *91*, 15–20. [CrossRef]
92. Wang, Z.; Bai, X.; Zhang, S.; Wang, Y.; Han, S.; Zhang, X.; Yan, Y.; Xiong, Z. User-Oriented AR Assembly Guideline: A New Classification Method of Assembly Instruction for User Cognition. *Int. J. Adv. Manuf. Technol.* **2021**, *112*, 41–59. [CrossRef]
93. Bhattacharya, B.; Winer, E.H. Augmented Reality via Expert Demonstration Authoring (AREDA). *Comput. Ind.* **2019**, *105*, 61–79. [CrossRef]
94. De Amicis, R.; Ceruti, A.; Francia, D.; Frizziero, L.; Simões, B. Augmented Reality for Virtual User Manual. *Int. J. Interact. Des. Manuf.* **2018**, *12*, 689–697. [CrossRef]
95. Liu, Y.; Li, S.; Wang, J. Assembly Auxiliary System for Narrow Cabins of Spacecraft. *Chinese J. Mech. Eng.* **2015**, *28*, 1080–1088. [CrossRef]
96. Makris, S.; Pintzos, G.; Rentzos, L.; Chryssolouris, G. Assembly Support Using AR Technology Based on Automatic Sequence Generation. *CIRP Ann.* **2013**, *62*, 9–12. [CrossRef]
97. Ong, S.K.; Wang, Z.B. Augmented Assembly Technologies Based on 3D Bare-Hand Interaction. *CIRP Ann.* **2011**, *60*, 1–4. [CrossRef]
98. Radkowski, R.; Herrema, J.; Oliver, J. Augmented Reality-Based Manual Assembly Support With Visual Features for Different Degrees of Difficulty. *Int. J. Hum. Comput. Interact.* **2015**, *31*, 337–349. [CrossRef]
99. Danielsson, O.; Syberfeldt, A.; Brewster, R.; Wang, L. Assessing Instructions in Augmented Reality for Human-Robot Collaborative Assembly by Using Demonstrators. *Procedia CIRP* **2017**, *63*, 89–94. [CrossRef]
100. Rentzos, L.; Papanastasiou, S.; Papakostas, N.; Chryssolouris, G. Augmented Reality for Human-Based Assembly: Using Product and Process Semantics. *IFAC Proc. Vol.* **2013**, *46*, 98–101. [CrossRef]
101. Schuster, F.; Engelmann, B.; Sponholz, U.; Schmitt, J. Human Acceptance Evaluation of AR-Assisted Assembly Scenarios. *J. Manuf. Syst.* **2021**, *61*, 660–672. [CrossRef]
102. Liu, X.; Zheng, L.; Shuai, J.; Zhang, R.; Li, Y. Data-Driven and AR Assisted Intelligent Collaborative Assembly System for Large-Scale Complex Products. *Procedia CIRP* **2020**, *93*, 1049–1054. [CrossRef]
103. Luxenburger, A.; Mohr, J.; Spieldenner, T.; Merkel, D.; Espinosa, F.; Schwartz, T.; Reinicke, F.; Ahlers, J.; Stoyke, M. Augmented Reality for Human-Robot Cooperation in Aircraft Assembly. In *2019 IEEE International Conference on Artificial Intelligence and Virtual Reality (AIVR)*; IEEE: Piscataway, NJ, USA, 2019; pp. 263–2633.
104. Chen, H.; Chen, C.; Sun, G.; Wan, B. Augmented Reality Based Visualization Method for Spacecraft Cable Assembly Process. *IOP Conf. Ser. Mater. Sci. Eng.* **2019**, *612*, 1–7. [CrossRef]
105. Rodriguez, L.; Quint, F.; Gorecky, D.; Romero, D.; Siller, H.R. Developing a Mixed Reality Assistance System Based on Projection Mapping Technology for Manual Operations at Assembly Workstations. *Procedia Comput. Sci.* **2015**, *75*, 327–333. [CrossRef]
106. Evans, G.; Miller, J.; Iglesias Pena, M.; MacAllister, A.; Winer, E. Evaluating the Microsoft HoloLens through an Augmented Reality Assembly Application. In *Degraded Environments: Sensing, Processing, and Display 2017*; SPIE: Anaheim, CA, USA, 2017; pp. 1–16.
107. Gavish, N.; Gutiérrez, T.; Webel, S.; Rodríguez, J.; Peveri, M.; Bockholt, U.; Tecchia, F. Evaluating Virtual Reality and Augmented Reality Training for Industrial Maintenance and Assembly Tasks. *Interact. Learn. Environ.* **2013**, *23*, 778–798. [CrossRef]
108. Neb, A.; Strieg, F. Generation of AR-Enhanced Assembly Instructions Based on Assembly Features. *Procedia CIRP* **2018**, *72*, 1118–1123. [CrossRef]
109. Li, B.; Dong, Q.; Dong, J.; Wang, J.; Li, W.; Li, S. Instruction Manual for Product Assembly Process Based on Augmented Visualization. In *2018 Chinese Automation Congress (CAC)*; IEEE: Piscataway, NJ, USA, 2018; pp. 3248–3253.
110. Danielsson, O.; Syberfeldt, A.; Holm, M.; Wang, L. Operators Perspective on Augmented Reality as a Support Tool in Engine Assembly. *Procedia CIRP* **2018**, *72*, 45–50. [CrossRef]
111. Novak-Marcincin, J.; Barna, J.; Torok, J. Precision Assembly Process with Augmented Reality Technology Support. *Key Eng. Mater.* **2014**, *581*, 106–111. [CrossRef]
112. Li, W.; Wang, J.; Jiao, S.; Wang, M.; Li, S. Research on the Visual Elements of Augmented Reality Assembly Processes. *Virtual Real. Intell. Hardw.* **2019**, *1*, 622–634. [CrossRef]
113. Funk, M.; Bächler, A.; Bächler, L.; Kosch, T.; Heidenreich, T.; Schmidt, A. Working with Augmented Reality? In *Proceedings of the 10th International Conference on PErvasive Technologies Related to Assistive Environments*; ACM: New York, NY, USA, 2017; pp. 222–229.
114. Qeshmy, D.E.; Makdisi, J.; Ribeiro da Silva, E.H.D.; Angelis, J. Managing Human Errors: Augmented Reality Systems as a Tool in the Quality Journey. *Procedia Manuf.* **2019**, *28*, 24–30. [CrossRef]
115. Urbas, U.; Vrabič, R.; Vukašinović, N. Displaying Product Manufacturing Information in Augmented Reality for Inspection. *Procedia CIRP* **2019**, *81*, 832–837. [CrossRef]
116. Thamm, S.; Huebser, L.; Adam, T.; Hellebrandt, T.; Heine, I.; Barbalho, S.; Velho, S.K.; Becker, M.; Bagnato, V.S.; Schmitt, R.H. Concept for an Augmented Intelligence-Based Quality Assurance of Assembly Tasks in Global Value Networks. *Procedia CIRP* **2021**, *97*, 423–428. [CrossRef]
117. Eschen, H.; Kötter, T.; Rodeck, R.; Harnisch, M.; Schüppstuhl, T. Augmented and Virtual Reality for Inspection and Maintenance Processes in the Aviation Industry. *Procedia Manuf.* **2018**, *19*, 156–163. [CrossRef]

118. Canepa-Talamas, D.; Nassehi, A.; Dhokia, V. Innovative Framework for Immersive Metrology. *Procedia CIRP* **2017**, *60*, 110–115. [CrossRef]
119. Chang, M.M.L.; Nee, A.Y.C.; Ong, S.K. Interactive AR-Assisted Product Disassembly Sequence Planning (ARDIS). *Int. J. Prod. Res.* **2020**, *58*, 4916–4931. [CrossRef]
120. Weidig, C.; Aurich, J.C. Systematic Development of Mobile AR-Applications, Special Focus on User Participation. *Procedia CIRP* **2015**, *28*, 155–160. [CrossRef]
121. Lee, H. Real-Time Manufacturing Modeling and Simulation Framework Using Augmented Reality and Stochastic Network Analysis. *Virtual Real.* **2019**, *23*, 85–99. [CrossRef]
122. van Lopik, K.; Sinclair, M.; Sharpe, R.; Conway, P.; West, A. Developing Augmented Reality Capabilities for Industry 4.0 Small Enterprises: Lessons Learnt from a Content Authoring Case Study. *Comput. Ind.* **2020**, *117*, 103208. [CrossRef]
123. Bagassi, S.; De Crescenzio, F.; Piastra, S. Augmented Reality Technology Selection Based on Integrated QFD-AHP Model. *Int. J. Interact. Des. Manuf.* **2020**, *14*, 285–294. [CrossRef]
124. Müller, T. Challenges in Representing Information with Augmented Reality to Support Manual Procedural Tasks. *AIMS Electron. Electr. Eng.* **2019**, *3*, 71–97. [CrossRef]
125. Yew, A.W.W.; Ong, S.K.; Nee, A.Y.C. Towards a Griddable Distributed Manufacturing System with Augmented Reality Interfaces. *Robot. Comput. Integr. Manuf.* **2016**, *39*, 43–55. [CrossRef]
126. Ong, S.K.; Wang, X.; Nee, A.Y.C. 3D Bare-Hand Interactions Enabling Ubiquitous Interactions with Smart Objects. *Adv. Manuf.* **2020**, *8*, 133–143. [CrossRef]
127. Li, W.K.; Ong, S.K.; Nee, A.Y.C. User-Oriented Augmented Reality Content Delivery and Management for Ubiquitous Manufacturing. *J. Manuf. Sci. Eng.* **2019**, *141*. [CrossRef]
128. Schumann, M.; Fuchs, C.; Kollatsch, C.; Klimant, P. Evaluation of Augmented Reality Supported Approaches for Product Design and Production Processes. *Procedia CIRP* **2021**, *97*, 160–165. [CrossRef]
129. Schmiedinger, T.; Petke, M.; von Czettritz, L.; Wohlschläger, B.; Adam, M. Augmented Reality as a Tool for Providing Informational Content in Different Production Domains. *Procedia Manuf.* **2020**, *45*, 423–428. [CrossRef]
130. Zhu, Z.; Liu, C.; Xu, X. Visualisation of the Digital Twin Data in Manufacturing by Using Augmented Reality. *Procedia CIRP* **2019**, *81*, 898–903. [CrossRef]
131. Syberfeldt, A.; Holm, M.; Danielsson, O.; Wang, L.; Brewster, R.L. Support Systems on the Industrial Shop-Floors of the Future—Operators' Perspective on Augmented Reality. *Procedia CIRP* **2016**, *44*, 108–113. [CrossRef]
132. Syberfeldt, A.; Danielsson, O.; Holm, M.; Wang, L. Dynamic Operator Instructions Based on Augmented Reality and Rule-Based Expert Systems. *Procedia CIRP* **2016**, *41*, 346–351. [CrossRef]
133. Caricato, P.; Colizzi, L.; Gnoni, M.G.; Grieco, A.; Guerrieri, A.; Lanzilotto, A. Augmented Reality Applications in Manufacturing: A Multi-Criteria Decision Model for Performance Analysis. *IFAC Proc. Vol.* **2014**, *47*, 754–759. [CrossRef]
134. Fast-Berglund, Å.; Gong, L.; Li, D. Testing and Validating Extended Reality (XR) Technologies in Manufacturing. *Procedia Manuf.* **2018**, *25*, 31–38. [CrossRef]
135. Riexinger, G.; Kluth, A.; Olbrich, M.; Braun, J.-D.; Bauernhansl, T. Mixed Reality for On-Site Self-Instruction and Self-Inspection with Building Information Models. *Procedia CIRP* **2018**, *72*, 1124–1129. [CrossRef]
136. Quandt, M.; Knoke, B.; Gorldt, C.; Freitag, M.; Thoben, K.-D. General Requirements for Industrial Augmented Reality Applications. *Procedia CIRP* **2018**, *72*, 1130–1135. [CrossRef]
137. Röltgen, D.; Dumitrescu, R. Classification of Industrial Augmented Reality Use Cases. *Procedia CIRP* **2020**, *91*, 93–100. [CrossRef]
138. Mourtzis, D.; Angelopoulos, J.; Zogopoulos, V. Integrated and Adaptive AR Maintenance and Shop-Floor Rescheduling. *Comput. Ind.* **2021**, *125*, 103383. [CrossRef]
139. Konstantinidis, F.K.; Kansizoglou, I.; Santavas, N.; Mouroutsos, S.G.; Gasteratos, A. MARMA: A Mobile Augmented Reality Maintenance Assistant for Fast-Track Repair Procedures in the Context of Industry 4.0. *Machines* **2020**, *8*, 88. [CrossRef]
140. Siew, C.Y.; Ong, S.K.; Nee, A.Y.C. A Practical Augmented Reality-Assisted Maintenance System Framework for Adaptive User Support. *Robot. Comput. Integr. Manuf.* **2019**, *59*, 115–129. [CrossRef]
141. Gattullo, M.; Scurati, G.W.; Fiorentino, M.; Uva, A.E.; Ferrise, F.; Bordegoni, M. Towards Augmented Reality Manuals for Industry 4.0: A Methodology. *Robot. Comput. Integr. Manuf.* **2019**, *56*, 276–286. [CrossRef]
142. Uva, A.E.; Gattullo, M.; Manghisi, V.M.; Spagnulo, D.; Cascella, G.L.; Fiorentino, M. Evaluating the Effectiveness of Spatial Augmented Reality in Smart Manufacturing: A Solution for Manual Working Stations. *Int. J. Adv. Manuf. Technol.* **2018**, *94*, 509–521. [CrossRef]
143. Mourtzis, D.; Vlachou, A.; Zogopoulos, V. Cloud-Based Augmented Reality Remote Maintenance Through Shop-Floor Monitoring: A Product-Service System Approach. *J. Manuf. Sci. Eng.* **2017**, *139*, 061011. [CrossRef]
144. Masoni, R.; Ferrise, F.; Bordegoni, M.; Gattullo, M.; Uva, A.E.; Fiorentino, M.; Carrabba, E.; Di Donato, M. Supporting Remote Maintenance in Industry 4.0 through Augmented Reality. *Procedia Manuf.* **2017**, *11*, 1296–1302. [CrossRef]
145. Fiorentino, M.; Uva, A.E.; Monno, G.; Radkowski, R. Natural Interaction for Online Documentation in Industrial Maintenance. *Int. J. Comput. Aided Eng. Technol.* **2016**, *8*, 56. [CrossRef]
146. Zhu, J.; Ong, S.K.; Nee, A.Y.C. A Context-Aware Augmented Reality Assisted Maintenance System. *Int. J. Comput. Integr. Manuf.* **2015**, *28*, 213–225. [CrossRef]

147. Fiorentino, M.; Uva, A.E.; Gattullo, M.; Debernardis, S.; Monno, G. Augmented Reality on Large Screen for Interactive Maintenance Instructions. *Comput. Ind.* **2014**, *65*, 270–278. [CrossRef]
148. Zhu, J.; Ong, S.K.; Nee, A.Y.C. An Authorable Context-Aware Augmented Reality System to Assist the Maintenance Technicians. *Int. J. Adv. Manuf. Technol.* **2013**, *66*, 1699–1714. [CrossRef]
149. Espíndola, D.B.; Fumagalli, L.; Garetti, M.; Pereira, C.E.; Botelho, S.S.C.; Ventura Henriques, R. A Model-Based Approach for Data Integration to Improve Maintenance Management by Mixed Reality. *Comput. Ind.* **2013**, *64*, 376–391. [CrossRef]
150. Fernández del Amo, I.; Erkoyuncu, J.; Vrabič, R.; Frayssinet, R.; Vazquez Reynel, C.; Roy, R. Structured Authoring for AR-Based Communication to Enhance Efficiency in Remote Diagnosis for Complex Equipment. *Adv. Eng. Inform.* **2020**, *45*, 101096. [CrossRef]
151. Vorraber, W.; Gasser, J.; Webb, H.; Neubacher, D.; Url, P. Assessing Augmented Reality in Production: Remote-Assisted Maintenance with HoloLens. *Procedia CIRP* **2020**, *88*, 139–144. [CrossRef]
152. Mourtzis, D.; Zogopoulos, V.; Vlachou, E. Augmented Reality Application to Support Remote Maintenance as a Service in the Robotics Industry. *Procedia CIRP* **2017**, *63*, 46–51. [CrossRef]
153. Mourtzis, D.; Angelopoulos, J.; Boli, N. Maintenance Assistance Application of Engineering to Order Manufacturing Equipment: A Product Service System (PSS) Approach. *IFAC-PapersOnLine* **2018**, *51*, 217–222. [CrossRef]
154. Manuri, F.; Pizzigalli, A.; Sanna, A. A State Validation System for Augmented Reality Based Maintenance Procedures. *Appl. Sci.* **2019**, *9*, 2115. [CrossRef]
155. Utzig, S.; Kaps, R.; Azeem, S.M.; Gerndt, A. Augmented Reality for Remote Collaboration in Aircraft Maintenance Tasks. In *2019 IEEE Aerospace Conference*; IEEE: Piscataway, NJ, USA, 2019; pp. 1–10.
156. De Crescenzio, F.; Fantini, M.; Persiani, F.; Di Stefano, L.; Azzari, P.; Salti, S. Augmented Reality for Aircraft Maintenance Training and Operations Support. *IEEE Comput. Graph. Appl.* **2011**, *31*, 96–101. [CrossRef] [PubMed]
157. Cachada, A.; Romero, L.; Costa, D.; Badikyan, H.; Barbosa, J.; Leitao, P.; Morais, O.; Teixeira, C.; Azevedo, J.; Moreira, P.M. Using AR Interfaces to Support Industrial Maintenance Procedures. In *IECON 2019—45th Annual Conference of the IEEE Industrial Electronics Society*; IEEE: Piscataway, NJ, USA, 2019; pp. 3795–3800.
158. Makris, S.; Karagiannis, P.; Koukas, S.; Matthaiakis, A.-S. Augmented Reality System for Operator Support in Human–Robot Collaborative Assembly. *CIRP Ann.* **2016**, *65*, 61–64. [CrossRef]
159. Wang, X.; Ong, S.K.; Nee, A.Y.C. Real-Virtual Interaction in AR Assembly Simulation Based on Component Contact Handling Strategy. *Assem. Autom.* **2015**, *35*, 376–394. [CrossRef]
160. Wang, Z.B.; Ong, S.K.; Nee, A.Y.C. Augmented Reality Aided Interactive Manual Assembly Design. *Int. J. Adv. Manuf. Technol.* **2013**, *69*, 1311–1321. [CrossRef]
161. Chen, C.; Tian, Z.; Li, D.; Pang, L.; Wang, T.; Hong, J. Projection-Based Augmented Reality System for Assembly Guidance and Monitoring. *Assem. Autom.* **2021**, *41*, 10–23. [CrossRef]
162. de Souza Cardoso, L.F.; Mariano, F.C.M.Q.; Zorzal, E.R. Mobile Augmented Reality to Support Fuselage Assembly. *Comput. Ind. Eng.* **2020**, *148*, 106712. [CrossRef]
163. Hietanen, A.; Pieters, R.; Lanz, M.; Latokartano, J.; Kämäräinen, J.-K. AR-Based Interaction for Human-Robot Collaborative Manufacturing. *Robot. Comput. Integr. Manuf.* **2020**, *63*, 101891. [CrossRef]
164. Lai, Z.-H.; Tao, W.; Leu, M.C.; Yin, Z. Smart Augmented Reality Instructional System for Mechanical Assembly towards Worker-Centered Intelligent Manufacturing. *J. Manuf. Syst.* **2020**, *55*, 69–81. [CrossRef]
165. Mourtzis, D.; Zogopoulos, V.; Xanthi, F. Augmented Reality Application to Support the Assembly of Highly Customized Products and to Adapt to Production Re-Scheduling. *Int. J. Adv. Manuf. Technol.* **2019**, *105*, 3899–3910. [CrossRef]
166. Tao, W.; Lai, Z.-H.; Leu, M.C.; Yin, Z.; Qin, R. A Self-Aware and Active-Guiding Training & Assistant System for Worker-Centered Intelligent Manufacturing. *Manuf. Lett.* **2019**, *21*, 45–49.
167. Valentini, P.P. Natural Interface for Interactive Virtual Assembly in Augmented Reality Using Leap Motion Controller. *Int. J. Interact. Des. Manuf.* **2018**, *12*, 1157–1165. [CrossRef]
168. Wang, X.; Ong, S.K.; Nee, A.Y.C. Real-Virtual Components Interaction for Assembly Simulation and Planning. *Robot. Comput. Integr. Manuf.* **2016**, *41*, 102–114. [CrossRef]
169. Gimeno, J.; Morillo, P.; Orduña, J.M.; Fernández, M. A New AR Authoring Tool Using Depth Maps for Industrial Procedures. *Comput. Ind.* **2013**, *64*, 1263–1271. [CrossRef]
170. Wang, Z.; Bai, X.; Zhang, S.; He, W.; Wang, Y.; Han, D.; Wei, S.; Wei, B.; Chen, C. M-AR: A Visual Representation of Manual Operation Precision in AR Assembly. *Int. J. Hum.–Comput. Interact.* **2021**, *37*, 1799–1814. [CrossRef]
171. Wang, X.; Ong, S.K.; Nee, A.Y.C. Multi-Modal Augmented-Reality Assembly Guidance Based on Bare-Hand Interface. *Adv. Eng. Inform.* **2016**, *30*, 406–421. [CrossRef]
172. Pilati, F.; Faccio, M.; Gamberi, M.; Regattieri, A. Learning Manual Assembly through Real-Time Motion Capture for Operator Training with Augmented Reality. *Procedia Manuf.* **2020**, *45*, 189–195. [CrossRef]
173. Chu, C.-H.; Ko, C.-H. An Experimental Study on Augmented Reality Assisted Manual Assembly with Occluded Components. *J. Manuf. Syst.* **2021**, *61*, 685–695. [CrossRef]
174. Kousi, N.; Stoubos, C.; Gkournelos, C.; Michalos, G.; Makris, S. Enabling Human Robot Interaction in Flexible Robotic Assembly Lines: An Augmented Reality Based Software Suite. *Procedia CIRP* **2019**, *81*, 1429–1434. [CrossRef]

175. Lampen, E.; Teuber, J.; Gaisbauer, F.; Bär, T.; Pfeiffer, T.; Wachsmuth, S. Combining Simulation and Augmented Reality Methods for Enhanced Worker Assistance in Manual Assembly. *Procedia CIRP* **2019**, *81*, 588–593. [CrossRef]
176. Mengoni, M.; Ceccacci, S.; Generosi, A.; Leopardi, A. Spatial Augmented Reality: An Application for Human Work in Smart Manufacturing Environment. *Procedia Manuf.* **2018**, *17*, 476–483. [CrossRef]
177. Michalos, G.; Karagiannis, P.; Makris, S.; Tokçalar, Ö.; Chryssolouris, G. Augmented Reality (AR) Applications for Supporting Human-Robot Interactive Cooperation. *Procedia CIRP* **2016**, *41*, 370–375. [CrossRef]
178. Syberfeldt, A.; Danielsson, O.; Holm, M.; Wang, L. Visual Assembling Guidance Using Augmented Reality. *Procedia Manuf.* **2015**, *1*, 98–109. [CrossRef]
179. Mura, M.D.; Dini, G.; Failli, F. An Integrated Environment Based on Augmented Reality and Sensing Device for Manual Assembly Workstations. *Procedia CIRP* **2016**, *41*, 340–345. [CrossRef]
180. Provost, J.; Ebrahimi, A.H.; Åkesson, K. Online Support for Shop-Floor Operators Using Body Movements Tracking. *IFAC Proc. Vol.* **2013**, *46*, 102–109. [CrossRef]
181. Serván, J.; Mas, F.; Menéndez, J.L.; Ríos, J. Assembly Work Instruction Deployment Using Augmented Reality. *Key Eng. Mater.* **2012**, *502*, 25–30. [CrossRef]
182. Chu, C.-H.; Liao, C.-J.; Lin, S.-C. Comparing Augmented Reality-Assisted Assembly Functions—A Case Study on Dougong Structure. *Appl. Sci.* **2020**, *10*, 3383. [CrossRef]
183. Konig, M.; Stadlmaier, M.; Rusch, T.; Sochor, R.; Merkel, L.; Braunreuther, S.; Schilp, J. MA 2 RA—Manual Assembly Augmented Reality Assistant. In Proceedings of the 2019 IEEE International Conference on Industrial Engineering and Engineering Management (IEEM), Macao, China, 15–19 December 2019.
184. Ojer, M.; Alvarez, H.; Serrano, I.; Saiz, F.A.; Barandiaran, I.; Aguinaga, D.; Querejeta, L.; Alejandro, D. Projection-Based Augmented Reality Assistance for Manual Electronic Component Assembly Processes. *Appl. Sci.* **2020**, *10*, 796. [CrossRef]
185. Sand, O.; Büttner, S.; Paelke, V.; Röcker, C. *SmARt.Assembly—Projection-Based Augmented Reality for Supporting Assembly Workers*; Springer: Cham, Switzerland, 2016; pp. 643–652.
186. Hořejší, P. Augmented Reality System for Virtual Training of Parts Assembly. *Procedia Eng.* **2015**, *100*, 699–706. [CrossRef]
187. Pusda, F.R.; Valencia, F.F.; Andaluz, V.H.; Zambrano, V.D. Training Assistant for Automotive Engineering Through Augmented Reality. In *Augmented Reality, Virtual Reality, and Computer Graphics*; Springer: Berlin/Heidelberg, Germany, 2019; pp. 146–160.
188. Marino, E.; Barbieri, L.; Colacino, B.; Fleri, A.K.; Bruno, F. An Augmented Reality Inspection Tool to Support Workers in Industry 4.0 Environments. *Comput. Ind.* **2021**, *127*, 103412. [CrossRef]
189. Dalle Mura, M.; Dini, G. An Augmented Reality Approach for Supporting Panel Alignment in Car Body Assembly. *J. Manuf. Syst.* **2021**, *59*, 251–260. [CrossRef]
190. Liu, S.; Lu, S.; Li, J.; Sun, X.; Lu, Y.; Bao, J. Machining Process-Oriented Monitoring Method Based on Digital Twin via Augmented Reality. *Int. J. Adv. Manuf. Technol.* **2021**, *113*, 3491–3508. [CrossRef]
191. Li, S.; Zheng, P.; Zheng, L. An AR-Assisted Deep Learning-Based Approach for Automatic Inspection of Aviation Connectors. *IEEE Trans. Ind. Inform.* **2021**, *17*, 1721–1731. [CrossRef]
192. Runji, J.M.; Lin, C.-Y. Markerless Cooperative Augmented Reality-Based Smart Manufacturing Double-Check System: Case of Safe PCBA Inspection Following Automatic Optical Inspection. *Robot. Comput. Integr. Manuf.* **2020**, *64*, 101957. [CrossRef]
193. Motoyama, Y.; Iwamoto, K.; Tokunaga, H.; Okane, T. Measuring Hand-Pouring Motion in Casting Process Using Augmented Reality Marker Tracking. *Int. J. Adv. Manuf. Technol.* **2020**, *106*, 5333–5343. [CrossRef]
194. Muñoz, A.; Martí, A.; Mahiques, X.; Gracia, L.; Solanes, J.E.; Tornero, J. Camera 3D Positioning Mixed Reality-Based Interface to Improve Worker Safety, Ergonomics and Productivity. *CIRP J. Manuf. Sci. Technol.* **2020**, *28*, 24–37. [CrossRef]
195. Bruno, F.; Barbieri, L.; Marino, E.; Muzzupappa, M.; D'Oriano, L.; Colacino, B. An Augmented Reality Tool to Detect and Annotate Design Variations in an Industry 4.0 Approach. *Int. J. Adv. Manuf. Technol.* **2019**, *105*, 875–887. [CrossRef]
196. Muñoz, A.; Mahiques, X.; Solanes, J.E.; Martí, A.; Gracia, L.; Tornero, J. Mixed Reality-Based User Interface for Quality Control Inspection of Car Body Surfaces. *J. Manuf. Syst.* **2019**, *53*, 75–92. [CrossRef]
197. Li, K.; Tian, G.Y.; Chen, X.; Tang, C.; Luo, H.; Li, W.; Gao, B.; He, X.; Wright, N. AR-Aided Smart Sensing for In-Line Condition Monitoring of IGBT Wafer. *IEEE Trans. Ind. Electron.* **2019**, *66*, 8197–8204. [CrossRef]
198. Álvarez, H.; Lajas, I.; Larrañaga, A.; Amozarrain, L.; Barandiaran, I. Augmented Reality System to Guide Operators in the Setup of Die Cutters. *Int. J. Adv. Manuf. Technol.* **2019**, *103*, 1543–1553. [CrossRef]
199. Holm, M.; Danielsson, O.; Syberfeldt, A.; Moore, P.; Wang, L. Adaptive Instructions to Novice Shop-Floor Operators Using Augmented Reality. *J. Ind. Prod. Eng.* **2017**, *34*, 362–374. [CrossRef]
200. Liu, C.; Cao, S.; Tse, W.; Xu, X. Augmented Reality-Assisted Intelligent Window for Cyber-Physical Machine Tools. *J. Manuf. Syst.* **2017**, *44*, 280–286. [CrossRef]
201. Doshi, A.; Smith, R.T.; Thomas, B.H.; Bouras, C. Use of Projector Based Augmented Reality to Improve Manual Spot-Welding Precision and Accuracy for Automotive Manufacturing. *Int. J. Adv. Manuf. Technol.* **2017**, *89*, 1279–1293. [CrossRef]
202. Franceschini, F.; Galetto, M.; Maisano, D.; Mastrogiacomo, L. Towards the Use of Augmented Reality Techniques for Assisted Acceptance Sampling. *Proc. Inst. Mech. Eng. Part B J. Eng. Manuf.* **2016**, *230*, 1870–1884. [CrossRef]
203. Mourtzis, D.; Siatras, V.; Zogopoulos, V. Augmented Reality Visualization of Production Scheduling and Monitoring. *Procedia CIRP* **2020**, *88*, 151–156. [CrossRef]

204. Hofmann, C.; Staehr, T.; Cohen, S.; Stricker, N.; Haefner, B.; Lanza, G. Augmented Go & See: An Approach for Improved Bottleneck Identification in Production Lines. *Procedia Manuf.* **2019**, *31*, 148–154.
205. Antonelli, D.; Astanin, S. Enhancing the Quality of Manual Spot Welding through Augmented Reality Assisted Guidance. *Procedia CIRP* **2015**, *33*, 556–561. [CrossRef]
206. Barbieri, L.; Marino, E. *An Augmented Reality Tool to Detect Design Discrepancies: A Comparison Test with Traditional Methods*; Springer: Cham, Switzerland, 2019; pp. 99–110.
207. Segovia, D.; Ramírez, H.; Mendoza, M.; Mendoza, M.; Mendoza, E.; González, E. Machining and Dimensional Validation Training Using Augmented Reality for a Lean Process. *Procedia Comput. Sci.* **2015**, *75*, 195–204. [CrossRef]
208. Segovia, D.; Mendoza, M.; Mendoza, E.; González, E. Augmented Reality as a Tool for Production and Quality Monitoring. *Procedia Comput. Sci.* **2015**, *75*, 291–300. [CrossRef]
209. Zhou, J.; Lee, I.; Thomas, B.; Menassa, R.; Farrant, A.; Sansome, A. Applying Spatial Augmented Reality to Facilitate In-Situ Support for Automotive Spot Welding Inspection. In Proceedings of the 10th International Conference on Virtual Reality Continuum and Its Applications in Industry—VRCAI '11, Hong Kong, China, 11–12 December 2011.
210. Alves, J.; Marques, B.; Dias, P.; Santos, B.S. *Using Augmented Reality and Step by Step Verification in Industrial Quality Control*; Springer: Cham, Switzerland, 2021; pp. 350–355.
211. Ong, S.K.; Yew, A.W.W.; Thanigaivel, N.K.; Nee, A.Y.C. Augmented Reality-Assisted Robot Programming System for Industrial Applications. *Robot. Comput. Integr. Manuf.* **2020**, *61*, 101820. [CrossRef]
212. Kokkas, A.; Vosniakos, G.-C. An Augmented Reality Approach to Factory Layout Design Embedding Operation Simulation. *Int. J. Interact. Des. Manuf.* **2019**, *13*, 1061–1071. [CrossRef]
213. Tzimas, E.; Vosniakos, G.-C.; Matsas, E. Machine Tool Setup Instructions in the Smart Factory Using Augmented Reality: A System Construction Perspective. *Int. J. Interact. Des. Manuf.* **2019**, *13*, 121–136. [CrossRef]
214. Ragni, M.; Perini, M.; Setti, A.; Bosetti, P. ARTool Zero: Programming Trajectory of Touching Probes Using Augmented Reality. *Comput. Ind. Eng.* **2018**, *124*, 462–473. [CrossRef]
215. Pai, Y.S.; Yap, H.J.; Singh, R. Augmented Reality–Based Programming, Planning and Simulation of a Robotic Work Cell. *Proc. Inst. Mech. Eng. Part B J. Eng. Manuf.* **2015**, *229*, 1029–1045. [CrossRef]
216. Zhang, J.; Ong, S.K.; Nee, A.Y.C. A Multi-Regional Computation Scheme in an AR-Assisted in Situ CNC Simulation Environment. *Comput. Des.* **2010**, *42*, 1167–1177. [CrossRef]
217. Wang, X.; Yew, A.W.W.; Ong, S.K.; Nee, A.Y.C. Enhancing Smart Shop Floor Management with Ubiquitous Augmented Reality. *Int. J. Prod. Res.* **2020**, *58*, 2352–2367. [CrossRef]
218. Mueller, F.; Deuerlein, C.; Koch, M. Intuitive Welding Robot Programming via Motion Capture and Augmented Reality. *IFAC-PapersOnLine* **2019**, *52*, 294–299. [CrossRef]
219. Mourtzis, D.; Angelopoulos, J.; Panopoulos, N. Collaborative Manufacturing Design: A Mixed Reality and Cloud-Based Framework for Part Design. *Procedia CIRP* **2021**, *100*, 97–102. [CrossRef]
220. Bottani, E.; Longo, F.; Nicoletti, L.; Padovano, A.; Tancredi, G.P.C.; Tebaldi, L.; Vetrano, M.; Vignali, G. Wearable and Interactive Mixed Reality Solutions for Fault Diagnosis and Assistance in Manufacturing Systems: Implementation and Testing in an Aseptic Bottling Line. *Comput. Ind.* **2021**, *128*, 103429. [CrossRef]
221. Blaga, A.; Militaru, C.; Mezei, A.-D.; Tamas, L. Augmented Reality Integration into MES for Connected Workers. *Robot. Comput. Integr. Manuf.* **2021**, *68*, 102057. [CrossRef]
222. Masood, T.; Egger, J. Adopting Augmented Reality in the Age of Industrial Digitalisation. *Comput. Ind.* **2020**, *115*, 103112. [CrossRef]
223. Longo, F.; Nicoletti, L.; Padovano, A. Smart Operators in Industry 4.0: A Human-Centered Approach to Enhance Operators' Capabilities and Competencies within the New Smart Factory Context. *Comput. Ind. Eng.* **2017**, *113*, 144–159. [CrossRef]
224. Novak-Marcincin, J.; Torok, J.; Janak, M.; Novakova-Marcincinova, L. Interactive Monitoring of Production Process with Use of Augmented Reality Technology. *Appl. Mech. Mater.* **2014**, *616*, 19–26. [CrossRef]
225. Rosales, J.; Deshpande, S.; Anand, S. IIoT Based Augmented Reality for Factory Data Collection and Visualization. *Procedia Manuf.* **2021**, *53*, 618–627. [CrossRef]
226. Wang, X.; Kim, M.J.; Love, P.E.D.; Kang, S.-C. Augmented Reality in Built Environment: Classification and Implications for Future Research. *Autom. Constr.* **2013**, *32*, 1–13. [CrossRef]
227. Macal, C.M.; North, M.J. Tutorial on Agent-Based Modelling and Simulation. *J. Simul.* **2010**, *4*, 151–162. [CrossRef]
228. Monostori, L.; Váncza, J.; Kumara, S.R.T. Agent-Based Systems for Manufacturing. *CIRP Ann.* **2006**, *55*, 697–720. [CrossRef]
229. Shen, W.; Norrie, D.H. Agent-Based Systems for Intelligent Manufacturing: A State-of-the-Art Survey. *Knowl. Inf. Syst.* **1999**, *1*, 129–156. [CrossRef]
230. Smithers, T.; Tang, M.X.; Tomes, N.; Buck, P.; Clarke, B.; Lloyd, G.; Poulter, K.; Floyd, C.; Hodgkin, E. Development of a Knowledge-Based Design Support System. *Knowl.-Based Syst.* **1992**, *5*, 31–40. [CrossRef]
231. Chen, S.-J.; Chen, L.-C.; Lin, L. Knowledge-Based Support for Simulation Analysis of Manufacturing Cells. *Comput. Ind.* **2001**, *44*, 33–49. [CrossRef]
232. Manivannan, S.; Lehtihet, A.; Egbelu, P.J. A Knowledge Based System for the Specification of Manufacturing Tolerances. *J. Manuf. Syst.* **1989**, *8*, 153–160. [CrossRef]

233. Havard, V.; Baudry, D.; Savatier, X.; Jeanne, B.; Louis, A.; Mazari, B. *Augmented Industrial Maintenance (AIM): A Case Study for Evaluating and Comparing with Paper and Video Media Supports*; Springer: Berlin/Heidelberg, Germany, 2016; pp. 302–320.
234. Bauer, P.; Fink, F.; Magaña, A.; Reinhart, G. Spatial Interactive Projections in Robot-Based Inspection Systems. *Int. J. Adv. Manuf. Technol.* **2020**, *107*, 2889–2900. [CrossRef]
235. Yang, X.; Fan, X.; Wang, J.; Yin, X.; Qiu, S. Edge-Based Cover Recognition and Tracking Method for an AR-Aided Aircraft Inspection System. *Int. J. Adv. Manuf. Technol.* **2020**, *111*, 3505–3518. [CrossRef]
236. Govindarajan, U.H.; Trappey, A.J.C.; Trappey, C.V. Immersive Technology for Human-Centric Cyberphysical Systems in Complex Manufacturing Processes: A Comprehensive Overview of the Global Patent Profile Using Collective Intelligence. *Complexity* **2018**, *2018*, 1–17. [CrossRef]
237. Pedersen, I. Radiating Centers: Augmented Reality and Human-Centric Design. In *2009 IEEE International Symposium on Mixed and Augmented Reality—Arts, Media and Humanities*; IEEE: Piscataway, NJ, USA, 2009; pp. 11–16.
238. Lopez, H.; Navarro, A.; Relano, J. An Analysis of Augmented Reality Systems. In *2010 Fifth International Multi-conference on Computing in the Global Information Technology*; IEEE: Piscataway, NJ, USA, 2010; pp. 245–250.
239. Gazzaneo, L.; Padovano, A.; Umbrello, S. Designing Smart Operator 4.0 for Human Values: A Value Sensitive Design Approach. *Procedia Manuf.* **2020**, *42*, 219–226. [CrossRef]
240. Zhou, F.; Duh, H.B.-L.; Billinghurst, M. Trends in Augmented Reality Tracking, Interaction and Display: A Review of Ten Years of ISMAR. In *2008 7th IEEE/ACM International Symposium on Mixed and Augmented Reality*; IEEE: Piscataway, NJ, USA, 2008; pp. 193–202.
241. Kitchenham, B.; Pearl Brereton, O.; Budgen, D.; Turner, M.; Bailey, J.; Linkman, S. Systematic Literature Reviews in Software Engineering—A Systematic Literature Review. *Inf. Softw. Technol.* **2009**, *51*, 7–15. [CrossRef]
242. Pfeil, K.; Masnadi, S.; Belga, J.; Sera-Josef, J.-V.T.; LaViola, J. Distance Perception with a Video See-Through Head-Mounted Display. In Proceedings of the 2021 CHI Conference on Human Factors in Computing Systems, Yokohama, Japan, 8–13 May 2021.
243. Ong, S.K.; Yuan, M.L.; Nee, A.Y.C. Augmented Reality Applications in Manufacturing: A Survey. *Int. J. Prod. Res.* **2008**, *46*, 2707–2742. [CrossRef]
244. Alvarez, H.; Aguinaga, I.; Borro, D. Providing Guidance for Maintenance Operations Using Automatic Markerless Augmented Reality System. In *2011 10th IEEE International Symposium on Mixed and Augmented Reality*; IEEE: Piscataway, NJ, USA, 2011; pp. 181–190.
245. Golding, A.R.; Lesh, N. Indoor Navigation Using a Diverse Set of Cheap, Wearable Sensors. In *Digest of Papers. Third International Symposium on Wearable Computers*; IEEE Computer Society: Washington, DC, USA, 1999; pp. 29–36.
246. Reid, G.B.; Nygren, T.E. The Subjective Workload Assessment Technique: A Scaling Procedure for Measuring Mental Workload. *Adv. Psychol.* **1988**, *52*, 185–218.
247. Dulle, F.W.; Minishi-Majanja, M.K. The Suitability of the Unified Theory of Acceptance and Use of Technology (UTAUT) Model in Open Access Adoption Studies. *Inf. Dev.* **2011**, *27*, 32–45. [CrossRef]

applied sciences

MDPI

Article

Design and Implementation of OPC UA-Based VR/AR Collaboration Model Using CPS Server for VR Engineering Process

Jeehyeong Kim and Jongpil Jeong *

Department of Smart Factory Convergence, Sungkyunkwan University, 2066 Seobu-ro, Jangan-gu, Suwon 16419, Korea; ghyeong2@naver.com
* Correspondence: jpjeong@skku.edu; Tel.: +82-31-299-4267

Abstract: In order to cope with the changing era of the innovative management paradigm of the manufacturing industry, it is necessary to advance the construction of smart factories in the domestic manufacturing industry, and in particular, the 3D design and manufacturing content sector is highly growthable. In particular, the core technologies that enable digital transformation VR (Virtual Reality)/AR (Augmented Reality) technologies have developed rapidly in recent years, but have not yet achieved any particular results in industrial engineering. In the manufacturing industry, digital threads and collaboration systems are needed to reduce design costs that change over and over again due to the inability to respond to various problems and demands that should be considered when designing products. To this end, we propose a VR/AR collaboration model that increases efficiency of manufacturing environments such as inspection and maintenance as well as design simultaneously with participants through 3D rendering virtualization of facilities or robot 3D designs in VR/AR. We implemented converting programs and middleware CPS (Cyber Physical System) servers that convert to BOM (Bill of Material)-based 3D graphics models and CPS models to test the accuracy of data and optimization of 3D modeling and study their performance through robotic arms in real factories.

Keywords: virtual reality; augmented reality; cyber physical system; OPC UA; CAD

Citation: Kim, J.; Jeong, J. Design and Implementation of OPC UA-Based VR/AR Collaboration Model Using CPS Server for VR Engineering Process. *Appl. Sci.* **2022**, *12*, 7534. https://doi.org/10.3390/app12157534

Academic Editors: Guido Tosello, Roque Calvo and José A. Yagüe-Fabra

Received: 10 June 2022
Accepted: 23 July 2022
Published: 27 July 2022

1. Introduction

Recently, innovations in the ICT (Information and Communications Technology) field have begun to rapidly change existing manufacturing methods into smart manufacturing methods or CPS (Cyber Physical Systems) [1]. These manufacturing methods can use IoT (Internet of Things) technologies and web-based services to communicate and interact with other products in a factory environment. Due to the COVID-19 pandemic, many manufacturers have come to seriously consider the problem of flexibly adjusting production according to environmental changes while promoting employee safety. Depending on the situation regarding changes in infectious diseases, the demand for certain products varies substantially by country or region. Specifically, manufacturing companies that rely on small-scale production of multiple varieties should come up with countermeasures such as quickly readjusting their production lines and educating their employees on new production manuals in line with changes in consumer demand. As telecommuting continues to increase, the need to clearly check the status of production plants online is increasing as well. For manufacturers to overcome the COVID-19 pandemic crisis, it has been stated that it is necessary to actively introduce robot automation that can promote the safety of employees and increase production efficiency, and to combine virtual convergence reality technologies such as Digital Twin and VR (Virtual Reality)/AR (Augmented Reality) technology.

Although the development of Digital Twin started from the Aerospace industry, the industry which is exploring the technology the most is the manufacturing industry. Digital Twins have been described as the key enablers of Industry 4.0 and Smart Manufacturing [2,3]. Any manufactured product goes through four main phases throughout its life cycle: design, manufacture, operation, and disposal. Smart manufacturers can leverage Digital Twins in all four phases of the product [4,5]. During the design phase, Digital Twin allows the designers to verify their product design virtually, which enables them to test different iterations of the product and choose the best one. An example of this is the car manufacturer Maserati, which used Digital Twin for optimizing car body aerodynamics using wind tunnel tests virtually, which are elaborate and expensive otherwise. Unlike traditional bike designing methods, which are based on the designers' knowledge and experience, the Digital Twin of a bicycle constantly collects the data from the physical space, which could be compared, analyzed, and used for designing or redesigning next-generation bicycles. With customer reviews and usage habits, designers can get a better understanding of customer requirements, which can be translated into better and improved features. Capturing customer preferences via a Digital Twin lets businesses know about the market trends which can be integrated with the customer usage data to see the effects on product performance. This allows businesses to take design-decisions and incorporate them in an informed way, thus making the process of integrating the customer feedback into the product to deliver customized products easier [6].

One of the most developed directions in manufacturing, especially in the scope of Industry 4.0, is the robotization and automatization of the production lines [7]. Digital Twin is playing a crucial role in this integration, as industrial robots are being programmed with mainly three methods, which are closely related to the twinning of the manufacturing equipment. Those three methods are: (i) Offline, a dedicated virtual environment (usually each robot brand has their own) for programming each aspect of the robotic cell for later deployment through the network to the physical robot; (ii) Online, which is being adapted by means of sensor information, usually being twinned in the dedicated virtual environment (e.g., ROS (Robot Operation System), and is able to directly affect the pre-programmed path and routine of the robotic systems; (iii) Manual, which is robot programming by the usage of a flex pendant, but with the introduction of VR/AR interfaces, it also uses a twin for manipulation near the virtual robot remotely [8]. As can be observed, combinations of those methods exploit the virtual twin of the robotic cell and are widely used across the various industrial sectors. Moreover, Digital Twin is used as a validation tool for HRC (Human–Robot Collaboration) safety standards, to evaluate the safety level of the system first, for example, using VR human avatars, before experiments with real operators in the actual system.

In this study, it is possible to visually check the converted 3D model in a virtual environment for product designers, purchasing teams, and laboratory staff of small and medium-sized enterprises that lack manpower and capital as well as large enterprises. Then, one develops a VR/AR-based product inspection environment that enables collaboration, which is the main goal in terms of the CAD data conversion process of the CPS-based VR/AR system. In addition, the research is focused on establishing a cooperative relationship between each system entity that observes, calculates, and manipulates physical phenomena through CPS.

The smart factory market is expanding with the development of technology following the fourth industrial revolution. Advancements in smart factories include factory automation, and various mechanical devices perform the process automatically. With the introduction of smart factories, field workers mainly carry out inspections and fault repairs, and field workers need a lot of prior learning to check and repair various mechanical devices. In the manufacturing industry, skilled workers are aging, and it is difficult to properly train unskilled workers due to the increased complexity of manufacturing facilities. Collaboration ideas are also emerging as the need for untact emerges due to the COVID-19 pandemic. In the manufacturing industry, collaborations often use AR and VR technologies, and as a result, related jobs are increasing worldwide.

To cope with the changing era of the fourth industrial revolution's innovative management paradigm, it is necessary to accelerate the construction of smart factories in the manufacturing industry, and the 3D design and manufacturing content sectors show high potential for growth. It is also necessary to introduce a standardized CPS communication platform technology to facilitate the construction of smart factories in the manufacturing industry [9]. OPC UA (Open Platform Communications Unified Architecture) has become an integral part of the construction of CPS communication platforms that must have an independent and standardized communication method between different species. OPC UA technology is considered to be important and valuable for data exchange and standardized data exchange for various industrial applications such as ERP (Enterprise Resource Planning)/MES (Manufacturing Execution System)/PDM (Product Data Management). To increase the reliability of BigData, OPC UA, a device-friendly communication, high value-added is created with the high reliability of simulation. OPC UA has real-time functionality in Ethernet environments with TSN (Time Sensitive Network), thus making it suitable for the high performance needs of the VR/AR network, and OPC UA-based CPS connects closed networks as standard networks in response to numerous industrial protocols to integrate OT and IT. Further, applications that require standardized data integration, such as ERP/MES, can search and collect data without having to install a separate network.

This study proposes a VR/AR collaboration design environment based on CPS communication platform technology and quality function development. This study makes the following contributions:

- We extract VR/AR rendering and data of 3D design files through data exchange and distribution through CPS servers with OPC UA and propose a model specialized in manufacturing collaboration.
- While referring to the OPC UA specification of international industry standards, the system is constructed by connecting both OPC UA modeling and CPS Node modeling with technical convergence with CPS.
- The CAD Data Import Interface allows designers to insert CAD data into the CPS and reads the CAD data interface to a VR Parser and converts it into a 3D object.
- BOM (Bill of Material) is extracted from CAD or migrated from legacy DB to refer the contents of the BOM to the related node in the Address Space of the CPS server to enhance the functionality of the physical engine based on the BOM.

The structure of this paper explains the necessity of collaboration in Industry 4.0 and the necessity of this study in Section 1. In Section 2, we examine the related studies of OPC UA-based CPS and VR/AR systems. In Section 3, we describe the composition and role of the VR/AR collaboration model and VR engineering process architecture proposed in this paper. In Section 4, we measure the address space conversion accuracy and optimization of the BOM of the actual factory equipment, and measure the rendering speed. Section 5 summarizes the proposed architecture, implementation, and test results.

2. Related Work

In this study, it is necessary to first understand the characteristic CPS factors and clarify the design review. Next, you need to understand VR/AR.

At the core of a CPS is the physical part based on mechatronic products [10] or intelligent mechatronic products. Such a system gets a label called CPS, which has the ability to communicate with other products [11]. The most important features of CPS are high autonomy, network communication, personalization capabilities, and general user affinity [12]. Moreover, CPS features dynamic reconfigurability over the entire product life cycle and real-time responsiveness to environmental changes. Initiatives such as German RAMI 4.0 (Reference Architectural Model for Industrie 4.0) [13] or US IIRA (Industrial Internet Reference Architecture) [14] aim to provide guidance for building IoX (Internet of Everything) devices for industrial use cases [15]. Each component within the CPS has a so-called management shell that associates components with other components of the reference architecture. The intended benefits of this linkage result in overall cost savings due to improved planning processes or utilization of feedback information from real CPS for identifying bottlenecks as well as future process and product optimization [16]. We describe ontology-based concepts with the aim of obtaining and providing knowledge feedback on product design and deriving rules for fault prevention [17]. Thus, design reviews of CPS can access on-site reports from previous generations of physical components as well as feedback data and services provided by product types and components. In this work, we will use an approach using CPS based on OPC UA to improve the design review process of CPS [18–23].

In the process of the switch to CPS, more data will be generated in the future, and this will help improve the physical environment. To bridge the gap between the physical and digital/virtual worlds, VR is a suitable technology for visualizing and interacting with geometric 3D models in virtual environments [24]. In general, when VR is used for design review and other engineering tasks, the goal is to achieve cost savings and improve or accelerate related processes. There are many approaches in the academic literature that aim to support multiple engineering applications through VR. VR makes it easy to verify design, and manufacturing costs can be reduced by replacing review processes involving physical prototypes and mock-ups with processes involving virtual prototypes [25]. Further, the product modifications can be implemented quicker if physical parts have not already been used. In addition to its general feasibility, it focuses on improving user interaction within a VR environment. Mohamed Tabaa et al. [26] have developed a hardware and software-based user interface that simplifies the interaction of VR review processes. Adrian Harwood et al. [27] investigated tools and methods for the CAD-like interaction and manipulation capabilities of 3D models. Research on VR-supported design reviews shows that participants are more likely to detect errors in 3D models within VR scenes than traditional CAD software approaches on PC screens. Further, conducting design reviews and interacting with 3D models is more intuitive and natural for non-CAD professionals due to their high immersion. In this work, we consider the design review of CPS and therefore connect the digital and physical worlds. In particular, the focus on the proposed implementation depends on the availability of accessing and interacting with digital content as well as representing object information in a collaborative and immersive environment.

2.1. OPC UA-Based CPS

OPC UA-based CPS presents the architecture of a Smart Factory CPS using the structure and concepts of the OPC UA Framework [28], which is an international smart factory standard, and it relates to the structure and method of dynamically constructing CPS models with OPC UA modeling.

OPC UA-based CPS is largely functionally composed of OPC UA Server and Factory CPS Model [29]. The CPS Server consists of an OPC CPS Node Generator, which automatically creates a systematic connection between the Factory CPS Model and the OPC UA Client and a variety of OPC FileA sensors and clients, along with an information exchange service for data exchange between CPS Connect OPC UA and external systems. The OPC UA Address Space is a variable system that changes the value of the modeled entire object node in real time. The OPC UA specification manages and delivers all values through the Address Space. Within the Centralized CPS Server, OPC UA Pub/Sub, OPC UA Monitored Item, Alarm/Event, and Historian, all data are exchanged and updated through the OPC UA Address Space.

The Information Exchange Service is responsible for exchanging and delivering data with external systems. It consists of OPC UA standard specifications and asynchronously distributes data to external brokers and clouds through OPC UA Pub/Sub. The OPC UA Monitored Item periodically registers the node value that the system and application that wants to exchange data from the outside and accordingly transmits data periodically. When registering with an OPC UA Monitored Item, attributes such as transmission period and data transfer filtering are registered along with the index value of the node to be obtained, and the OPC UA Monitored Item transmits the data for each session accordingly [30]. OPC

UA Alarm/Event distributes event data by simultaneously generating the alarm/Event internally and externally whenever that event occurs, in accordance with the attributes of the Alarm/Event defined in OPC UA Node Modeling [31]. The CPS Connect OPC UA is responsible for updating and changing all data in the OPC UA Address Space and for transferring internal alarms/events from the Information Exchange Service to CPS Node Control, CPS Logic Control, and Product Process Control inside the Factory CPS Model. The OPC CPS Node Generator automatically generates the properties and codes required for CPS Node Control, CPS Logic Control, and Product Process Control, all of which are registered in OPC UA Address Space and OPC UA Client by parsing OPC UA Modeling [32]. The OPC UA Historian records all updated information in the Address Space. The recording method can be used by selecting a relational database, a NoSQL database, an XML, a binary file, etc.

The OPC UA-based CPS architecture allows CPS to be configured with only OPC UA Modeling and continuous system operation without the need for a system redesign or program changes or downtime [33]. By embracing the international standard OPC UA Spec, numerous facilities, machinery, and robot products can be accommodated in a single protocol. Managing data in real time through Address Space increases the efficiency of horizontal data distribution and exchange operations, significantly reduces work time and costs, and solves the problems of data loss, system performance degradation, and real-time processing delay.

2.2. Industrial VR Technology

VR is one of the key technologies of digital transformation. The powerful and immersive hardware and software systems that are currently on the market allow one to visualize complex systems used in technology, nature, physics, chemistry, and anatomy in a realistic virtual environment. This allows us to fully immerse ourselves in the observed virtual environment and interact with graphic objects and systems, thus creating new possibilities for describing and communicating complex and abstract system behavior to non-experts in a concrete and understandable manner. VR technology has yet to have a breakthrough technology that can be recognized in the industry, except for large companies in the automobile and aircraft industries, and small and medium-sized enterprises are particularly facing difficulties building VR systems and applications, introducing them into business processes, and operating them economically. This is mainly due to the lack of standardized interfaces between CAD systems and VR systems and the substantial difficulty involved in implementing VR applications. At present, this data integration remains largely divided, especially when there is a need for more than just geometry to be transferred outside the software environment of CAD system providers, and the "Industrial Utilization of Virtual Reality in Product Design and Manufacturing" survey [34] suggests one of the key issues in VR. CAD data and related design data in the product design process cannot be directly sent to VR applications, so basic CAD data must be converted to VR data. For companies characterized by frequent and rapid CAD data changes in development processes, use of VR technology is currently difficult and almost unimaginable in terms of efficiency, time, and economy. To utilize VR potential in engineering and industrial environments, there is an urgent need for solutions that promote the implementation and setup of VR applications—particularly for small and medium-sized enterprises—and enable intelligent use of this technology. In addition, AR technologies, including object recognition, deep point cloud scanning, gesture manipulation, and VR have gradually been introduced in recent years, and technologies related to the fusion of VR and AR have been explored [35].

2.3. CAD Data Conversion Process for VR/AR Systems

VR refers to an artificial environment that feels like reality. It also refers to a computer graphic system technology that allows a user to immerse himself in a 3D virtual environment using interface equipment, and the core of VR is the concept of realism, which allows users to exist in a space or virtual place.

AR is a technology that superimposes a real image or a three-dimensional virtual image of a background and displays it as a single image, and this is called Augmented Reality. At present, the main method is one that involves receiving a real-time image through an image input device mounted on a mobile device, then overlapping, displaying, and providing information on a 3D object.

The advantage of VR is that it can give a direct feeling of space to a space that does not actually exist, so it is easy to grasp the situation without going directly to the site. Further, visual elements have the advantage of being able to be observed and felt more directly than other senses, so they can be accurately adapted and highly understood, and they can thus be used conveniently by users. It is also possible to prevent problems caused by construction errors in actual sites by simulating them virtually through collaboration, and it has the advantage of reducing costs such as labor costs by reducing redesign.

The business processes of most industrial companies that develop and manufacture technology systems and products are CAD-based [36–40]. CAD models are the primary targets for recording and representing technical product data over the entire product lifecycle [41]. All lifecycle processes and tasks are based on CAD models to create, process, and store design, simulation, manufacturing, production, logistics, training, and maintenance data. Many standardized SETP and IGES data formats have been defined for this purpose specifically. These standardized data formats are handled by the most common engineering tools over the entire product lifecycle, except for VR engines. Over the past years, the development and application of VR technology has mainly been driven by the gaming and entertainment industries. Graphical requirements in these areas are completely different from those in the engineering domain. For example, the data content of a game object is very superficial compared to that of an engineering object. For game objects, the surface shape is important for VR applications, but for engineering VR applications, complete metadata, including material and tolerance information for parts, is needed. This has led to common VR engines and VR data interfaces and processors that can be implemented for game design to focus on processing only minor surface data and reducing metadata. To solve this problem, it is necessary to automatically process the data transfer process from the CAD environment to the VR environment.

Numerous studies have been conducted over the past few years with the goal of addressing this CAD-VR data conversion problem and reducing the gap between CAD and VR systems [34]. The focus is on transmitting CAD data to VR using a PDM [37]. CAD-VR data transformation workflows were defined using the VRML (Virtual Reality Modeling Language) standard [42], and approaches for VR-based collaborative platforms were defined using the STEP (Standard for the Exchange of Product Model Data) and JT (Jupiter Tesla) standards [43]. However, this approach focuses on the data transformation of a particular CAD system that uses a particular CAD data format [36,44,45]. Specifically, it only considers VR applications [46] and data used in VR-Cave, and it does not apply to HMD (Head Mounted Display)-based VR solutions.

3. OPC UA-Based VR/AR Collaboration Model

This research aims to reduce design costs, which have changed often due to various problems and the demands of manufacturing products. This study provides participants with VR/AR with 3D rendering virtualization to improve design efficiency. Through data exchange and distribution through CPS servers applying OPC UA, an international standard technology, VR/AR rendering and data of 3D design files are extracted and a platform specialized for manufacturing collaboration is developed. This provides an editor that allows users to write AR content themselves by utilizing VR modeling, and it supports re-reflection of product operation and operation results through simulators in design and rapid rendering through a central virtual rendering server. From design to future maintenance, it provides collaboration functions over the entire life of the product, holds data, and enables AR inspection and maintenance using 3D rendering models to collect and analyze data that is difficult to obtain using existing research methods through VR to visualize BI (Intelligence Business).

3.1. VR/AR Collaboration Model

Figure 1 depicts the main components of the VR/AR collaboration model using CPS by dividing the VR/AR layer, Visualization layer, and CPS layer.

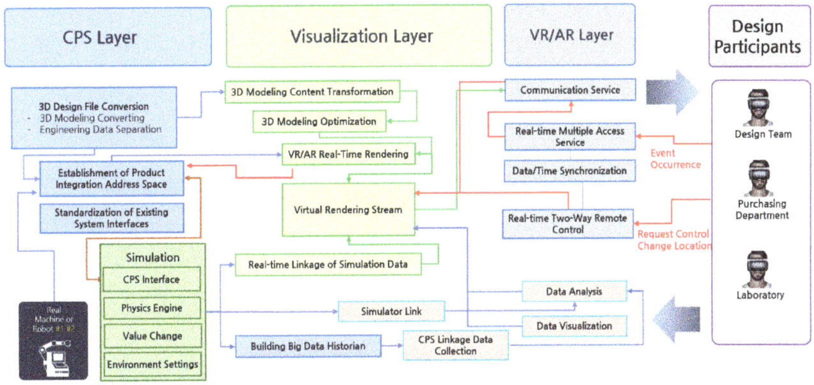

Figure 1. VR/AR collaboration model construction process.

The CPS Layer is the central source data store of this system and serves as middleware. The standard of OPC UA, an international standard technology for smart factories used for Digital Twin technology in factories, was applied here. OPC UA is not just a protocol, but it standardizes all the authentication, exchange, storage, alarms, and distribution required for data exchange. The CPS service has a real-time memory structure called Address Space, which serves the machine's information modeling and attribute values in real time. In this system, real-time data generated by a simulator or an actual machine is delivered to a VR/AR rendering service. It also exchanges data with standard specifications on all existing systems that require data interworking and exchange. In big data analysis, all information data generated in this cooperative design is stored and used to generate big data. The simulation, which is characteristic of this service, calculates real-time data, such as the actual machine operating based on the physical engine, stores and analyzes it, and lastly returns the analysis results to the simulator to improve the accuracy and virtuality of the simulation. Further, by visualizing and serving all data, participating designers provide services that enhance design capabilities through visualized data analysis. Moreover, by providing a management module, it offers systematic services such as all of the user management, security management, authentication management, access management, and history management services necessary to service this 3D collaboration design system.

The visualization layer optimizes 3D modeling to convert models extracted from 3D design files into VR/AR, converts them into stream images from virtual rendering servers, and simultaneously transmits simulator values to VR/AR in real-time interworking with CPS. It synchronizes the time and data of simultaneous multi-access users; detects the location, control, and event occurrence of design participants; and communicates in both directions with the VR/AR rendering server in real time to service the video stream without delay. Further, the voices of participating designers are synchronized in real time through VOIP.

The VR/AR Layer based on 3D design modeling and physical engine is used for product development, and the contents necessary for product inspection are developed and applied to deliver inspection results to CPS systems. The CPS system delivers it to the existing system required for existing inspection and quality. Based on the 3D design modeling and physical engine used for product development, it provides contents and services that allow customers to perform maintenance and defect repair through an AR service that allows them to watch video transmitted through AR goggles and Live Cam.

Figure 2 shows the system configuration diagram. It is a structure in which several engines and servers such as real-time rendering and big data analysis engines are combined with existing CPS servers and VR/AR collaboration servers. Digital Twin is not essential for untact manufacturing VR/AR collaborative AI systems, but the most effective configuration is to build real-time Digital Twins for factory machines and robots. In the OPC UA-based CPS server and VR/AR collaboration system, the real-time data of a Digital Twin is an important system for data analysis and prediction. Collaborators and fielders are connected to each other with videos through communication, and the fielders' images are shared with collaborators. If a Digital Twin is used, VR models and CPS data are overlaid over real-time images with an AR function.

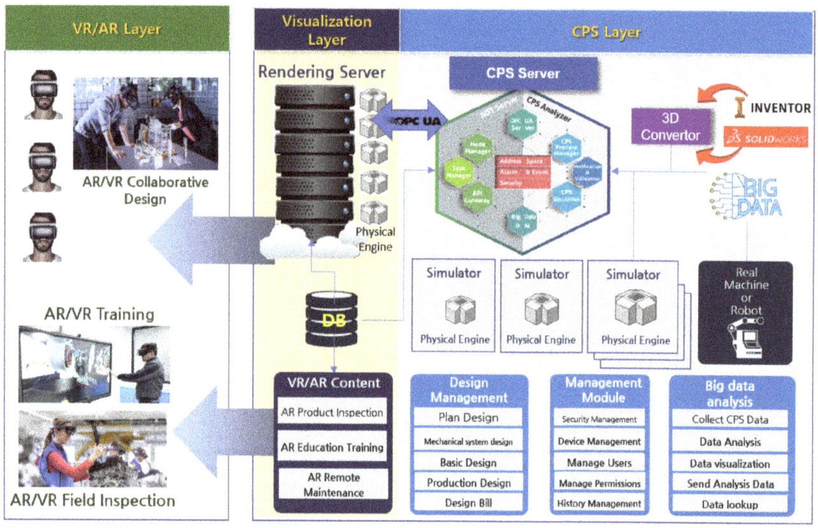

Figure 2. VR/AR collaborative model.

3.2. VR Engineering Process

The system maps the real-time values of the existing CPS server Address Space to the AR Object in the manner shown in Figure 3. The existing collaboration system has focused on object recognition and VR matching using VR/AR technology alone, and it only raises the actual value of AR through some data interfaces. Overseas, only PTC products receive actual data through Kepware's CPS Server. Figure 3 shows that rendering is possible through 3D design files by exchanging data through the CPS server to which OPC UA, an international standard technology, is applied. CPS maps from OPC UA Server to Address Space to exchange equipment information in real time without delay. As a central address space, the CPS system is built from multiple OPC UA Servers in the field as one huge central address space, and with one connection, the entire factory information can be mapped to the product inspection system in real time without delay.

Figure 3. OPC UA information space of VR engineering.

The existing CPS played the role of collecting data from all equipment in the plant. In addition to the existing role, the proposed CPS plays a role of linking data to the 3D model through the OPC UA protocol. CPS performs the role of middleware including object information and value of the 3D model object delivered to the rendering server by applying technology based on the OPC UA Framework, which is an international table. The data stored in the CPS is utilized not only in the 3D model but also in the rendering server. In addition to communication, with computational control as the core concept, the physical world objects that coexist with humans and objects such as sensors, actuators, and embedded systems are fused into the cyber world, and the data communicated with the simulator are calculated and the results are redistributed. Using the OPC UA specification of international industrial standards, the system is constructed as depicted in Figure 4, with the smart factory optimized and both OPC UA modeling and CPS Node modeling connected with technological convergence with CPS. It converts CAD 3D binary files such as Solidworks into OPC UA Address Space data. If Edge is needed, data are collected from simulators, real machines, robots, etc., and data are delivered to CPS.

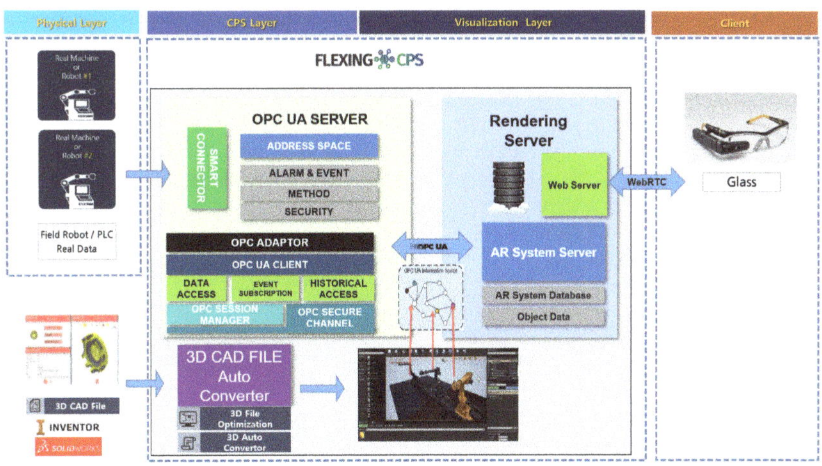

Figure 4. System architecture of VR engineering.

The design file is treated to the modeling and separation conversion of the design file, and the process proceeds as depicted in Figure 5. The CAD Data Import Interface provides an interface that allows designers to insert CAD data into applications and reads CAD Data Interface to VR Parser and converts it into 3D objects. For visualization, if a designer places 3D objects to collaborate and then performs virtual collaboration space work, CAD models are input to create virtual space objects through a 3D Converter, and designers can freely assemble and decompose objects in virtual space. It is also possible to easily create a virtual collaboration space so that it can be displayed to collaborators.

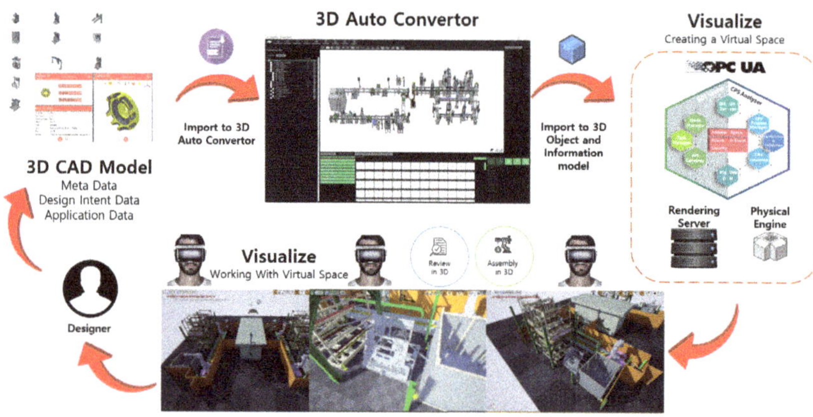

Figure 5. Model separation from 3D design files.

Further, as shown in Figure 6, the contents of the BOM are referred to the relevant Node in the Address Space of the CPS server through extraction of the BOM from CAD or migration from legacy DB to enhance the functionality of the physical engine based on the BOM. Simulator advancements can be achieved with physical engine and Address Space. For the basic process of design management, all changes to the design for each stage are recorded in the DB, and then the rendering server lists the version history by referring to the DB. A blueprint for each stage of each VR/AR version can be recalled. It is possible to manage the version of each design level and real-time management of changes to each VR/R-based design. Products produced through VR collaboration design can be inspected using AR technology. When selecting a part list, the presence or absence of the relevant part, specifications, etc., are displayed to check for abnormalities.

Physical engines can simulate behavior without manufacturing products by applying optimized physical engines of certain real-world devices. If important design modifications—such as those involving parts, materials, and appearances—are made by sending data to CPS in real-time simulation, the appropriateness or stability of the change can be verified.

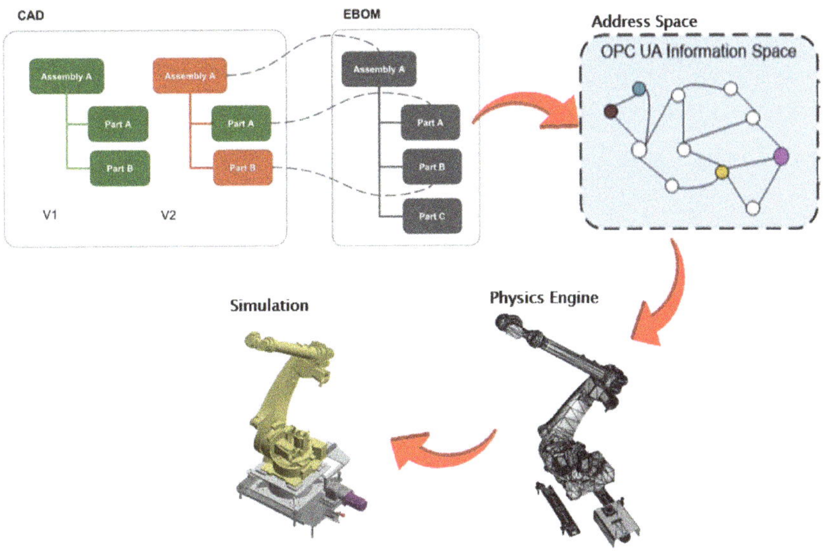

Figure 6. Simulation of CPS interworking.

4. Implementation and Results

4.1. System Configuration

In this paper, OPC UA Foundation's SDK and C language were used. The basic functions of CPS server and Smart Connect were implemented as shown in Table 1 using the source of the CPS server part of Flexing Server [29], which developed the CPS server. Smart Connect is part of the CPS server, and each distributed system is implemented as a CPS server and OPC UA server protocol that can have a centralized address space.

Table 1. Implementation of CPS Server.

Classification	CPS Server
Operating system	Windows 10 Pro
Specification	AMD Ryzen9 3940X 16-Core, Mem 128 G, SSD 256 GB
Purpose	Receive BOM data from VR Parser and configure Address Space on CPS Server in real time

The Smart Connector needs to generate NodesetXML to be applied to the CPS Server based on data received from the Socket (.iam file to open node). It reads element ID information from an XML file, collects relevant BOM information, and generates objects from the model. CAD Socket Class—serves as a socket server that receives data from the VR Parser. Figure 7 shows the details of the process. When VR Parser converts the .iam file parsed by the Socket Client into the Json format and sends the data to the CAD Socket, tcpListenerThread sends the received data to the NodeSetXmlGenerator after checking the received data. TCP Listener was implemented with Newtonsoft.json NuGet package with C-based NetCore 3.1 version, and NodeSetXML file generation was studied according to the OPC UA standard.

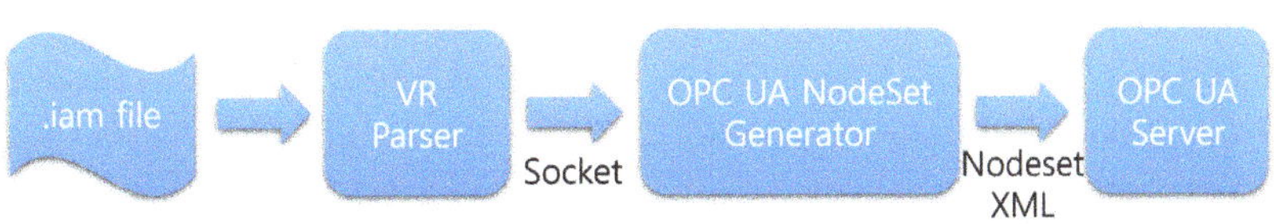

Figure 7. Smart connector structure.

It is possible to change the OPC Node in real time, collect process device data, and change the .iam file data. Finally, we study custom scripts and VR Parsers to parse the corresponding Json file and extract relevant information before re-generating the model by implementing the C class in the Unity SDK. Table 2 details the implementation environment.

Table 2. VR Parser Classification.

Classification	VR Parser
Operating system	Windows 10 Pro
Specification	AMD Ryzen 3 3100 4-Core, Mem 32 GB, SSD 256 GB
Purpose	Purpose of extracting BOM data from CAD files

To parse the equipment or CAD File data into OPC UA, VR Parser is UI-verifiable and visualizes 3D models. Users can browse models; search among engineering models, specifications, and configurations in a VR environment; check them through the UI as shown in Figure 8; and change the OPC Node in real-time and process the device data.

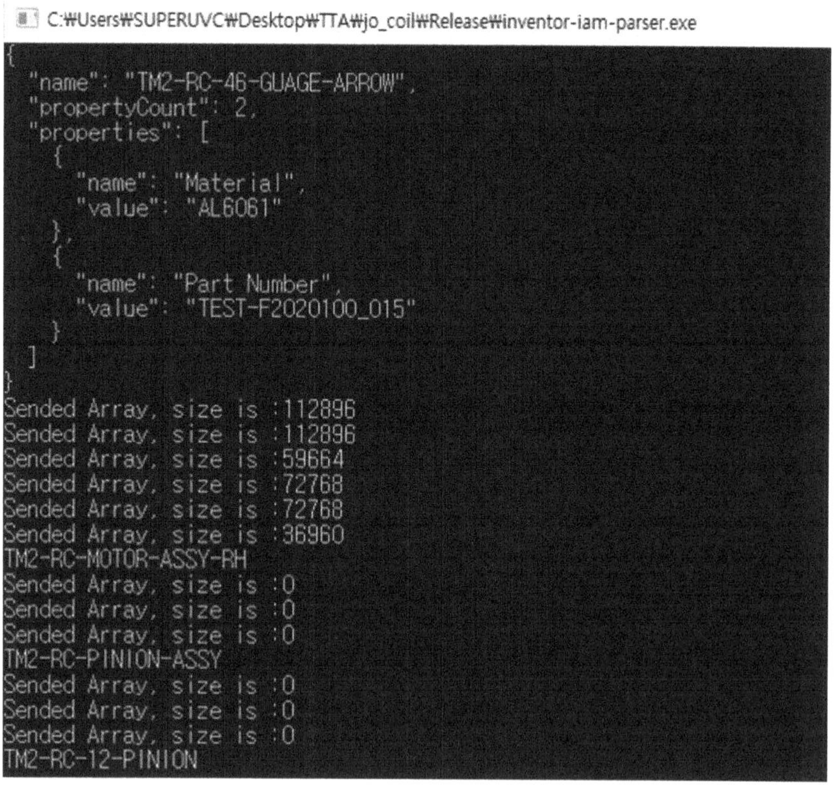

Figure 8. Smart connector and VR parser.

Moreover, high-spec servers are required for 3D model rendering, and the server configuration environment is as described in Table 3. There are performance limitations to performing rendering of 3D models on user devices, and 3D models are generated and streamed to devices through predefined 3D modeling information. This allows for the designed product to be simulated by utilizing the AR function provided by Vuforia Engine, a physical engine, as if it were operating in a real environment. The Vuforia EnginE includes image recognition, QR recognition, and AR object viewer functions, which are key AR technologies, and it is optimized for Unity engines.

Table 3. Implementation of Rendering Server.

Classification	Rendering Server
Operating system	Windows 10 Pro
Specification	AMD Ryzen 9 3900X, Mem 64GB, SSD 4TB
Purpose	3D model rendering and streaming

4.2. Implementation

Company T, located in Pyeongtaek, Korea, manufactures assembly line automation facilities and specialized equipment (piston insertion robots, valve finishing devices). In fact, there are several facilities and various robots available for use, but in this study, as shown in Figure 9, one 7-axis robot arm was used to model a physical factory environment.

Figure 9. Automated facilities of Company T's.

Data collection was required for designing the 3D VR production of the robot arm, and a 2D information product drawing, automated mechanical equipment 3D CAD information, and photo base mapping were conducted. Modeling data, including layers of components of each automated mechanical device and layers of each component, were collected. In the drawing of the acquired automation machine, the storage extension conversion was carried out through an inverter, and the layer was parted in the design program to form a part-oriented layer. Upon completion of the extension conversion, it was loaded into a real-time engine after being converted to general data such as IGS, OBJ, and 3DS. Further, by applying the Surface Name, a preparation step for material and baking was prepared. After defining the material of the robot arm based on an image, VR content sets on the same scale as actual equipment are configured, and graphic performances such as unnecessary polygons and high-calculated material improvement are optimized. CAD design files of real-world physical automation machines are converted to 3D objects. CAD design files exist in various formats, but they are basically conducted based on commonly used inventory. As shown in Figure 10, the 3D CAD design file for the automation machine includes data on the BOM and the 3D shape.

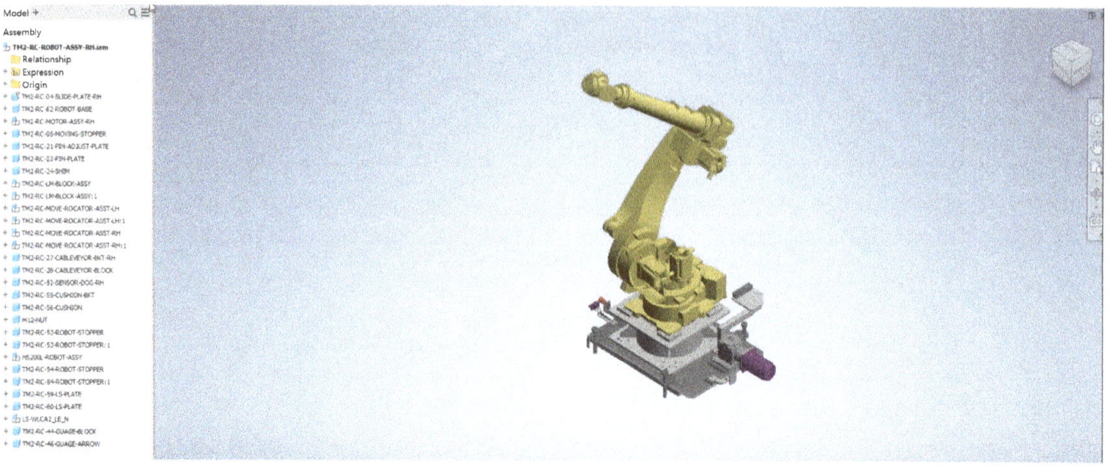

Figure 10. .iam Design File with Inventor.

The implemented VR engineering process was implemented in a total of four projects as shown in Figure 11. FBXExport wraps it and delivers it in DLL format to use FBX-related API in Unity3D project. CAD Convertor creates and displays a mesh on the screen based on DLLs such as CAD and FBX, and delivers it to the cloud server. VR Application uses libaray such as AR Core, Intelloid, and WebRTC to build a platform in AR environment and deliver it to users. The VR Projects Parser uses VB.net to transmit a program written through an actual Inventor to CAD Convertor through socket communication. The reason why it is written in VB.net is that the API provided by Inventor is provided through VB.net, and it is transmitted as a loopback for sharing between processes.

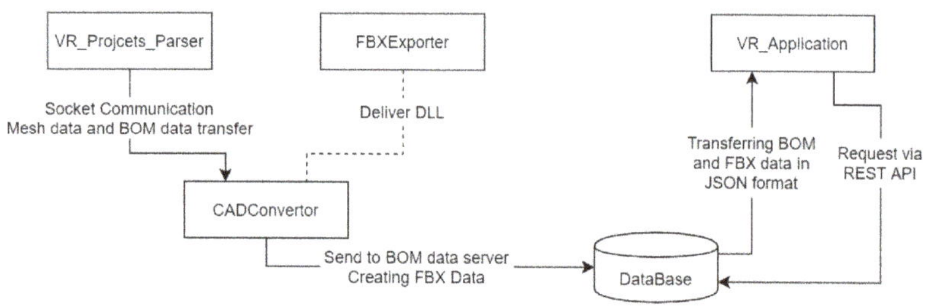

Figure 11. VR engineering project configuration.

A CAD Converter is a necessary tool for converting these 3D CAD design files into 3D objects in a VR environment. The CAD Converter role is performed by the VR Parser developed in this study. The .iam file should be loaded as shown in Figure 12 to read the information in the CAD file. When all are loaded, BOM and 3D objects are created.

By implementing the BOM data and metadata in CAD design files as the Address Space of CPS, a VR Parser can transmit BOM data to an OPC UA-based CPS server. When the transmission is completed, the OPC UA client program can access the CPS server to verify that the structure of the BOM data is created in the Address Space. Further, when accessing the CPS server, it can be confirmed that the structure of the BOM data is generated as shown in the figure. As shown in Figure 13, the BOM structure of the CAD file was typically generated as the Address Space in CPS, and the entire part information and the operable part information were implemented in the Address Space.

Figure 12. Creating a 3D object in a VR Parser.

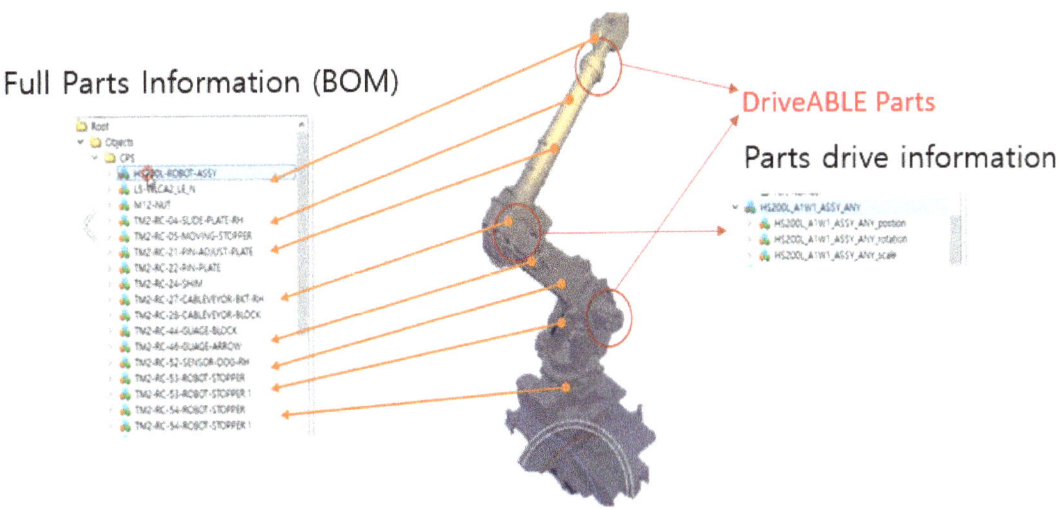

Figure 13. Address Space implementation of 3D objects.

To optimize the FBX model file capacity, the CAD converter can extract 3D objects as FBX models. When EXPORT is performed, it is automatically extracted as an FBX model. The capacity is optimized for extraction and default is 50%. In Figure 14, it can be observed that the FBX 3D model is made normally.

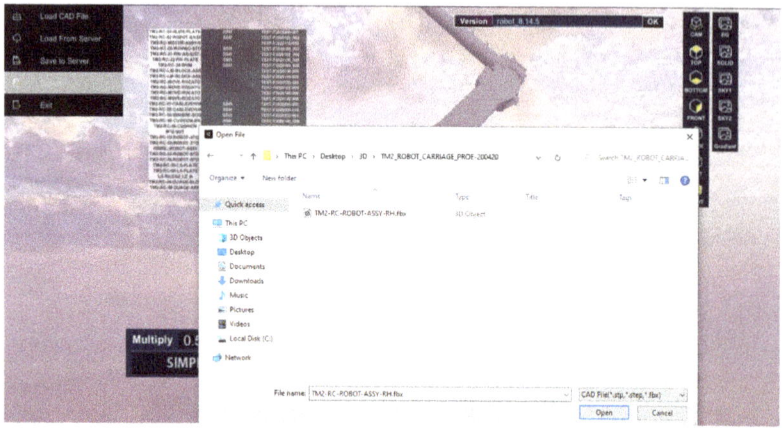

Figure 14. FBX model file capacity optimization.

In the CAD file, the robot arm was simulated by adding a physical engine to confirm the physical force of the model converted into a 3D object. For the simulation, a physical engine that calculates the physical force for the load through two robot arm colliders has been added. The collider is a component for collision detection and can add physical functions with scripts. Unity Physics and Havok physics were added to each automated machine, as shown in Figure 15.

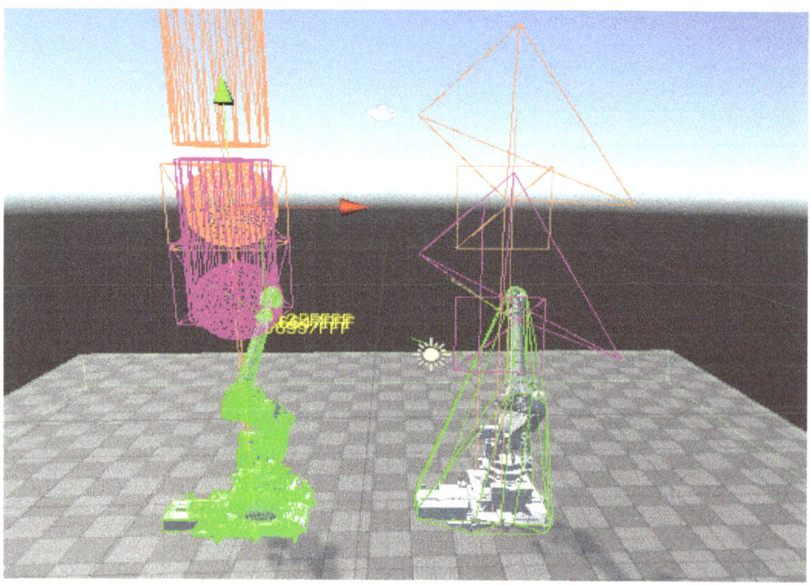

Figure 15. Physical engine simulation of unity physics and Havok physics.

After importing the designed .iam file into the VR Parser, the design problem is checked through load simulation, and the stress–strain rates of specific parts due to the action of vertical stress resulting from the load in the size and shape, and the structure of the designed file is checked as well. It expresses the collider of each other object and expresses the movable range of the object. When a condition exceeding the yield point on the diagram occurs, the problematic part is expressed through the VR Parser, and the user can check the wrongly designed part through this simulation.

4.3. Results

BOM data is extracted from CAD files created by inventory through VR Parser, converted to Json, and delivered to the CPS Server. At this time, the BOM data of the CAD file modified in the inventory is well transmitted to the CPS server, and the BOM data from the VR Parser checks the accuracy of the configuration of the address space of the CPS Server in real time. Further, the FBX file extraction function for external program interworking compresses the data in the CAD file to measure the capacity optimization (capacity reduction) value. The test environment was constructed as depicted in Figure 16.

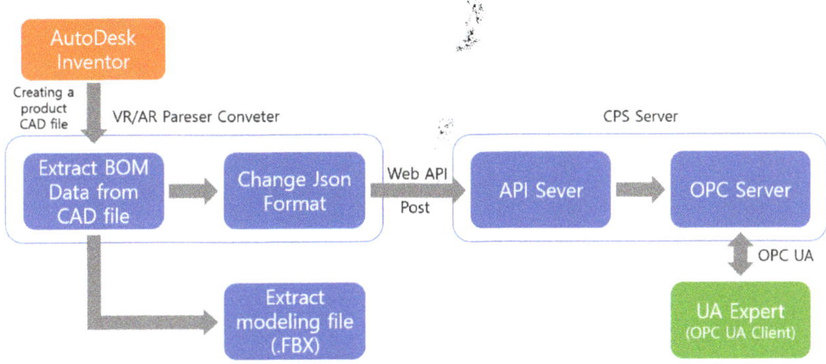

Figure 16. Experiment diagram.

The total number of CAD files to be converted is 5, and the total number of data objects is 77. After running the inventory, the CAD file should be opened to check the CAD BOM data. The VR Parser should then be run to load CAD files. As shown in Figure 17, UA Expert is a representative client that supports the OPC UA protocol with software from a company called Unified Automation. The BOM data of the inventory, the number of nodes converted to OPC UA, Node name, Node tree structure, and Node properties can be verified by running the most commonly used internationally test client that supports OPC UA functions such as data access, alarms, history, and UA method calls.

This procedure is executed by the number of CAD files, and the accuracy converted to the expression of (1) is calculated.

$$\text{Conversion accuracy} = \frac{\text{Number of Objects Converted to Address Space}}{\text{CAD File Total Objects}} \times 100 \quad (1)$$

As a result of the data conversion test, the accuracy of address space conversion in BOM was 100%, as shown in Table 4, thus indicating the reliability of BOM product inspection. This shows that real and bidirectional configuration results are visualized in real time in collaborative design and that interaction between users is possible.

Table 4. Translation accuracy of Address Space in BOM.

Number	CAD File Total Objects	Number of Objects Converted to Address Space	Results
1	29	29	100%
2	28	28	
3	3	3	
4	14	14	
5	3	3	
Sum	77	77	-

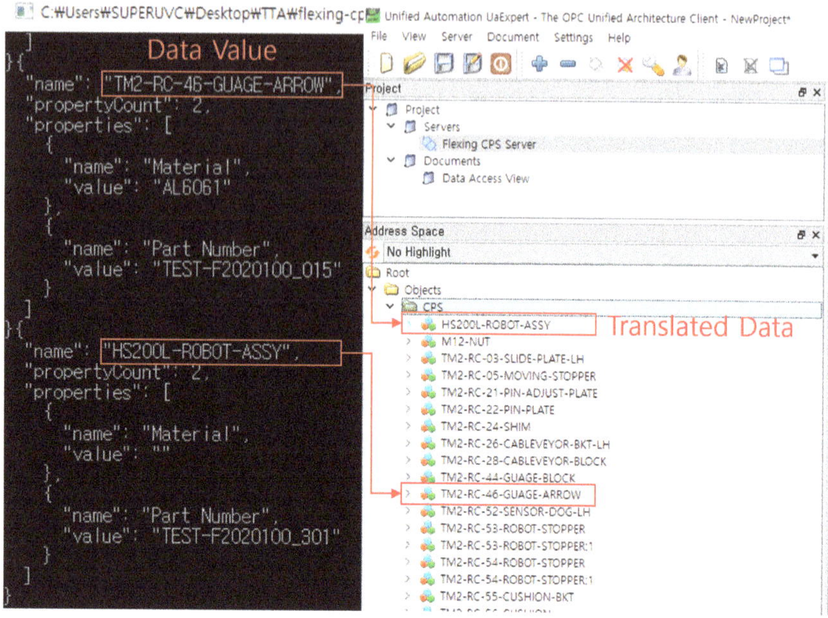

Figure 17. Data conversion of CPS server.

For CAD data, it should be converted to a data format that can be processed by the VR engine. The Unity engine can only handle two types of data: FBX and OBJ. Almost all CAD systems have data preprocessing for generating FBX and OBJ data, but the generated FBX and OBJ do not have the quality to be used directly as VR data. The FBX and OBJ data formats visualize the 3D shape of the object using polygon elements, and the number of polygons used is critical to the 3D graphic quality of the visualization object. The higher the number of polygons, the more realistic is the 3D object and the 3D graphics quality is higher. However, having a large number of polygons increases data volume and leads to performance problems in VR applications. Since this study requires optimization to ensure a balance between graphic data quality and VR application performance, the optimization value is calculated by calculating the capacity size of the extracted FBX file and the existing CAD file using Equation (2) by importing and exporting CAD files from the VR Parser. Figure 18 shows the optimization results.

$$\text{Optimization Value} = \frac{\text{File Size After Optimization}}{\text{CAD File Size}} \times 100 \tag{2}$$

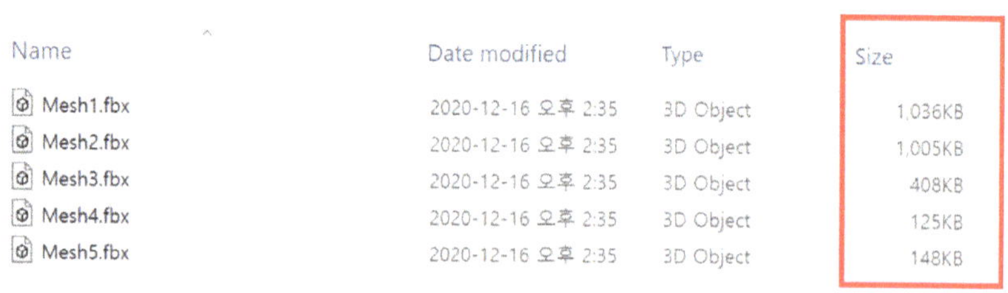

Figure 18. Data optimization size.

The results of the optimization test comparing the size of the CAD design file, as presented in Table 5, show an average optimization value of 37.4%. We confirmed an improvement of 12.6% over the optimization default of 50%, which can be expected to improve the performance of the same VR application and the quality of 3D graphics in the IT infrastructure.

Table 5. CAD Design File Optimization (Reduction).

Number	CAD File Size (KB)	File Size after Optimization (KB)	Results
1	2004	1036	51.7%
2	2214	1005	47.3%
3	1103	408	37.0%
4	790	125	15.8%
5	421	148	35.2%
Sum	-	-	37.4%

To measure the graphic quality through optimization, the influence and quality of rendering were tested. It should be verified that real-time VR rendering content can be provided by importing CAD files from VR Parser and then exporting them to convert the extracted FBX files into 3D formats. The test was conducted as shown in Figure 19, while focusing on the frame rendering speed and the resolution of the 3D content.

Based on the 3D format-converted 2D files, 3.5.99 frames per second was used to measure the number of frames per second of VR image rendered through CAD converter, and the number of frames per second was checked after moving to the left, right, and top of VR images. Further, the resolution of the rendered VR image was confirmed.

Rendering to CAD files in a VR system can result in high transmission load, slow rendering speed, and poor 3D control due to the generation of large amounts of polygons, so it should be verified that it is efficient for VR systems. In existing studies [47] aiming to simplify CAD files, it was measured at 75 frames when optimization was performed at 50%. Converting CAD data into FBX, in most cases, decomposing a single object into a sub-object, damaging the object structure, which was feared to degrade performance, but the frame measurement results of the model provided as VR Parser are shown in Table 6. Full view from parent of 1, the overall foreground was 86 frames per second, Full foreground from left end 2, the preliminary foreground was 88 frames per second, and Full foreground from right end of 3, the overall foreground was 83 frames, showing an average result of 85.6 frames. It also provides a high resolution of 2K for resolution.

Figure 19. Frame rendering speed.

Table 6. Frame count and resolution per second.

Classification	Measurement Location	Frames per Second	Resolution
1	Full view from parent	86	2K (2048 × 1080)
2	Full foreground from left end	88	2K (2048 × 1080)
3	Full foreground from right end	83	2K (2048 × 1080)
Average Number of Frames per Second	-	85.6	2K (2048 × 1080)

5. Conclusions

In this paper, we proposed a VR/AR collaboration model and VR engineering process using CPS applied with international standard technology. Using VR/AR rendering of product design 3D drawings in the manufacturing industry, a collaborative model was presented to allow participants to participate in design at the same time, and an application was developed to check the virtualization model with data collected through OPC UA-based CPS. The programs used in machine design and product design currently use 100% foreign software. However, the current foreign software is specialized in design, but collaborative design does not provide functionality, so the actual design process of reviewing and reflecting opinions, and then simulating them, can be replaced with domestic programs to cover the growth of foreign design programs. Further, domestic manufacturers will find problems later by starting products or main production after design is completed, and it will take a long time and high cost due to the method of re-reflecting and re-production. However, sufficient simulation and design review using the product inspection system can dramatically reduce this time and cost, thus contributing to the competitiveness of manufacturing companies and actively using it for smart factories in the manufacturing industry.

In this study, we developed a converter that converts design CAD files and BOMs into graphic models and CPS models, and that successfully loads BOM data of 3D CAD files from the VR Parser and sends BOM data to CPS server to construct Address Space. We also successfully extracted 3D objects converted from CAD files with capacity-optimized FBX. A CPS server, a middleware operating under the OPC UA protocol, was established to link OPC UA–BOM Address Space with OPC UA–BOM-based real-time values. In addition, when converting to 3D VR content using CAD drawings, it was implemented in the same way as the measured equipment, and errors were determined in CPS through the calculated load value of the load in the physical engine as the value generated by the simulator.

Here, time and competency were both insufficient to implement all of the proposed VR/AR collaboration models. In the future, we will develop a VR/AR collaborative design client to create virtual characters for multi-user access, VR and motion capture wearers, and build a WebRTC-based backend (server and service) to stream high-end content and voice communication between users using VoIP.

Author Contributions: Conceptualization, J.K.; methodology, J.K.; validation, J.K. and J.J.; formal analysis J.K.; investigation, J.K.; resources, J.K.; data curation, J.K.; writing—original draft preparation, J.K.; writing—review and editing, J.K.; supervision, J.J.; project administration, J.K.; funding acquisition, J.J. All authors have read and agreed to the published version of the manuscript.

Funding: This research was supported by the MSIT (Ministry of Science and ICT), Korea, under the ITRC (Information Technology Research Center) support program (IITP-2022-2018-0-01417) supervised by the IITP (Institute for Information & Communications Technology Planning & Evaluation). Also, this work was supported by the National Research Foundation of Korea (NRF) grant funded by the Korea government (MSIT) (No. 2021R1F1A1060054).

Institutional Review Board Statement: Not applicable.

Informed Consent Statement: Not applicable.

Data Availability Statement: Not applicable.

Acknowledgments: This research was supported by the MSIT (Ministry of Science and ICT), Korea, under the ICT Creative Consilience Program (IITP-2022-2020-0-01821) and the ITRC (Information Technology Research Center) support program (IITP-2022-2018-0-01417) supervised by the IITP (Institute for Information & communications Technology Planning & Evaluation).

Conflicts of Interest: The authors declare no conflict of interest.

References

1. García, M.V.; Pérez, F.; Calvo, I.; Moran, G. Developing CPPS within IEC-61499 based on low cost devices. In Proceedings of the 2015 IEEE World Conference on Factory Communication Systems (WFCS), Palma de Mallorca, Spain, 27–29 May 2015; pp. 1–4.
2. Qi, Q.; Tao, F. Digital twin and big data towards smart manufacturing and industry 4.0: 360 degree comparison. *IEEE Access* **2018**, *6*, 3585–3593. [CrossRef]
3. Shao, G.; Helu, M. Framework for a digital twin in manufacturing: Scope and requirements. *Manuf. Lett.* **2020**, *24*, 105–107. [CrossRef] [PubMed]
4. Liu, M.; Fang, S.; Dong, H.; Xu, C. Review of digital twin about concepts, technologies, and industrial applications. *J. Manuf. Syst.* **2021**, *58*, 346–361. [CrossRef]
5. Boschert, S.; Rosen, R. Digital twin—the simulation aspect. In *Mechatronic Futures: Challenges and Solutions for Mechatronic Systems and Their Designers*; Hehenberger, P., Bradley, D., Eds.; Springer International Publishing: Cham, Switzerland, 2016; pp. 59–74.
6. Liu, Q.; Zhang, H.; Leng, J.; Chen, X. Digital twin-driven rapid individualised designing of automated flow-shop manufacturing system. *Int. J. Prod. Res.* **2019**, *57*, 3903–3919. [CrossRef]
7. Kuts, V.; Cherezova, N.; Sarkans, M.; Otto, T. Digital Twin: Industrial robot kinematic model integration to the virtual reality environment. *J. Mach. Eng.* **2020**, *20*, 53–64. [CrossRef]
8. Guerra-Zubiaga, D.; Kuts, V.; Mahmood, K.; Bondar, A.; Nasajpour-Esfahani, N.; Otto, T. An approach to develop a digital twin for industry 4.0 systems: Manufacturing automation case studies. *Int. J. Comput. Integr. Manuf.* **2021**, *34*, 933–949. [CrossRef]
9. Adwernat, S.; Wolf, M.; Gerhard, D. Optimizing the design review process for cyber-physical systems using virtual reality. *Procedia CIRP* **2020**, *91*, 710–715. [CrossRef]
10. Abramovici, M.; Gebus, P.; Savarino, P. *Engineering Smarter Produkte und Services Plattform Industrie 4.0 STUDIE*; Acatech–Deutsche Akademie der Technikwissenschaften (Hrsg): München, Germany, 2018.
11. Neugebauer, R.; Denkena, B.; Wegener, K. Mechatronic systems for machine tools. *CIRP Ann.* **2007**, *56*, 657–686. [CrossRef]
12. Baheti, R.; Gill, H. Cyber-physical systems. *Impact Control Technol.* **2011**, *12*, 161–166.
13. Spec, D. *91345: Referenzarchitekturmodell Industrie 4.0 (RAMI4. 0)*; Deutsches Institut für Normung (DIN) eV: Berlin, Germany, 2016.
14. Löwen, U.; Braune, A.; Diesner, M.; Hüttemann, G.; Klein, M.; Thron, M.; Manger, T.; Okon, M.; Kochseder, R. *Exemplification of the Industrie 4.0 Application Scenario Value-Based Service following IIRA Structure.*; Federal Ministry for Economic Affairs and Energy (BMWi): Berlin, Germany, 2017.
15. Di Martino, B.; Li, K.C.; Yang, L.T.; Esposito, A. Internet of Everything. *Internet Things (Technol. Commun. Comput.)* **2018**, *103*, 130.
16. Biesinger, F.; Meike, D.; Kraß, B.; Weyrich, M. A digital twin for production planning based on cyber-physical systems: A Case Study for a Cyber-Physical System-Based Creation of a Digital Twin. *Procedia CIRP* **2019**, *79*, 355–360. [CrossRef]
17. Brandmeier, M.; Bogner, E.; Brossog, M.; Franke, J. Product design improvement through knowledge feedback of cyber-physical systems. *Procedia CIRP* **2016**, *50*, 186–191. [CrossRef]
18. Pérez, F.; Irisarri, E.; Orive, D.; Marcos, M.; Estevez, E. A CPPS Architecture approach for Industry 4.0. In Proceedings of the 2015 IEEE 20th Conference on Emerging Technologies & Factory Automation (ETFA), Luxembourg, 8–11 September 2015; pp. 1–4.
19. Kang, W.; Kapitanova, K.; Son, S.H. RDDS: A real-time data distribution service for cyber-physical systems. *IEEE Trans. Ind. Inform.* **2012**, *8*, 393–405. [CrossRef]
20. Rajhans, A.; Bhave, A.; Ruchkin, I.; Krogh, B.H.; Garlan, D.; Platzer, A.; Schmerl, B. Supporting heterogeneity in cyber-physical systems architectures. *IEEE Trans. Autom. Control* **2014**, *59*, 3178–3193. [CrossRef]
21. Vicaire, P.A.; Hoque, E.; Xie, Z.; Stankovic, J.A. Bundle: A group-based programming abstraction for cyber-physical systems. *IEEE Trans. Ind. Inform.* **2011**, *8*, 379–392. [CrossRef]
22. Stojmenovic, I. Machine-to-machine communications with in-network data aggregation, processing, and actuation for large-scale cyber-physical systems. *IEEE Internet Things J.* **2014**, *1*, 122–128. [CrossRef]
23. García, M.V.; Pérez, F.; Calvo, I.; Morán, G. Building industrial CPS with the IEC 61499 standard on low-cost hardware platforms. In Proceedings of the 2014 IEEE Emerging Technology and Factory Automation (ETFA), Barcelona, Spain, 16–19 September 2014; pp. 1–4.
24. Shafto, M.; Conroy, M.; Doyle, R.; Glaessgen, E.; Kemp, C.; LeMoigne, J.; Wang, L. Modeling, simulation, information technology & processing roadmap. *Natl. Aeronaut. Space Adm.* **2012**, *32*, 1–38.
25. Rosen, R.; Von Wichert, G.; Lo, G.; Bettenhausen, K.D. About the importance of autonomy and digital twins for the future of manufacturing. *IFAC-Papersonline* **2015**, *48*, 567–572. [CrossRef]

26. Tabaa, M.; Chouri, B.; Saadaoui, S.; Alami, K. Industrial communication based on modbus and node-RED. *Procedia Comput. Sci.* **2018**, *130*, 583–588. [CrossRef]
27. Harwood, A.; Wenisch, P.; Revell, A. A real-time modelling and simulation platform for virtual engineering design and analysis. In Proceedings of the 6th European Conference on Computational Mechanics (ECCM 6) and 7th European Conference on Computational Fluid Dynamics (ECFD 7), Glasgow, UK, 11–15 June 2018; p. 504.
28. International Electrotechnical Commission. *OPC Unified Architecture—Part 1: Overview and Concepts*; IEC62541; International Electrotechnical Commission: Geneve, Switzerland, 2010.
29. Jo, G.; Jang, S.H.; Jeong, J. A Study on CPPS Architecture integrated with Centralized OPC UA Server. *J. Inst. Internet Broadcast. Commun.* **2019**, *19*, 73–82.
30. Fernbach, A.; Kastner, W.; Mätzler, S.; Wollschlaeger, M. An OPC UA information model for cross-domain vertical integration in automation systems. In Proceedings of the 2014 IEEE Emerging Technology and Factory Automation (ETFA), Barcelona, Spain, 16–19 September 2014; pp. 1–8.
31. Palm, F.; Grüner, S.; Pfrommer, J.; Graube, M.; Urbas, L. Open source as enabler for OPC UA in industrial automation. In Proceedings of the 2015 IEEE 20th Conference on Emerging Technologies & Factory Automation (ETFA), Luxembourg, 8–11 September 2015; pp. 1–6.
32. Lu, H.; Zhifeng, Y. Research on key technology of the address space for OPC UA Server. In Proceedings of the 2010 IEEE 2nd International Conference on Advanced Computer Control, Shenyang, China, 27–29 March 2010; Volume 3, pp. 278–281.
33. Rohjans, S.; Piech, K.; Lehnhoff, S. UML-based modeling of OPC UA address spaces for power systems. In Proceedings of the 2013 IEEE International Workshop on Inteligent Energy Systems (IWIES), Vienna, Austria, 14 November 2013; pp. 209–214.
34. Berg, L.P.; Vance, J.M. Industry use of virtual reality in product design and manufacturing: A survey. *Virtual Real.* **2017**, *21*, 1–17. [CrossRef]
35. Wang, B.; Xie, J.; Wang, X.; Liu, S.; Liu, Y. A New Method for Measuring the Attitude and Straightness of Hydraulic Support Groups Based on Point Clouds. *Arab. J. Sci. Eng.* **2021**, *46*, 11739–11757. [CrossRef]
36. Jeschke, S.; Kobbelt, L.; Dröge, A. *Exploring Virtuality*; Springer: Berlin/Heidelberg, Germany, 2014.
37. Mihelj, M.; Novak, D.; Beguš, S. *Virtual Reality Technology and Applications*; Springer: Berlin/Heidelberg, Germany, 2014.
38. Liu, D.; Dede, C.; Huang, R.; Richards, J. *Virtual, Augmented, and Mixed Realities in Education*; Springer: Berlin/Heidelberg, Germany, 2017.
39. The Diverse Potential of VR & AR Applications. Available online: https://www.statista.com/chart/4602/virtual-and-augmented-reality-software-revenue/ (accessed on 6 April 2016).
40. Virtual Reality Ist Vor Allem Beim Gaming Attraktiv. Available online: https://de.statista.com/infografik/7430/virtual-reality-ist-vor-allem-beim-gaming-attraktiv/ (accessed on 5 January 2017).
41. Bellalouna, F. VR-Based Design Process of Industrial Products. In Proceedings of the International Conference on Competitive Manufacturing (COMA'19), Stellenbosch, South Africa, 30 January–1 February 2019; pp. 239–245.
42. Lorenz, M.; Spranger, M.; Riedel, T.; Pürzel, F.; Wittstock, V.; Klimant, P. CAD to VR—A Methodology for the Automated Conversion of Kinematic CAD Models to Virtual Reality. *Procedia CIRP* **2016**, *41*, 358–363. [CrossRef]
43. Stelzer, R.; Steger, W.; Petermann, D. The VR Session Manager: A Tool to Co-Ordinate a Collaborative Product Development Process in a Virtual Environment. In Proceedings of the International Design Engineering Technical Conferences and Computers and Information in Engineering Conference. American Society of Mechanical Engineers, Chicago, IL, USA, 12–15 August 2012; Volume 45011, pp. 1517–1525.
44. Kim, S.; Weissmann, D. Middleware-based integration of multiple CAD and PDM systems into virtual reality environment. *Comput.-Aided Des. Appl.* **2006**, *3*, 547–556. [CrossRef]
45. Raposo, A.; Corseuil, E.T.; Wagner, G.N.; dos Santos, I.H.; Gattass, M. Towards the use of cad models in VR applications. In Proceedings of the 2006 ACM international conference on Virtual reality continuum and its applications, Hong Kong, China, 14–17 June 2006; pp. 67–74.
46. Coburn, J.Q.; Freeman, I.; Salmon, J.L. A review of the capabilities of current low-cost virtual reality technology and its potential to enhance the design process. *J. Comput. Inf. Sci. Eng.* **2017**, *17*, 031013. [CrossRef]
47. Tang, Y.; Gu, H. CAD model's Simplification and Conversion for Virtual Reality. In Proceedings of the 2010 IEEE Third International Conference on Information and Computing, Wuxi, China, 4–6 June 2010; Volume 4, pp. 265–268.

applied sciences

MDPI

Article

Performance Analysis of a Repairable Production Line Using a Hybrid Dependability Queueing Model Based on Monte Carlo Simulation

Ferdinando Chiacchio *, Ludovica Oliveri, Soheyl Moheb Khodayee and Diego D'Urso

Department of Electrical, Electronic and Computer Engineering, University of Catania, 95125 Catania, Italy
* Correspondence: chiacchio@dmi.unict.it

Abstract: Due to the augmented complexity of the factory on the one hand and the increased availability of information on the other hand, nowadays it is possible to design models of production lines able to consider the state of the health of the production system. Such models must combine both the deterministic and the stochastic behaviours of a system, with the former accounting for the mechanics and physics of the industrial process and the latter for randomness, including reliability of the production systems and the unpredictability of the maintenance and of the manufacturing lines. This study proposes the application of a Hybrid Dependability Modelling based on Monte Carlo simulation to estimate the performances of a repairable production line modelled with a queueing G/G/1 system. The model proposed is characterized by random interarrival and service times and by the wearing and dynamic aging phenomena of the machine tools that depend on the working and operating conditions.

Keywords: industry 4.0; hybrid dependability modelling; production scheduling; dynamic failure rate; discrete event simulation; time-driven simulation

Citation: Chiacchio, F.; Oliveri, L.; Khodayee, S.M.; D'Urso, D. Performance Analysis of a Repairable Production Line Using a Hybrid Dependability Queueing Model Based on Monte Carlo Simulation. *Appl. Sci.* **2023**, *13*, 271. https://doi.org/10.3390/app13010271

Academic Editors: José A. Yagüe-Fabra, Guido Tosello and Roque Calvo

Received: 3 November 2022
Revised: 15 December 2022
Accepted: 20 December 2022
Published: 26 December 2022

1. Introduction

Industry 4.0 is playing a crucial role in the digital transformation of manufacturing/production and related industries as it provides new paradigms for the management of the enterprises. Cyber–physical systems form the basis of Industry 4.0 (e.g., "smart machines"), and they are becoming predominant also in small and medium enterprises (SMEs), such as in several phases of the manufacturing and production processes [1].

The success of any industry is tightly interconnected to its suppliers and to the response of the market. Under this viewpoint, Industry 4.0 can represent a great opportunity to improve the effectiveness of the value chain [2] because the collaboration among the actors of the supply chain can be facilitated by the digitalization of the Smart Factories. In fact, Smart Factories become able to exchange real-time information and increase the transparency among each other [3] thanks to their advanced equipment constituted by hi-tech robotics and artificial intelligence, IoT sensors, cloud computing, data capture and analytics, digital fabrication (including 3D printing), software-as-a-service and other new marketing models.

The efficiency of a production system is, indeed, one of the main drivers of competitiveness. The quantity of different products and the uncertain demand require an increased flexibility in terms of manufacturing which, in turn, requires the ability to adapt to the market demand and supply uncertainties [4].

In the context of the SME, more flexible and less expensive tools, supported by a growing number of new technologies, are being preferred to traditional enterprise information systems, such as ERP and MES [5,6]. They can be easily integrated with modern control and embedded software systems, allow individuals to monitor and process data control via IoT (the Internet of Things) and Internet and improve the overall management and scheduling of industrial processes, from the production to the maintenance.

From the field-level of a production department, it is now possible to capture many important signals that can be collected and used to evaluate the status of the system. Moreover, the diffusion of data analytics tools is helping the organizations in the understanding of meaningful information from the produced data. Nowadays, it looks possible to code models able to capture information data of operational behaviours that link environmental conditions, workloads and cumulative aging of materials which were not available before. As highlighted by [7], systems are generally working under time-varying operating conditions that can influence the failure time of systems. In the modelling of a queueing system [8], the trivial hypothesis of a dyadic state (working, failed) is not realistic because industrial and electromechanical equipment are deteriorating from wearing and ageing effects. As discussed in [9], one of the main challenges is the unpredictability of the manufacturing lines that can be tackled with novel tools, including digital twin and simulation, of the Industry 4.0 generation.

In this context, in order to better assess the main performance and dependability attributes of the system, including reliability, availability and maintainability [10], the development looks promising in the investigation of a methodology that can combine a realistic model of a production line with up-to-date and more detailed information data of the equipment. A general dependability framework has been proposed in [11], and in this research paper, it has been modified in order to integrate it into a dynamic reliability problem [12]: a non-exponential queueing system subjected to non-exponential failure and repair and characterized by dynamic working conditions (process scheduling and workloads).

Due to the numerous physical and stochastic interdependencies typical of a multi-state system, the choice for the use of the simulation paradigm over any other mathematical formalism appears the most suitable. The simulation model is fed by the mechanical differential equations of the machine tool wearing which are integrated numerically. Moreover, the proposed simulation environment, coded within the electronic spreadsheet, can be easily interconnected (using many plugins, such as OPC-DA, Modbus, ODBC, etc.) to a production line to get real-time data from the field level (operational variables), communicate the diagnosis of its status (evaluation of aging) and control the operations (assessing its remaining useful life) as a cyber physical system.

This paper is organized as follows: In Section 2, a thorough literature review about repairable queuing systems modelling is presented in order to depict the state of the art and to identify the space of intervention. Section 3 presents the theoretical background of reliability and repairable queueing systems to better frame the concepts in place—what are their main limitations and how they can be combined in a simulation of a dynamic reliability problem. The flowchart algorithm of the Hybrid Monte Carlo simulation able to solve the challenge of a generic repairable queueing system with a dynamic reliability problem is thus presented. This represents one of the main novelties with respect to the state of the art described in the next section. With the aid of a case study, Section 4 presents the hybrid simulation model and the design of the experiment that consists of different working scenarios under a corrective maintenance policy. This latter includes, among everything else, some validation scenarios that are used to test the correctness of the simulation model. Finally, conclusions are reported in Section 5.

The proposed paper is thus providing a theoretical framework to evaluate the performance of a production line characterized by wearing and aging which can be coded according to different types of production hypotheses. Moreover, it can be coded to connect the electronic spreadsheet to a production environment and achieve more realistic information from the field, providing to the production stakeholders a powerful tool that works not just offline but also in real-time.

2. Literature Review

Production scheduling, maintenance scheduling and quality management are three shop floor level policies that aim to minimize the costs of a manufacturing business. Job rejections, machines' down time, waste, manpower, energy, raw material and delay costs are all the consequences of a mishandled management of the production activity. Although these three policies have been treated as separate problems in the past [13], comparing the three different maintenance policies, the interdependencies between them and the competitive markets together have shifted the recent studies to analyse joint considerations for which, based on the objectives, different models are proposed [14]. In the literature, the interest of research and practitioners for finding optimal values of floor-shop variables is demonstrated by many recent papers that undertake this goal by proposing different mathematical, engineering and operative research solutions that are often conceived to solve the specific use case under investigation.

A non-linear programming approach to model a joint production and maintenance scheduling (JPMS) with multiple preventive maintenance services and determine the best sequence of jobs is proposed in [15]. The results demonstrate the effectiveness of the proposed algorithm that remains valid under the hypothesis of constant failure and repair rates. Conversely, the objective to determine the maintenance plan of deteriorating production machines subject to a production scheduling system is discussed in [16–18], with the main assumptions of a Markovian process. In [19], a model that includes two coupled Markovian queues is presented, with one queue that represents the decoupling inventory and the other the order backlog.

The Weibull distribution characterizes the failure behaviour of the single station problem proposed in [20] where an optimisation problem for the job scheduling with, at most, 18 items is solved, assuming preventive maintenance. As it can be seen, the resolution approach to these problems is with meta-heuristic methods, with a dependability model that is set with a well-defined probability density function of failure/repair.

Another class of problems for queueing systems deals with minimizing loss (of materials or time) either by focusing on the properties of the queue or service station. If such models are coupled with a stochastic dependability model, they can provide important key performance indexes of a production system, identifying bottlenecks and weaknesses, and other relevant hints that can reveal and improve the effectiveness of the production operations, from the production scheduling to the maintenance policies. This latter category, known as "queue with unreliable service station" or "repairable queue", is the main focused by this manuscript.

The resolution of unreliable queues is an interdisciplinary field between queueing model, dependability and operational research. The main measures of a queueing system such as waiting time, queue length, station occupancy, etc., are the objectives of such analyses. Analytical methods, heuristic methods and simulations are the main methodologies used to model and solve this class of problems.

Analytical methods include generating function, stochastic decomposition, renewal processes and Quasi-Birth–Death (QBD) models. For instance, in [21] a mathematical model for the optimal quality and process control of a queueing production system is proposed to maximize expected profit per unit time. Unscheduled and preventive maintenance of a single unreliable station processing N jobs are analysed [22] using arguments from renewal theory which stick to a mathematical model to obtain a closed form solution. An interesting evolution of such modelling is presented by [23] which focuses on condition-based maintenance to increase the system-level performance of a degradation system described with a Markovian model. A transition matrix is used in their solutions with the limitation that this analytical model can be applied only to a few operational scenarios.

In the general category of the problem of a queue with unreliable service stations, there is a specific type of problems called repairable queues with degradation. In this case, it can be assumed that the service station can fail during the production activities either because of a fault or due to a degradation of its workload capacity. Such type of scenario is very common and difficult to analyse with a pure analytical approach. The research papers [8] adopt an analytical explicit solution for an M/M/1 of a repairable queueing model, and [24] shows an approach to model the failure rate of the service station that depends on the age of the components, without considering when the system is in idle or in a busy mode. Arrival rate of the jobs and occurrence of failures follow a Poisson process; service time and maintenance time are exponentially distributed.

The attention over the working conditions which can affect the system performance is another important setting for this type of modelling. In [25], the authors examined an M/M/1 retrial queue with an unreliable server in an exogenous random environment and obtained the approximate orbit size distribution and mean queueing performance measures using matrix–analytic methods (level dependent QBD). In addition, [26] considered a variable failure rate for a M/M/1 queueing system with repairable service station with different parameters in idle time and busy time. They used transform method (QBD) to obtain the steady-state availability and the generating function method to obtain steady-state mean queueing length. Moreover, [7,27,28] focus on a single component system: [7] assumed time-varying operational conditions that are modelled by means of a continuous-time Markov process, whereas [27] combined a continuous-time Markov chain with queueing theory to obtain a model for the optimization of the repairing costs. In another work, [29] studied the M/M/2 repairable queueing system with varying failure rate at different times and used transform method (QBD) in steady-state period to obtain the probability of the states, the steady-state availability and the steady-state queueing length. In [30], authors presented and used an analytical method based on a decomposition approach that applies to a multi-stage manufacturing system. The queueing system was modelled with a Markov process and the degradation with a fixed decreasing yield that does not depend on the physics of the system and of the manufacturing process.

As it can be noticed, the previous papers narrow the scenarios to the assumption that the distribution of failure, repair, interarrival and service are random and can be modelled with the Poisson distribution. This represents an important limitation for models which aim to analyse real production environment. Moving towards a more realistic modelling, an evolution model for the failure rate is proposed in [31] to evaluate the optimal integrated predictive maintenance strategy on the manufacturing system with a gamma process to characterize the performance degradation. Although this kind of modelling can be defined as hybrid because it combines at once mechanics and stochastic behaviours of the system, the main limitation of the previous manuscript is that the degradation follows a specific distribution, set as input of the model. A different viewpoint is presented in a recent work [32] that focuses on a joint model of production, quality control and preventive maintenance for a serial–parallel multistage production system. The complexity of the problem is higher than single stage queue and, as in [33], the novelty proposed is to evaluate the machine structure importance measure and productivity when selecting the machine to put in maintenance. In this case, the problem is modelled and solved by means of Monte Carlo simulation, but only a single quality attribute for modelling the machine deterioration is considered without taking into account the active contribution of operational conditions.

Therefore, according to the literature review performed, the main limitations addressed in this manuscript are to enable a general modelling of repairable G/G/1 queue with a repairable service station subjected to wear and aging. These, depending on the jobs scheduling, the environmental and the working conditions, will be integrated into the mechanical equations of physics for the modelling of the system dynamic failure rate. Moreover, although the working station can be in a failure or degraded state [28], the manufacturing system can, eventually, continue its work at a reduced capability and, thus, increase the number of items waiting in the queue. Therefore, to model the proposed system, a Monte Carlo hybrid simulation that combine Discrete-Event and Time-Driven simulation is presented.

3. Theoretical Background and Methodology

This section illustrates the theoretical background on which the proposed methodology is built upon, with the aim to provide the readers with the required relevant know-how to understand the hybrid simulation methodology for the study of a general repairable queuing system. With reference to the queuing (Section 3.1) and reliability (Section 3.2) theories, the main hypotheses that the hybrid modelling aims to overcome with respect to traditional resolutions are presented.

3.1. Queueing Theory

The study of queueing systems allows one to evaluate various metrics to assess the performance and the quality of a service such as the mean number of items in the system (queue length), the mean waiting time, the service occupancy, etc. [34]. A queuing system consists of items (i.e., customers or jobs in a queue) and servers.

Kendall Notation (A/B/C/D/E/F) is used to synthetize the hypotheses of a queuing model [35] where A is the arrival process, B is the service process, C is the service mechanism, D is the system capacity, E is the population and F is the queue discipline.

The simplest queueing model (Figure 1) is when, for each item, both interarrival and service time are independent among each other (i.i.d), and it is possible to model the corresponding probability distributions with the negative exponential distributions, assuming, respectively, a constant rate $\lambda = 1/MTA$ (i.e., the Mean Time of Arrive) and a constant service rate $\mu = 1/MTS$ (i.e., the Mean Time of Service).

This modelling obeys the following hypotheses:

- **Hp1:** Interarrival of items can be modelled with a random variable characterized by a mean interarrival time.
- **Hp2:** The service time can be modelled with a random variable characterized by a mean service time.

These hypotheses bring about a special type of Continuous Time Markov Chain (CTMC), the well-known birth and death process [36] in which jumps exist only among the neighbouring states and each state of the Markov Chain represents the number of items in the queue (Figure 2).

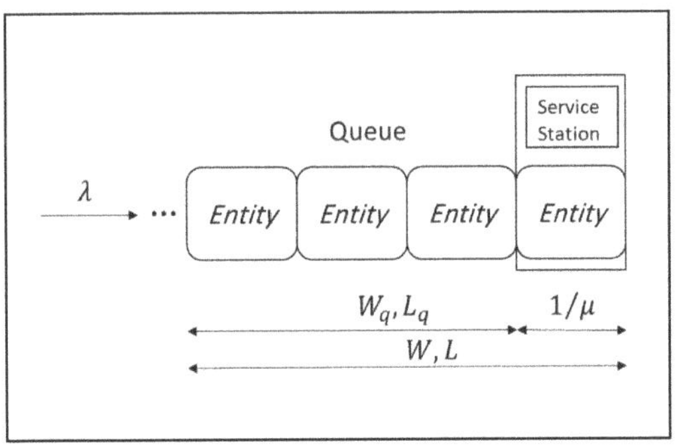

Figure 1. M/M/1 queueing system diagram representation.

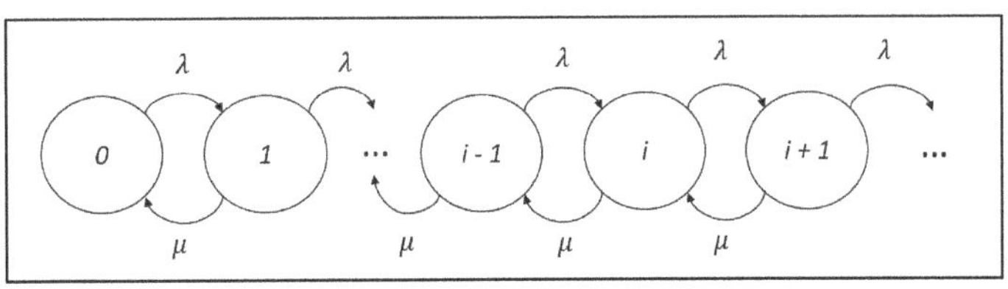

Figure 2. Birth and Death process.

In this process, a Birth is an arrival of a new item into the queueing system, with the rate of λ , and a Death is a departure of a served item, with the rate of μ. Under this assumption, the amount of time θ_i the system spends until another birth or death happens is exponentially distributed with rate $\vartheta = \lambda + \mu$, and the sojourn probability in the state ith can be computed as shown in Equation (1):

$$P\{\theta_i \leq t\} = 1 - e^{-\vartheta t}, t \geq 0 \tag{1}$$

For a long run, it is possible to evaluate the steady-state system probabilities:

$$\rho = \frac{\lambda}{\mu} \tag{2}$$

$$\pi_n = (1-\rho)\rho^n \tag{3}$$

$$L = \frac{\lambda}{\mu - \lambda} \tag{4}$$

$$W = L/\lambda \tag{5}$$

where

- ρ is the probability of being served;
- π_n is the probability of having n items in the queueing system;
- L is the mean number of items in the queue;
- W is the mean waiting time in the queue: $W = L/\lambda$.

As it can be understood, this simple queuing model is very unlikely in real case scenarios for which the complexity of the corresponding processes cannot really be captured with those assumptions. Therefore, the study of a generic G/G/k must be carried on with other computational models such as the hybrid simulation model proposed in Section 4.

3.2. *Reliability (Dynamic) Theory*

Reliability is the probability that a system will operate for the entire duration of a mission, T_m, in the mode for which it was designed, given that it was working at the beginning of the mission. As known, the mathematic formulation of this concept is based on the definition of a random variable θ that measures the probability of an item to survive from time 0 up to time T:

$$\mathrm{R(T)} = \mathrm{P}\cdot(\theta > \mathrm{T}) \tag{6}$$

When a small-time interval is considered, $\xi_2 - \xi_1 = \varepsilon$, the probability that the system fails is called failure rate $h(t)$ [37], and the following relationships hold [38]:

$$h(t) = -\frac{1}{\mathrm{R}(t)}\frac{d}{dt}\mathrm{R}(t) \tag{7}$$

which brings to the general equation of the system reliability the following:

$$\mathrm{R}(t) = e^{-\int_0^T h(t)\,dt} \tag{8}$$

Dynamic reliability studies those problems for which the failure rate is dynamic, which varies according to the physical conditions in which the system operates. The concept of failure rate is involved in most resolution methods of reliability engineering such that, in many industrial applications, it is assumed to analyse a system during its useful-time period, assuming three main hypotheses, which simplify the resolution of a problem:

- **Hp3:** During the useful-time period, there is an approximately constant failure rate with respect to time, where only random failures can occur.
- **Hp4:** The behaviour of a component can be defined by means of a dyadic operational state (on/off).

- **Hp5:** The component works under well-defined operative conditions, similar to those observed by the original manufacturer.

With these assumptions, the failure rate is constant and independent from time and other external factors. Therefore, Mean Time To Failure (MTTF), generally provided by the component manufacturer [39], is given, and Equation (8) can be written as follows:

$$R(t) = e^{-\frac{t}{MTTF}} \tag{9}$$

Repairable systems represent the majority of industrial applications. In these cases, a stochastic model is characterized by another parameter, the system downtime, referred to as the total time required to repair it, during which the system is not available or operating [37]. A stochastic model of repairable systems is often following the next hypothesis:

- **Hp6:** Downtime can be modelled with a random variable characterized by its own expected value.

In this case, it is possible to model the repair distribution with the negative exponential distribution, characterized by Mean-Time-To-Repair that models the average time it takes to recover from the failure [37].

For a repairable system, it is possible to calculate the percentage of time that the component is operational, the Availability (A), and this represents the probability that a system is available at a given point of time (point availability) [37,38,40]:

$$A = \frac{average\ up\ time}{average\ up\ time + average\ down\ time} = \frac{MTTF}{MTTF + MTTR} \tag{10}$$

3.3. Simulation Algorithm of the Hybrid Dynamic Queueing System

The production process of an industrial product is subjected to interdependent physical and aleatory phenomena that cannot be caught by Hypotheses Hp1–Hp6. Indeed, these provide a rigid framework of operations which are not applicable to real cases that are characterized by complex working and environmental conditions that can modify the performance of a system and affect its wear. More generally, a system can be subjected to continuous changes during its mission, and the stochastic modelling requires coupling the aleatory behaviour of the system (linked to the uncertainties of the features of workloads and resistance of the materials) with the physical equations that describe its mechanics and its process. This type of modelling [33], based on the knowledge of the mathematical relationships that make the dynamic failure rate, changes over time [39]; it aims to relax the framework defined by the Hypotheses Hp1–Hp6. The conception and resolution of such models is complex and can be undertaken using simulation techniques. Under this modelling framework, numerical integration is used to integrate the dynamics equations of the system, and Monte Carlo simulation is applied to take into account the contribution of the random variables; the resulting model is defined as hybrid.

The simulation paradigm is required to model and evaluate complex systems for which a closed analytical solution is difficult to model and compute. From a simulation perspective, systems can be categorized as discrete or continuous. The difference between a discrete and continuous system is the change in their variables that may happen in discrete points in time or continuously.

Simulation models are analysed by numerical methods which employ computational procedures to solve mathematical models. Within the scope of the simulation, the model is solved by means of a "run" or "iteration" [41] which constitutes one of the possible realizations of the system's evolution. The simulation model must keep track of the results generated during each run and, at the end of the simulation, it computes the average. Therefore, in order to achieve an accurate result, simulation can be time consuming.

Figure 3 shows the flowchart of the Monte Carlo simulation (MCS) used as a general algorithm for combining a queueing system with a dynamic reliability problem. As said, it is important to highlight that this algorithm is able to combine together a Discrete Event

Simulation and Time-Driven (Continuous) Simulation engines by adding the complexity of a repairable queueing system constituted by a simple machine tool that can get broken and repaired, overcoming the limitations of [39] and [24]. In fact, the Block.4, as explained in the section of the case study, is the simulation component that manages the scheduling of the production performed by the service station. The discrete simulation engine has to consider when the manufacturing station is unavailable and, thus, shifts or extends the processing of a work item accordingly.

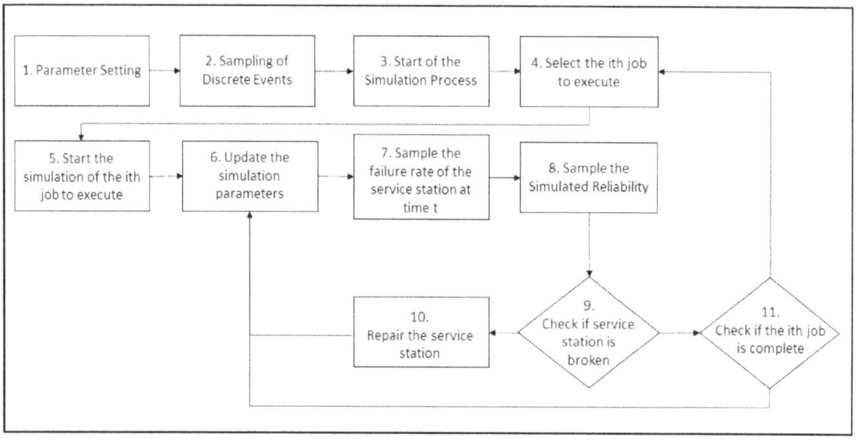

Figure 3. Flowchart of the Monte Carlo simulation algorithm.

Let us illustrate the algorithm proposed.

The first block (Block.1) concerns the setting of the parameters which characterize the use case scenario. Basically, it should be possible to configure the variables and the related distributions of the queueing system (e.g., the machine tool) and of the dynamic reliability problem.

In the Block.2, the simulation engine samples the interarrival (t_{mi}) and service (t_{sm}) time of each job which will enter in the queueing system, so as to build the table of the Discrete Events.

In particular, for the ith job, it is possible to compute:

$$ti_{mi} = F^{-1}(\cdot, rni) \tag{11}$$

$$ti_{sm} = R^{-1}(\cdot, rmi) \tag{12}$$

where F^{-1} is the inverse of the distribution function of the time arrival, and R^{-1} is the inverse of the distribution function of the service time; finally, rn_i and rm_i are two real random number in [0,1] extracted using a uniform distribution.

In the Block.3, the simulation start time is set so that the actual time for starting the processing of the next job, selected in block 4 according to the FIFO policy (which at the beginning of the simulation is the Job 1) is computed accordingly (at the beginning of the simulation t = 0). The relevant time points computed are stored in a memory table, called the Event-Driven Table.

From Block.5 to Block.11, the continuous Time-Driven simulation is explained. These blocks perform the update of the dynamic failure rate for every Dtsim and, in Block.9, verify if the service station (e.g., the machine tool) becomes broken. The operations performed by these blocks are stored in a memory table, called the Time-Driven Table which contains the result of a simulation-run. If the Block.9 assesses that the queuing system breaks, the algorithm gives the control to Block.10 to compute the restoration time using the probability function of repair characterizing the model. Therefore, the simulation parameters are updated in Block.6, and the control is given back to the Block.8 for continuing the process simulation of the continuous system. During the restoration time window, the service station of the queueing system stops working on the items enqueued in the queue; therefore, the time points of the Event-Driven Table are update accordingly, assuming an increase in the time of their occurrence with a value equal to the restoration time required to repair the service station. If the Block.9 does not detect any failure, the Time-driven simulation can proceed until the ith job is completed in Block.11. The process of simulation continues starting a new job, and this process repeats until all the items are completed.

4. Case Study and Hybrid Modelling

The system under analysis is characterized by a production line where items are enqueued and processed one by one in a service station. It is assumed that both the interarrival time and the service time of the system items are random. Moreover, the service station can be failure prone. It is assumed that smart sensors can monitor the working temperature of operation which, in this model, will affect the dynamic failure rate of the service station. Figure 4 shows a schema of the system, where it is possible to identify within the bold contour the main components of the model, namely the enqueueing conveyor and the service station. Each item (or job) is characterized by its own time for being processed (time of service), whereas they are enqueued one by one by the picker robot (interarrival time).

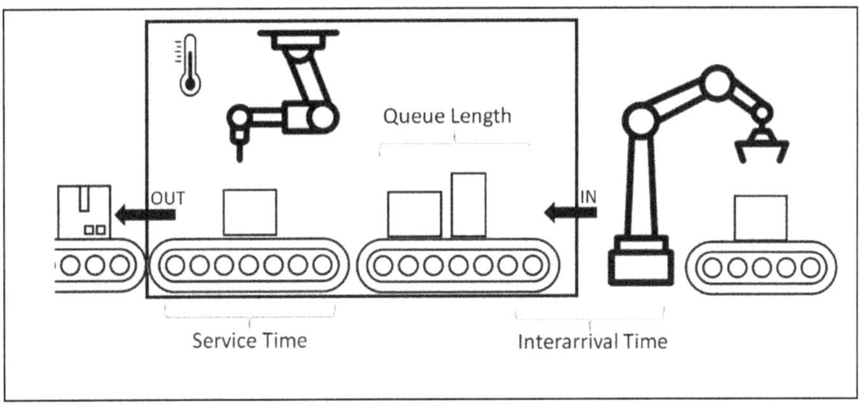

Figure 4. Generic production line of an automated factory.

The main assumptions for this generic production line are thereby listed:

- Interarrival Time follows a rectified Normal Distribution (see Table 1 parameters);
- Service Time follows a rectified Normal Distribution (see Table 1 parameters);
- Mean Time to Failure follows a Weibull Distribution (see Table 2 parameters);
- Mean Time to Repair follows a Uniform Distribution (see Table 2 parameters);
- The machine tool is subjected to aging that depends on the operating temperature, following the Equations (14) and (15).

Table 1 shows the main parameters of the queueing model. It was assumed that the interarrival time and the service time follow two rectified Normal distributions to avoid negative time values and overcome Hypotheses 1 and 2.

Table 1. Normal distribution parameters characterizing the automated factory.

Variable	Symbol	Value	Measuring Unit
Expected service time	ts_{mean}	5	h
Standard deviation service time	σ_{ts}	1	h
Expected interarrival time	ti_{mean}	10	h
Standard deviation between jobs	σ_{ti}	2	h

Table 2. Main parameters for the failure and repair distributions of the machine tool.

Distribution	Variable	Symbol	Value	Measuring Unit
Failure Event (Weibull)	Expected Mean Life	α	100	h
	Shape Factor	β	3	-
Repair Event (Uniform)	Min Time to Repair	TTR_{min}	1	h
	Max Time to Repair	TTR_{max}	8	h

The failure of the service station is modelled using a Weibull distribution characterized by a shape factor, $\beta = 3$, and an expected machine tool life $\alpha(T)$ of 100 hours which represents the mean time to failure of the service machine if it operates in the same fixed operational conditions in which it was tested by the Original Equipment Manufacturer (OEM). Therefore, the failure rate can be written as follows:

$$h(t,T) = \frac{\beta}{\alpha(T)}\left(\frac{t}{\alpha(T)}\right)^{\beta-1} \tag{13}$$

The time to repair for the machine tool follows a uniform distribution time between a maximum repair and a minimum repair time (TTR_{min}, TTR_{max}). Table 2 resumes the main data of the failure and repair model of the machine tool.

In order to model a daily variation of the temperature during the season of the year, the equation used to model the working temperature is the following:

$$T(t) = Tm + A_T \sin\left(\frac{2\pi}{T}t\right) \tag{14}$$

where Tm = 20 °C is the mean operation temperature, T = 24 h is the oscillation period of the temperature, A_T = 5 °C is the amplitude of the oscillation, and t is the independent time variable.

As it was said, the machine tool is subjected to aging that depends also on the operating temperature. Under this assumption, the expected machine tool life takes the following form:

$$a(T) = \frac{a_0}{\frac{T}{T_0}} \tag{15}$$

where a_0 is the expected lifetime at the nominal working temperature T_0.

4.1. Simulation Algorithm of the Hybrid Dynamic Queueing System

In a simulation model, variables can be deterministic (and non-random) or stochastic (and random). A Monte Carlo simulation (MCS) is basically a stochastic simulation method and a sampling technique in which random sampling is used. In this paper, the Monte Carlo simulation (MCS) core algorithm has been codified so as to combine together Discrete-Event and a Time-Driven simulation. In this way, MCS can change dynamically the parameters characterizing each run of the system evolution during the mission time. In order to understand the hybridization of the simulation, let us recall first the generic algorithms for the queue and reliability independently before merging the two of them.

For a queueing system, one run of the simulation will result in a sample path that models the number of items in the queue during the time. The interarrival $Z_i(t)$ and service time $S_i(t)$, characterizing a generic i^{th} item entering the queue, are simulated by sampling the inverse of the distribution probability functions, Z^{-1} and S^{-1}, characterizing the two processes. This allows one to construct a sample path of the system evolution that, as shown in Figure 5 [42], depicts for each discrete event time-point the number of items enqueued in the system. As it can be understood, the advantage of the Monte Carlo simulation is to overcome the limitation of the analytical "Little's Law" which is valid only for the Poisson distribution and to generalize the modelling with the possibility to simulate the behaviour of a general queueing system.

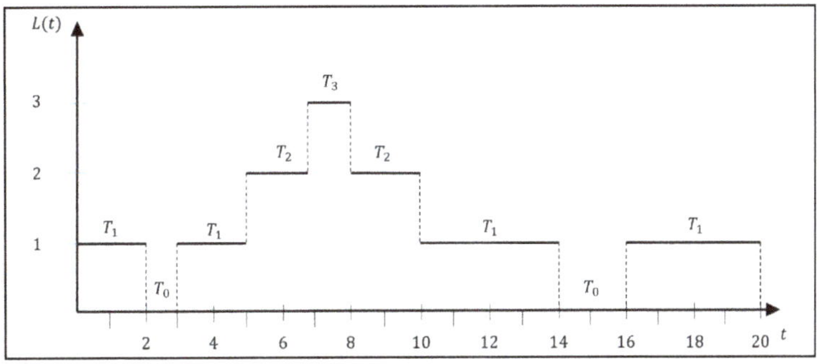

Figure 5. Number of entities in the system versus time.

For the reliability of a system, the MCS corresponds to performing an experiment for a certain mission time (T_M) in which a large number of identical stochastic systems, each one behaving differently, runs virtually instead of performing real physical tests.

The simulation algorithm monitors whether a system fails before the completion of the mission and, in that case, records the time of the failure. This procedure corresponds to the observation of several realizations of the system process. Based on the amount of observations fallen into each range, simulation results are examined [41].

In a reliability problem, on the one hand, a set of possible states can be considered for a component (e.g., functioning, degraded and failed, as shown in Figure 6a), and based on its failure distribution, random numbers are generated for the duration that the component remains in each state; on the other hand, an infinite set of other states can be modelled considering how the failure rate h(t) changes over time, depending on the age of the component and on the boundary conditions in which it operates (Figure 6b).

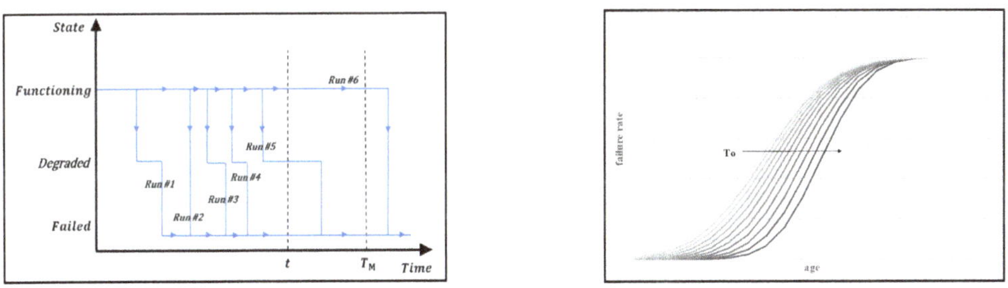

(**a**) (**b**)

Figure 6. Instance runs for failure analysis of a component with state (**a**) and failure rate (**b**).

Both these approaches overcome the hypothesis HP4 of dyadic states. With regard to the first, the sojourn time into a state (duration) is computed by the inverse distribution function F^{-1} that characterizes the jump towards another state. To obtain the unreliability for a given time (t), the number of runs that have jumped into a failure state before the mission time can be divided by the total number of runs [43]. Figure 6a shows an example of the sample path generated by 6 runs of a simulation where the transitions among the three different states of the components occur at different times. For the example of Figure 6a, the component has reached the failed state in four out of the total six runs before "t"—therefore, the unreliability of the system, F(t) = 4/6.

With regard to the second, with reference to Figure 6b, modelling the failure rate evolution enables one to calculate the instantaneous reliability R(T) as the probability that the component survives over the next times step ΔT. If the jump from one time step to another can be written with a mathematical functional (F, t, $\cdot$), which is a combination of dynamic failure rates and deterministic (mechanical) system equations (with this latter taking into account the dynamic change of the component's failure behaviour and the working conditions), it is possible to build a hybrid distribution function that overcomes the limitation of the hypotheses HP3, HP4 and HP5. This type of modelling reconfigures a reliability problem into a Dynamic Reliability problem where is a mathematical functional that depends on the stochastic distribution function of the failure rate F, the time variable t and $(\cdot)$ which represents all the other physical parameters included in the mechanical relationship. Moreover, for this type of problem, the numerical integration of the functional is carried on through adopting a time-step integration Δt_{sim}.

For the proposed case study, the time-step integration, Δt_{sim}, has been set to 1 h and the dynamic failure rate functional that depends on the environmental temperature can be written as follows:

$$\Delta_i = \frac{\beta}{\alpha_i}\left[\left(\frac{t}{a_i}\right)^{\beta-1} - \left(\frac{t-\Delta t_{sim}}{\alpha_i}\right)^{\beta-1}\right] \tag{16}$$

$$\alpha_i = \begin{cases} \alpha_i = \alpha_0 \frac{T_0}{T_i}\,, & if\ T_i > T_0 \\ \alpha_0\,, & otherwise \end{cases} \tag{17}$$

$$T_i = Tm + A_T \sin\left(\frac{2\pi}{T} i\right) \tag{18}$$

The numerical integration of the functional can be computed using the following formula:

$$_{i+1} = {}_i + \Delta_i \tag{19}$$

In order to overcome the hypothesis HP6, the restoration of a component can be modelled by sampling the generic repair distribution which will give the time-point at which the repair transition occurs and bring the component back to the functioning state (i.e., as good as new with a corrective maintenance policy). Therefore, in the repairable queue system, whenever the manufacturing station goes into the failed state, it is possible to assume that during the unavailable state, the queueing system cannot continue its processing (and thus decreases the number of items in the queue), but, eventually, it can be subject only to the arrival of new items. Only when the manufacturing station gets restored, the provision of service will start again. By assuming a general probability distribution of repair, it is possible to write the general discrete equation of the equipment unavailability that can be expressed as the ratio of the total accidental shutdown time over the total task time [31]:

$$U_{i+1} = \frac{\mathbb{E}\left(\tau_u^{min}\right)\mathbb{E}\left(\sum_0^{i+1}{}_{i+1}\right) + \mathbb{E}\left(\tau_p^{min}\right)}{i * \Delta\mathrm{T} + \mathbb{E}\left(\tau_p^{min}\right)} \tag{20}$$

where $\mathbb{E}\left(\tau_u^{min}\right)$ is the expected value o on the jobs scheduling for the minimal repair duration, $\mathbb{E}\left(\tau_p^{min}\right)$ is the expected value of the planned maintenance duration, $\mathbb{E}\left(\sum_0^{i+1}{}_{i+1}\right)$ is the the expected number of failures in the (i + 1)th timestep.

4.2. Simulation Execution of the Hybrid Monte Carlo Simulation Algorithm

This section describes the results of single iteration of the simulation process for the use case described in Section 4.1.

As explained in the previous section, the Discrete Event Table is initialized in Block.2 at the beginning of each iteration. For instance, Table 3 shows the main data-points event times of the *i*th jobs of a generic iteration, which are sampled using the inverse functions of Equations (11) and (12).

From Block.4, the continuous simulation starts, and Table 4 shows the evolution of the system that, among other things, depends on the physical mechanical equations of the system. The instantaneous dynamic failure rates are updated at any ΔT.

Table 3. Discrete Events Table of the generic *i*th run of the simulation of the items.

Job	T_{mi} (h)	t_{sm} (h)	Measuring Unit
1	0	94	h
2	10	89	h
3	46	87	h
...	...	...	h
N	...	...	h

Table 4. Time-Driven Table of the generic iteration (run) of the simulation process.

SimTime	T(t) [°C]	α(T) [h]	Δ		Random Number	Failure Test	Job Complete
0	20.0	100.0	-	0.000	0.3834	False	False
1	21.3	93.9	1.09×10^{-5}	1.086×10^{-5}	0.3831	False	False
2	22.5	88.9	2.14×10^{-5}	3.222×10^{-5}	0.9560	False	False
3	23.5	85.0	3.42×10^{-5}	6.644×10^{-5}	0.1671	False	False
...	...	...	...	...	...	...	False
80	24.3	82.2	9.13×10^{-5}	2.52×10^{-2}	0.0126	True	False
87	...	...	...	...	...	False	False
...	...	...	...	...	...	...	False
101	...	...	...	...	...	...	True

Let's explain the meaning of each column for a better understanding of the paper.

- The column SimTime represents the time of simulation that increases according to the integration time step (Δtsim).
- The column Δ is the instantaneous failure rate of the machine tool, computed using Equation (16).
- The column α(T) is the expected mean life of the machine tool, computed using Equation (17).
- The column T(t) is the working temperature, computed using Equation (18).
- The column is the integral of the instantaneous failure rate functional.
- The column rand_num samples a uniform real random number in [0,1] to perform the failure test.

Therefore, in the failure test column is indicated the result performed in Block.9 to verify whether the machine tool of the service station becomes broken (Failure_Test = True) at that time point:

$$\text{Failure Test} = \begin{cases} \text{False}, & \text{if } e^{-} < \text{rand\_num} \\ \text{True}, & \text{if } e^{-} \geq \text{rand\_num} \end{cases} \tag{21}$$

As shown in Table 3, it is possible to see that the Job 1 would require 94 hours to complete (see the t_{sm} column). From the evolution of the Time-Driven Table (Table 4), it is possible to notice that at time 80 h the machine tool becomes broken (see column "Failure Test" → True), and it resumes after 7 h. Therefore, the next row of the Time-Driven Table has the value SimTime = 87 h. In the end, the Job 1 completes at time SimTime = 101 h (see column "Job Complete" True) so that the algorithm executes in Block.4 the next job. During that interval, the machine has gone under repair, Block.10; the simulation parameters are updated (Block.6), and the next time of failure for the service station is computed (Block.7 to Block.9).

Otherwise, if the service station is functioning, the simulation process keeps verifying when the ith job has completed in Block.11. If it is not completed, the parameters are updated in the Block.6; otherwise, in Block.12, the simulation engine checks whether the jobs scheduled in Table 3 are completed. If there are other jobs to process, the simulation starts back from the Block.4; otherwise, the run of the simulation is complete.

4.3. Design of Experiment and Simulation Results

The case study has been analysed according to the design experiment of Table 5. The simulation scenarios 0 and 1 model a queueing system without failures, assuming interarrival and service times according to a negative exponential and a normal distribution; they are needed in order to validate the coded MCS and compare the results with closed analytical equations.

Table 5. Design of experiments.

Scenario						
Parameter Description	***Symbol***	**Measuring Unit**	**0**	**1**	**2**	**3**
pdf of interarrival time	I	-	neg exp		norm	
mean interarrival time	ti_m	h	10	10	10	10
std. deviation of interarrival time	σ_{ti}	h	10	2	2	2
pdf of service time	Z	-	neg exp		norm	
mean service time	ts_m	h	5	5	5	5
std. deviation of service time	σ_{ts}	h	5	1	1	1
pdf of failure time	Λ		#N/A		Weibull	
service station expected life time	a_0	h	-		100	100
shape factor	β		-		3	3
pdf of repair time	Φ		#N/A		Uniform	
maintenance strategy			#N/A		Corrective	
max restoration period	TTR_{max}	h	-		8	8
min restoration period	TTR_{min}	h	-		1	1

In scenarios 2 and 3, on the other hand, the restoration of the system and the evolution of the failure rate due to aging (scenario 2) and due to aging, wear and variations of the operating conditions (scenario 3) are modelled.

Figure 7 shows the comparison between the simulative process and the analytic benchmark of scenarios 0. With 10^3 runs of the simulation, it is possible to observe that the two models converge. In particular, the simulated mean awaiting in queue time is W_0 = 10.12 h, and the analytic mean awaiting queue time W_{ref} = 10.0 h (mean error $\Delta e = \frac{W_0 - W_{ref}}{W_{ref}} = 1.12$ %).

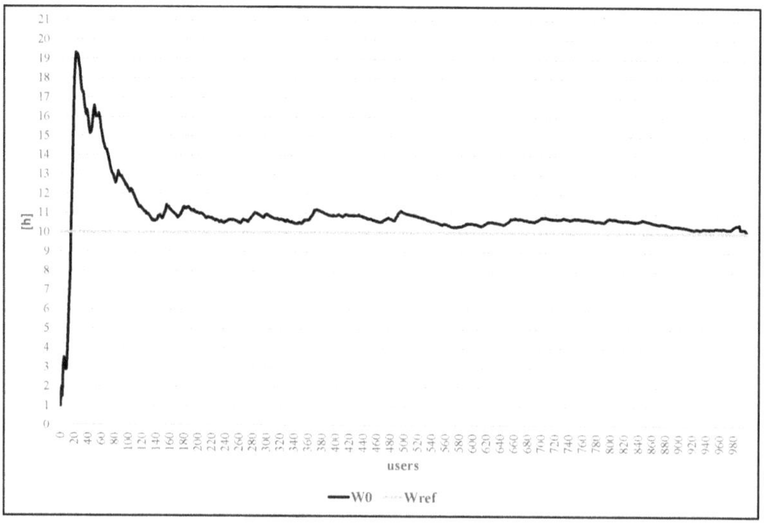

Figure 7. Scenario 0—Comparison between simulated progressive mean waiting time (W_0) and analytic value (W_{ref}).

The simulation results of scenario 1 show that the mean waiting time is 5 h. With regard to this scenario, inferential statistics enables one to define an upper bound for mean waiting time in queue according to the following equation [44]:

$$E(W_q) \leq \frac{\sigma_{ti}^2 + \sigma_{ts}^2}{ti_m(1-\rho)} = 7\text{ h} \tag{22}$$

where $p = ts_m/ti_m$ is the level of service utilization; σ_{ti} and σ_{ts} are the standard deviations, respectively, of interarrival and service time; ti_m is the mean interarrival time. Both the previous results confirm the goodness of the MCS coded.

Figure 8 shows the Monte Carlo simulation process of the progressive mean waiting time of Scenarios 1–3. The reported trends allow a person to conclude that the average waiting time in queue increases with the increased complexity of the operative scenario.

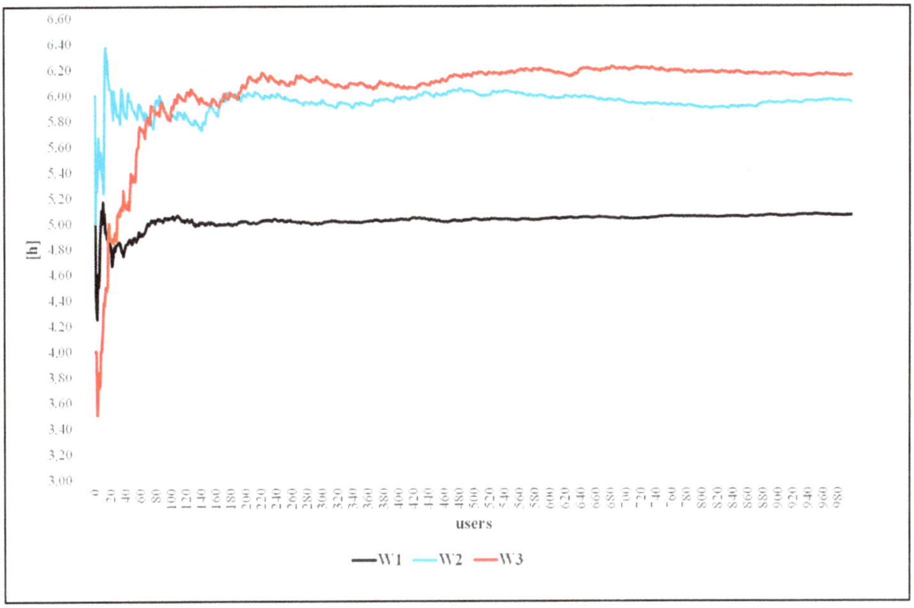

Figure 8. The Monte Carlo simulation process of progressive mean waiting in queue (W).

In fact, Scenario 1 which characterizes a system that cannot fail (max reliability of the system) shows the shortest mean waiting time in the queue. In Scenarios 2 and 3, the mean waiting time increases due to the increasing unavailability of the service station caused by the modelling of the failure rate. In particular for Scenario 2, this result is mainly given by the aging of the service station; whereas in Scenario 3, this effect is more emphasized by the contribution of the operating temperature.

The results given by the simulation shows that the average waiting in queue time is, respectively, W1 = 5.0 h; W2 = 5.94 h; W3 = 6.12 h. These cannot be compared with an analytical benchmark because, as explained, the modelling and evaluation of such scenarios are not doable with analytical methods. Finally, only for the Scenarios 2 and 3, the model has been used to evaluate the system's availability, the mean waiting time in queue and the mean number of jobs in queue for different level of service utilization r = [0.5; 0.6; 0.7; 0.8; 0.9; 0.95; 0.99]. As shown in Table 6, this subset of scenarios confirms that the performances of the queueing system decrease when operational conditions and aging are considered, and this is more emphasized in Scenario 3 (W_i: Mean waiting time in the queue; A_i: Availability of the machine tool; L_i: Mean items in the queueing system).

Table 6. Performances of the system (ts_{mean} = [5, 6, 7, 8, 9, 9.5, 9.9]; σ_{ts} = 0.2 ts_{mean}).

r	W_2	W_3	A_2	A_3	L_2	L_3
	[h]	[h]				
0.5	5.9	6.03	86.9%	84.8%	0.57	0.60
0.6	7.4	7.68	86.9%	84.8%	0.71	0.79
0.7	9.2	9.51	86.9%	84.8%	0.94	0.96
0.8	11.8	13.54	86.9%	84.8%	1.17	1.36
0.9	96.0	188.35	86.9%	84.8%	9.37	18.05
0.95	451.9	472.04	86.9%	84.8%	41.38	43.21
0.99	708.0	811.60	86.9%	84.8%	62.77	70.85

5. Conclusions and Remarks

The importance of a more realistic modelling and evaluation of the manufacturing and production processes is becoming a key component in the success of modern enterprises which are called upon to be agile and reactive to the market demand. The revolution brought with Industry 4.0 is helping to automate and digitalize such processes allowing us to collect the relevant information to identify bottlenecks and reveal opportunities to increase the throughput of the system processes.

In this paper, the performance evaluation of a production system has been addressed by the implementation of a simulation model of a repairable G/G/1 queueing system with a corrective maintenance policy. The model has been coded adopting a hybrid dynamic dependability approach able to link the aging and the wearing of the queuing system to the change of the working conditions, using the mechanical equations of the wearing physical process. The rationale behind the proposed approach is that the real-time information produced by the smart factories are now significant and help to extract, and thus model, with more precision the characteristics of the general stochastic processes and to find out the relationship between the working conditions and the wearing of the system. The simulation proposed represents a powerful tool in comparison with the analytical theoretical queueing models of the referred literature because it has been implemented in order to combine the physics of the system with the stochastic processes of the queuing theory and reliability. This has allowed us to consider in the model not only the randomness of the queue and of the machine reliability but also how the operational conditions affect these two processes. Moreover, a simple—yet valuable—corrective maintenance is analysed, but the simulation tool could be eventually improved to model, also, a preventive maintenance policy.

As expected, the results of the analysis performed into the G/G/1 with corrective maintenance show that the throughput of the system (mean waiting in queue) improves with the increasing of the system reliability; but, on the other hand, it demonstrates the importance of such advanced modelling tools that are able to take into account the influences caused by the working conditions on real production scenarios. The Monte Carlo simulation model has been coded using the electronic spreadsheet that—by means of the industrial connectors plug-in (ODBC, Modbus, IoT, OPC, etc.)—can be easily integrated into the manufacturing process. This demonstrates, on the one hand, that it is possible to code a complex simulation scenario (Event Driven with Continuous input data) in this software environment and that it can be used to analyse the system performance or to evaluate how to improve the weak elements of a production process, even before any engineering refactoring. Under this viewpoint, in future works, the model can be improved in order to handle other types of maintenance policies and compare the performance and the costs with respect to the corrective maintenance policy.

From the management point of view, the paper demonstrates that it is possible to develop a simulation model that can allow one to estimate the performance of the production line with tools which are not too complex or expensive, having a good knowledge of the business processes and of mathematical reliability and queuing theory. The proposed model, moreover, offers the possibility to modify the probability distributions of the queueing and reliability processes which can be customized ad hoc, depending on the production line modelled. This allows a person not only to estimate the performance of the system as it is, but also to analyse what modifications can be brought to the production line in order to understand the effects and the possible improvements.

Of course, the proposed approach presents some limitation. For instance, the Electronic Spreadsheet is not general enough to model any production scenario which could require an ad hoc model of the system and tuning of the model. Therefore, further research should identify other tools and generalize the proposed modelling in order to improve the effectiveness of the Monte Carlo simulation hybrid approach discussed.

Author Contributions: Conceptualization, F.C. and D.D.; methodology, F.C.; software, F.C. and S.M.K.; validation, D.D. and F.C.; formal analysis, F.C.; investigation, L.O.; resources, D.D.; data curation, S.M.K. and L.O.; writing—original draft preparation, S.M.K. and L.O.; writing—review and editing, D.D.; visualization, F.C.; supervision, F.C.; project administration, D.D.; funding acquisition, D.D. All authors have read and agreed to the published version of the manuscript.

Funding: This paper belongs to a research path funded by University of Catania (PIA.CE.RI. 2020–2022 Linea 2—Progetto Interdipartimentale GOSPEL—Principal investigator prof. A. Costa—Codice 61722102132).

Conflicts of Interest: The authors declare no conflict of interest. The funders had no role in the design of the study; in the collection, analyses, or interpretation of data; in the writing of the manuscript, or in the decision to publish the results.

Nomenclature

CPS	Cyber Physical Systems
CTMC	Continuous Time Markov Chains
ERP	Enterprise Resource Planning
FIFO	First-In First-Out
IoT	Internet of Things
JPMS	Joint Production and Maintenance Scheduling
L	Mean number of items in queue
MES	Manufacturing Execution System
MCS	Monte Carlo Simulation
MTA	Mean Time of Arrive
MTS	Mean Time of Service
MTTF	Mean Time to Failure
MTTR	Mean Time to Repair
ODBC	Open Database Connectivity
OEM	Original Equipment Manufacturer
OPC	Open Platform Communications
OPC-DA	Open Platform Communications Data-Access
QBD	Quasi-Birth–Death
SMEs	Small and Medium Enterprises
TTR	Time to Repair
W	Mean waiting time in queue

References

1. Cotrino, A.; Sebastián, M.A.; González-Gaya, C. Industry 4.0 Roadmap: Implementation for Small and Medium-Sized Enterprises. *Appl. Sci.* **2020**, *10*, 8566. [CrossRef]
2. Tjahjono, B.; Esplugues, C.; Ares, E.; Pelaez, G. What does Industry 4.0 mean to Supply Chain? *Procedia Manuf.* **2017**, *13*, 1175–1182. [CrossRef]
3. Geissbauer, R.; Vedsø, J.; Schrauf, S. A Strategist's Guide to Industry 4.0: Global Businesses Are about to Integrate Their Operations into a Seamless Digital Whole, and Thereby Change the World. 2016. Available online: https://www.strategy-business.com/article/A-Strategists-Guide-to-Industry-4.0 (accessed on 2 November 2022).

4. Müller, M.; Lehmann, M.; Kuhn, H. Measuring sequence stability in automotive production lines. *Int. J. Prod. Res.* **2020**, *59*, 7336–7356. [CrossRef]
5. Maresova, P.; Soukal, I.; Svobodova, L.; Hedvicakova, M.; Javanmardi, E.; Selamat, A.; Krejcar, O. Consequences of Industry 4.0 in Business and Economics. *Economies* **2018**, *6*, 46. [CrossRef]
6. Amaral, A.; Peças, P. A Framework for Assessing Manufacturing SMEs Industry 4.0 Maturity. *Appl. Sci.* **2021**, *11*, 6127. [CrossRef]
7. Hu, J.; Shen, J.; Shen, L. Periodic preventive maintenance planning for systems working under a Markovian operating condition. *Comput. Ind. Eng.* **2020**, *142*, 106291. [CrossRef]
8. Lam, Y.; Zhang, Y.L.; Liu, Q. A geometric process model for M/M/1 queueing system with a repairable service station. *Eur. J. Oper. Res.* **2006**, *168*, 100–121. [CrossRef]
9. Rolo, G.R.; Rocha, A.D.; Tripa, J.; Barata, J. Application of a Simulation-Based Digital Twin for Predicting Distributed Manufacturing Control System Performance. *Appl. Sci.* **2021**, *11*, 2202. [CrossRef]
10. Avižienis, A.; Laprie, J.-C.; Randell, B. Dependability and Its Threats: A Taxonomy. *Build. Inf. Soc.* **2008**, *156*, 91–120. [CrossRef]
11. Chiacchio, F.; Iacono, A.; Compagno, L.; D'Urso, D. A general framework for dependability modelling coupling discrete-event and time-driven simulation. *Reliab. Eng. Syst. Saf.* **2020**, *199*, 106904. [CrossRef]
12. Liu, J.; Zio, E. System dynamic reliability assessment and failure prognostics. *Reliab. Eng. Syst. Saf.* **2017**, *160*, 21–36. [CrossRef]
13. Rebaiaia, M.L.; Ait-Kadi, D. Maintenance policies with minimal repair and replacement on failures: Analysis and comparison. *Int. J. Prod. Res.* **2020**, *59*, 6995–7017. [CrossRef]
14. Govil, M.K.; Fu, M.C. Queueing theory in manufacturing: A survey. *J. Manuf. Syst.* **1999**, *18*, 214–240. [CrossRef]
15. Mokhtari, H.; Mozdgir, A.; Abadi, I.N.K. A reliability/availability approach to joint production and maintenance scheduling with multiple preventive maintenance services. *Int. J. Prod. Res.* **2012**, *50*, 5906–5925. [CrossRef]
16. Bajestani, M.A.; Banjevic, D.; Beck, J.C. Integrated maintenance planning and production scheduling with Markovian deteriorating machine conditions. *Int. J. Prod. Res.* **2014**, *52*, 7377–7400. [CrossRef]
17. Ghaleb, M.; Taghipour, S.; Sharifi, M.; Zolfagharinia, H. Integrated production and maintenance scheduling for a single degrading machine with deterioration-based failures. *Comput. Ind. Eng.* **2020**, *143*, 106432. [CrossRef]
18. van Dijkhuizen, G.C.; van Harten, A. Two-stage generalized age maintenance of a queue-like production system. *Eur. J. Oper. Res.* **1998**, *108*, 363–378. [CrossRef]
19. Fiems, D.; De Cuypere, E.; De Turck, K.; Claeys, D. Performance Analysis of Hybrid MTS/MTO Systems with Stochastic Demand and Production. *Mathematics* **2020**, *8*, 1925. [CrossRef]
20. Wang, S.; Liu, M. A branch and bound algorithm for single-machine production scheduling integrated with preventive maintenance planning. *Int. J. Prod. Res.* **2013**, *51*, 847–868. [CrossRef]
21. Hsu, L.-F.; Tapiero, C.S. An economic model for determining the optimal quality and process control policy in a queue-like production system. *Int. J. Prod. Res.* **1990**, *28*, 1447–1457. [CrossRef]
22. Hsu, L.-F.; Tapiero, C.S. Maintenance of an unreliable M/G/1 queue-like job shop. *Queueing Syst.* **1987**, *2*, 333–349. [CrossRef]
23. Ejaz, I.; Alvarado, M.; Gautam, N.; Gebraeel, N.; Lawley, M. Condition-Based Maintenance for Queues with Degrading Servers. *IEEE Trans. Autom. Sci. Eng.* **2019**, *16*, 1750–1762. [CrossRef]
24. Lv, S.; Zhu, L. Single Server Repairable Queueing System with Variable Service Rate and Failure Rate. *IEEE Access* **2020**, *9*, 1233–1239. [CrossRef]
25. Cordeiro, J.D.; Kharoufeh, J.P. The UnreliableM/M/1 Retrial Queue in a Random Environment. *Stoch. Model.* **2012**, *28*, 29–48. [CrossRef]
26. Lv, S.; Li, J. The M/M/NRepairable Queueing System with Variable Breakdown Rates. *Discret. Dyn. Nat. Soc.* **2013**, *2013*, 1–10. [CrossRef]
27. Srivathsan, S.; Viswanathan, S. A queueing-based optimization model for planning inventory of repaired components in a service center. *Comput. Ind. Eng.* **2017**, *106*, 373–385. [CrossRef]
28. Kalidass, K.; Kasturi, R. A queue with working breakdowns. *Comput. Ind. Eng.* **2012**, *63*, 779–783. [CrossRef]
29. Lv, S.; Li, J.; Yue, D.; Xiao, X. The M/M/2 repairable queueing system. In *Proceedings of the 2007 IEEE International Conference on Control and Automation, ICCA, Guangzhou, China, 17–20 October 2007*; Institute of Electrical and Electronics Engineers Inc.: Piscataway, NZ, USA, 2007; pp. 2071–2075.
30. Colledani, M.; Tolio, T. Integrated quality, production logistics and maintenance analysis of multi-stage asynchronous manufacturing systems with degrading machines. *CIRP Ann.* **2012**, *61*, 455–458. [CrossRef]
31. He, Y.; Gu, C.; Chen, Z.; Han, X. Integrated predictive maintenance strategy for manufacturing systems by combining quality control and mission reliability analysis. *Int. J. Prod. Res.* **2017**, *55*, 5841–5862. [CrossRef]
32. Cheng, G.; Li, L. Joint optimization of production, quality control and maintenance for serial-parallel multistage production systems. *Reliab. Eng. Syst. Saf.* **2020**, *204*, 107146. [CrossRef]
33. Famoso, F.; Brusca, S.; D'Urso, D.; Galvagno, A.; Chiacchio, F. A novel hybrid model for the estimation of energy conversion in a wind farm combining wake effects and stochastic dependability. *Appl. Energy* **2020**, *280*, 115967. [CrossRef]
34. Worthington, D.; Bunday, B.D. An Introduction to Queueing Theory. *J. Oper. Res. Soc.* **1996**, *47*, 1423. [CrossRef]
35. Smith, J.M. *Introduction to Queueing Networks Theory Practice*; Springer: Berlin/Heidelberg, Germany, 2018; pp. 1–556.
36. Haviv, M. *Queues: A Course in Queueing Theory*; Springer: New York, NY, USA, 2013; Volume 191, ISBN1 978-1-4614-6764-9. ISBN2 978-1-4614-6765-6.

37. Dhillon, B.S. *Applied Reliability and Quality-Fundamentals, Methods and Procedures*; Springer: Berlin/Heidelberg, Germany, 2006; ISBN 978-1-84628-498-4. Available online: https://link.springer.com/book/10.1007/978-1-84628-498-4 (accessed on 2 November 2022).
38. Ushakov, I.A. *Handbook of Reliability Engineering*; John Wiley & Sons: New York, NY, USA, 1994.
39. D'Urso, D.; Chiacchio, F.; Borrometi, D.; Costa, A.; Compagno, L. Dynamic failure rate model of an electric motor comparing the Military Standard and Svenska Kullagerfabriken (SKF) methods. *Procedia Comput. Sci.* **2021**, *180*, 456–465. [CrossRef]
40. Misra, K.B. *Handbook of Advanced Performability Engineering*; Springer: Berlin/Heidelberg, Germany, 2021; ISBN 978-3-030-55732-4. Available online: https://link.springer.com/book/10.1007/978-3-030-55732-4 (accessed on 2 November 2022).
41. Radziwill, N. *Discrete and Continuous Simulation: Theory and Practice*; Taylor and Francis Group: London, UK, 2014; Volume 21, ISBN 9780429073472. [CrossRef]
42. Banks, J.; Carson, J.S.; Barry, L.; Nelson, D.M.N. *Discrete-Event System Simulation: Pearson New International Edition PDF EBook*; Pearson Education: London, UK, 2013.
43. Zio, E. *The Monte Carlo Simulation Method for System Reliability and Risk Analysis*; Springer Series in Reliability Engineering; Springer: Berlin/Heidelberg, Germany, 2013. [CrossRef]
44. Kouvatsos, D.D. A Maximum Entropy Queue Length Distribution for the G/G/1 Finite Capacity Queue. In *Proceedings of the 1986 ACM SIGMETRICS Joint International Conference on Computer Performance Modelling, Measurement and Evaluation, New York, NY, USA, 1 May 1986*; ACM Digital Libraries: New York, NY, USA, 1986; pp. 224–236. [CrossRef]

applied sciences

MDPI

Article

Assisted-Driven Design of Customized Maintenance Plans for Industrial Plants

Néstor Rodríguez-Padial, Marta M. Marín and Rosario Domingo *

Department of Construction and Manufacturing Engineering, Universidad Nacional de Educación a Distancia (UNED), C/Juan del Rosal 12, 28040 Madrid, Spain; nesrodriguez@motril.uned.es (N.R.-P.); mmarin@ind.uned.es (M.M.M.)
* Correspondence: rdomingo@ind.uned.es

Abstract: Current production systems that respond to market demands with high rates of production change and customization use complex systems. These systems are machines with a high capacity for communication, sensing and self-diagnosis, although they are susceptible to failures, breakdowns and a loss of reliability. The amount of data they provide as a productive system and, individually, as a machine can be treated to improve customized maintenance plans. The objective of this work, with an operational scope, is to collect and exploit the knowledge acquired in the industrial plant on failures and breakdowns based on its historical data. The acquisition of the aforementioned data is channeled through the human intellectual capital of the work groups formed for this purpose. Once this knowledge is acquired and available in a worksheet format according to the Reliability-Centered Maintenance (RCM) methodology, it is implemented using Case-Based Reasoning algorithms in a Java application developed for this purpose to carry out the process of RCM, accessing a base of similar cases that can be adapted. This operational definition allows for the control of the maintenance function of an industrial plant in the short term, with a weekly horizon, to design a maintenance plan adjusted to the reality of the plant in its current operating context, which may differ greatly from the originally projected plan or from any other plan caused by new production requirements. This new plan designed as such will apply changes to the equipment, which make up the production system, as a consequence of the adaptation to the changing market demand. As a result, a computer application has been designed, implemented and validated that allows, through the incorporation of RCM cases already successfully carried out on the productive system of the plant, for the development of a customized maintenance plan through an assistant, which, in a conductive way, guides the plant maintenance engineer through their design process, minimizing human error and design time and leveraging existing intellectual capital.

Keywords: case-based reasoning; failure mode; effect and criticality analysis; knowledge-based system; nearest neighbor; reliability-centered maintenance

Citation: Rodríguez-Padial, N.; Marín, M.M.; Domingo, R. Assisted-Driven Design of Customized Maintenance Plans for Industrial Plants. *Appl. Sci.* **2022**, *12*, 7144. https://doi.org/10.3390/app12147144

Academic Editors: Dino Musmarra and Abílio Manuel Pinho de Jesus

Received: 28 March 2022
Accepted: 13 July 2022
Published: 15 July 2022

1. Introduction

The uncertainty of demand has led production systems to be increasingly complex. In this context, the production requires a strategic alignment with global company objectives and a high capacity on the part of plant operations and technology to respond to market needs [1] and flexible equipment with many configurations for their machines [2]. This can affect the availability of the machines and therefore their maintenance, which is why adequate management of the information is necessary to facilitate decision making. In fact, the academic literature shows tools that facilitate efficient maintenance management, such as Erozan [3], the selection of an effective maintenance strategy, such as Srivastava et al. [4], specific solutions for predictive maintenance in machining processes, such as Jimenez-Cortadi et al. [5], or a devotion to synchronizing manufacturing and maintenance activities [6]. All this shows that the problem of selecting the best maintenance strategy is conditioned by a high complexity and that multi-criteria decision systems must be used, as Aghaee et al. [7] points out.

On the other hand, the reliability is presented as principal focus for the maintenance [8]. However, in this current era of industry 4.0, according to Vrignat [9], lean manufacturing policies are considered insufficient and must be complemented with other policies of sustainability. The same author also highlights the need to incorporate new maintenance strategies with prognostics and health management to make decisions in time, creating opportunities for scheduled maintenance intervention and avoiding unexpected failure. Therefore, a system for decision making regarding the design of customized maintenance plans could attend to an industry demand.

Although there is a recent concern among plant experts regarding expanding the capabilities of the Reliability-Centered Maintenance (RCM) method in order to improve it, according to the work of Melani [10], these improvements involve the replacement of the RCM method by a sequential model of methodologies of Functional Analysis of Operability, Hazard and Operability Study (HAZOP), Analysis of Failure Modes, Effects and Criticalities, Failure Mode, Effect and Criticality Analysis (FMECA) and Analytic Network Process (ANP). Similarly, Ma et al. [11] suggests performing a data-driven RCM process, which basically replaces the FMECA with a quantitative analysis on possible maintenance effects and replaces the original RCM decision logic with a quantitative decision model based on MonteCarlo Simulation. Both of the proposed models involve an enormous effort in preparing all the documentation and studies to carry out this philosophy, resulting in an increase in the effort and time devoted to the application of a classic RCM process.

In addition to the methodologies to select the best maintenance action according to the RCM process, Condition-Based Maintenance (CBM) technologies must be integrated with Computer Maintenance Management Systems (CMMS) to achieve the maximum capacity of the three aforementioned systems, as indicated by López-Campos et al. [12], and include unified functional knowledge, as pointed out by Song [13]. This even includes TPM, as proposed Kumtekar et al. [14], and Failure Modes and Effects Analysis, as proposed by Mkalaf et al. [15].

This work presents a system for decision making regarding the design of customized maintenance plans within a production plant whose general objective is the proposal of a system that assists an expert in the decision making for the design of maintenance plans tailored to the real productive context of an industrial plant, as based on the alignment of the company's strategic objectives and tactical and maintenance operations, according to a customized concept.

As a result of the previously applied strategic definition, a productive area of the plant was selected to implement the improvements [16]. This area was taken as the starting point for the tactical definition, in which sections of the machine tree were analyzed in detail within the cited area according to the works of Rodríguez-Padial et al. [17,18]. The present operational definition, in synthesis, will try to solve a reliability problem posed by the proven success methodology of Reliability-Centered Maintenance (hereafter RCM) and driven by Case-Based Reasoning algorithms (hereafter CBR) to offer an optimized maintenance solution adapted to a new problem presented, such as a new case. On the other hand, Ruschel [19] considers it essential that the CBR methodology improves the various models used in decision making, seeking to anticipate maintenance actions before the failure occurs.

This maintenance solution will have a maintenance type or class format to be applied. Its frequency and specific instructions are to be developed in the future work order that will be programmed in the Computer Aided Maintenance Management software (hereafter CMMS).

As the state of the art on the integration of RCM and CBR methodologies, it is worth highlighting, as a background, the work of Cheng et al. [20], who developed an intelligent system called IRCMA with a similar approach, although with a great difference in terms of the approach regarding the treatment of the RCM process. Fundamentally, the big difference is that it tries to apply the CBR paradigm only in the FMECA part of the RCM methodology, making it possible to recover cases for this part, which is more difficult to document and consumes more time by experts, according to Candea et al. [21], in subsequently applying the decisional flowchart of the RCM process in the new context of the problem through consultation. In this way, it is possible to, on the one hand, group the advantages of the CBR algorithm—recovering similar cases of FMECA, achieving substantial savings in the time spent by the expert and maximizing human capacities in terms of remembering similar cases when there is an abundant base of cases—and, on the other hand, correctly apply the RCM criteria regarding the classification of the maintenance policy to be carried out, not by recovering a similar CBR case. This is intended to use only the recovery of cases in the FMECA part of the method, closely related to the functional failure and its root cause, a fact that depends more on the type of equipment but not on the maintenance policy, which depends more on the context where said team works by using the original RCM criteria established by Moubray. There is a remarkably small bibliography in which both RCM and CBR methodologies are integrated, as indicated by Kobbacy [22], exclusively evidencing the IRCMA reference already cited Cheng et al. [20] until 2012. In this sense, a current bibliographic search shows that there are no other related publications to date.

The objective of this work of operational definition, as a final part of the proposed global system, is to assist in the design of customized maintenance plans focused on reliability in a driven way—that is, to guide the maintenance expert in an ideal way according to RCM method. In this way, human error is minimized, as Rahman et al. [23] suggested, and an adequate level of excellence is ensured in the applied methodology. On the other hand, the time spent by the maintenance expert in case recovery work, analysis and the adaptation of the same case to the new case or the reliability problem presented is drastically reduced, since it is carried out automatically by means of a software application developed for this purpose. A fundamental advantage resides in the fact of having a large number of historical cases, and the expert will only consider the most similar k-cases, kNN, where k is the number of similar cases that will be displayed by the application. All of this translates the operational objectives by integrating the RCM method with the CBR methodology. Therefore, the objective is the operational definition, allowing the expert or leader of the RCM group to be efficiently conducted in the RCM method with great savings in devoted time, specifically for the part of the method where failure modes, effects and criticalities—FMECA—are analyzed for the organization's business.

2. Methodology

In the design of this decision support system and the planning of customized maintenance processes of any industrial plant, it has been decided to use the RCM method due to its proven effectiveness in the industrial environments where it is implemented and which are equipped with a knowledge-based system through CBR algorithms to endow the RCM process with faster decision making by a group of RCM experts in its implementation. In parallel, being a driving-assisted system, human errors caused by forgetfulness or carelessness in the correct implementation of the RCM method are minimized.

Finally, the system will be a computer support, as an independent application that implements the RCM method according to CBR algorithms, in such a way that results in a driven RCM method, assisting the expert from its inception to completion. The following sections very briefly describe the RCM and CBR methodologies.

2.1. RCM Method

The Reliability-Centered Maintenance (RCM) method was originally designed for the aviation industry by Nowlan and Heap [24], Anderson and Neri [25] and Smith [26], and, later, an extension of the RCM concept was used by Moubray [27] for a generic industrial environment. The concept is based on designing a maintenance plan based on the operational reliability of the plant. In summary, the functions required by the systems are defined; then, a Failure Modes, Effects and Criticalities Analysis (FMECA) is used, where they consider the possible functional failures being listed and link the different associated failure modes, their effects and their criticalities. The critical elements of the analyzed productive system are highlighted and located. The information is collected in an information worksheet, as shown in Figure 1, adapted from Waeyenbergh and Pintelon [28]. The next step indicates the use of a decision diagram, which helps in deciding the type of maintenance to apply, whose structure is shown in Figure 1 and detailed in Figure 2, the latter adapted from Saniuk et al. [29]. This diagram helps in recording the decision worksheet.

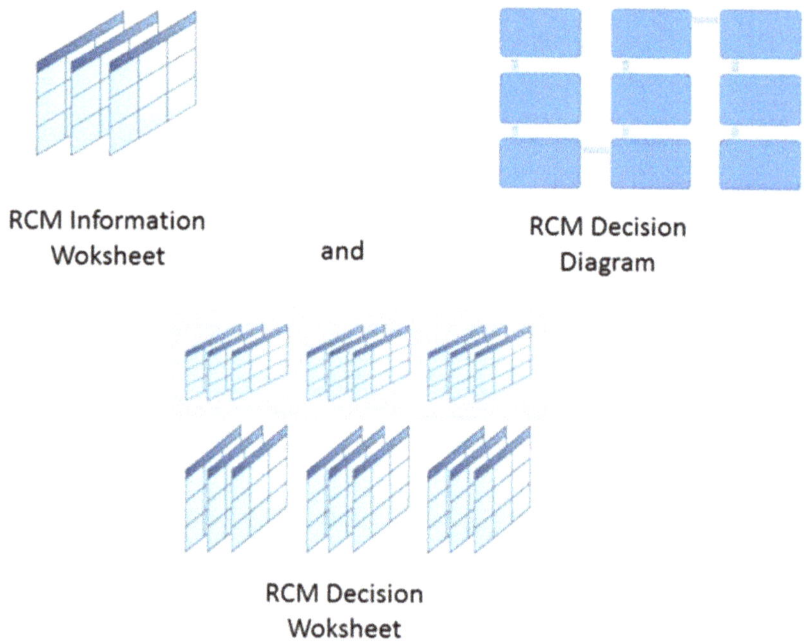

Figure 1. RCM approach, structured with information worksheets, a decision diagram and a decision worksheet; adapted from [28].

Although RCM is a proven, successful method focused on increasing the plant reliability, strictly applied, it is a complex and costly method, justifiable in aeronautical and high-risk industries, where high reliability requirements are imperative, but too expensive for the general industry, where the maintenance problem is more economical than reliability. These disadvantages can be overcome by replacing the strict application of the method with a more flexible and customized approach to the maintenance to be applied.

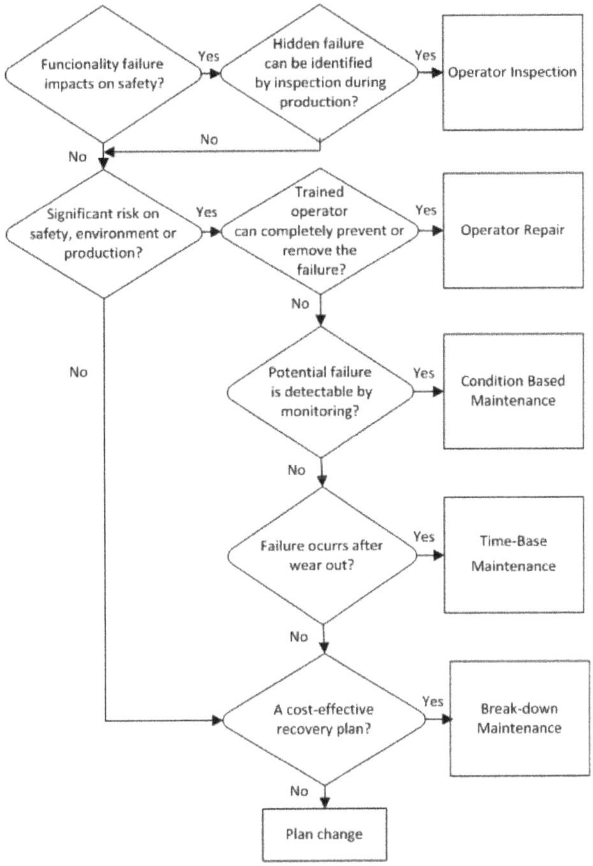

Figure 2. Detail of the RCM decision diagram; adapted from [29].

2.2. CBR Method

The Case-Based Reasoning (CBR) method was originally pioneered by Schank [30] from his research work on dynamic memory [31], in which he discovered that memory recalls solved cases as similarity patterns to solve new problems posed. In this way, he defines reasoning based on cases as "the resolution of problems using or adapting the solutions of old problems" [32]. Case-Based Reasoning (CBR) is an artificial intelligence paradigm that combines problem solving and learning, as established by Watson [33]—that is, Case-Based Reasoning (CBR) is a method to describe a problem posed where various technologies can be used for its resolution. In this work, jCOLIBRÍ, developed by Díaz-Agudo [34], has been used as a technology for the algorithmic implementation of the Case-Based Reasoning (CBR) methodology in the Java programming language [35].

In summary, the Case-Based Reasoning (CBR) method is based on the hypothesis of human reasoning, where similar problems present similar solutions; therefore, it is about solving problems by adapting solutions to problems previously solved successfully and by describing them in a similar way to the new problem raised. In this sense, it is possible to differentiate two spaces: the space of the problem posed, called the description space, and the solution space. The CBR method is described as a cyclical process, as seen in Figure 3, with four activities: Recover, Reuse, Review and Retain. Once a new problem arises, a new case is created in the description space, and the Retrieve activity makes it possible to obtain from the case base those problems that are most similar to the new case described.

Once the most similar case is obtained, called the recovered case, the second activity is called Reuse—that is, the solution of the recovered case is used as an approximation of the solution of the new problem posed. This proposed solution, if necessary, is revised or better adapted to the problem initially raised, obtaining an adapted case with a new solution. If this solution is validated as good, the last activity consists of retaining the case as a new learned case and incorporating it as a new case added to the case base. As can be seen, the CBR method seeks to increase knowledge by incorporating new cases each time the CBR cycle is invoked.

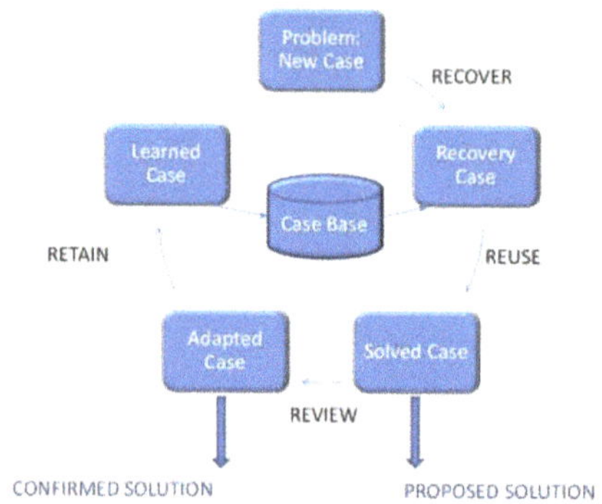

Figure 3. Standard cycle of the CBR method adapted by Watson [33].

Of the four activities, attention is focused on the recovery activity to establish that this methodology allows for the use of different technologies to recover cases due to similarity. In this work, the Nearest Neighbor algorithm (hereafter NN) has been used to model the similarity between the new case and the existing cases in order to obtain the recovered case. kNN is the most widely used recovery method in the CBR methodology [36]; this has been evaluated as the most accurate method compared to other methodologies such as k-means clustering (k-means), FCM (Fuzzy C-Means) and SOM (Self-Organizing Map). According to Watson [33], the similarity can be assessed as described in the following equation:

$$\text{Similarity}(\text{T}, \text{S}) = \sum \text{f}(\text{Ti}, \text{Si}) \cdot \text{wi}, \tag{1}$$

The similarity between an objective case, T, and a case of the base of cases, S, is evaluated with the similarity function, f, for the attribute, i, extending it from the first to the last, n, of each case. As can be seen in "(1)", it is a sum weighted by the weight, wi, that marks the importance of each attribute, i. Extending this assessment between the objective case, T, and all the cases of the case base, Sj, of those existing, m, a normalized and ordered list can be obtained from the highest to the lowest, thus configuring from the most similar cases to those least close to the target. Finally, only a group of them will be taken from the entire list, the first k most similar to the objective. In this work, a value of k = 3 has been considered, according to the recommendation of an interval comprised of 1 to 4 according to the study by Salton and McGrill [37]. It has been shown in some cases that the choice of the value of k influences the efficiency of the CBR process. Thus, a possible improvement of the method to increase the efficiency of the recovery activity could be to not tacitly decide a value of k, if not to calculate it automatically based on the optimization of the statistical disparity applied to the base of cases, avoiding human intervention and its associated error for choosing the value of k [38].

2.3. Implementation of the CBR Method with jCOLIBRI

The driven design system for Reliability-Centered Maintenance (hereafter driven RCM) has been implemented in the Java programming language using specialized CBR libraries, for which the jCOLIBRI environment has been selected, according to Recio-García [39]. In particular, jCOLIBRI version 2 or jCOLIBRI 2 has been used as a framework for the development and construction of CBR systems developed by Recio-García [40] in Java. In this work, the development layer in Java has been used, whose architecture allows for the representation of cases and their handling by means of a specific library of CBR methods to be used in the dedicated CBR application that is the object of this work, called AIRCM.

The division of the application is modularized in the pre-cycle, cycle and post-cycle periods. The pre-cycle and post-cycle periods contain methods in charge of managing the resources required by the main cycle methods. There are two main methods of the cycle: (1) Recovery and Selection Methods. From this group of methods, the closest neighbor (NN) recovery method has been used, with global and local similarity metrics. The local metrics use the similarity function, f, of "(1)" to calculate the similarity value between the different attributes that describe the target case and those of the comparable cases in the case base, while the global metric calculates the weighted mean (wi) of the similarity values returned by the local metrics. Once the cases are retrieved, methods are used that select a group of the retrieved cases to present them to the end user. The simplest approach consisting of returning the k best cases has been used, the combination of which brings together the retrieval method of the k most neighbors next or kNN. (2) Reuse and Review Methods, which include basic methods for copying the solution of a case to the query or for copying the values from the attributes to the description of the solution.

Finally, the structure of the cases is defined in a series of classes that enable all the representations of the cases.

3. Application: Results and Discussion

The Case-Based Reasoning (CBR) and Reliability-Centered Maintenance (RCM) methodologies have been integrated with the ultimate objective of obtaining a customized design of maintenance plan adjusted to the real needs of industrial plants and assisting the person in charge of maintenance in its design. This integration of both methodologies, implemented in the Java programming language, results in the obtention of an independent computer application for the driven RCM model called Artificial Intelligence Reliability-Centered Maintenance (AIRCM) and designed for the effect to be applied in any industrial environment.

In particular, the AIRCM application has been conceived in two parts. The first part implements the FMECA, typical of the RCM method, with the CBR method using the jCOLIBRI environment and the second part, where the maintenance policy has been implemented according to the decisional flow chart in Figure 2, using the operating context of the new case raised. Therefore, the maintenance policy adopted for the recovered case is discarded. This is because the maintenance policy depends more on the operational context where the equipment is located—that is, its effects and not the failure mode—as this is more related to the type of equipment and the maintenance policy with the productive context.

Every Knowledge-Based System (KBS), such as the CBR method, must have access to and the ability to modify a database where the different historical cases validated as successful are stored. This case base is made up of the worksheets (Figure 1) of FMECA information once the classic RCM method has been successfully applied in a productive environment. These record sheets have been considered as already existing and will suppose that the input information for a record in the database is a base of cases that have been interconnected to the AIRCM application that is the object of this work. Since the data sheets are usually arranged in spreadsheets, it is easy to dump the information and its structure into a new spreadsheet that will make up the case record and therefore the case base that the AIRCM application needs. This case base is a Comma-Separated Values (CSV) file. Figure 4 shows the values of the CSV file in tabular form, each column corresponding to the attributes specified in Table 1.

#IDP	S	I	E	FF	IDS	F
1	EW	RESMAS	AXIS	Fit format out of range	1	Fit to format width
2	EW	RESMAS	AXIS	Do not center the axis with each other	2	Center all the axes with respect to each other on the machine axis
3	EW	RESMAS	AXIS	Do not center any axis on the rest	3	Center all the axes with respect to each other on the machine axis
4	EW	RESMAS	FOLDERS	Do not fold	4	Lateral folding of the package v=15 folded/min
5	EW	RESMAS	FOLDERS	Do not fold	5	Lateral folding of the package v=15 folded/min
6	EW	RESMAS	FOLDERS	Do not fold	6	Lateral folding of the package v=15 folded/min
7	EW	RESMAS	GLUE	Do not dose the glue through any nozzle	7	Dosage of glue for gluing the package dc/dt=15 lines/minute x 3 nozzles
8	EW	RESMAS	GLUE	Do not dose the glue through any nozzle	8	Dosage of glue for gluing the package dc/dt=15 lines/minute x 3 nozzles
9	EW	RESMAS	GLUE	Do not dose the glue through any nozzle	9	Dosage of glue for gluing the package dc/dt=15 lines/minute x 3 nozzles
10	EW	RESMAS	GLUE	Do not dose the glue through any nozzle	10	Dosage of glue for gluing the package dc/dt=15 lines/minute x 3 nozzles
11	EW	RESMAS	GLUE	Do not dose the glue through any nozzle	11	Dosage of glue for gluing the package dc/dt=15 lines/minute x 3 nozzles
12	EW	RESMAS	GLUE	Do not dose the glue through any nozzle	12	Dosage of glue for gluing the package dc/dt=15 lines/minute x 3 nozzles
13	EW	RESMAS	GLUE	Do not dose the glue through any nozzle	13	Dosage of glue for gluing the package dc/dt=15 lines/minute x 3 nozzles
14	EW	UNWINDER	EDGE TRIMMER	Do not cut	14	Cut the width of the package wrapping paper format. (vp=4 bar, ph=2 bar
15	EW	UNWINDER	EDGE TRIMMER	Do not cut accurately	15	Cut the width of the package wrapping paper format. (vp=4 bar, ph=2 bar
16	EW	UNWINDER	EDGE TRIMMER	Do not cut accurately	16	Cut the width of the package wrapping paper format. (vp=4 bar, ph=2 bar
17	EW	UNWINDER	EDGE TRIMMER	Do not cut accurately	17	Cut the width of the package wrapping paper format. (vp=4 bar, ph=2 bar
18	EW	I/0 PALLETIZER	PALLETIZER	During loading, the pallet turns as it passes through the intermediate belt	18	Automatic loading and positioning of empty pallets
19	EW	I/0 PALLETIZER	PALLETIZER	It does not position the pallet correctly	19	Automatic loading and positioning of empty pallets
20	EW	I/0 PALLETIZER	PALLETIZER	It does not position the pallet correctly	20	Automatic loading and positioning of empty pallets
21	EW	I/0 PALLETIZER	PALLETIZER	Do not transfer the loaded pallet	21	Transfer loaded pallets by belt to the transport line
22	EW	I/0 PALLETIZER	PALLETIZER	Do not transfer the loaded pallet	22	Transfer loaded pallets by belt to the transport line
23	EW	RESMAS	PADS	Not adequacy of the minimum pressure according to the optimum operating value	23	The pads must be adjusted to the height of the package and introduce a pressure between a minimum and a maximum to achieve adequate quality in the wrapped package. 1st PRESS: p1{min=1,max=2.5} bar, 2nd PRESS: p2{min=0,max=1.5} bar, 3rd PRESS: p3{min=1,max=2} bar
24	EW	RESMAS	PADS	Not adequacy of the minimum pressure according to the optimum operating value	24	The pads must be adjusted to the height of the package and introduce a pressure between a minimum and a maximum to achieve adequate quality in the wrapped package. 1st PRESS: p1{min=1,max=2.5} bar, 2nd PRESS: p2{min=0,max=1.5} bar, 3rd PRESS: p3{min=1,max=2} bar
25	EW	RESMAS	BRAKES	It does not perform the braking in the time required for the manoeuvre. Excessive braking time	25	Perform motor braking act in the maneuver
26	EW	RESMAS	TORQUE LIMITER	It does not perform the braking in the time required for the manoeuvre. Excessive braking time	26	Perform an adaptation during start and stop between the accto. (motor) and the drive gear of the load transmission (rams and conveyor belts)
27	EW	RESMAS	DEPLOY BARS	Do not adjust the unwinding length within the allowable tolerance	27	Pre-unwind the wrapping to the length of the format (P) according to GRM specifications calculation: P=A+H-130 (mm)
28	EW	RESMAS	DEPLOY BARS	Do not adjust the unwinding length within the allowable tolerance	28	Pre-unwind the wrapping to the length of the format (P) according to GRM specifications calculation: P=A+H-130 (mm)
29	EW	RESMAS	DEPLOY BARS	Do not center the unfolded packaging on the bands	29	Pre-unwind the wrapping to the length of the format (P) according to GRM specifications calculation: P=A+H-130 (mm)
30	EW	RESMAS	DEPLOY BARS	Do not center the unfolded packaging on the bands	30	Pre-unwind the wrapping to the length of the format (P) according to GRM specifications calculation: P=A+H-130 (mm)
31	EW	RESMAS	LV DISTRIBUTION PANEL	The alarm system for Circuit Breaker	31	Magneto-Thermal Protection of the electrical installation of the machine. Protection settings according to motors to be protected
32	EW	RESMAS	LV DISTRIBUTION PANEL	Alarm System for Jumping Electric Torque Limiter	32	Magneto-Thermal Protection of the electrical installation of the machine. Protection settings according to motors to be protected
33	EW	UNWINDER	LOADING OF COILS	Partial cargo handling assistance	33	Assist in the loading of packaging reels to the operator in the form of ergonomic loading
34	EW	DESPALLETIZER	PROTECTION	Uncontrolled rise	34	Package Feeding Lift Platform Rise
35	EW	DESPALLETIZER	PROTECTION	Uncontrolled rise	35	Package Feeding Lift Platform Rise

Figure 4. Records of the CSV file that make up the Case Base. Attributes {IDP}.

Table 1. Attributes of the Case Base. CSV Database File Column Headers.

Attributes	Description	Typology
IDP	Problem Identifier	Sequential value of cases. Integer
S	Machine Tree Section	Alphanumeric code that identifies the section
I	Installation inside the machine tree section	Alphanumeric code that identifies the installation
E	Equipment within the machine tree installation	Alphanumeric code that identifies the equipment
FF	Functional Failure	Description of the failure that makes the partially required or total required function. Character string
IDS	Solution Identifier	Sequential value of cases. Integer number
F	Equipment Required Function	Description of the required function and its nominal operating values. Character string

Table 1. *Cont.*

Attributes	Description	Typology
FM	Failure Mode	Description of the failure modes that can trigger the failure. Character string
FE	Failure Effects	Description of the effects that affect the safety, production and economy of the plant. Character String
NPR	Risk Priority Number	Real value that indicates the priority regarding the risk evaluated by the failure. It is the product of the probability of occurrence (O), the severity (S) and the detection (D). Real number
PT	Proposed Task	Description of the maintenance actions to be carried out. Character string
MC	Maintenance Classification	Maintenance policy to be carried out. Listing: Periodic search for hidden faults, operator-based maintenance, condition-based maintenance, time-based maintenance, unplanned (corrective) maintenance, redesign
II	Initial Interval	Frequency of the application of systematic preventive maintenance actions. Integer number
R	Maintenance Responsible	Group, person or workshop responsible for the correct execution of the planned maintenance tasks

This section has been divided into three phases, according to chronological order: the preparation of the case base used the design and implementation of the AIRCM application and the use of it to solve a new case raised, enabling the applicability of the driven RCM method through running the AIRCM application.

3.1. Case Base

For this work, a case base has been prepared with a total of 35 cases (Figure 4) by means of the easy dumping of data from the worksheets corresponding to 35 problems that actually occurred and were successfully solved under the classic RCM method on a specific machine within the productive area as a result of a previous strategic definition.

The machine is identified with the Section code "EW", the result of a previous tactical definition. This machine is divided according to a machine tree in hierarchical format (Section, Installation, Equipment) to locate the zone where the fault occurs.

3.2. Design and Implementation of Conductive RCM: Application

Once a database with structured attributes has been established, the implementation of the AIRCM computer application begins, using the Java programming language under the eclipse environment. In the first part of the code, the jCOLIBRI environment has been used to implement the CBR method with the specific libraries designed for this purpose. In the second part, the code has been implemented through conditional sentences to establish the maintenance policy according to the RCM decision diagram for the operational context of the new case presented. To implement the code, the Integrated Development Environment (IDE), eclipse [41] has been used, since jCOLIBRI presents a framework prepared for this environment through a perspective intended for this purpose, as can be seen in Figure 5.

The design has been divided into four phases: configuration, pre-cycle, cycle and post-cycle. In the configuration, a mapping has been established between the attributes of the case base of the CSV file and the computational variables of the computer application. The correspondence between the attributes of the case base, Table 1, and the computational variables of the application, the central red box in Figure 5, which distinguish between the descriptive variables of the problem "d." and the descriptive variables of the solution "s.", is noted. The complete mapping can be seen in Table 2. This is done with the assistance of jCOLIBRI assigning headers through a drop-down of variables. Regarding persistence or memory storage, the location of the case base must be indicated by its memory address, as can be seen in Figure 5 on the file that contains RCM_EW.csv. The last part of the configuration is based on the third XML file that can be seen inside the config folder of the same figure, called similarityConfig.xml, which is generated from the jCOLIBRI wizard, as can be seen in Figure 6. In this figure, the similarity of the method used has been configured—in this case, Nearest Neighbors (NN). It can be seen that the variables to consider for the comparison of cases are the three inherent locations within the machine tree (Section, Installation, Equipment) and that of the problem description (Functional Failure). Likewise, these variables have been combined with each other to obtain a global mean according to the chosen variable, weighting them to give more emphasis to those variables that have been considered more important, according to "(1)". In this case, it has been decided to weight the functional failure variable with 70%, considering it the most important, since it is what the application tries to solve; however, the equipment and its location within a machine must be taken into account to resolve a possible tie between similar descriptions. In this case, it involves a 30% weighting, each variable of the section, installation and equipment being distributed equally at 10%. The local similarity function used for all the variables is the MaxString function in order to compare the similarity between the texts of the new case raised (objective) and all the available cases in the case base.

In the pre-cycle and post-cycle phases, cases have been loaded before executing the cycle and the resources have been released once the cycle has been executed, respectively, through connectors to the case base. However, the corresponding cycle phase contains the logical functioning flow of the CBR method (Figure 3) and is executed through queries. This cycle presents two stages. The first one applies the flow of CBR on the FMECA part of the RCM method; this is a recovery of failure modes within the base of cases, comparing them with the new problem raised through consultation. The second stage attempts to reapply the maintenance actions, re-evaluating the Risk Priority Number (NPR), the maintenance task instruction, the maintenance policy or class, the interval and the person assigned.

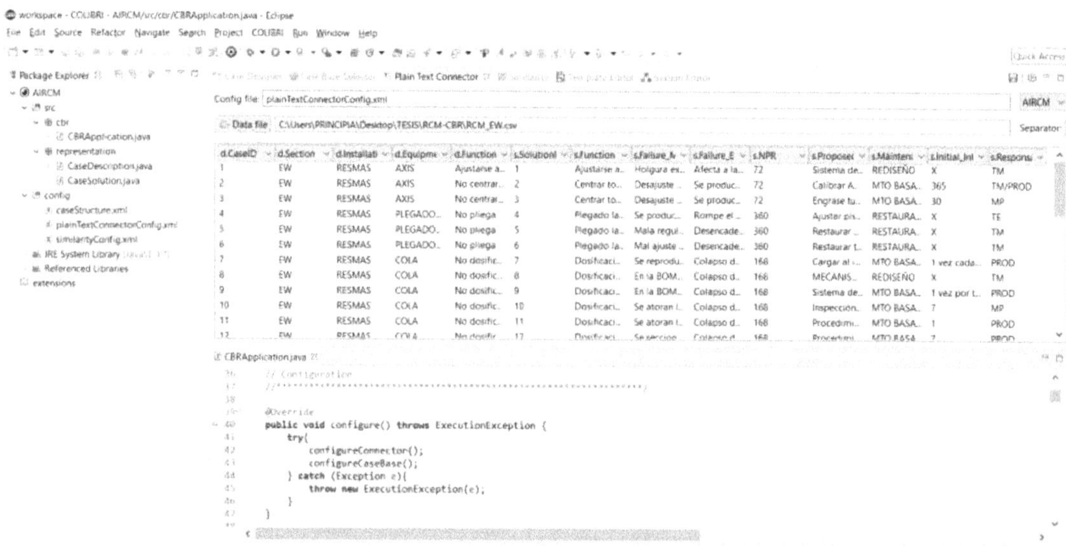

Figure 5. Java development environment in the eclipse IDE from the jCOLIBRI perspective.

Table 2. Attributes Mapping of the Case Base and Variables implemented in the AIRCM Application.

Attributes Base Case (CSV)	AIRCM Variables	Domain or Space
E	Equipment	Description
FF	Functional_Failure	Description
I	Installation	Description
IDP	CaseID	Description
S	Section	Description
F	Function	Solutions
FE	Failure_Effect	Solutions
FM	Failure_Mode	Solutions
IDS	SolutionID	Solutions
II	Initial_Interval	Solutions
MC	Maintenance_Classify	Solutions
NPR	NPR	Solutions
PT	Proposed_Task	Solutions
R	Responsible	Solutions

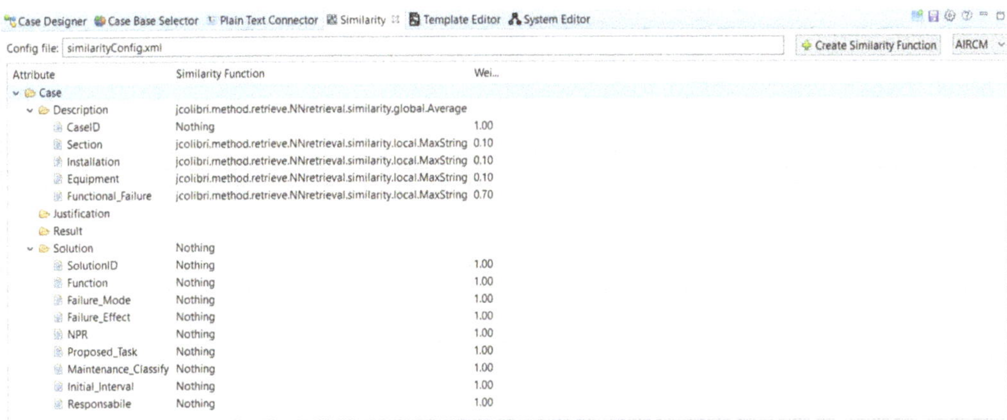

Figure 6. Configuration of the similarity function of the NN method under the jCOLIBRI perspective.

3.2.1. First Stage—CBR-Based FMECA

Recover: At this stage of the Java source code using jCOLIBRI libraries, the NN method has been implemented. In the first place, the similarity configuration seen in the configuration phase (Figure 6) has been obtained to later use it in the retrieval of cases, according to a query introduced in the cycle method as an input parameter. This query corresponds to the description of the problem of the new case raised. In the second instance of the code, the collection of all retrieved cases has been stored in a variable, where a score has been added for each case based on its similarity to the query.

Reuse: In this stage of the code, the selection of the k closest cases has been implemented due to their similarity to the new case. In this work, it has been decided to select the k = 3 most similar cases, considered as the ideal value within the range [1,9], as justified previously. In this way, the user has been allowed to choose between the three most similar cases, making himself the final decision maker on the similarity of the new case presented with respect to the case base. In the first sentence, the selected cases variable stores the three closest cases in order of the similarity of the cases retrieved in the previous eval collection, while in the second sentence, the choice variable stores the three cases selected but orders them in a table to be presented on the screen to the application user by using a pop-up window.

However, this window, in addition to presenting the three most similar cases in descending order, allows the user to decide which case is finally selected from the three presented, allowing the user to decide which one best suits the problem, as seen in the third block of the code using the Buy or Quit conditional.

Review: At this stage of the code, the user has been allowed to review the selected case by creating a case using the bestCase variable and copying the selected one (choice). The revision itself is based on modifying the values contained in the attributes of the identifier, the newCaseID description and the newSolutionID solution of the bestCase case in such a way that a new case is prepared with sequential identification to the last case of the case base.

3.2.2. Second Stage—Driven RCM within the CBR Cycle Review Activity

In this second stage, being within the review activity within the CBR cycle, the code has been implemented to modify the values of those attributes that involve the operational context by the RCM method—that is, instead of retrieving the values of the attributes of the entire solution, only those associated with the FMECA are retrieved, and the rest of the attribute values will be modified (revised) by the user according to the RCM decision-making methodology through the decision diagram (Figure 2), allowing the operational context to be reconsidered again when designing the solution of the new case raised. The aforementioned decision diagram has been implemented in Java code through conditional sentences so that the solution can be reviewed through established questions—hence the name of the conductive RCM method. The attributes whose values have been reviewed through the implementation of the driven RCM are: Risk Priority Number (NPR) (through its occurrence, severity and detection factors), Proposed Task (PT), Initial Interval (II) and Responsible (R). Finally, the new maintenance policy or Maintenance Classification (MC) has been applied, obeying the RCM decision diagram and leading the user through the questions to obtain the maintenance action to be applied in this context.

The final part of the cycle is the Retain activity. In this case, the code only presents a single sentence. Where the new case chosen and reviewed, bestCase has been stored in the case base by adding one more instance in the RCM_EW.CSV file.

3.3. *Use of the Application to Resolve a New Failure Case by Assistance*

Starting from a situation in which the complete planning and design assistance system has been launched (that is, a previous strategic definition has been applied) and obtaining as a result a productive area of interest to the management, for that productive area, a previous tactical definition by middle managers is later applied, resulting in a machine (section) to which the RCM design must be applied in order to improve its reliability and maintainability. At this moment, instead of applying a classic design using the RCM method, the use of driven RCM is made possible through the AIRCM application designed in a previous section for this purpose, with the advantages of minimizing the time spent by the users responsible for the design and the human errors inherent in handling extensive case databases.

At this point, the AIRCM application has been launched on the machine identified as section "EW", the result of a previous tactical definition. Once the case base has been prepared, the data needed as input information will be fourfold, as can be seen in Figure 7. The last three variables locate the area where the problem to be analyzed has occurred, while the textual description of the problem has been entered with the first input variable. This has been prepared in this way because users (machine operators) who request maintenance intervention must submit a work request, PT, which is nothing more than the start of a Work Order (OT) of the CMMS application used by the maintenance department. It takes advantage of the existing equivalence between the description of a work request and the functional failure of a device, attribute FF, in Table 1.

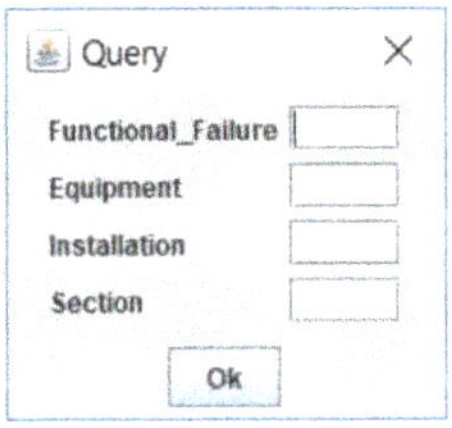

Figure 7. Initial window for input information running on Windows 10.

The input information of the new problem posed in the query window of Figure 8 has been introduced, describing the problem as a Functional Failure: "They do not center the axes"; for the AXIS equipment; for the Equipment: "AXIS"; within Installation: "RESMAS"; and Section: "EW".

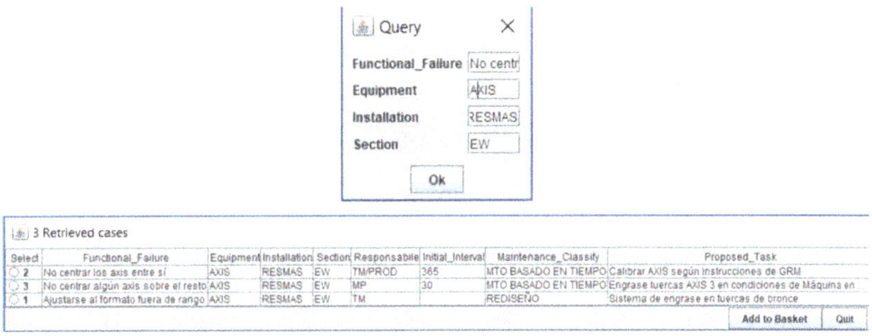

Figure 8. Problem data input (above) and output window with the three similar cases.

The three recovered cases similar to the problem raised have been obtained and ordered in descending order of similarity. As can be seen in the table in Figure 8, the case identified as ID = 2 of the Cases Base is the most similar to the problem posed. As can be seen, the cases ID = 3 and ID = 1 will be the second and third most similar among those found in the case base. However, the user of the application can choose the case that best suits the three cases presented, since it also allows him to see other variables such as failure modes or failure effects, which complete the FMECA. In this way, it is possible to recover similar cases and carry out an analysis of failure modes, effects and criticalities faster and easier than doing it again, in addition to adding the existing knowledge of the previous analysis of proven success.

Once the user has chosen the case that best suits the needs of the new case or problem raised, he selects it himself by clicking on the Select attribute of the three cases presented on the screen. In this work, he has chosen the first one, ID = 2, since it is the most similar to the problem posed. Obtaining as an output the recovered case ID = 2, the review stage begins, assigning the case ID = 36 as an additional case to the 35 existing cases in the case base.

Continuing with the review process, the second stage of the driven RCM process begins immediately by reviewing the contextual information of the new problem, as shown in Figure 9. With the occurrence, severity and undetectable data entered, the new Risk Priority Number (NPR) is automatically calculated.

The second revision stage comprises redefining the new maintenance policy or class by following the RCM diagram in a guided way—that is, the user has been guided through the questions in the diagram, as seen in Figure 10, where the user answers the questions and the maintenance policy is obtained as an output. In this work, the maintenance classification obtained is Maintenance by Operator, thus completing the solution of the chosen case (see the last case in Figure 11).

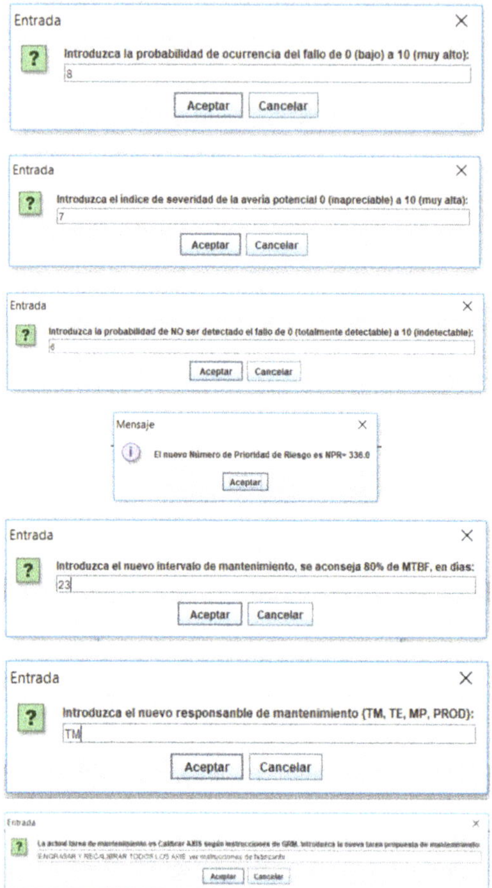

Figure 9. Data review windows for the new chosen case using questions from the operational context.

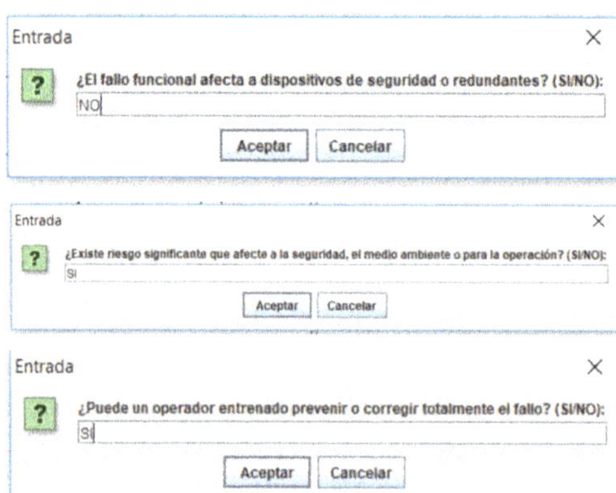

Figure 10. Data review window for the new chosen case using questions of operational context. Second stage: driven RCM to apply the maintenance policy.

#IDP	S	I	E	FF	IDS	F
33	EW	UNWINDER	LOADING OF COILS	Partial cargo handling assistance	33	Assist in the loading of packaging reels to the operator in the form of ergonomic loading
34	EW	DESPALLETIZER	PROTECTION	Uncontrolled rise	34	Package Feeding Lift Platform Rise
35	EW	DESPALLETIZER	PROTECTION	Uncontrolled rise	35	Package Feeding Lift Platform Rise
36	EW	RESMAS	AXIS	Do not center the axis with each other	36	Center all the axes with respect to each other on the machine axis

Figure 11. Case Base in the RCM_EW.CSV File with the new Case ID = 36 REGISTERED.

Finally, the retention activity is checked by verifying the persistence (that is, verifying that the new case is stored in the database)—in this case, in the RCM_EW.CSV file—and that a new case is added with the description identifiers and solutions IDP = 36 and IDS = 36, respectively, as evidenced in Figure 11. The newly added case is also highlighted in Figure 11, where it can be seen how the case number 36 has been added within the case identifier attribute as a consecutive number to the last existing case in the original case base. It is verified that all the registered attributes of case 36 contain all the values obtained in the solution process of the new case, according to the conductive process in Figures 8–10.

3.4. Achieved Development

A maintenance model focused on driven RCM has been developed in such a way that it assists the user responsible for the customized design of the maintenance plans of any industrial plant, making three great advantages possible: the first is the considerable saving of time involved in carrying out an FMECA; the second the minimization of human error regarding the treatment of an extensive series of historical cases; and the third advantage is that the AIRCM application created for this purpose is used as training for the inexperienced maintenance user, where they can be fully guided in the process, accessing all the historical cases stored and successfully resolved. In this way, an AIRCM application is available, which can be used as a training system through case simulation, providing a didactic component.

Once the AIRCM application has been validated, different input cases are entered and the corresponding solution is obtained. The advantages and benefits of the proposed driven RCM model have been successfully tested by raising a series of new cases in the EW section, where there is a base of success stories of RCM programs already implemented.

3.5. Results Analysis

A conductive RCM model has been developed in such a way that it assists the user responsible for the customized design of maintenance plans for an industrial plant, thus enabling three major advantages: the first is the considerable time savings involved in carrying out a Failure Modes, Effects and Criticalities Analysis (FMECA); the second is the minimization of human error in the treatment of an extensive series of historical cases; and the third advantage is that the AIRCM application created for this purpose is used as training for the inexperienced user of maintenance, where he can be totally guided in the process, accessing all the historical cases stored and successfully resolved. In this way, he has an AIRCM application that can be used as a training system through case simulation, providing a didactic component.

Once the AIRCM application has been validated, different input cases are entered and the corresponding solution is obtained. The advantages and benefits of the proposed conductive RCM model have been successfully tested by proposing a series of new cases in the EW section, where there is a database of successful cases of Reliability-Centered Maintenance (RCM) programs already implemented. The new cases have enabled the simulation in the AIRCM application environment, obtaining the results shown in Table 3. The total number of simulated cases for each of the input variables {INSTALLATION, EQUIPMENT and FUNCTIONAL FAILURE} is shown with X.

Table 3. Simulations carried out for the validation of the AIRCM application.

Simulation	Installation	Equipment	Functional Failure	Success (%)
1	X			100
2		X		100
3			X	100
4	X	X		100
5		X	X	100
6	X		X	100
7	X	X	X	100

It should be noted that the robustness of the proposed conductive RCM model has been verified, leaving some values corresponding to the input variables empty, as evidenced in Table 3 for simulations 1 to 3, where only one input variable has been reported, for simulations 4 to 6, reporting two input variables, and for simulation 7, reporting all the input variables. For all the simulations carried out, it has been observed that the retrieved cases most similar to the problem posed are correctly displayed as results, even with a lack of input information, thus concluding the validation process. In the same way, the goodness of the system has been verified, introducing some input cases as a query to the problem. This only informs the FUNCTIONAL FAILURE (FF) input field in the query, approximately as what would be done in a real case, leaving the rest of the fields—EQUIPMENT, INSTALLATION and SECTION—empty, without input information. That is, it is about describing the problem of an existing case by means of a similar description to observe the response of the system. Table 4 highlights this type of simulation to show the results, such as similar recovered cases, concluding that there is a good system response.

Table 4. Simulations of raised cases and similar cases recovered.

Posed Problems	Query Description (FF)	First Recovered Case (Description FF)	IDs
1	Packaging is folded	Not centering the unfolded packaging on the bands	29, 30, 2
2	Magneto-thermal trigger	The alarm system for magneto-thermal failure	31, 14, 15
3	Pallet turns	Not transferring the loaded pallet	21, 22, 19
4	Paper length is not adjusted by spreader bars	Not centering the unfolded packaging on the bands	29, 30, 27

The use of the application developed in a maintenance design process in an industrial plant in operation has allowed us to observe the following improvements in the results.

Empirically, a notable reduction in the time devoted to applying the Reliability-Centered Maintenance (RCM) methodology has been achieved. Considering that a case like the one presented in this work requires a series of meetings of the RCM group created expressly, in addition to a documentary analysis of the specifications of the assets subject to maintenance and their operational context, the development of the Mode Analysis process of Failure, Effects and Criticalities (FMECA) and the correct application of the Reliability-Centered Maintenance (RCM) flowchart, a reduction of the four hours dedicated to the classic RCM process by the person responsible for the maintenance design is obtained—only thirty minutes is required to use the conductive RCM process through the AIRCM application developed and used in this work. This does not account for the time that has had to be devoted by the members of the RCM group in the different meetings organized.

Another advantage achieved with the execution of the AIRCM application has been the management of the knowledge acquired regarding the resolution of problems that have occurred in the plant. Starting from successfully executing the application of the conductive RCM program a vast number of times, a large number of effectively resolved historical cases will be obtained, which implies a large amount of information, which will be considered as stored intellectual capital. When a large number of cases are available in a database, it becomes difficult and arduous to manipulate it through direct consultation by a maintainer. This phenomenon is mitigated by similar case retrieval processes implemented in the application itself, so case retrieval is handled very precisely, and this problem is solved.

Special emphasis is placed on the use of the developed application as a knowledge manager in the broadest term, since, by hosting a large base of success stories regarding the solution of real plant problems, it enables its use a simulator to train novice or inexperienced users in the maintenance plan design processes.

4. Conclusions

The successful integration of both methods—Reliability-Centered Maintenance and Case-Based Reasoning (RCM-CBR)—for the scope of the improvement in the optimal management of the maintenance of assets of an industrial plant is the aim pursued in this work. Although the individual benefits of both methods have been shown to improve the efficiency of industrial processes, the union of both pointing to the single direction of designing maintenance plans produces a combined effect in terms of their efficiency, their precision and the minimization of the time devoted to obtaining it. An additional advantage of redirecting the time dedicated to customizing the maintenance plan has been achieved, since, by saving time in the use of the RCM method by using a driven process and the automatic recovery of solutions to the problem posed, it allows the person responsible for the design of the maintenance plan to invest his time in customizing, by adapting recovered cases, the new proposed solution, in addition to correctly monitoring the execution of the proposed solution.

As a conclusion of this operational definition, within the global process of assisting in the decision making and planning of maintenance processes, an independent application in the Java programming language has been developed and used to conduct the complete RCM process with the use of Case-Based Reasoning (CBR) algorithms in order to recover and explore similar failure cases from a base of historical cases that have occurred and whose solutions were successfully applied in equivalent industrial plants. This has meant an exploitation of the knowledge of historical problems and their solutions that occurred in industrial plants, and this has also made it possible to extend and pool the human intellectual capital inherent in all industries, such as experts from various areas; the middle management of maintenance, production, process and maintenance engineers; workshop personnel; mechanics and electricians; etc. This philosophy of the integration of both paradigms (RCM-CBR) has made it possible to group the different previously described agents that are involved in the reliability problems of the industrial plant in order to reduce the time spent applying a classic RCM process. This time saving is a competitive advantage as soon as the implementation of the corrective actions become part of the maintenance management program. The effects of their application are obtained earlier compared to the classic application of the RCM method, channeling the analysis effort into the following critical areas for improvement. Finally, this last part is much faster, more operational and its objectives are aligned with the two preceding objectives—tactical and strategic, respectively—whose joint effect has resulted in increasing the overall efficiency of the industrial company.

As a result of this operational definition, a solution has been established as a maintenance action or task to be applied in response to a reliability problem detected for the equipment from a chosen section of the previous tactical definition. This solution, carried out iteratively on the section under analysis, allows for the creation of a complete maintenance plan for it to be programmed in the CMMS. It should be noted that this methodology is based on a continuous improvement scheme, which allows it to be applied frequently according to the launch of the complete system together with the previous strategic and tactical definitions, or independently when a loss of reliability is detected.

Although the term "adaptive maintenance" (AdM) has not been coined, Burggräf [42] suggested that the maintenance function should not only restore or maintain the operating condition of productive equipment and used the term "adaptive remanufacturing" (AdR) to suggest that the evolution towards intelligent and strategic maintenance should improve productive resources by increasing the life cycles of assets, monitoring their performance levels and considering the technical, economic and ecological aspects. It is considered relevant for future work to add this new concept of adaptive maintenance to this operational definition, where, in addition to re-establishing the required function of the high-impact asset to be maintained, the availability is ensured and the required function is increased—for example, when there is an increase in the speed of a machine due to the demands of the changing production context. In summary, it is suggested that scheduled maintenance interventions provide opportunities for the improvement or adaptation of the originally projected required functions of production systems in an increasingly demanding context.

Author Contributions: Conceptualization, N.R.-P. and R.D.; methodology, N.R.-P.; validation, N.R.-P.; formal analysis, N.R.-P., M.M.M. and R.D.; investigation, N.R.-P.; resources, N.R.-P., M.M.M. and R.D.; writing—original draft preparation, N.R.-P. and R.D.; writing—review and editing, N.R.-P., M.M.M. and R.D.; supervision, M.M.M. and R.D.; project administration, M.M.M. and R.D.; funding acquisition, M.M.M. and R.D. All authors have read and agreed to the published version of the manuscript.

Funding: This research was funded by the Spanish Ministry of Science, Innovation and Universities through the RTI2018-102215-B-I00 project and by the College of Industrial Engineers of UNED, grant number 2022-ETSII-UNED-08.

Institutional Review Board Statement: Not applicable.

Informed Consent Statement: Not applicable.

Data Availability Statement: Not applicable.

Conflicts of Interest: The authors declare no conflict of interest. The funders had no role in the design of the study; in the collection, analyses or interpretation of data; in the writing of the manuscript; or in the decision to publish the results.

References

1. Sardana, D.; Terziovski, M.; Gupta, N. The impact of strategic alignment and responsiveness to market on manufacturing firm's performance. *Int. J. Prod. Econ.* **2016**, *177*, 131–138. [CrossRef]
2. Lazarova-Molnar, S.; Mohamed, N. Reliability Assessment in the Context of Industry 4.0: Data as a Game Changer. *Procedia Comput. Sci.* **2019**, *151*, 691–698. [CrossRef]
3. Erozan, I. A fuzzy decision support system for managing maintenance activities of critical components in manufacturing systems. *J. Manuf. Syst.* **2019**, *52*, 110–120. [CrossRef]
4. Srivastava, P.; Khanduja, D.; Agrawal, V.P. Agile maintenance attribute coding and evaluation based decision making in sugar manufacturing plant. *Opsearch* **2020**, *57*, 553–583. [CrossRef]
5. Jimenez-Cortadi, A.; Irigoien, I.; Boto, F.; Sierra, B.; Rodriguez, G. Predictive maintenance on the machining process and machine tool. *Appl. Sci.* **2020**, *19*, 224. [CrossRef]
6. Ashayeri, J. Development of computer-aided maintenance resources planning (CAMRP): A case of multiple CNC machining centers. *Robot. Comput. Integr. Manuf.* **2007**, *23*, 614–623. [CrossRef]
7. Aghaee, A.; Aghaee, M.; Fathi, M.R.; Shoabin, S.; Sobhani, S.M. A novel fuzzy hybrid multi-criteria decision-making approach for evaluating maintenance strategies in petrochemical industry. *J. Qual. Maint. Eng.* **2021**, *27*, 351–365. [CrossRef]
8. Hoffmann Souza, M.L.; da Costa, C.A.; de Oliveira Ramos, G.; da Rosa Righi, R. A survey on decision-making based on system reliability in the context of Industry 4.0. *J. Manuf. Syst.* **2020**, *56*, 133–156. [CrossRef]
9. Vrignat, P.; Krtaz, F.; Avila, M. Sustainable manufacturing, maintenance policies, prognostics and health management: A literature review. *Reliab. Eng. Syst. Saf.* **2022**, *218*, 108140. [CrossRef]
10. Melani, A.H.A.; Murad, C.A.; Netto, A.C.; de Souza, G.F.M.; Nabeta, S.I. Criticality-based maintenance of a coal-fired power plant. *Energy* **2018**, *147*, 767–781. [CrossRef]
11. Ma, Z.; Ren, Y.; Xiang, X.; Turk, Z. Data-driven decision-making for equipment maintenance. *Autom. Constr.* **2020**, *112*, 103103. [CrossRef]
12. López-Campos, M.; Crespo Márquez, A.; Gómez Fernández, J.F. The integration of open reliability, maintenance, and condition monitoring management systems. In *Advanced Maintenance Modelling for Asset Management*; Springer: Cham, Switzerland, 2018; pp. 43–78.
13. Song, M. Optimization of RCM/CBM maintenance strategy by using unified functional knowledge. In Proceedings of the DHRTC Technology Conference, Kolding, Denmark, 25 November 2021.

14. Kumtekar, R.; Swapnil, K.; Rane, S. Integration of TPM, RCM, and CBM: A practical approach applied in shipbuilding industry. In *System Assurances*; Academic Press: Cambridge, MA, USA, 2022; pp. 389–402.
15. Mkalaf, K.A.; Abdulwahhab, A.; Al-Sabbagh, A.K. Failure modes and effects analysis: An integrated approach for optimizing maintainability and redesign for the concrete pump. In Proceedings of the 8th International Conference on Industrial Technology and Management (ICITM), Cambridge, UK, 2–4 March 2019.
16. Rodríguez-Padial, N.; Marín, M.; Domingo, R. Strategic framework to maintenance decision support systems. *Procedia Eng.* **2015**, *132*, 903–910. [CrossRef]
17. Rodríguez-Padial, N.; Marín, M.; Domingo, R. An approach to integrating tactical decision-making in industrial maintenance balance scorecards using principal components analysis and machine learning. *Complexity* **2017**, *2017*, 3759514. [CrossRef]
18. Rodríguez-Padial, N.; Marín, M.; Domingo, R. An approach to evaluate tactical decision-making in industrial maintenance. *Procedia Manuf.* **2017**, *13*, 1051–1058. [CrossRef]
19. Ruschel, E.; Alves Portela Santos, E.; de Freitas Roucha Loures, E. Industrial maintenance decision-making: A systematic literature review. *J. Manuf. Syst.* **2017**, *45*, 180–194. [CrossRef]
20. Cheng, Z.; Jia, X.; Gao, P.; Wu, S.; Wang, J. A framework for intelligent reliability centered maintenance analysis. *Reliab. Eng. Syst. Saf.* **2008**, *93*, 806–814. [CrossRef]
21. Candea, G.; Kifor, S.; Constantinescu, C. Usage of case-based reasoning in FMEA-driven software. *Procedia CIRP* **2014**, *25*, 93–99. [CrossRef]
22. Kobbacy, K. Application of artificial intelligence in maintenance modelling and management. *IFAC Proc. Vol.* **2012**, *45*, 54–59. [CrossRef]
23. Rahman, A.; Pasaribu, E.; Nugraha, Y.; Khair, F.; Soebandrija, K.E.N.; Wijaya, D.I. Industry 4.0 and society 5.0 through lens of condition based maintenance (CBM) and machine learning of artificial intelligence (MLAI). In Proceedings of the IOP Conference Series: Materials Science and Engineering, Chennai, India, 16–17 September 2020; p. 852.
24. Nowlan, F.S.; Heap, H.F. *Reliability Centered Maintenance*; United Airlines Publications: San Francisco, CA, USA, 1978.
25. Anderson, R.T.; Neri, L. *Reliability Centered Maintenance: Management and Engineering Methods*; Elsevier: London, UK, 1990.
26. Smith, A.M. *Reliability Centered Maintenance*; McGraw-Hill: New York, NY, USA, 1993.
27. Moubray, J. *Reliability-Centered Maintenance*, 2nd ed.; Industrial Press: New York, NY, USA, 1997.
28. Waeyenbergh, G.; Pintelon, L. A framework for maintenance concept development. *Int. J. Prod. Econ.* **2002**, *77*, 299–313. [CrossRef]
29. Saniuk, A.; Jasiulewicz-Kaczmarek, M.; Samolejova, A.; Saniuk, S.; Lenort, R. Environmental favourable foundries through maintenance activities. *Metalurgija* **2015**, *54*, 725–728.
30. Schank, R.; Abelson, R. *Scripts, Plans, Goals and Understanding: An Inquiry into Human Knowledge Structures*; Erlbaum: Hillsdale, NJ, USA, 1977.
31. Schank, R. *Dynamic Memory: A Theory of Reminding and Learning in Computers and People*; Cambridge University Press: New York, NY, USA, 1983.
32. Riesbeck, C.K.; Schank, R. *Inside Case-Based Reasoning*; Erlbaum: Northvale, NJ, USA, 1989.
33. Watson, I. Case-based reasoning is a methodology not a technology. *Knowl. Based Syst.* **1999**, *12*, 303–308. [CrossRef]
34. Díaz-Agudo, B.; González-Calero, P.A.; Recio-García, J.A.; Sánchez-Ruiz-Granados, A.A. Building CBR Systems with jCOLIBRI. *Sci. Comput. Program.* **2007**, *69*, 68–75. [CrossRef]
35. Java. Available online: https://www.oracle.com/es/java (accessed on 28 June 2021).
36. Watson, I. *Applying Case-Based Reasoning: Techniques for Enterprise Systems*; Morgan Kaufmann: San Francisco, CA, USA, 1997.
37. Salton, G.; McGill, M.J. *Introduction to Modern Information Retrieval*; McGraw-Hill: New York, NY, USA, 1983.
38. Chieh-Yuan, T.; Chuang-Cheng, C. Developing a significant nearest neighbour search method for effective case retrieval in a CBR system. In Proceedings of the International Association of Computer Science and Information Technology-Spring Conference, Singapore, 17–20 April 2009; IEEE: Los Alamitos, CA, USA, 2009; pp. 262–266.
39. Recio-García, J.A.; Díaz-Agudo, B.; González-Calero, P.A. *jCOLIBRI 2 Case-Based Reasoning Framework. Document Version 1.2. Group for Artificial Intelligence Applications*; Universidad Complutense de Madrid: Madrid, Spain, 2008.
40. Recio-García, J.A. jCOLIBRI: Una Plataforma Multi-Nivel Para la Construcción y Generación de Sistemas de Razonamiento Basado en Casos. Ph.D. Thesis, Departamento de Ingeniería del Software e Inteligencia Artificial, Facultad de Informática, Universidad Complutense de Madrid, Madrid, Spain, 2008.
41. Eclipse Foundation. Available online: https://www.eclipse.org (accessed on 28 June 2021).
42. Burggräf, P.; Adlon, T.; Müller, K.; Föhlisch, N.; Dackweiler, J.; Fölling, C. Adaptive Remanufacturing–Methodology towards an intelligent maintenance strategy for production resources. *Procedia CIRP* **2021**, *98*, 330–335. [CrossRef]

applied sciences

MDPI

Article

A Robust Scheduling Framework for Re-Manufacturing Activities of Turbine Blades

Lei Liu * and **Marcello Urgo**

Department of Mechanical Engineering, Politecnico di Milano, 20133 Milano, Italy; marcello.urgo@polimi.it
* Correspondence: lei.liu@polimi.it

Abstract: Refurbished products are gaining importance in many industrial sectors, specifically high-value products whose residual value is relevant and guarantee the economic viability of the re-manufacturing at an industrial level, e.g., turbine blades for power generation. In this paper, we address the robust scheduling scheme of re-manufacturing activities for turbine blades. Parts entering the process may have very different wear states or presence of defects. Thus, the repair process is affected by a significant degree of uncertainty. The paper investigates the uncertainties and discusses how they affect the scheduling performance of the re-manufacturing system. We then present a robust scheduling framework for the re-manufacturing scheduling strategies, policies, and methods. This framework is based on a wide variety of experimental and practical approaches in the re-manufacturing scheduling area, which will be a guideline for the planning and scheduling of re-manufacturing activities of turbine blades. A case study approach was adopted to examine how re-manufacturers design their scheduling strategies.

Keywords: turbine blades; re-manufacturing; uncertainty; robust scheduling

Citation: Liu, L.; Urgo, M. A Robust Scheduling Framework for Re-Manufacturing Activities of Turbine Blades. *Appl. Sci.* **2022**, *12*, 3034. https://doi.org/10.3390/app12063034

Academic Editors: Roque Calvo, José A. Yagüe-Fabra and Guido Tosello

Received: 17 February 2022
Accepted: 15 March 2022
Published: 16 March 2022

1. Introduction and Industrial Motivation

In the past few years, increasing attention has been devoted towards enhancing the sustainability of manufacturing processes by reducing the consumption of resources and key materials, energy consumption and environmental footprint, while also reducing costs and increasing competitiveness in the global market. Re-manufacturing, which can be defined as "the rebuilding of a product to specifications of the original manufactured product using a combination of reused, repaired and new parts" [1], is a form of product recovery process entailing the repair or replacement of worn out components to obtain re-manufactured products with the same characteristics as new products. The re-manufacturing paradigm is aimed at supporting sustainability challenges in strategic manufacturing sectors for high-value products whose residual value is high, such as aeronautics, automotive, electronics, consumer goods, and mechatronics [2].

Re-manufacturing processes, compared to the original manufacturing processes, entail higher degrees of uncertainty, complexity and dynamics, due to the unpredictable and variable conditions of the used parts to be processed. This significantly affects process and production planning, as well as the requirements driving the design of systems operating re-manufacturing activities. Thus, low production efficiency, unstable product quality, frequent abnormal production accidents and high product rework rates are typical characteristics of re-manufacturing systems [3]. To address these operating scenarios, smart approaches are needed to match production management and control decisions with the updated state of production processes, resources and requirements.

Industry 4.0, which is one of the most trending topics in manufacturing area, considers smart manufacturing as its central element, and relies on the adoption of digital technologies such as Internet of Things (IoT), cloud services, big data and analytics, to gather data in real time and to analyze it, providing useful information to the manufacturing system [4].

To hedge against unexpected events in the manufacturing system, predictive and reactive approaches can take advantage of advanced sensor technologies providing monitoring capability and artificial intelligence supporting analyses and decisions [5]. Taking advantage of the described scenario, smart management approaches for re-manufacturing systems are required to support management, planning and scheduling, within the circular economy paradigm [2].

Grounding on this perspective, this paper focuses on scheduling approaches for re-manufacturing activities, able to cope with uncertainty affecting both processing times and steps. The aim is to devise a robust schedule to mitigate the impact of these uncertain events. The industrial application addressed is gas turbine blade, an extremely complex and high-value product.

The paper is organized as follows: Section 2 reviews relevant literature; Section 3 describes the addressed re-manufacturing environment entails the characteristics and operating environment of turbine blades, re-manufacturing process and associated uncertainties; Section 4 presents the proposed robust scheduling framework, while a case study is reported in Section 5. Finally, Section 6 provides final considerations and conclusions.

2. Literature Review

Re-manufacturing, as one of the most important sustainable economy paradigms, has drawn a lot of attention due to its advantages on cost-effectiveness, energy-saving and emission-reduction. Tolio et al. [2] revises system level problems, methods and tools to support the re-manufactuing paradigm and highlights the main challenges and opportunities towards a new generation of advanced re-manufacturing systems. Goodall et al. [6] reviews the work on tools and methods which have been developed to support the decision process of assessing and evaluating the viability of conducting re-manufacturing, and evaluate how they have met the requirements of the decision stage.

Turbine blades re-manufacturing environment is an emerging industrial sector where many advanced tools and technologies are proposed. Rickli et al. [7] described a framework for re-manufacturing systems that aims to take advantage of additive manufacturing processes to remanufacture end–of–life cores. Huang et al. [8] proposes a re-manufacturing scheme design method based on the incomplete reconstruction of used part information to solve the uncertain and highly personalized problems in re-manufacturing. Nevertheless, re-manufacturing environment for turbine blades is way more complicated than the traditional manufacturing environment, and it involves enormous decision-making practices due to the uncertainties incurred by the different demand, inventory, processing routes and times, which depend on the condition of the returned products [9]. However, among the literature on the tools and approaches for re-manufacturing environment, 48% focus on strategic-level and 34% on tactical-level, with only 5% focusing on operational-level, evenly, only 36% of the studies address uncertainties which is a highly important factor companies face in product recovery management [10,11].

Production planning and scheduling, which play important roles in the organization of manufacturing activities and directly affect the overall performance of manufacturing, is also extensively studied in the re-manufacturing environment. Morgan and Gagnon [12] classified the re-manufacturing scheduling into disassembly versus integrated scheduling, single versus multiple products scheduling and reviewed the relevant literature. Kin et al. [13] proposed a conceptual methodology for reconditioning process sequence planning based on the ranking of the defects and the precedence relationships which consider the criticality of the defects. Zhang et al. [14] addressed the re-manufacturing scheduling problem by adopting a simulation-based optimization framework and proposed a genetic algorithm to optimize two objective functions. Liu and Urgo [9] proposed a approximate branch and bound algorithm for 2-machine permutation flow shop scheduling re-manufacturing activities for the repair of turbine blades, with a value-at-risk performance investigated.

In the re-manufacturing of turbine blades, two main sources of uncertainty are processing routes and times, which depend on the condition of returned products. To our best knowledge, no general framework exists to support the scheduling of re-manufacturing activities with uncertain processing routes and times.

3. Re-Manufacturing of Turbine Blades

Gas turbine (Figure 1), in which burning of an air-fuel mixture produces hot gases that spin a turbine to produce power, is one of the most widely-used power generating technologies. Turbine blades (Figure 2), whose individual price is close to a middle-class car (i.e., about ten-thousand euros), is one of the most important and expensive parts in a gas turbine. To maximize the performance of a gas turbine, they are constituted by multiple stages, each of them equipped with specially-designed turbine blades. According to the specific OEM (Original Equipment Manufacturer), the number of stages can vary, as an example, a F-class turbine [15] needs about 400 blades for 3 to 6 stages. Blades belonging to the same stage of the turbine are usually manufactured/re-manufactured in batch to guarantee homogeneous characteristics and the balancing of the portion of the rotor for the stage.

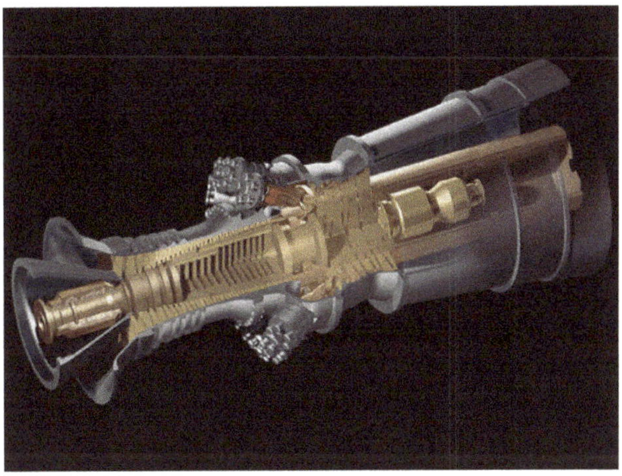

Figure 1. GE H-series power generation gas turbine [16].

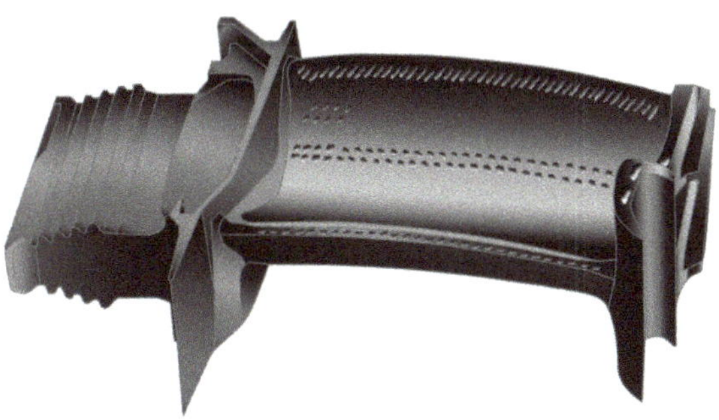

Figure 2. New turbine blade [17].

During the functioning of the turbine, blades are exposed to an extremely strenuous environment (high temperatures, stresses, vibration and corrosion) constituting a major cause for wear and failures of the blades (Figure 3) and, consequently, reduced energy conversion efficiency, and potentially disruptive failures of the whole turbine.

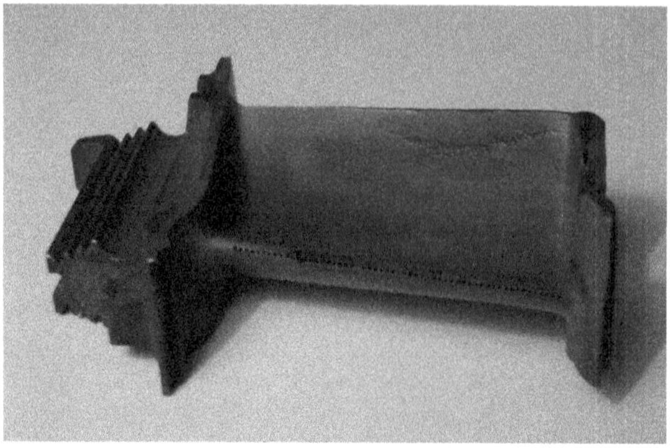

Figure 3. Damaged blades resulting from exposure to high temperature and stress [17].

Thus, planned maintenance operations are enforced for gas turbines, requiring all the blades in the rotor to be disassembled, inspected and, if needed, re-manufactured or replaced with new ones.

3.1. Re-Manufacturing Processes

Taking the economy cost into consideration, extending the service life of blades through re-manufacturing is a feasible and attractive choice, since their replacement with a new one is far more expensive.

The re-manufacturing of turbine blades generally involves multiple processes and technologies, e.g., visual inspection, non-destructive testing, machining, additive manufacturing, grinding, heat-treatments, coating, etc. The whole process can be summarized in the following main steps:

1. Blades are disassembled at the production site and shipped to the OEM to be refurbished/renewed.
2. Upon arrival at the OEM, blades are inspected to assess the severity of wear and damages. Blades that cannot be repaired are discarded and the supply/production of new blades is issued. On the contrary, for reparable blades, the re-manufacturing process starts, whose process parameters depends on the assessed damage level.
3. The re-manufacturing process consists of removing the damaged (e.g., cracked) parts of the blades, restore the removed material through welding/additive technologies, machining/grinding the blade to reconstruct the desired shape. During the material removal phase, as well as at the end of the processing, non-destructive tests are operated to verify the complete removal of the defects.
4. The blades undergo the definition of small-size features (e.g., internal cooling ducts, specific shapes in the terminal side or in the coupling with the rotor) and are successively coated with high-resistance materials.
5. Finally, the blades belonging to the same stage are assembled together and balanced. Hence they are shipped back to the operating site of the turbine where they will be re-assembled and be ready to go into production again.

In this paper the focus is on the core part of the re-manufacturing process, i.e., only the phases operating the removal of the defects, the reconstruction of the original shape through an additive process, and the machining of the blade to obtain the desired final shape. While the blades go through these phases, due to the different and unpredictable degree of wear and severity of the damages on the parts, rework activities and non-destructive testing operations could be needed, as shown in Figure 4.

Thus, depending on their characteristics, wear and possible presence of damages, different blades will undergo different sequences of operations. Figure 4 presents an example of the repairing process for two different batches of turbine blades. The process flow of the first batch of blades (i.e., stage 1) is described with a black arrow, while a blue dashed arrow is used for the second batch blades. O_i denotes the operations belongs to defects removal, material additive, machining and NDT process steps respectively. Although both stage 1 and stage 2 blades undergo the four main process steps, the detailed operations could differ according to the specific stage they belong to.

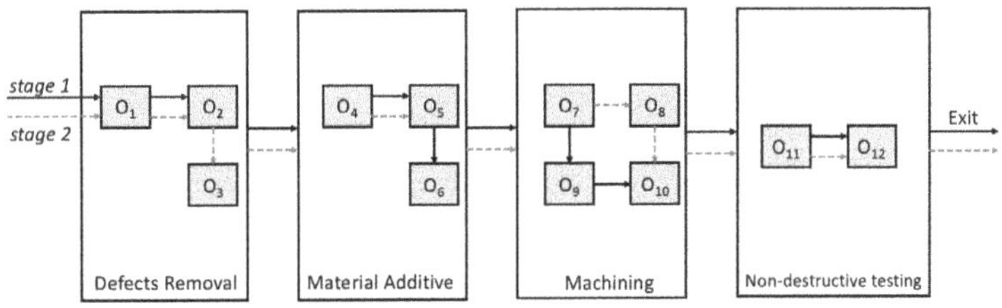

Figure 4. Re-manufacturing process example for two batches of turbine blades.

3.2. *Uncertainty in Re-Manufacturing Processes*

One of the peculiar requirements of re-manufacturing processes is the capability to cope with the intrinsic uncertainty affecting the parts and products to process. This drives the need for re-manufacturing plants able to operate in variable conditions with high degree of efficiency awhile guaranteeing the quality of the processed products [2,18].

The main sources of uncertainty to manage can be summarized in the following classes:

1. A major source of uncertainty is embedded in the characteristics and volumes of the parts to be processed. Used parts arrives from different use-phase conditions and, thus, could be worn differently. Moreover, grounding on their characteristics, the number of parts whose re-manufacturing is feasible or attractive could be different, thus impacting on the actual volumes of parts to be re-manufactured.
2. The re-manufacturing process plan could vary for different specimens of the same products. Thus, according to the characteristic of the item (wear, presence of damages, etc), different process steps arranged in different sequences could be needed. This impact on the management of the re-manufacturing system due to the need of implementing different routines among the available resources.
3. Grounding on the characteristics of each specific part, re-manufacturing process can differ in terms of the characteristics and parameters of re-manufacturing operations. This a clear impact on the execution of the processes (forces, velocity, processing times, etc.)

The described sources of uncertainty also fits the case of re-manufacturing turbine blades. As stated in Section 3, blades are repaired in batches, consisting of a set of blades belonging to the same stage of the turbine. Blades belonging to different stages, as well as blades in different positions in the same blade, could be exposed too different stresses and environmental conditions and, thus, result in different degrees of wear. Moreover, after the initial inspection phase, a subset of the blades results not repairable and must be discarded. Thus, the number of blades to be processed as well as their characteristics are not known in advance.

Moreover, repairable blades with a higher degree of damages will likely entail longer processing times and different process parameters respect to the less severely damaged blades. The severity of the damages can be partially estimated in the initial inspection phase but, while the re-manufacturing process is operated, deviations from what initially estimated could emerge. These deviations can just affect the processing time of specific operation or, in some cases, require different sequence of operations to be executed, e.g., non-destructive testing and rework (Figure 5). Thus, processing times as well as the steps to be operated in the system entail a certain degree of uncertainty.

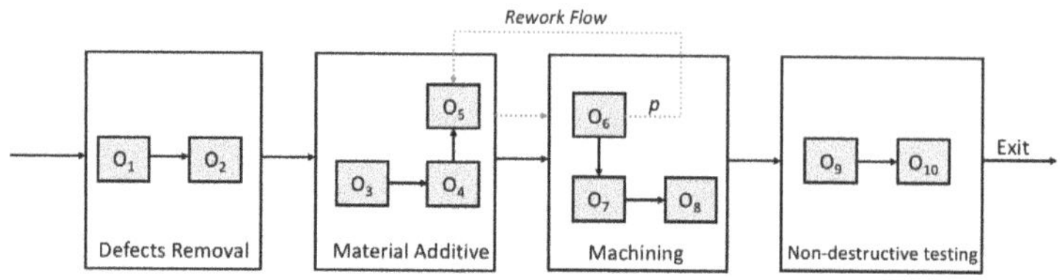

Figure 5. Repairing process example for a batch with rework.

The sources of uncertainty described above primarily affects the planning and scheduling of re-manufacturing operations, responsible of matching the requirements of the customers (delivery dates) as well as defining planned execution of operations matching the characteristics and constraints within the re-manufacturing system. Thus, it would not be possible to obtain a realistic schedule of re-manufacturing operations without being able of taking into consideration the complexity and uncertainty of the described re-manufacturing environment [19], entailing the need of the development and adoption of robust scheduling approaches [20,21].

4. Robust Scheduling Framework

Robust scheduling approaches focus on constructing preventive schedules to minimize the effects of disruptions on the performance measure and try to ensure that the predictive and realized schedules do not differ drastically, while maintaining a high level of schedule performance [20]. In this section, a resource constrained project scheduling (RCPSP) approach is used to formalize the re-manufacturing process of turbine blades, different modelling approaches are used to describe relevant sources of uncertainty, proper robustness measures are introduced to support the devising of robust schedules.

4.1. Shop Scheduling Model

Grounding on the description of the re-manufacturing process in Section 3.1, several batches of turbine blades are repaired in the same shop, different batches may need different sets of operations, sharing the same resource (e.g., workers, machines). The scheduling of batches of turbine blades can be modeled through a resource constrained project scheduling problem (RCPSP) with limited renewable resources [22]. The structure of the processes to be operated is defined by precedence relations among the activities, the associated processing times can be modeled through random variables to consider the associated uncertainty. The aim of the scheduling approach is the minimization of the makespan, supporting the optimized utilization of the resources [23].

Each activity represents a processing operation of a whole batch of blades in a specific process phase. Activities cannot be interrupted, hence, a non-preemptive schedule is pursued. The set of precedence relations are usually given as a directed acyclic graph, where nodes represent activities while an edge (u, v) models a precedence relation enforcing u to finish before v is allowed to start. The graph contains two dummy activities, the source node S and the sink node T, modeling the start and finish of the whole set of activities. A graph representing the processing of k batches of blades is reported in Figure 6. A set of batches $(1, 2, 3, \ldots, k)$ of turbine blades which come from different of customer orders and stages need to be re-manufactured, each horizontal chain from source node S to sink node T represents the processing steps of one batch of turbine blades. Different batches undergo different process steps while with many equal operations $(A, B, C, D, \ldots)$ consuming same resources.

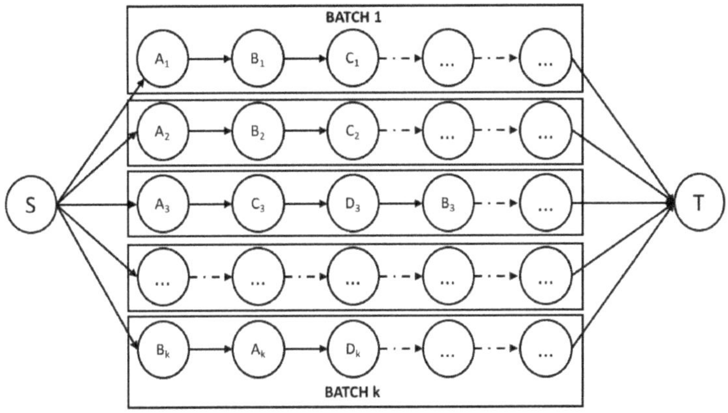

Figure 6. RCPSP model for re-manufacturing process of turbine blades.

The set of renewable resources represents production resources in the shop (e.g., experienced workers, machines). Each resource has a maximum capacity per time slot while each activity is associated to a requested amount of each resource for each time slot (greater or equal to zero). The dummy source and sink activities require no resource.

With respect to the objective to be pursued, the minimization of the makespan (i.e., the completion time of the dummy sink activity) is addressed in this paper. Nevertheless, pursuing a robust approach entails the need of addressing the uncertainty affecting the scheduling problem in the objective function. The following subsections are going to address this aspect.

4.2. Modeling Uncertainty

Grounding on the description of uncertainties in the re-manufacturing process of turbine blades in Section 3.2, different classes of unexpected events must be modeled. Specifically, with respect to turbine blades, the effects of these classes of uncertainties affects processing times and the sequence of process steps.

4.2.1. Uncertainty Affecting Processing Times

The uncertainty affecting processing times can be modeled in different ways, according to the available amount of knowledge and data. The simpler approach is the definition of an interval so that the processing time of an operation p_i can assume values within an interval $[p^{L_i}, p^{U_i}]$, i.e., $p_i \in [p^{L_i}, p^{U_i}]$. The probability distribution associated to the interval can be the uniform one [24].

Pursuing this approach, only the possible minimum and maximum values of the processing times should be determined, grounding on the available knowledge or on historical production data [25]. In the re-manufacturing of turbines, the blades are repaired in batches and the processing time uncertainty associated to a batch depends on two main factors as anticipated in Section 3.2:

1. The number of blades in a batch depends on their level of damage. Extremely worn blades cannot be refurbished and must be discarded. A rejection rate table based on the damage level of each type of blade is defined. Thus, each batch of blades to be processed is assigned a level of damage (high, medium, low) sampled from a discrete distribution. Hence, the rejection rate is modeled as an uniform distribution defined by a lowest and highest rejection rates for that class of blades.
2. The processing time of an operation on a single blade is modeled as an interval whose boundaries can be determined grounding on historical data. Thus, the distribution of the processing time of a batch of blades can be obtained through a convolution

operation between the distribution of the number of blades in the batch to be re-manufactured and the distribution of the processing time of a single blade.

If more detailed information and knowledge are available, uncertainty related to processing times can be modeled by independent random variables with an associated general probability distribution, either discrete or continuous [26]. Differently from the approach described above, additional hypotheses are required for the definition of the probability distributions, namely, their type and the associated parameters. The main rule here is trying to limit hypotheses that cannot be fully supported by the available data. Thus, simple distributions (e.g., triangular) are preferred [27].

4.2.2. Uncertainty Affecting the Process Steps

In re-manufacturing, the characteristics of the parts to be processed, e.g., their level of wear, could entail the need of operating different processes. A typical situation is the need of executing a rework.

The execution of rework activities can be modeled in different ways, grounding on the specific characteristic of the process and re-manufacturing environment as well as with respect to the available data. A first approach is incorporating possible rework activities in the standard process, thus, modeling the distribution of processing time to take into consideration the possibility of longer times due to the need to operate a rework. Although simple, this approach is clearly giving up in modeling the actual sequencing of operations. In fact, rework activities are usually operated after a verification of the standard process. Thus, they are usually operated at a later temporal stage. Putting work and rework operations in the same scheduled activity is not correctly modeling the fact that, in the meanwhile, other activities can be processed. A second approach is hypothesizing that all batches need to be reworked, which is a really common occurrence in the re-manufacturing of turbine blades. Thus, an additional set of activities is considered, for each batch of blades to be processed). These are then added to the original RCPSP introduced in Section 4.1, considering the associated processing times as well as precedence constraints.

Another approach considers that not all the possible rework activities are always operated. Thus, their occurrence is modeled through a Bernoulli distribution while the associate processing time is still defined by a probabilistic distribution, thus:

1. A RCPSP model is defined considering both work and rework activities.
2. The processing time of rework activities is obtained through the convolution of the occurrence and processing time distributions described above.

4.3. *Robustness Measures*

A robust schedule, which is defined as a schedule that is insensitive to unforeseen disturbances [20], minimizes the effect of uncertainties on the primary performance measure of the schedule when implemented. Most of the robust scheduling approach consider the expected realized performance of the schedule, while it is really limited since minimizing the expected value fails in estimating the quality of the schedule in a stochastic point of view [28]. In this paper, we consider risk based robustness measures, specifically, the min-max regret, value-at-risk and conditional value-at-risk of the realized schedule are investigated.

4.3.1. Minimizing the Maximum Regret of the Objective Function

The min-max regret scheduling approach is a risk-aversion method focusing on the worst case scenarios, rather than considering all the possible realizations. It takes into account the regret, i.e., the deviation of an outcome from the best possible one in each specific scenario. Thus, this is not an absolute measure of performance of the solutions, but relative to the best available performance for a specific scenario [29].

The described re-manufacturing process of turbine blades with uncertain processing times and rework activities is modeled as a resource constrained project scheduling (RCPSP) problem (Section 4.1), grounding on the modeling of the uncertainty related to processing times and rework activities as described in Section 4.2. For this class of problems, the optimization of the maximum regret can be addressed considering extreme scenarios only [30]. In the case under study, this is pursued considering extreme scenarios for the processing times p_i, i.e., with $p_i = p_i^{min}$ or $p_i = p_i^{max}$ for all i and exploiting the algorithm proposed in [31].

4.3.2. Minimizing the Value-at-Risk/Conditional Value-at-Risk of the Objective Function

The minimization of the maximum regret has the advantage of only requiring knowledge about the extreme scenarios and, consequently, the simple knowledge of the extreme values of the domains for uncertain variable is enough. At the same time, focusing the optimization on worst-case scenarios, that may be unlikely to occur, tends to be too conservative. To cope with this limitation, different risk measures can be used to guide the optimization, namely, the value-at-risk (VaR) and the conditional value-at-risk (CVaR) [32]. These indicators take into consideration the whole distribution of uncertain variables and, thus, are able to consider the impact of uncertain events both in terms of their effect and occurrence probability. The use of both the VaR and CVaR was initiated in the financial area [32] but their popularity in robust scheduling area is rapidly increasing [21,33,34].

In the case of the deterministic RCPSP, a solution to the RCPSP is a schedule s, i.e., a vector of starting times $(s_0, s_1, \ldots, s_n, s_{n+1})$, with each activity duration is a constant. While for the scheduling problem under study, the decision maker does not know which exact information of activity duration, and yet a number of sequencing decisions need to be made. Hence, the execution of the project with uncertain activity durations is a dynamic decision process, and a schedule solution which denoted as a vector x, is a policy, which defines actions at the start of the project and at the completion times of activities. A vector of random variables $y = \{p_1, \ldots, p_n\}$ models the random processing times, governing by a probability measure P on Y and independent of scheduling decision x. The probability distribution of the makespan, $f_{C_{max}}(x, y)$, depends on the values of x and y. For a given schedule x, the resulting cumulative density function (cdf) for the makespan is defined as:

$$F_{C_{max}}(x, \zeta) = P(f_{C_{max}}(x, y) \leq \zeta | x) \tag{1}$$

Then, the value-at-risk α (VaR_α) of C_{max}, associated with a schedule decision x, denoted as $\zeta_\alpha(x)$, is defined according to the following:

$$\zeta_\alpha(x) = min\{\zeta | F_{C_{max}}(x, \zeta) \geq \alpha\} \tag{2}$$

Further, the $\alpha - CVaR$ of (1) associated to a schedule x is the mean of the $\alpha-$tail distribution (3) [35].

$$F_X^\alpha(z) = \begin{cases} 0, & \text{when } z < \zeta_\alpha(X) \\ \frac{F_X(z) - \alpha}{1 - \alpha}, & \text{when } z \geq \zeta_\alpha(X) \end{cases} \tag{3}$$

By considering the proposed Equation (2), VaR can be specified as the risk measure on the random $f(x, y)$, and minimizing ζ corresponds to seeking the schedule solution with the smallest possible VaR measure for a specified α value, i.e., for confidence level α, VaR_α is the $(1 - \alpha)$- quantile of the makespan distribution which is the largest value that ensures that the probability of obtaining a makespan less than this value is lower than $1 - \alpha$ [33].

CVaR at confidence level α, i.e., $CVaR_\alpha$, is defined as the expected value of makespan smaller than the $(1 - \alpha)$-quantile of the probability distribution of makespan, i.e., VaR_α [36]. CVaR is the expected value of the makespan for the worst α% cases with a value greater than the VaR.

The main difficulty in pursuing robustness through the minimization of the VaR/CVaR lies in the need to calculate the distribution of the objective function, e.g., the makespan. In scheduling problems affected by uncertainty, thus entail the capability of dealing with the correlation among all the possible paths in the network of activities (Section 4.1) [37,38]. To overcome this difficulty, a Markovian Activity Network(MAN) approach can be used to support the analytical estimation of this distribution. Basic MANs require the processing times of the activities to follow exponential distributions [39] but can be extended to cope with general distributions, approximated by phase-type distributions [40]. Grounding on this, the distribution of objective function based on the completion times of the activities (e.g., the makespan), enables the use of risk measures to address robustness of scheduling.

5. Case Study

The proposed robust scheduling framework has been preliminary tested in an industrial environment to support the scheduling of the re-manufacturing activities for turbine blades. In this case study, four turbine stages are considered. Each stage consists of a set of identical blades that are thus processed as a batch. The steps of the repair process are the same for all the batches, i.e., defects removal, material additive, machining, and non-destructive testing (Section 3). Within these steps, 9–12 operations in total are executed, with some differences among the different types of batches. Moreover, 2/3 operations in the process may need a successive rework. The initial state of the turbine blades is modeled in terms of their damage level, defined for each stage grounding on historical data, reported in Table 1. Starting from the damage level, a rejection rate can be defined (the probability for a blade to be too damaged to be repaired), as well as the distribution of the processing times for rework operations (Section 4.2.1). Each activity is expected to require a single renewable resource from a resource set with four different resources representing machines and human workers involved.

Table 1. Damage level table.

Damage Level	Rejection Rate LB (%)	Rejection Rate UB (%)
Heavy	60	90
Medium	30	60
Light	0	30

We will show how to appropriately use the framework to provide relevant information to the decision-makers, and help them develop robust schedules in their turbine blades re-manufacturing plants.

A resource constrained project scheduling model (RCPSP) is used to formalize the described scheduling problem (Section 4.1). A total of four batches are considering, resulting in 41 activities to schedule (including the start and complete dummy activity). The objective of the approach is the minimization of the maximum regret of the makespan (Section 4.3.1). The processing time of each activity is modeled with an interval derived from historical data and convolution operations (Section 4.2.1). Rework activities are described through a Bernoulli distribution modeling the probability to occur (Section 4.2.2). Thus, the integrated uncertainty is described by a set of scenarios, obtained by considering the possible occurrences of processing times and reworks for all the activities. For each scenario, a processing time vector $[p_0, p_1, \ldots, p_{40}]$ is used to describe the processing time for all the activities. Since only extreme scenarios need to be examined in the maximum regret minimization model [30], $2^{(41-2)}$ scenarios in total need to be evaluated in this case study, the scenario relaxation algorithm with accelerated convergences for robust resource-constrained project scheduling problem proposed in [23] has been adopted to support the definition of robust schedules.

To demonstrate the benefits of this robust scheduling framework, an alternative scheduling approach is used for comparison, namely, one only considering the expected values of processing times and a probability of 50% rework occurrence, ignoring the disturbance and volatility in the re-manufacturing process, is compared with the proposed framework.

Due to the large amount of possible scenarios, it is unfeasible to test the schedule obtained through the alternative approach on all of them. To overcome this limitation, a subset of the total number of scenarios is sampled and, for each of them, the performance obtained with the schedule considered (i.e., the makespan). These values are then used to assess the value of the stochastic solution (VSS), i.e., the value of exploiting stochastic information to support a robust schedule. It is calculated as the average deviation between the performance of the robust approach and the one just considering expected values (Equation (4)) over the considered set of scenarios [41].

$$VSS = \frac{1}{|S|} \sum_{s \in S} (EVS_s - RVS_s) \tag{4}$$

In Equation (4), EVS denotes the value of the solution considering expected values of the stochastic variables, and RVS is the one of the robust schedule. Moreover, $|S|$ is the number of scenarios considered, and s denotes a specific scenario in the set.

The proposed robust scheduling framework has been tested on 50 randomly generated problem instances. Each instance represents the re-manufacturing of a turbine with all its stages. Considering the results obtained for the whole set of instances, the relative value of the VSS is on average 5.4%. Since the absolute value of the VSS depends on the test instance, a scatter plot of the relative values of the VSS is presented in Figure 7, showing that for the vast majority of them, values are positive. This supports the value of the proposed robust scheduling framework in comparison with the expected value solution (EVS).

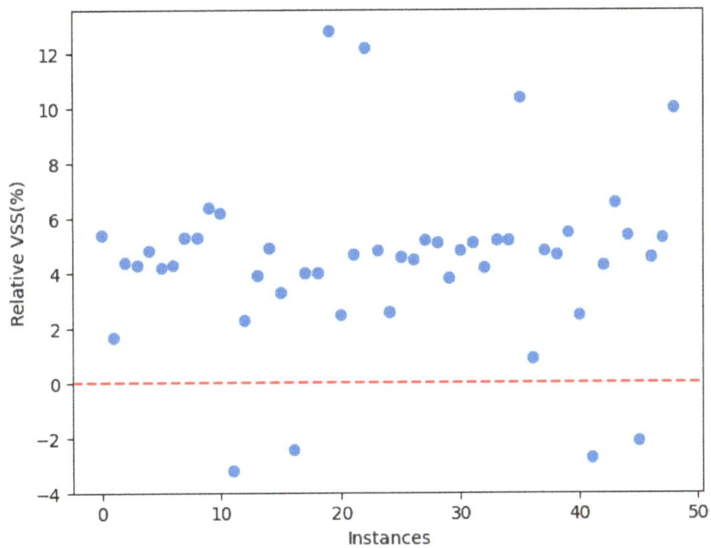

Figure 7. Relative value of VSS.

To further investigate the performance of the proposed robust approach, an instance has been selected, and the objective values of the two approaches with respect to 20 randomly generated scenarios. The results are shown in Figure 8 (left), showing that, for most of the scenarios, the robust scheduling framework performs better, i.e., leads to a smaller makespan. Furthermore, the results obtained in the described scenarios has been evaluated in terms of the difference between the makespans obtained. The results are reported in Figure 8 (right), showing that the proposed robust approach performs better on average (the 0.25-quantile of the difference EVS-RVS is positive). Moreover, the top whisker of the plot shows that, in extremely unfavorable cases, the protection provided by the robust schedule is extremely valuable.

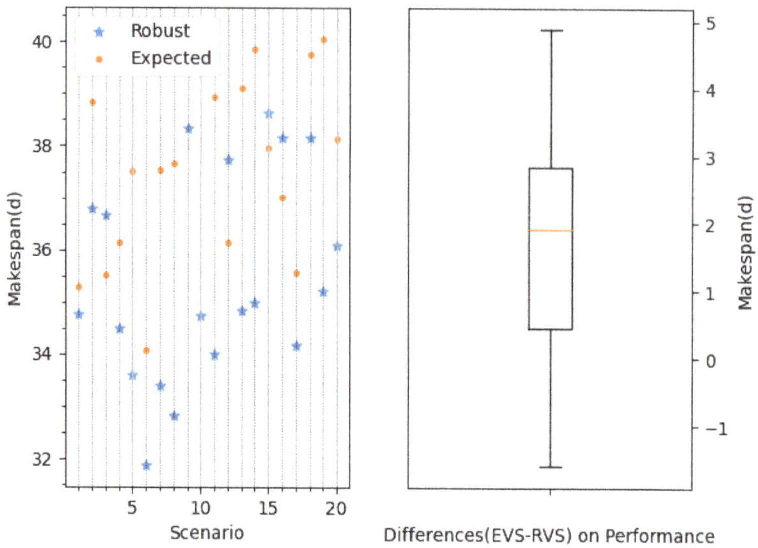

Figure 8. Performances under two approaches.

In addition, an extreme scenario has been analyzed, i.e., a worst-case scenario where all the processing times have the maximum possible value and all rework operations occur. The resulting execution of the robust schedule and the one obtained considering the expected values of uncertain variables are represented in Figure 9 where the blue one is the robust schedule and red the expected value one. The figure provides an aggregate representation, showing the aggregate processing of stages, and a detailed one, showing the detailed processing of all the re-manufacturing activities. It is clear from the Gantt in Figure 9 that, upon the occurrence of an extreme scenario, the robust schedule guarantees a smaller makespan.

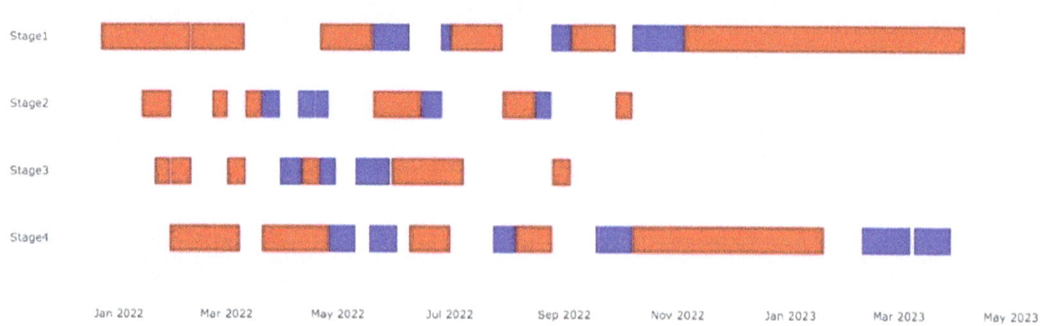

Figure 9. *Cont.*

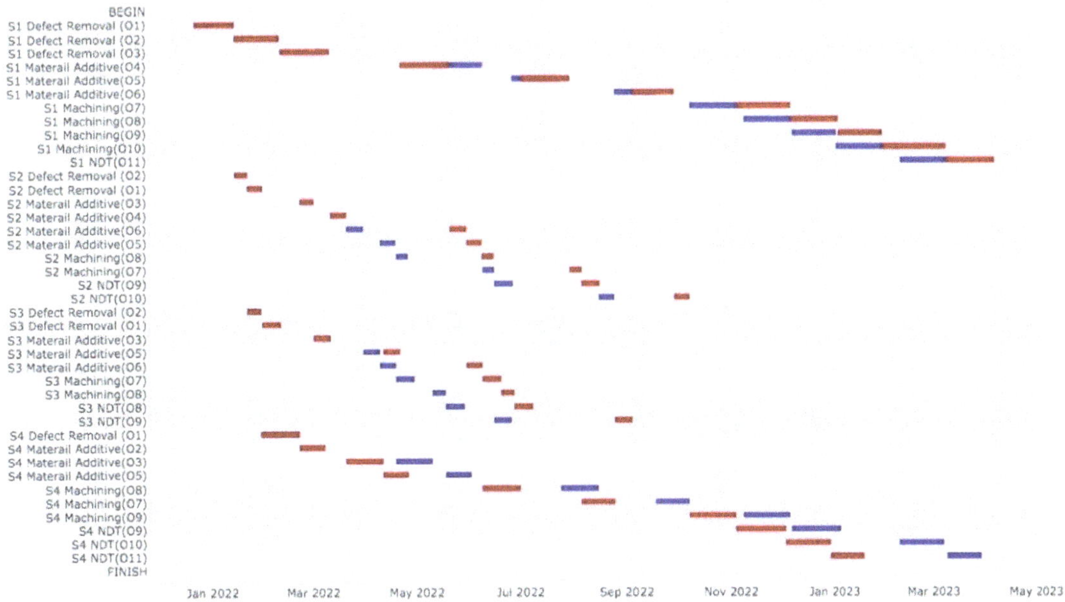

Figure 9. Gantt Chart.

6. Conclusions

In this paper, a robust scheduling framework is proposed to support re-manufacturing activities for gas turbine blades. The re-manufacturing problem and the relevant sources of uncertainty are described and formalized. To pursue the definition of robust schedules, different robustness measures, i.e., minimax regret and VaR/CVaR, are exploited. A case study, which can be a guideline for supporting the planning and scheduling of re-manufacturing activities of turbine blades, is presented.

Although robust scheduling approaches have the potential to bring significant advantages in the minimization of risks, it may lead to non-optimal solutions in a subset of scenarios. To overcome this limitation, proactive-reactive approaches [42] could serve and will be investigated in further research activities. Moreover, only the core repair process of the considered re-manufacturing process has been addressed in this paper. Future work will also address the extension to a wider portion of the whole re-manufacturing process, including inspection, disassembly, additional repair steps and re-assembly.

Author Contributions: The authors contributed equally to this work. All authors have read and agreed to the published version of the manuscript.

Funding: This research was partially funded by DIGIMAN4.0 project ("DIGItal MANufacturing Technologies for Zero-defect Industry 4.0 Production", http://www.digiman4-0.mek.dtu.dk/, accessed on 31 Decmber 2021). DIGIMAN4.0 is a European Training Network supported by Horizon 2020, the EU Framework Program for Research and Innovation (Project ID: 814225).

Institutional Review Board Statement: Not applicable.

Informed Consent Statement: Not applicable.

Data Availability Statement: Not applicable.

Acknowledgments: We thank Ansaldo Energia for the support in the definition of the requirements in relation to the planning and scheduling of remanufacturing activities for turbine blades.

Conflicts of Interest: The authors declare no conflict of interest.

References

1. Johnson, M.R.; McCarthy, I.P. Product recovery decisions within the context of extended producer responsibility. *J. Eng. Technol. Manag.* **2014**, *34*, 9–28. [CrossRef]
2. Tolio, T.; Bernard, A.; Colledani, M.; Kara, S.; Seliger, G.; Duflou, J.; Battaia, O.; Takata, S. Design, management and control of demanufacturing and remanufacturing systems. *CIRP Ann.* **2017**, *66*, 585–609. [CrossRef]
3. Liu, C.; He, K.; Zhang, Y.; Liu, C. Research on data mining technology for the connotation and measurement of uncertainty for reassembly dimensions. *Int. J. High Perform. Syst. Archit.* **2018**, *8*, 13–21. [CrossRef]
4. Frank, A.G.; Dalenogare, L.S.; Ayala, N.F. Industry 4.0 technologies: Implementation patterns in manufacturing companies. *Int. J. Prod. Econ.* **2019**, *210*, 15–26. [CrossRef]
5. Magruk, A. Analysis of Uncertainties and Levels of Foreknowledge in Relation to Major Features of Emerging Technologies—The Context of Foresight Research for the Fourth Industrial Revolution. *Sustainability* **2021**, *13*, 9890. [CrossRef]
6. Goodall, P.; Rosamond, E.; Harding, J. A review of the state of the art in tools and techniques used to evaluate remanufacturing feasibility. *J. Clean. Prod.* **2014**, *81*, 1–15. [CrossRef]
7. Rickli, J.L.; Dasgupta, A.K.; Dinda, G.P. A descriptive framework for additive remanufacturing systems. *Int. J. Rapid Manuf.* **2014**, *4*, 199–218. [CrossRef]
8. Huang, W.; Jiang, Z.; Wang, T.; Wang, Y.; Hu, X. Remanufacturing scheme design for used parts based on incomplete information reconstruction. *Chin. J. Mech. Eng.* **2020**, *33*, 1–14. [CrossRef]
9. Liu, L.; Urgo, M. Scheduling Remanufacturing Activities for the Repair of Turbine Blades: An Approximate Branch and Bound Approach to Minimize a Risk Measure. In *Selected Topics in Manufacturing*; Springer: Berlin/Heidelberg, Germany, 2022; pp. 41–59.
10. Rizova, M.I.; Wong, T.; Ijomah, W. A systematic review of decision-making in remanufacturing. *Comput. Ind. Eng.* **2020**, *147*, 106681. [CrossRef]
11. Guide, V.D.R., Jr. Production planning and control for remanufacturing: Industry practice and research needs. *J. Oper. Manag.* **2000**, *18*, 467–483. [CrossRef]
12. Morgan, S.D.; Gagnon, R.J. A systematic literature review of remanufacturing scheduling. *Int. J. Prod. Res.* **2013**, *51*, 4853–4879. [CrossRef]
13. Kin, S.T.M.; Ong, S.; Nee, A. Remanufacturing process planning. *Procedia Cirp* **2014**, *15*, 189–194. [CrossRef]
14. Zhang, R.; Ong, S.; Nee, A.Y. A simulation-based genetic algorithm approach for remanufacturing process planning and scheduling. *Appl. Soft Comput.* **2015**, *37*, 521–532. [CrossRef]
15. Ducker, M. The Fall of the F-Class Turbine. 2015. Available online: https://www.power-eng.com/emissions/policy-regulations/the-fall-of-the-f-class-turbine/#gref (accessed on 19 January 2022).
16. GE H Series Power Generation Gas Turbine. Available online: https://en.wikipedia.org/wiki/Gas_turbine (accessed on 8 March 2022).
17. Wikipedia. Turbine Blade. 2009. Available online: https://en.wikipedia.org/wiki/Turbine_blade (accessed on 19 December 2021).
18. Behdad, S.; Williams, A.S.; Thurston, D. End-of-Life Decision Making with Uncertain Product Return Quantity. *J. Mech. Des.* **2012**, *134*, 100902. doi: 10.1115/1.4007394 [CrossRef]
19. Bermejo, J.; Calinescu, A.; Efstathiou, J.; Schirn, J. Dealing with uncertainty in manufacturing: The impact on scheduling. In Proceedings of the Thirty-Second International Matador Conference, Manchester, UK, 10–11 July 1997; Springer: Berlin/Heidelberg, Germany, 1997; pp. 149–154.
20. Jorge Leon, V.; David Wu, S.; Storer, R.H. Robustness measures and robust scheduling for job shops. *IIE Trans.* **1994**, *26*, 32–43. [CrossRef]
21. Urgo, M.; Váncza, J. A branch-and-bound approach for the single machine maximum lateness stochastic scheduling problem to minimize the value-at-risk. *Flex. Serv. Manuf. J.* **2019**, *31*, 472–496. [CrossRef]
22. Nudtasomboon, N.; Randhawa, S.U. Resource-constrained project scheduling with renewable and non-renewable resources and time-resource tradeoffs. *Comput. Ind. Eng.* **1997**, *32*, 227–242. [CrossRef]
23. Artigues, C.; Demassey, S.; Neron, E. *Resource-Constrained Project Scheduling: Models, Algorithms, Extensions and Applications*; John Wiley & Sons: Hoboken, NJ, USA, 2013.
24. Sotskov, Y.N.; Lai, T.C.; Werner, F. Measures of problem uncertainty for scheduling with interval processing times. *OR Spectr.* **2013**, *35*, 659–689. [CrossRef]
25. Liao, W.; Fu, Y. Min–max regret criterion-based robust model for the permutation flow-shop scheduling problem. *Eng. Optim.* **2019**, *52*, 687–700. [CrossRef]
26. Paprocka, I. The model of maintenance planning and production scheduling for maximising robustness. *Int. J. Prod. Res.* **2019**, *57*, 4480–4501. [CrossRef]
27. Law, A.M.; Kelton, W.D.; Kelton, W.D. *Simulation Modeling and Analysis*; McGraw-Hill: New York, NY, USA, 2000; Volume 3.
28. Tolio, T.; Urgo, M.; Váncza, J. Robust production control against propagation of disruptions. *CIRP Ann.* **2011**, *60*, 489–492. [CrossRef]
29. Aissi, H.; Bazgan, C.; Vanderpooten, D. Min–max and min–max regret versions of combinatorial optimization problems: A survey. *Eur. J. Oper. Res.* **2009**, *197*, 427–438. [CrossRef]
30. Averbakh, I. Computing and minimizing the relative regret in combinatorial optimization with interval data. *Discret. Optim.* **2005**, *2*, 273–287. [CrossRef]

31. Artigues, C.; Leus, R.; Nobibon, F.T. Robust optimization for resource-constrained project scheduling with uncertain activity durations. *Flex. Serv. Manuf. J.* **2013**, *25*, 175–205. [CrossRef]
32. Rockafellar, R.T.; Uryasev, S. Conditional value-at-risk for general loss distributions. *J. Bank. Financ.* **2002**, *26*, 1443–1471. [CrossRef]
33. Atakan, S.; Bülbül, K.; Noyan, N. Minimizing value-at-risk in single-machine scheduling. *Ann. Oper. Res.* **2017**, *248*, 25–73. [CrossRef]
34. Urgo, M. A branch-and-bound approach to schedule a no-wait flow shop to minimize the CVaR of the residual work content. *Comput. Ind. Eng.* **2019**, *129*, 67–75. [CrossRef]
35. Sarykalin, S.; Serraino, G.; Uryasev, S. Value-at-risk vs. conditional value-at-risk in risk management and optimization. In *State-of-the-art Decision-Making Tools in the Information-Intensive Age*; Informs: Catonsville, MD, USA, 2008; pp. 270–294. [CrossRef]
36. Dixit, V.; Tiwari, M.K. Project portfolio selection and scheduling optimization based on risk measure: A conditional value at risk approach. *Ann. Oper. Res.* **2020**, *285*, 9–33. [CrossRef]
37. Dodin, B. Bounding the project completion time distribution in PERT networks. *Oper. Res.* **1985**, *33*, 862–881. [CrossRef]
38. Dodin, B. Determining the optimal sequences and the distributional properties of their completion times in stochastic flow shops. *Comput. Oper. Res.* **1996**, *23*, 829–843. [CrossRef]
39. Kulkarni, V.G.; Adlakha, V. Markov and Markov-regenerative PERT networks. *Oper. Res.* **1986**, *34*, 769–781. [CrossRef]
40. Angius, A.; Horváth, A.; Urgo, M. A Kronecker Algebra Formulation for Markov Activity Networks with Phase-Type Distributions. *Mathematics* **2021**, *9*, 1404. [CrossRef]
41. Felberbauer, T.; Gutjahr, W.J.; Doerner, K.F. Stochastic project management: Multiple projects with multi-skilled human resources. *J. Sched.* **2019**, *22*, 271–288. [CrossRef]
42. Manzini, M.; Demeulemeester, E.; Urgo, M. A predictive–reactive approach for the sequencing of assembly operations in an automated assembly line. *Robot. Comput.-Integr. Manuf.* **2022**, *73*, 102201. [CrossRef]

applied sciences

MDPI

Article

Value Chain Comparison of Additively and Conventionally Manufactured Multi-Cavity Tool Steel Inserts: An Injection Molding Industrial Case Study for High-Volume Production

Mandaná Moshiri [1,*], Mohsin Raza [2], Mohamed Sahlab [2], Ali Ahmad Malik [3], Arne Bilberg [2] and Guido Tosello [1,*]

1 Department of Civil and Mechanical Engineering, Technical University of Denmark, Produktionstorvet, 2800 Kongens Lyngby, Denmark

2 Institute of Technology and Innovation, University of Southern Denmark, Alsion, 6400 Sønderborg, Denmark

3 School of Engineering & Computer Science, Oakland University, Rochester, MI 48309, USA

* Correspondence: manmos@dtu.dk (M.M.); guto@dtu.dk (G.T.)

Featured Application: In this work, a methodology for the quantitative assessment of additive manufacturing and conventional manufacturing technology value chains is presented for the production of injection molding tools.

Abstract: The development of injection molding tools is an expensive, time-consuming, and resource-intensive process offering little to no flexibility to adapt to variations in product design. Metal additive manufacturing can be used to produce these tools in a cost-effective way. Nevertheless, in an industrial context, effective methods are missing for the selection of the most suitable technology for the given tooling project. This paper presents a method to compare process chains based on additive and conventional subtractive technologies for the manufacturing of metal tooling for injection molding. The comparison is based on a technology focused-performance analysis (TFPA) through computer simulation performed using Tecnomatix Plant Simulation developed by Siemens Digital Industries Software combined with a customized cost–benefit economic analysis tool. The analysis of the technology comparison highlights potential bottlenecks for production, such as the printing phase and the heat treatment. It also gives a deeper understanding of the technology maturity level of conventional milling machines against laser powder bed fusion machines. The result is that the total costs for an insert made by AM and CM are indeed rather similar (the cost difference between the two tooling process chains is lower than 5%). The cost analysis reveals major costs drivers in the production of high-performance molding tools, such as the cutting tools employed for the milling steps and their changeover frequency. The industrial case of a 32-cavity mold insert for plastic injection molding is used to perform the study, develop the analysis, and validate the results.

Keywords: value chain; additive manufacturing; subtractive manufacturing; cost comparison; plant simulation; technology comparison; industry 4.0

Citation: Moshiri, M.; Raza, M.; Sahlab, M.; Malik, A.A.; Bilberg, A.; Tosello, G. Value Chain Comparison of Additively and Conventionally Manufactured Multi-Cavity Tool Steel Inserts: An Injection Molding Industrial Case Study for High-Volume Production. *Appl. Sci.* **2022**, *12*, 10410. https://doi.org/10.3390/app122010410

Academic Editors: Joamin Gonzalez-Gutierrez and Mirco Peron

Received: 29 July 2022
Accepted: 10 October 2022
Published: 15 October 2022

1. Introduction

Metal additive manufacturing (MAM) technologies like laser powder bed fusion (LPBF) represent laser-based techniques employed to develop products using powdered metals. It is used to manufacture prototypes during product development, as well as for applications with small batch sizes and with frequent design changes [1]. Thanks to its flexibility, speed, and cost effectiveness [2], the technique is also being investigated for the development of injection molding tools [3]. The conventional methods of producing these tools are through subtractive manufacturing, which is time-consuming and relatively expensive. Furthermore, additive manufacturing (AM) offers opportunities for innovative modifications in the product design that conventionally are not possible in an effective and efficient way.

The choice of injection mold tooling using AM can bring several advantages, such as flexible development, design freedom (to improve the product's performance), and reduced assembly effort, but also some disadvantages such as higher machine and material cost and the skills gap in using sophisticate machinery and software [4,5].

Hence, the manufacturing of high-performance tooling can currently be achieved either with conventional machining (CM), i.e., subtractive technologies such as milling, or AM [6]. As both methodologies present advantages and disadvantages, a detailed analysis must be conducted to identify the consequences of selecting a method over the other throughout the process and value chains.

Such a comparison of subtractive and additive manufacturing is already available in literature; for example, in [7], Quinlan et al. compared AM against conventional manufacturing processes. To describe AM, they introduced the statement of "complexity for free". The metaphor describes the fact that, in the case of AM, an increased complexity in product design does not correspond to an increase in cost of production, which is contrary to conventional manufacturing. Furthermore, the research presents a framework for a direct comparison of LPBF and conventional machining (computer numerical controlled (CNC) milling). However, a detailed split of the costs and time contributions to the full value chain has not been fully investigated and, currently, a research gap in this topic remains. The lack of such a perspective makes the decision of the manufacturing route to follow (AM or CM) not as informed as it should be, without a clear understanding of the costs and lead time involved, together with the general productivity of the specific technology chosen.

In the research presented in this paper, AM and CM value chains are compared quantitatively and qualitatively for producing mold inserts for plastic injection molding, with the goal to provide a relevant example of the costs, lead time, and production efficiency involved in the two manufacturing routes, together with a proven methodology to conduct such a comparison. The AM and CM produced mold inserts are equally capable and can withstand the same number of molding cycles; therefore, the AM products are not considered as a prototype or a preventive maintenance tool (as in [8]) or a spare part (as in [9,10]), but a fully functional and finished product, totally comparable to the tools produced by CM. At first, a detailed technology comparison is presented, followed by the economic analysis. Such a step-by-step analysis is particularly relevant to better understand the consequence in terms of the technologies involved and the economic impact of the choice of the two streams on the final product. The article is structured as follows: Section 1 ('Introduction') presents the research topic and explains how the work addresses the current research gap in terms of quantitative comparison of tooling value chains; Section 2 ('Related Research') discusses the relevant state-of-the-art in the fields of virtual models and simulation of production and cost modelling techniques of additive manufacturing; Section 3 ('Materials and Methods') describes the details of the analyzed tool insert and study case; Section 4 ('Technology comparison') and Section 5 ('Economic analysis') present the application and the results of the process chain simulation and of the cost analysis, respectively; Section 6 ('Results') discusses the cost drivers of both value chains and their implications in terms of productivity; finally, Section 7 ('Conclusions') summarizes the paper and provides indications for future research.

2. Related Research

The use of virtual models in a computer simulation to simulate production floor is not novel in the literature. Factory simulation allows experiments to be conducted efficiently to test different scenarios and process capabilities, reducing the need for costly and time-consuming physical experiments. Such physical experiments are also difficult to conduct directly online [11]. To run a simulation of the production parameters, one of the most popular tools is Tecnomatix Plant Simulation by Siemens Digital Industries Software, which is a discrete simulation program, event-controlled, and object-oriented [11–13].

Many researchers have used the tool to optimize manufacturing systems, for example, Klos et al. [13], who analyzed resource availability, user allocation, and throughput in a conventional manufacturing factory floor [14,15]. Vaclav et al. presented a simulation for planning an assembly system, with the goal to improve workflow and remove interruptions [16]. Tecnomatix has also been successfully used to evaluate machine utilization and production efficiency [17] and energy-related considerations towards the creation of a digital factory [18]. Other applications include a detailed analysis of production process and the detection of bottlenecks [19]. Moreover, Trebuna et al. used a simulation tool to early compare different production plans, verifying the accuracy of the different solutions before realization [20].

Regarding cost models, especially for AM, different examples can be found in the literature, as discussed in [21]. Most of them focused on a single aspect of the AM process (typically the part generation by the machine), without necessarily covering a holistic view of the full process chain from design to deployment. In the field of injection molding tool inserts' manufacturing, a cost model and comparison between the costs of AM- and CM-based process chains was presented in [22]. Fera et al. analysed previous cost models for AM and presented a new model named MiProCAMAM (mixed production cost allocation model for additive manufacturing) [23]. The model consists of five phases: preparation, build job, setup, building, and removal. The model calculates the production costs starting from the process and the geometry of the part, considering the design freedom given by AM. Process times and performance are evaluated by estimating the build time. A similar structure is presented in this paper.

Other techniques can be applied to conduct cost estimations and to collect the data required for the technology and cost model. For example, Baldinger et al. [24] identified four classes of data required for such a cost analysis. These are qualitative—intuitive (based on experience), qualitative—analogical (based on historical data), quantitative—parametric (defined from a variable), and quantitative—analytical (using enterprise resource planning). Qualitative data are easier to use and are helpful in early decision-making; however, they can sometimes lack accuracy. Intuitive data rely on the experience of skilled specialists, but they make it possible to run an early evaluation when little or no stored data exist. Intuitive and analogical data are the types mainly used in this research. Extra input data were obtained from the literature data, where necessary and available. Obviously, a proper and reliable data collection plan is vital for the success of such a project, as the quality of the input data will affect the quality of the simulations and the results of the cost model. The great variety of applications of AM shows the importance of having a reliable method for an accurate manufacturing methods' comparison, including a more detailed business model and cost–benefit calculation tools, as reported in the review in [25].

3. Materials and Methods

For running this evaluation, a dedicated industrial case study was selected, as presented in Figure 1. This is the design of a mold insert for plastic injection molding used to manufacture consumer goods made in acrylonitrile butadiene styrene or ABS [26].

The insert is made of maraging steel grade 300 (1.2709), a very common material in the tooling industry. The insert can be manufactured with both CM and AM. With the higher design freedom allowed by AM, it was demonstrated that the number of cavities in the insert could be increased from 16 to 32, thanks to a better thermal management supplied by the conformal cooling channels [26].

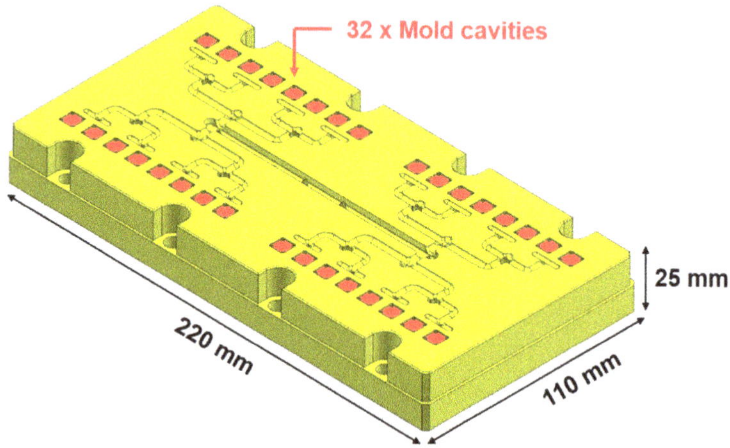

Figure 1. Mold insert for plastic injection molding (adapted with permission from from [26]).

To run the technology analysis and the production performance, the production process was simulated using discrete simulation, while the detailed cost–benefit analysis was performed with a custom-built tool based on an Excel spreadsheet that also incorporates the performance of the insert during the injection molding process.

The data used in this paper come partially from the literature and were partially collected by interviewing knowledgeable experts in the company where the research was conducted.

4. Technology Comparison

The simulation is used to obtain a full technology comparison between the CM and AM process chains. The dynamic simulation helped to define lead times, identify bottlenecks, and understand resource utilization. Different scenarios were defined and simulated to achieve this, allowing opportunities for improvements. The methodology used in the simulation presented is inspired by [27].

As a first step for this analysis, the process flow for the fabrication of the selected tool had to be mapped. The simplified process flow schematic of CM and AM process chains is presented in Figures 2 and 3, respectively.

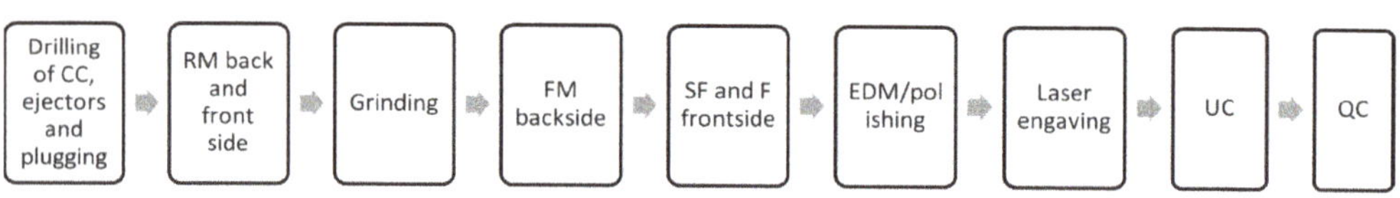

Figure 2. Simplified scheme of the conventional manufacturing process chain.

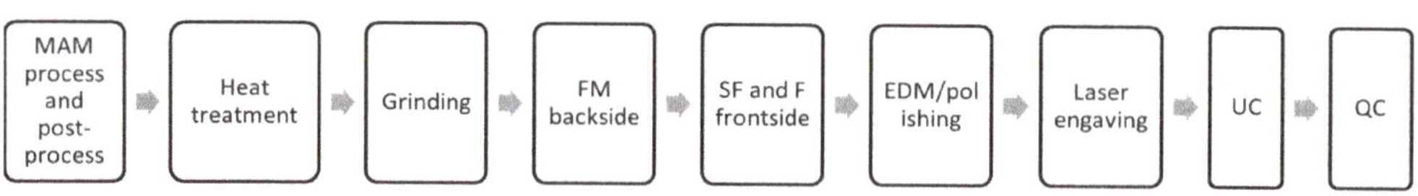

Figure 3. Simplified scheme of the additive manufacturing process chain.

In the CM flow, the first step consists of drilling in the metal block the holes for cooling channels (CCs) and ejector pins and plugging one of the ends, followed by rough milling (RM) and grinding with three-axis CNC. Finish-milling (FM), semi-finishing (SF), and finishing (F) operations are conducted with five-axis CNC, in order to achieve the specified shape and dimensional tolerances. To achieve the required texture and shape for the plastic product, the next step consists of electrical discharge machining (EDM) or manual polishing. Laser engraving is performed to mark the surface of the cavity for traceability. The last two steps are ultrasonic cleaning (UC) of the insert, to remove all residual lubricants, oil, and chips, before proceeding to quality control (QC) via 3D scanning.

The AM process chain starts with the MAM process and relative post-processes (i.e., powder cleaning, cutting from build plate, and support removal) and heat treatment. The next steps are then very similar to the CM workflow. However, at the start of each step, some additional time may be required to select the suitable cutting tools, as the printed part is near-net-shape and has different surface characteristics as compared with the CM-generated components.

To build up the model and run the simulation, it was necessary to collect, among others, data regarding the setup time [28], processing time [29], availability (the ratio between the time the machine is running successfully and the total time in which it could be running) [30], MMTR (mean time to repair, i.e., the time necessary to repair a machine) [31], number of workers, and any special observations.

The data were, as mentioned, collected through (i) conversations with experienced users, (ii) historical data, and (iii) literature research. An average was calculated when differences were noticed. The data were collected for each step of the manufacturing flow for both the CM and AM process chains. The processing and setup time data collected to run the simulations are presented in Table 1.

Table 1. Time data for CM and AM manufacturing flow (N/A: not applicable).

Process Step	Conventional Manufacturing		Additive Manufacturing	
	Processing (min)	Setup (min)	Processing (min)	Setup (min)
AM process	N/A	N/A	3240	120
AM post-processes	N/A	N/A	130	N/A
Heat treatment	N/A	N/A	2880	N/A
Drilling CC/ejectors	180	60	N/A	N/A
Plugging	60	N/A	N/A	N/A
Rough mill back	180	30	N/A	N/A
Rough mill front	180	90	N/A	N/A
Grinding	30	30	180	180
Finish mill back	105	75	180	120
Semi finish front	540	60	540	60
Finish front	540	60	540	60
EDM	480	60	480	60
Laser engraving	60	15	60	15
UC	40	N/A	40	N/A
QC	22	5	22	5

In the considered case, there are twenty-two workers in the factory, of which fifteen are dedicated to the CM area, two on AM, four on EDM, and one for the UC and QC. The number of workers is proportional to the number of machines involved.

The setup of the simulation model used is presented schematically in Figure 4, with a 3D view taken from Tecnomatix Plant Simulation. It consists of four main manufacturing areas: CM (three three-axis milling; four five-axis milling machines; and one laser engraver, UC, and QC), AM (two one-laser LPBF machines and one furnace), EDM (four wire cutters and four EDMs), and grinding (two grinding machines).

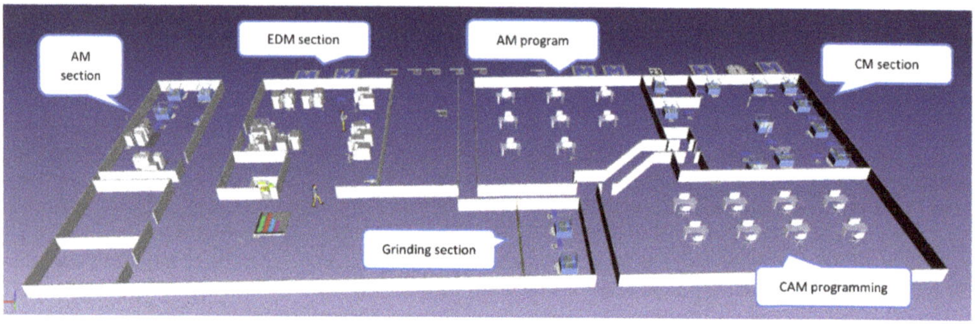

Figure 4. Three-dimensional model setup used for the manufacturing flow simulation (from Tecnomatix Plant Simulation).

Three scenarios were considered to evaluate the impact on the production throughput of selecting AM or CM as the manufacturing method:

- Scenario 1: one-to-one comparison of the AM and CM flow, with only one machine per type active and performing one step of the process chain;
- Scenario 2: an AM-to-CM comparison that includes all of the machines available listed above in the model setup, in order to consider the full factory capabilities;
- Scenario 3: focus on the AM process chain flow performance where a third newer machine is added, equipped with four lasers.

Each scenario was run to calculate the number of inserts manufactured in 1 year (considered as 250 days of active work), with the machine running 24 h a day, 7 days a week. A few assumptions were necessary to run the simulation, for example, the heat treatment is a parallel process to the printing and can fit eight inserts at the same time, while the UC can clean four parts at the same time.

4.1. Scenario 1—AM-to-CM Comparison—One Machine per Type Active

The results collected from running this production simulation for one year are presented in Figures 5 and 6 for CM and AM respectively. In this comparison, with the CM workflow, 595 inserts can be produced in one year, while 105 inserts can be produced with the AM process chain. The analysis of the resource shows that the LPBF machine represents the process bottleneck, as this step takes up to 54 h. Hence, the AM machine waiting times block the workflow productivity.

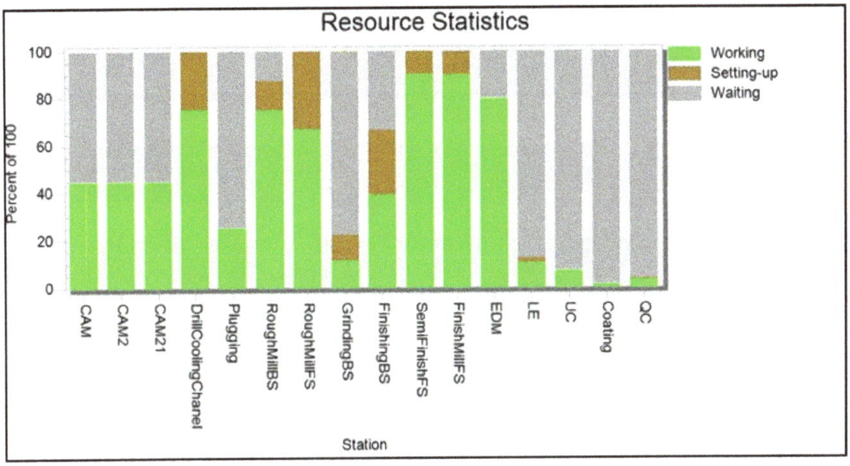

Figure 5. Resource statistics simulation for the CM workflow. The number of inserts produced in one year is 595.

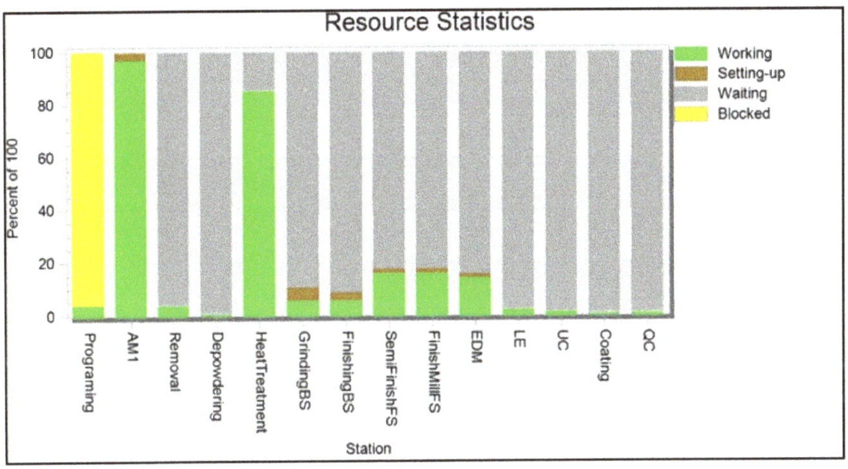

Figure 6. Resource statistics simulation for the AM workflow. The number of inserts produced in one year is 105.

4.2. Scenario 2—AM-CM Comparison—All Machines Active (Full Factory Capability)

The results of the simulation for one year of production are presented in Figures 7 and 8 for CM and AM respectively. The second scenario, in which all machines present at the production site are used, and thus two AM machines are employed, confirms the earlier conclusion that the printing step was the bottleneck; that is, its reduction by adding one extra AM machine almost doubles the throughput for AM (210 inserts). The productivity increase for CM is only marginal (two extra inserts produced compared with scenario 1, or a 0.3% increase). The AM step continues to be a bottleneck in this scenario too, as its occupancy is the highest, blocking the successive steps. An interesting aspect is that the next bottleneck in the flow can already be identified in the heat treatment because of the long time it takes and only one machine being available. One of the reasons that, in the CM manufacturing line, there is not such a big breakthrough compared with AM, despite the addition of an extra grinding and five-axis CNC, is that there are two consecutive bottlenecks: semi-finishing and finishing. These should be removed at the same time to further increase throughput.

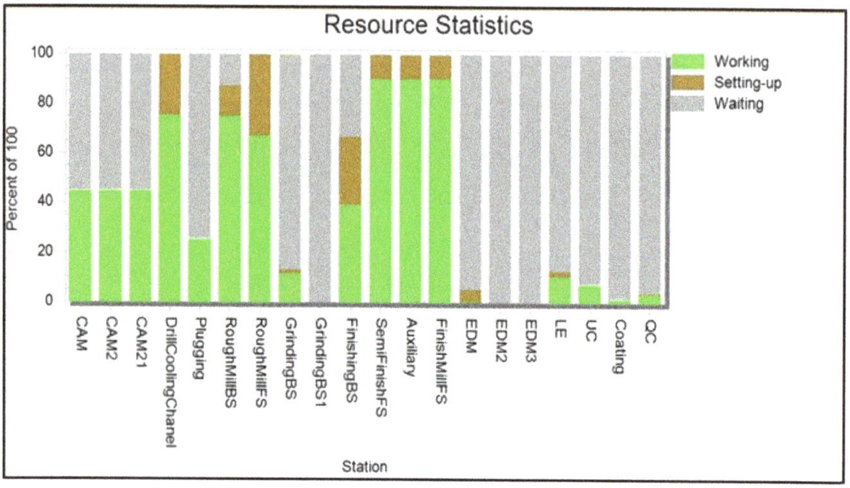

Figure 7. Resource statistics simulation for the CM workflow. The number of inserts produced in one year is 597.

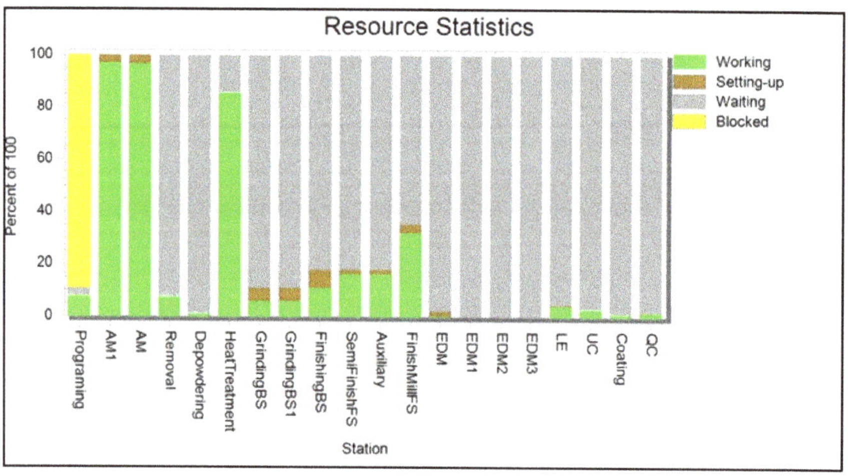

Figure 8. Resource statistics simulation for the AM workflow. The number of inserts produced in one year is 210.

4.3. Formatting of Mathematical Components

Only the AM workflow is considered in this scenario. One additional AM machine is included, this time equipped with four lasers, thus being considerably faster (Figure 9). The bottleneck of the printing step is decreased further, creating a breakthrough in throughput, which rises to 589 inserts per year. As foreseen in the previous simulation, the bottleneck is now definitely represented by the heat treatment, which is working continuously. No more inserts can be produced without reducing the heat treatment procedure and/or purchasing more furnaces. The presence of bottlenecks at the beginning of the manufacturing process means that the laser engraving, UC, and QC equipment are very underutilized.

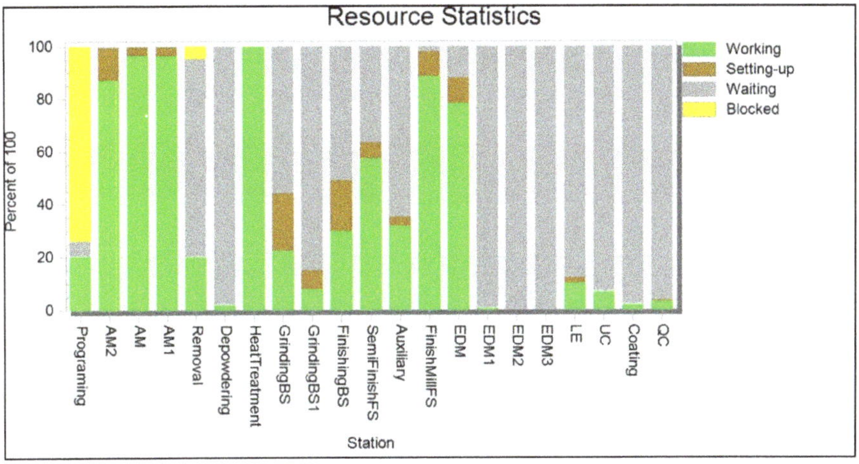

Figure 9. Resource statistics simulation for the AM workflow. The number of inserts produced in one year is 589.

5. Economic Analysis

The aim of this section is to understand the financial impact of choosing CM or AM routes to tooling fabrications, which is done with a time and cost breakdown for labour and machines for each step. All of the data for each step presented in the previous Figures 2 and 3 were collected.

All costs and prices refer to the Danish market. The labour rate is considered to be 650 DKK/h, so a round figure of 88 €/h is used (1 € = 7.4 DKK).

For cost calculation, the AM step is divided into four phases: preparation, setup, build, and removal.

- The ***preparation*** phase includes the time required to prepare the CAD design and for job preparation, which takes an average of 2 h, depending on the complexity of the design. This phase is thus estimated to cost 176 €;
- The ***setup*** costs include the machine preparation time, addition of powder, and other general manual preparation, which takes an average of 1 h. The machine then needs to start heating up the build plate and remove the oxygen from the chamber, which will also take approximately 1 h. The AM machine hourly rate needs to be considered for this step and, for a relatively old ($\approx$10 years) and robust machine (including its depreciation), the final cost is of 55 €/h. Thus, the cost of this phase is 143 €;
- The ***build*** phase needs 54 h for printing the test insert, at the same hourly rate calculated previously. The electricity consumption cost (based on Danish industrial electricity) is 0.0688 €/h, with machine consumption being 3.23 kWh. It is required now to consider the product: the mass of the geometry calculated from the volume of the insert and the density of the material (8100 kg/m^3) is multiplied by the cost per kilo of the powder (88 €/kg). The final price of 3392 € is derived by multiplying and summing all of the costs involved;
- The final ***removal*** phase includes cutting of the parts from the build plate, cleaning leftover powder, and the removal of supports. Overall, it takes approximately 2 h. The heat treatment takes 48 h. The furnace consumption is 15 kWh at the same energy cost. Summing all the activities and costs, the final price is 225 €.

All conventional processes also need to be considered for the CM flow, as well as for the finishing of the AM chain, which uses only five-axis CNC. The machine costs used here are approximations calculated from the external hourly rate for one hour of milling (approximately 101.35 €/h). It is important to consider the cutting tool costs, i.e., drill bits, required to manufacture the inserts. For the three-axis CNC, three tools are required, each costing 402 €, plus a drill bit set of 134 €. An important aspect to consider is how many times the tools are changed per insert produced. Mold makers prefer to change the tools as often as possible, in order to ensure the highest quality of the products. This means that a set of tools can be used to manufacture one or two inserts only. If each set of tools is used for two inserts, then the final figure for tools' costs for three-axis manufacturing is 675.68 €. A similar calculation was performed for the five-axis machine operations, which require five tools that each cost 402 €, with the final price being 103,151 €. An extra cost of 402 € needs to be added for the CM manufacturing flow to take into account the price of the initial block of steel. Given the processing time reported in Table 1, the final costs are 4193 € for the AM finishing steps and 8017 € for the full CM production stream.

The two products (made with the AM and CM process chain, respectively) converged at this point, and not relevant cost variation can be found for the last steps of the chain. The EDM steps cover both spark erosion, used to obtain the desired shape and surface of the cavity, and wire-cutting, employed to refine the holes in the insert. The EDM spark erosion also requires the design and manufacture of copper electrodes by milling that, together with CAM programming, takes around 5.5 h. The material cost is calculated considering between 8 and 10 electrodes per insert, with each block of copper costing 67 €, giving a total value of 608 €. The total cost associated with the EDM then becomes 3118 €.

Regarding the quality control step, again it is the same for both AM and CM. Roughly 5 min is necessary to manually prepare the products for the 3D scanner (machine cost of 34 €/h), which takes an average of 17 min to complete. Ultrasonic cleaning (machine cost of 40 €/h) takes 40 min. The machine costs are calculated by dividing the initial purchase price by the number of depreciation years multiplied by the number of working hours per year. In total, the cleaning and quality control phase costs 44 €.

The total costs involved are summarized in Table 2 and in the graph in Figure 10.

Table 2. Total AM and CM costs' overview.

Process	Additive Manufacturing	Conventional Manufacturing
AM printing	3936 €	-
CNC roughing	-	3824 €
CNC semi and finishing	4193 €	4193 €
EDM and QC	3162 €	3162 €
Total costs	11,290 €	11,179 €

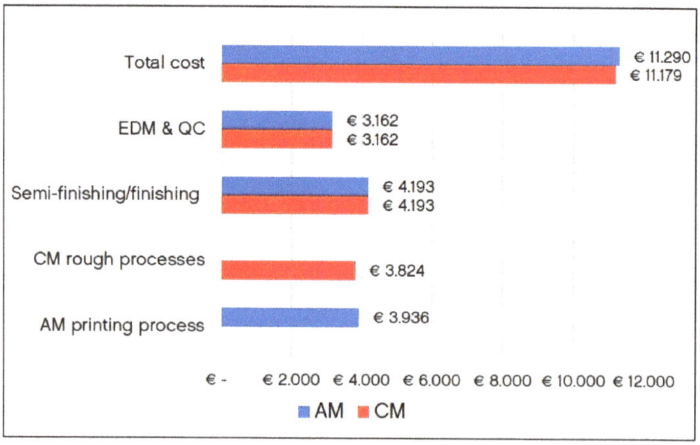

Figure 10. Total AM and CM cost comparison (each set of tools used for two inserts).

6. Results and Discussion

The result is that the total costs for an insert made by AM and CM are comparable, while the common thinking is that AM products are more expensive to produce [7]. The fact that they are comparable can be related to the cost associated with the cutting tools for the machining step, which represent a big proportion of the final price, and often their impact is either not included or considerably underestimated.

If we consider that cutting tools are changed for every insert, the calculation will actually show a total cost of 12,304 € for AM and 12,868 € for CM. In this case, the AM product is cheaper than the CM product, thanks to the production of a near-net shape component. It is, therefore, essential to strictly monitor the wear and the changeover of cutting tools to avoid unexpected costs.

Another consideration is the amount of time dedicated to labor and machine time for each stream. With a one-laser LPBF machine for AM, the total lead time is longer than CM (167 h against 88 h), but 89% of the AM time is taken by the machines working unsupervised. On the CM flow, only 68% is only machine time. Another consideration is the use of a newer LPBF machine for the AM chain, for example, equipped with four lasers, which would make the printing approximately 3.5 to 4 times faster.

It is important to consider that the final AM and CM product is not the same. AM allows to produce a mold insert that has double the number of cavities, and thus improves the performance in the injection molding phase. As a rough evaluation, it was known that the CM insert (16 cavities) has a cycle time of 9.1 s, while the AM insert (32 cavities) has a cycle time of 10.1 s. Considering an insert lifetime of 10 million shots, this corresponds to 117 continuous days for the AM insert and 105 days for the CM insert. With a difference 12 days, 160 million more molded elements can be produced with the AM insert. That also means that, during the same production time, e.g., 100 days, the AM and CM tool inserts will have potentially produced close to 27.4 and 15.2 million parts, respectively, with a potential productivity increase of 80.2% during the same production time of the AM inserts compared with the CM inserts.

7. Conclusions

This research presents an investigation into the process chains based on AM and CM manufacturing paths for the production of tooling, in particular multi-cavity mold inserts for plastic injection molding. A detailed technology and economic analysis was conducted to understand the impact of the choice.

One of the main conclusions of this project is the renewed highlighted importance of keeping a holistic vision of the complete process chains, in order to give an objective evaluation, potentially also including the performance of the manufactured tools in their intended final application. As demonstrated in this research, often (but not always), manufacturing a mold insert additively may incur higher production costs, but such a higher initial cost is completely paid back by the enhanced final performance, in this case represented by the productivity of the tool in terms of injection molded plastic components within the same production time.

The technological analysis demonstrated the production capabilities of the two streams, identifying the bottlenecks, and the results of the simulation showed how to potentially eliminate them (i.e., using newer AM machine with a higher number of lasers, adding more furnaces).

The aim of this study was not to recommend one technology path over the other, but rather to analyze in detail both process chains. The analysis helped to highlight the relatively high importance of costs that are often neglected (such as the change in the cost of cutting tools due to wear). Here, it was demonstrated that they play a considerable part in the final cost. However, a challenge faced during this exploration was in the data collection itself, as some of the data used in the comparison came from the literature or, alternatively, from user experience. The quality and truthfulness of the comparison, when looking at a real factory environment, will increase if only historical quantitative data are used. Still, the relevance of the data highly depends on the level of digitalization of the considered factory. Potential future research on this field could start from a work similar to [32] to develop a tool specialized for mold components in order to (i) compare the CM and AM process chains, as well as to (ii) support the engineers in the selection of the best method for each tool component during the design of the injection mold.

As a next investigation, it would be worth analyzing more in depth the design process and programming phase, which are executed at the computer, for preparing the design and the machines' programs to perform the job, as different types of software and skills level are required, and different levels of design complexities are involved for the two manufacturing flows. In the current work, the design approach for the AM and CM component was in fact the same. However, it has been demonstrated in previous research that, by adopting a design with the AM (DfAM) approach, a better trade-off between performance and manufacturing costs can be achieved [33].

Lastly, a final observation should be made in regard to the process technology maturity and the product complexities, whose considerations also represented a challenge of this investigation. AM- and CM-based process chains were compared in this analysis as if they were producing the exact same product. This was, however, not the case, as demonstrated by the final performance. The AM design could have been even further improved, for example, by topology optimization, as investigated in a similar case study by Sinico et al. in [34], to reduce weight and, consequently, the printing time. A similar investigation with a focus on the possible workflow to adopt for redesigning components to fully exploits the AM technology potential was conducted by Dalpadulo et al. in [35]. In addition to topology optimization, another DfAM area that could be further investigated for application areas, such as injection molding, is the possibility of part consolidation. A design method to explore this area was proposed by Kim et al. in [36].

The comparison was, despite the mentioned differences, carried out in the same way. The processes also compared CNC and LPBF, which have very different levels of technological advancement and industrial maturity. LPBF might reach the same level as CNC in the next decade, improving its performance further.

Author Contributions: Conceptualization, M.M., M.R., M.S. and G.T.; methodology, M.M., M.R., M.S., A.A.M. and A.B.; software, M.M., M.R., M.S. and A.A.M.; validation, M.M., M.R., M.S. and A.A.M.; investigation, M.M., M.R. and M.S.; resources, M.M.; data curation, M.M., M.R. and M.S.; writing—original draft preparation, M.M.; writing—review and editing, M.M., M.R., M.S., A.A.M., A.B. and G.T.; supervision, M.M., A.B. and G.T.; project administration, M.M.; funding acquisition, G.T. All authors have read and agreed to the published version of the manuscript.

Funding: The project received funding from the European Union's Horizon 2020 Marie Sklodowska-Curie grant agreement No. 721383, for the Precision Additive Metal Manufacturing (PAM2) project (https://pam2.eu/, accessed on 12/10/2022).

Institutional Review Board Statement: Not applicable.

Informed Consent Statement: Not applicable.

Data Availability Statement: Not applicable.

Acknowledgments: The authors are grateful to all of the people in the manufacturing company who supported this research by sharing their valuable experience and advice.

Conflicts of Interest: The authors declare no conflict of interest. The funders had no role in the design of the study; in the collection, analyses, or interpretation of data; in the writing of the manuscript; or in the decision to publish the results.

References

1. Cicconi, P.; Mandolini, M.; Favi, C.; Campi, F.; Germani, M. Metal Additive Manufacturing for the Rapid Prototyping of Shaped Parts: A Case Study. *Comput.-Aided Des. Appl.* **2021**, *18*, 1061–1079. [CrossRef]
2. Blakey-Milner, B.; Gradl, P.; Snedden, G.; Brooks, M.; Pitot, J.; Lopez, E.; Leary, M.; Berto, F.; Plessis, A. Metal additive manufacturing in aerospace: A review. *Mater. Des.* **2021**, *209*, 110008. [CrossRef]

3. Altaf, K.; Abdul-Rani, A.-M.; Raghavan, V. Prototype production and experimental analysis for circular and profiled conformal cooling channels in aluminium filled epoxy injection mould tools. *Rapid Prototyp. J.* **2013**, *19*, 220–229. [CrossRef]
4. Lindemann, C.; Jahnke, U.; Moi, M.; Koch, R. Analyzing product lifecycle costs for a better understanding of cost drivers in additive manufacturing. In Proceedings of the 23rd Annual Internatio Solid Freeform Fabrication Symposium—An Additive Manufacturing Conference, Austin, TX, USA, 6–8 August 2012; pp. 177–188.
5. Diegel, O.; Nordin, A.; Motte, D. *A Practical Guide to Design for Additive Manufacturing*; Springer: Singapore, 2019. [CrossRef]
6. Park, H.S.; Dang, X.P.; Nguyen, D.S.; Kumar, S. Design of advanced injection mold to increase cooling efficiency. *Int. J. Precis. Eng. Manuf.* **2020**, *7*, 319–328. [CrossRef]
7. Quinlan, H.E.; Hasan, T.; Jaddou, J.; Hart, A.J. Industrial and Consumer Uses of Additive Manufacturing: A Discussion of Capabilities, Trajectories, and Challenges. *J. Ind. Ecol.* **2017**, *21*, 15–20. [CrossRef]
8. Lolli, F.; Coruzzolo, A.M.; Peron, M.; Sgarbossa, F. Age-based preventive maintenance with multiple printing options. *Int. J. Prod. Econ.* **2022**, *243*, 108339. [CrossRef]
9. Sgarbossa, F.; Peron, M.; Lolli, F.; Balugani, E. Conventional or additive manufacturing for spare parts management: An extensive comparison for Poisson demand. *Int. J. Prod. Econ.* **2021**, *233*, 107993. [CrossRef]
10. Cantini, A.; Peron, M.; De Carlo, F.; Sgarbossa, F. A decision support system for configuring spare parts supply chains considering different manufacturing technologies. *Int. J. Prod. Res.* **2022**, 1–21. [CrossRef]
11. Kokareva, V.V.; Malyhin, A.N.; Smelov, V.G. Production Processes Management by Simulation in Tecnomatix Plant Simulation. *Appl. Mech. Mater.* **2015**, *756*, 604–609. [CrossRef]
12. Arunkumar, Y.; Patil, R.; Mohankumar, S. Discrete Event Simulation for Increasing Productivity in Digital Manufacturing. *SSRN Electron. J.* **2012**, *1*, 36–40. [CrossRef]
13. Siemens PLM. Tecnomatix. Available online: https://www.plm.automation.siemens.com/global/en/products/manufacturing-planning/plant-simulation-throughput-optimization.html (accessed on 13 July 2022).
14. Kłos, S.; Trebuna, P. The impact of the availability of resources, the allocation of buffers and number of workers on the effectiveness of an assembly manufacturing system. *Manag. Prod. Eng. Rev.* **2017**, *8*, 40–49. [CrossRef]
15. Pekarčíková, M.; Trebuňa, P.; Liaposhchenko, O.; Petrik, M. Increasing the efficiency of logistics flow in production system by the software tool tecnomatix process simulate. In Proceedings of the 3rd EAI International Conference on Management of Manufacturing Systems, Dubrovnik, Croatia, 6–8 November 2018. [CrossRef]
16. Václav, Š.; Lecký, Š.; Senderská, K.; Mareš, A. Simulation as a support tool in assembly systems planning. *MATEC Web Conf.* **2017**, *137*, 04008. [CrossRef]
17. Fedorko, G.; Honus, S.; Badiarová, S. Analysis of the utilization of machinery in the production process using computer simulation. *MATEC Web Conf.* **2017**, *134*, 00012. [CrossRef]
18. Stoldt, J.; Schlegel, A.; Putz, M. Enhanced integration of energy-related considerations in discrete event simulation for manufacturing applications. *J. Simul.* **2016**, *10*, 113–122. [CrossRef]
19. Kliment, M.; Popovič, R.; Janek, J. Analysis of the production process in the selected company and proposal a possible model optimization through PLM software module Tecnomatix Plant Simulation. *Procedia Eng.* **2014**, *96*, 221–226. [CrossRef]
20. Trebuňa, P.; Kliment, M.; Edl, M.; Petrik, M. Creation of simulation model of expansion of production in manufacturing companies. *Procedia Eng.* **2014**, *96*, 477–482. [CrossRef]
21. Kadir, A.Z.A.; Yusof, Y.; Saidin Wahab, M. Additive manufacturing cost estimation models—A classification review. *Int. J. Adv. Manuf. Technol.* **2020**, *107*, 4033–4053. [CrossRef]
22. Moshiri, M.; Loaldi, D.; Zanini, F.; Sgaravato, D.; Carmignato, S.; Tosello, G. Analysis of an as-built metal additively manufactured tool cavity insert performance and advantages for plastic injection moulding. *J. Manuf. Proc.* **2021**, *61*, 369–382. [CrossRef]
23. Fera, M.; Fruggiero, F.; Costabile, G.; Lambiase, A.; Pham, D.T. A new mixed production cost allocation model for additive manufacturing (MiProCAMAM). *Int. J. Adv. Manuf. Technol.* **2017**, *92*, 4275–4291. [CrossRef]
24. Baldinger, M.; Levy, G.; Schönsleben, P.; Wandfluh, M. Additive manufacturing cost estimation for buy scenarios. *Rapid Prototyp. J.* **2016**, *22*, 871–877. [CrossRef]
25. Savolainen, J.; Collan, M. How Additive Manufacturing Technology Changes Business Models?—Review of Literature. *Addit. Manuf.* **2020**, *32*, 101070. [CrossRef]
26. Sinico, M.; Ranjan, R.; Moshiri, M.; Ayas, C.; Langelaar, M.; Witvrouw, A.; van Keulen, F.; Dewulf, W. A mold insert case study on Topology Optimized design for Additive Manufacturing. In Proceedings of the 30th Annual International Solid Freeform Fabrication Symposium—An Additive Manufacturing Conference, Austin, TX, USA, 12–14 August 2019.
27. Bangsow, S. *Tecnomatix Plant Simulation*; Springer International Publishing: Cham, Switzerland, 2016. [CrossRef]
28. Dennis, P. *Lean Production Simplified: A Plain Language Guide to the World's Most Powerful Production System*; Productivity Press: New York, NY, USA, 2002.
29. Nash, M.A.; Poling, S.R. *Mapping the Total Value Stream*; Productivity Press: New York, NY, USA, 2008. [CrossRef]
30. Borris, S. *Total Productive Maintenance*; McGraw-Hill: New York, NY, USA, 2006.
31. Agustiady, T.K.; Cudney, E.A. *Total Productive Maintenance: Strategies and Implementation Guide*; CRC Press Taylor & Francis Group: Boca Raton, FL, USA, 2015.
32. Westerweel, B.; Basten, R.J.; van Houtum, G.J. Traditional or additive manufacturing? Assessing component design options through lifecycle cost analysis. *Eur. J. Oper. Res.* **2018**, *270*, 570–585. [CrossRef]

33. Sbrugnera Sotomayor, N.A.; Caiazzo, F.; Alfieri, V. Enhancing Design for Additive Manufacturing Workflow: Optimization, Design and Simulation Tools. *Appl. Sci.* **2021**, *11*, 6628. [CrossRef]
34. Sinico, M.; Ranjan, R.; Moshiri, M.; Ayas, C.; Langelaar, M.; van Keulen, F.; Dewulf, W.; Witvrouw, A. Improving the manufacturability of metal AM parts. *Mikroniek* **2019**, *5*, 23–28.
35. Dalpadulo, E.; Gherardini, F.; Pini, F.; Leali, F. Integration of Topology Optimisation and Design Variants Selection for Additive Manufacturing-Based Systematic Product Redesign. *Appl. Sci.* **2020**, *10*, 7841. [CrossRef]
36. Kim, S.; Moon, S.K. A Part Consolidation Design Method for Additive Manufacturing based on Product Disassembly Complexity. *Appl. Sci.* **2020**, *10*, 1100. [CrossRef]

applied sciences

MDPI

Article

Ontology-Based Production Simulation with OntologySim

Marvin Carl May *, **Lars Kiefer**, **Andreas Kuhnle** and **Gisela Lanza**

wbk Institute of Production Science, Karlsruhe Institute of Technology (KIT), Kaiserstr. 12, 76131 Karlsruhe, Germany; uufrm@student.kit.edu (L.K.); andreaskuhnle@t-online.de (A.K.); gisela.lanza@kit.edu (G.L.)

* Correspondence: marvin.may@kit.edu; Tel.: +49-1523-950-2624

Abstract: Imagine the possibility to save a simulation at any time, modify or analyze it, and restart again with exactly the same state. The conceptualization and its concrete manifestation in the implementation OntologySim is demonstrated in this paper. The presented approach of a fully ontology-based simulation can solve current challenges in modeling and simulation in production science. Due to the individualization and customization of products and the resulting increase in complexity of production, a need for flexibly adaptable simulations arises. This need is exemplified in the trend towards Digital Twins and Digital Shadows. Their application to production systems, against the background of an ever increasing speed of change in such systems, is arduous. Moreover, missing understandability and human interpretability of current approaches hinders successful, goal oriented applications. The OntologySim can help solving this challenge by providing the ability to generate truly cyber physical systems, both interlocked with reality and providing a simulation framework. In a nutshell, this paper presents a discrete-event-based open-source simulation using multi-agency and ontology.

Keywords: ontology; production simulation; multi-agent; digital twin

Citation: May, M.C.; Kiefer, L.; Kuhnle, A.; Lanza, G. Ontology-Based Production Simulation with OntologySim. *Appl. Sci.* **2022**, *12*, 1608. https://doi.org/10.3390/app12031608

Academic Editors: Roque Calvo, José A. Yagüe-Fabra and Guido Tosello

Received: 4 January 2022
Accepted: 26 January 2022
Published: 3 February 2022

Publisher's Note: MDPI stays neutral with regard to jurisdictional claims in published maps and institutional affiliations.

1. Introduction

Product differentiation and customer satisfaction today demonstrate a shift from a technical product focus towards customized, unique products. Increasing this product individualization leads to changes and increases in the complexity of production systems and amplifies the necessity for more flexible production systems [1,2]. Changes in production directly affect both modeling and the simulation of production systems, as these are carried out with the purpose of providing analyses and insights into the up-to-the-minute, real production system [3]. Most notably, to provide insights into complex systems and make decisions, for instance, regarding production changes [4]. To meet this new demand for flexibility of production system simulations [3], this paper presents an ontology-based simulation enabling the transfer of ontology advantages to simulation. The ontology's ability to dynamically map multi-dimensionality representations [5] offers new possibilities for simulation [6]. In this paper, the conceptualization of a simulation based on a knowledge graph as an instantiated ontology is presented. This is extended by introduction of the manifestation, the Owlready [7]-based open source solution OntologySim, which fully integrates an ontology into the simulation and offers a visualization via web development. The multi-agent-based OntologySim, thus, provides an ideal basis for the application of a digital twin. This research aims to present a combination of ontology with manufacturing simulation, which is available as an open-source solution to the general public and is distinguished from previous solutions by its flexibility and storability.A digital twin, as an up-to-the-minute representation of the real system [3], is instantiated based on the OntologySim as a digital master. Hence, the current production system shall be mapped in detail to the simulation, the ontology-based model that is both describing the state as a knowledge graph and generally modeling the ontological structure of the regarded system, as shown in Figure 1. The underlying concept of this storable and expandable ontology-based simulation is discussed in more detail in Section 4.1.

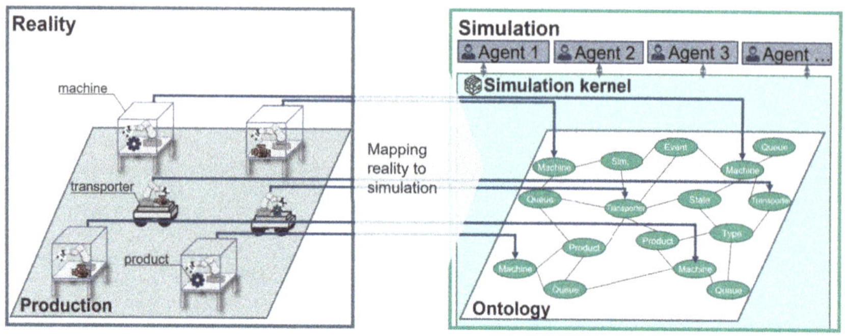

Figure 1. Outline of an ontology-based simulation model.

The paper is organized as follows. In Section 2, Foundation, the terms ontology and simulation are introduced. Section 3, Related Work, categorizes existing simulations and ontology-based simulations in the realm of production and beyond. In Section 4, Proposed Ontology-based Simulation, the unique features of an ontology-based simulation approach are presented and extended in Sections 5–7, introducing the manifestation OntologySim and the individual components and functionalities in more detail. This paper finishes with a discussion in Section 8 and a Summary and Outlook in Section 9.

2. Foundation

Given the novelty of the approach of using ontologies as means for simulating production systems and the necessity to elaborate both topics individually, this section presents an introduction to ontologies in Section 2.1, classical simulation in the context of production systems in Section 2.2, and a delimitation of simulations based on an ontology in Section 2.3.

2.1. Ontology

The term ontology was initially coined in philosophy and describes the semantic representation of existence. For today's problems, the term ontology is often associated with the term knowledge-base [8]. For the exact definitions of an ontology, there are multiple versions available, but, for this application, the following definition is quite appropriate [9]:

The formal description of an ontology is

$$O = (C, R, A, Top),$$

where C is the set of all concepts, R the set of all assertions, and A the set of all axioms. Top defines the top level of the hierarchy. R contains the two subsets H and N, where H is the "set of all assertions in which the relation is a taxonomic relation" [9], and N "a set of all assertions in which the relation is a non-taxonomic relation" [9]. An instantiated ontology can best described with the term knowledge graph, where concepts refer to vertices, and relations refer to assertions that signify the type and structure of the relation between two concepts.

The strengths of an ontology compared to object-oriented languages (OOL) are "reuseability, interoperability, flexibility, consistency and quality checking and reasoning" [10]. In the field of OOL, there are only UML diagrams that offer a similarly good basis for reuse and interoperability. However, implemented applications are often incompatible, even if stated with the same UML diagram. Furthermore, reasoning, by design achievable with ontologies, offers classification, consistency checking, and knowledge extraction, which can only be achieved with significantly more effort in OOL [10].

To take full advantage of an ontology, five design principles have been established [8]. These are named and briefly explained below:

- *Clarity* pursues the goal to design definitions objectively and formally and to prefer fully consistent definitions whenever possible.
- *Coherence* is the property of being logical and consistent.
- *Extendability* allows the ontology to be extended or modified regarding new relations and vocabularies.
- *Minimal encoding bias* means that the chosen representation goes beyond the current implementation to ensure knowledge-sharing over the current use-case.
- *Minimal ontological commitment*: An ontology should have few requirements and restrictions on the modeled world to represent specialized and individual circumstances.

These design principles are also applied to the OntologySim, with a focus on "minimal ontological commitment" and "minimal encoding bias". The goal of the OntologySim is to represent diverse production types and characteristics in the best possible way to be able to map complex and individualized production plants with all their characteristics to an ontology. One classification of ontologies is based on a three-step approach that classifies an ontology based on the scope of concepts that are included: (1) general ontologies, including cross-divisional concepts, (2) top level ontologies, focused on a particular application, and (3) domain ontologies, that translate known class and data models [11]. Another classification can be based on the purpose of an ontology, which can be limited to communication enablers, where agents derive their communication convention from a commonly accepted ontology, the purpose of automatic reasoning to identify new concepts and relations or the purpose of representation through knowledge reusability [12].

Typical languages for the description and exchange of ontologies in the wake of the Semantic Web Initiative [13] are the Resource Description Framework (RDF), initially developed to describe metadata by describing resources with characteristics and links to other resources, and the Extensible Markup Language (XML), developed to annotate and describe data and documents. Their power can be vastly increased by using schemata, i.e., ontology languages, such as the Resource Description Framework Schema (RDFS), a set theory-based formalism for ontologies, and the Web Ontology Language (OWL) family, based on description logic and set theory [7,14].

In the following, two classification approaches for ontologies are presented in more detail, firstly, the distinction between static versus dynamic ontologies [15]. Secondly, from a programming perspective, the distinction between the SPARQL Protocol And RDF Query Language (SPARQL), a graph-based query language for the Web Ontology Language (OWL) [16], and ontology-oriented programming for OWL [7,17], as well as traditional Application Programming Interfaces (API), is made.

A static ontology assumes that the world is static, so only queries are possible, and "inference, classification or dynamic class creation" are not allowed [17]. Examples of static ontologies are given by Kalyanpur et al. [18] and Goldman [19]. Dynamic ontologies allow the creation of classes and instances during runtime [17]. They often use concepts that include state, state transition, and process and are inspired by finite state machines and Petri nets [10].

An overview of the classification of an ontology is visualized in Figure 2.

In the structure and access of an ontology, three types have been established, SPARQL, the traditional OWL API, and object-oriented OWL. The differences between the two types are explained using the Java OWL API and Owlready. SPARQL will not be discussed further in the following because of the following significant disadvantages. SPARQL is an RDF query language based on SQL (Structured Query Language), but, since it is not based on OWL and queries have to be written for each access [17,20], it is significantly slower than the alternatives. OWL API and Owlready use the OWL2 standard based on the W3C specification [21]. OWL2 is an extension to OWL concerning "richer data types, data ranges, qualified cardinality restrictions, asymmetric, reflexive, and disjoint properties" [21]. The RDF (Resource Description Framework) sets the standard grammar for formal "association among resources" [10] and is based on an XML schema [10]. OWL extends the RDF concerning the "ability to express more information about the characteristics of properties and to define classes by grouping those instances that meet these characteristics" [10].

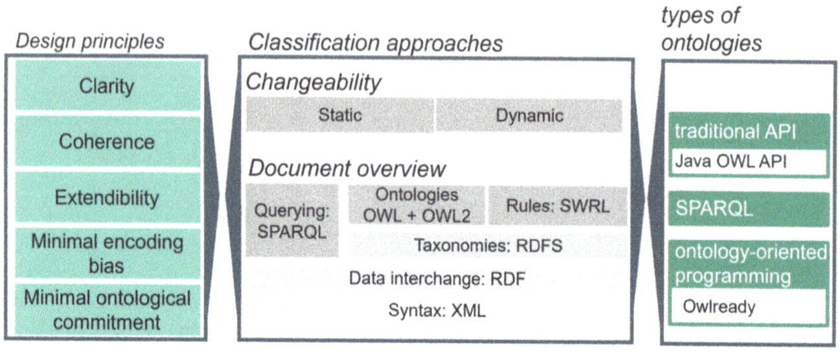

Figure 2. Summary of the OntologySim foundation, based on [8,10].

Traditional APIs define unified methods and classes to modify and customize the ontology. The Java-based OWL API allows loading and saving of different syntaxes, but with an independent internal syntax. The OWL API is widely used in Protégé-4 [22], SWOOP [23], and NeOnToolkit [24,25], which are tools capable of modeling ontologies, sometimes even graphically.

In ontology-oriented programming, the entities of the ontology classes, properties, and individuals are considered as classes, attributes, and instances in the object model [17]. The basic structure of Owlready is based on a SQLite3 database, which stores the optimized RDF quadstore [7]. If required, the ontology entities are loaded in Python, and a modification of the Python objects leads to an automatic update of the quadstore [7]. The advantages of ontology-oriented programming are lower programming effort and easy readability of the code [17]. Furthermore, Owlready combines the agility of object-oriented programming with the expressiveness of an ontology with good access speeds through relational databases [26]. These reasons were decisive for the use of Owlready2.

A comparison of the resource efficiency between OWL API and Owlready shows that Owlready requires less memory and has faster loading times than the OWL API for smaller applications (with about 60,000 classes). For larger ontologies (5 million triples), the loading time for OWL API is 100% faster. In terms of listening time, OWL API is significantly faster, although the gap increases exponentially with larger applications [17]. For this reason, when designing an ontology, care must be taken to ensure that the ontology remains compact, as the resource overhead increases exponentially, especially for listening for very large ontologies. For the application of the simulation in the presented OntologySim framework, for instance, with 5 machines, Owlready has 73 classes and 1225 objects, and it, thus, uses significantly fewer classes and edges than the benchmark test. For this reason, the problem regarding large, slow ontologies is of less relevance in this case. Owlready was introduced in 2017 and has been successfully used in biomedical informatics [17], for football games [27], and production simulation [28]. However, the differences between Liu [28] and OntologySim are discussed in Section 3.

2.2. Simulation

The term simulation refers to mapping a system to analyze its dynamic processes in a simplified replica (of the target system) and to obtain transferable knowledge [29]. When modeling the simulation, different approaches can be used either individually or in combination. There is a distinction between Discrete Event Simulation (DES), which has event-based step sizes, System Dynamics, which has a top-down approach over system changes per time, and Agent-based Simulation, which executes decision processes based on agents. Agent-based simulation can be further divided into multi-agent-based and single agent-based systems [30]. For the application in complex systems, multi-agent-based can be advantageous because of the decentralized and robust structure [31–33]. The OntologySim is a discrete event-based, multi-agent system. The exact classification of the simulation is discussed in Section 4.

Besides the modeling type, a simulation consists of the following components according to the VDI3633 [29]: a simulation kernel, data management (see Figure 3), user interface, and interfaces to external databases. The simulation kernel contains the model world and necessary elements, particularly events to ensure automatic execution [29]. The last level, the interfaces to external databases, is not included in this simulation because the OntologySim attempts to represent as many applications as possible, and this interlocking with a particular external database conflicts with the many individual external databases prevailing in real production systems. Hence, a config file or owl file specifying the required information to be extracted before or while execution and API interfaces serve as external interfaces. For more detail, the reader is referred to Section 7.

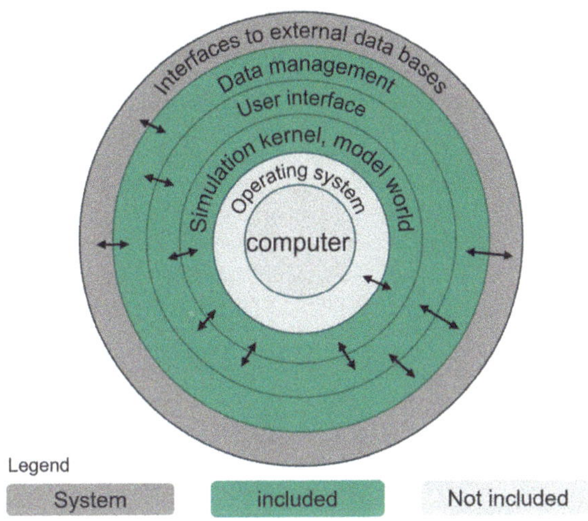

Figure 3. Components of the OntologySim simulation tool (DES), based on [29].

2.3. Term Ontology-Based Simulation

Ontology-based simulation as a term is used differently throughout the literature as the degree of ontology application is not clearly defined. The proposed classification scheme, as shown in Figure 4, is based on References [30,34] and has been adapted and integrated with respect to simulations based on an ontology. The Ontology Integration Level (OIL) for simulations continuously increases from production capacity to ontologies that store all subcategories and simulation specifications. It allows the classification of ontologies that are interwoven with an ontology-based simulation in the area of production simulations.

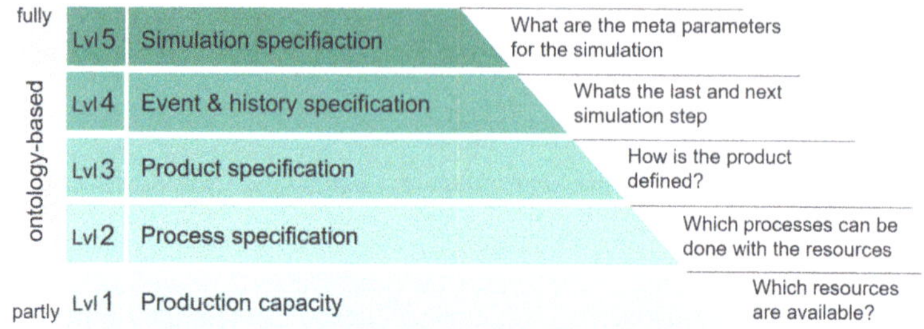

Figure 4. Examining the Ontology Integration Level for ontology-based production simulation.

The definition of each level is as follows:

- *Lvl1: Production capacity:* The ontology contains information about the capacity and structure of the ontology, such as machines, transporters, and queues.
- *Lvl2: Process specification:* The next level adds information regarding processes, setup behavior and resource change.
- *Lvl3: Product specification:* An ontology enables products to be produced in simulation. This means that production plans are included in the ontology and the current processing step of the product is known in the ontology at any time.
- *Lvl4: Event & history specification:* Events and data from the past lay the foundation for processes and allow the determination of the next simulation step.
- *Lvl5: Simulation specification:* The highest ontology integration level contains all information of the previous levels and focuses on all further necessary information, such as current time steps or work orders, to be contained in the ontology. Fully ontology-based simulations allow saving the ontology, and, thus, the simulation core and current status, at any time without information loss. This is possible when both the data management and the simulation kernel are integrated into the ontology.

According to this definition, OntologySim is a fully ontology-based simulation in ontology integration level 5.

3. Related Work

As introduced in Section 2.3, different integration levels for ontologies in simulation and production exist. The literature in this domain is reviewed and extended with other approaches of ontology-based simulations in Section 3.1. Then, available simulations for production systems are reviewed in Section 3.2.

3.1. Literature Review

In the literature, there are many approaches to integrate ontologies in production and simulation. The majority of approaches to date focus on using the ontology as a memory or database to build a simulation model. This is similar to handling the simulation data management with the help of an ontology, yet, the simulation itself remains unchanged. Based on a grounded theory literature review [35], a large set of papers dealing with both simulation and ontology or both digital twin and ontology in the domain of production research were identified. By manual analysis a total of 20 papers, the most relevant research approaches were selected (cf. Table 1), with priority given to approaches with higher Ontology Integration Levels and the requirement of including ontologies from a research perspective in production organization. In general, the existing approaches can be clustered as follows:

- *Concepts/schemata for production simulation:*
  This category is characterized by the fact that schemes and concepts for a simulation are presented, but the active use has not yet been implemented or has only been carried out for an example case. Some of the concepts differ greatly in scope and

structure of the ontology. Thus, multi-agent approaches were designed, such as that by Karageorgos et al. [36], who define an agent-based approach to support logistics and production planning; and Mönch and Stehli [37], who store data based on domain-related predicates, such as machine structure and task-related predicates, such as scheduling. Additionally, there are concepts for the description of the factory layout [38] by ontological means. Furthermore, there are databases, such as ONKI [39], available, which hold production data sets and ontology schemata for production.

- *Real world application (CPS, MES):*
  These ontology approaches are characterized by the fact that a virtual representation of a production plant and a Manufacturing Execution System (MES) is extended by an ontology to maintain flexibility. In these approaches, the ontology often serves as a classic knowledge database. Examples are the use of ontologies for MES and OPC-UA interfaces [40] and a digital twin for a Cyber-physical Production system (CPS) [28]. A different approach is followed by the virtual factory data model presented by Terkaj [41], which aims at representing factory objects, i.e., from products to machines, virtually, in a static way.
- *External data source to facilitate easier production simulation start:*
  These papers pursue the integration of different external data sources to create a uniform simulation data model through an ontology. It is noticeable for this type of concept that multiple ontologies are used for modeling [42–44]. The simulation models are either converted into executable models, as in Silver et al. [42], or the ontology serves as a knowledge database adjunct to the simulation model [43,44]. Benjamin et al. [43] introduce an ontology-driven framework based on scheduling, simulation and optimization ontology. Du et al. [44] implement a framework, where multiple databases are integrated into multiple ontologies and then combined to one core ontology model. A hybrid approach focuses on modeling external data sources, a static virtual model of the factory and its history, presented by Terkaj and Urgo [45], likewise uses ontologies and extends so far static representations with the system's history and evolution. Thus, these can be classified as a digital shadow [3]. An extension of this approach is described by Terkaj et al. [46], who use this continuously synchronized, ontology-based virtual model to enable "in situ" simulation of future system behavior. Hence, the approach aims at enabling a foresighted digital twin [3].
- *Fully ontology-based simulations:*
  A similar approach to the OntologySim is only provided by the paper of Warden et al. [47]. Here, an attempt was made to create a simulation for transport logistics utilizing several ontologies, which were divided into different layers. The goals regarding "scalable, portable and reusable domain models [...] fell short [...] and turned out to yield more pain than gain" [47].

The categorization of different literature examples is presented in Table 1.

Table 1. Literature review.

	Ontology System						Real World		Simulation			
Approach by	**Ontology Schema**	**Multi Agent System**	**Multiple Ontology**	**Ontology as Data Storage**	**Application Framework**	**External Data Source**	**Cyber-Physical-System (CPS)**	**Manufacturing Execution System (MES)**	**Ontology Integration Level**	**Intervention**	**Visualization**	**Main Goal**
Concepts / schemata for production simulation												
Terkaj and Urgo [45]	●	○	○	●	●	●	○	○	4	○	○	Performance
Terkaj et al. [41]	●	○	○	●	○	○	○	○	3	○	○	Virtual model
Viljanen et al. [39]	●	○	○	○	○	○	○	○	-	○	○	Database
Mönch and Stehli [37]	●	●	○	●	●	○	○	○	3	○	○	MAS
Cheng et al. [6]	●	●	○	○	○	○	○	○	3	○	○	Concept
Scholz and Schabus [38]	●	○	○	●	○	○	○	○	1	○	○	Schema
Gurjanov et al. [48]	●	○	○	○	○	○	○	○	2	○	○	Product
Karageorgos et al. [36]	●	●	○	●	○	○	○	○	3	○	○	MAS
Fumagalli et al. [49]	●	○	○	●	○	○	○	◐	3	○	○	MES
Mazzola et al. [50]	●	○	○	○	○	●	○	○	2	○	○	Concept
Guizzardi and Wagner [51]	●	○	○	○	○	○	○	○	4	○	○	DES
External data source to facilitate easier production simulation start												
Du et al. [44]	●	●	●	●	●	●	○	○	3	○	●	AnyLogic
Terkaj et al. [46]	●	○	○	●	●	●	○	○	4	○	●	Speed
Benjamin et al. [43]	●	●	●	●	●	●	○	○	3	○	○	Framework
Silver et al. [42]	●	◐	●	○	●	●	○	○	3	○	○	Parsing
Real world application (CPS, MES)												
Liu et al. [28]	●	○	○	●	○	○	●	●	3	○	●	digital twin
Ansari et al. [52]	●	○	○	●	●	○	●	○	2	○	○	Problem
Katti et al. [40]	●	○	○	●	○	○	●	●	2	○	○	OPC UA
Chen and Tu [53]	●	○	○	●	○	○	◐	◐	1	○	●	CPS
Fully ontology-based simulations												
Warden et al. [47]	●	●	○	●	●	○	○	○	4	○	○	Scheduling

Legend: ● regarded ◐ rudimentary regarded ○ not regarded

All in all, there is still a research gap regarding fully ontology-based simulations that do not only use the ontology as a data storage. The use of the ontology as a control element, event handler, and data base has been little considered so far. However, promising advantages of continuing the ontology integration are flexibility [47], correct data through interlinkage with the real system [28], and extendability [45] among others. Thus, the need for researching and introducing concepts on Ontology Integration Level 5 arises.

3.2. Simulation Programs

Table 2 compares existing simulation systems on the market. For the comparison, a Python-based discrete-event simulation and, for simplification, two widely used commercial products, AnyLogic and Siemens Plant Simulation (PLM), are used. The selection of the 3 simulation tools from the available simulation is based on the similar use in the production environment and the high popularity. Given the regarded scope of the comparison, commercial alternatives, such as Arena- or Petri net-based production system simulation, do not differ greatly. While commercial applications include visualizations, open source approaches, such as SimPy [54,55], seldom provide this kind of user-friendliness. Ease of use and first start is typically provided by graphical editors [56] or configurations, thus providing little flexibility. Furthermore, given DES cannot be interrupted, the simulation core changed, and then continued, a feature dubbed intervention. Furthermore, KPIs are typically self defined or calculated, omitting international standards, such as the ISO-22400-2:2014 standard [57].

Table 2. Comparison of a selection of existing simulation applications.

	User-Friendliness				Availability	Modularity	
Simulation	**Visualization**	**Ease of Use**	**First Start**	**KPI Usage**	**Open Source**	**Intervention**	**Flexibility**
SimPy [54]	○	◐	●	◐	●	○	◐
AnyLogic [56]	●	◐	●	◐	○	○	◐
Siemens PLM	●	◐	●	◐	○	○	◐

Legend: ● implemented ◐ partly implemented ○ not possible

In comparison to traditional approaches, the main advantages of the OntologySim lie in the combination of a storable, changeable DES on open source basis with visualizations. In the following chapter, the structure and advantages of the OntologySim are explained in more detail.

4. Proposed Ontology-Based Simulation

In this chapter, the OntologySim idea is explained in more detail. Furthermore, the OnologySim is classified according to VDI 3633 [29], and the unique selling propositions, that fit the previously outlined research gaps, are presented.

4.1. OntologySim Conceptualization

The main requirement in the conception and implementation of OntoloySim is to implement a flexible, modular simulation that can be saved and reloaded at any time. The process of changing and loading the simulation enables a direct linkage and runtime adjustement to the reality, which is shown graphically in Figure 5. Thereby, OntologySim enables a truly interlinked Digital Twin.

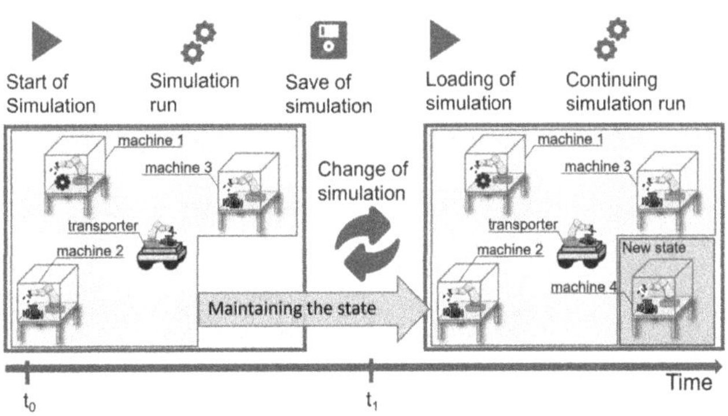

Figure 5. Main idea of OntologySim.

After configuration and the start of the simulation, the OntologySim can be saved at any time, with or without interruption. Since large parts of the simulation core are directly connected and modeled within the ontology and the fact that the ontology holds all relevant information. After saving the ontology in an "owl" file, changes can be made within or outside of the executable OntologySim, such as adding a machine or AGV (Automated Guided Vehicle). Loading the ontology guarantees the seamless continuation of the simulation at any time.

4.2. Unique Selling Proposition (USP)

Figure 6 summarizes the main points of differentiation from other simulations. The individual contents of the framework are explained in more detail in the various chapters: Sections 5–7.

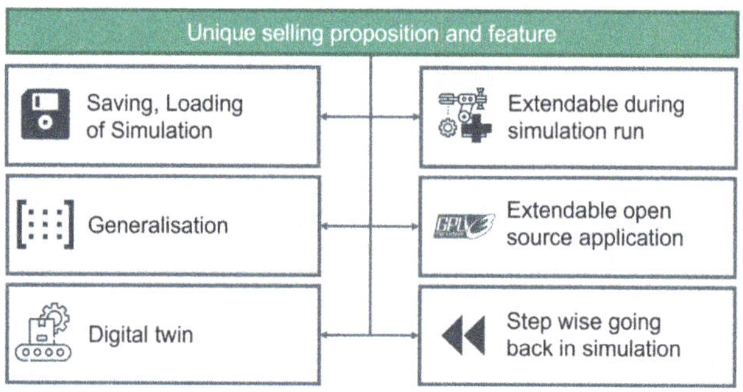

Figure 6. Unique position of the simulation.

The USP is based on the following advantages over existing discrete-event simulations:

- *Saving, Loading of Simulation:*
  Interrupting the simulation during runtime enables new use cases and possibilities, which are shown in Figure 7. When saving the simulation, the current state of the ontology is stored in a single OWL file. This saved data is sufficient to restart the simulation, as all essential information is contained in the ontology. In particular, the following new use cases are enabled:
  - The stored image can be run on other computational units and enables the easy extraction of information data.
  - After saving a defined state, several strategies, multi-agents, etc., can simulate different simulation runs, and the best agent configuration can be selected to facilitate research in this domain [31,55,58].
  - Furthermore, time saving is achieved because the initial run-up and warm-up of a simulation into a static state can be skipped, as a static stimulation can be loaded, that is "identical" to the real system state to increase inference accuracy required for real-world use cases [59,60].
  - A real world digital twin is enabled, as up-to-the-minute virtual models can be started with the exact same initial situation [3].

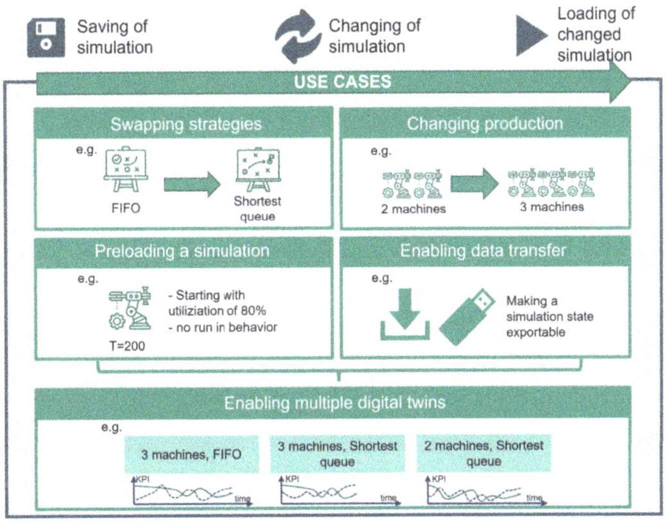

Figure 7. Use cases for saving and loading simulation.

- *Extendable during simulation run:*
  Saving and loading the simulation provides the basis for the simulation to be extended during a run. In addition to changing the strategy, adjustments can also be made to the manufacturing system which affect resources, such as machines, transporters, or processes. Thus, for example, transporters can be removed or added, or process times can be changed. It is also possible to add data and information that is not needed currently for the simulation. By adding supplementary nodes, it is possible to link external (real-world) data or to model the current description in detail. An example would be the addition of installation space sizes for the product, novel products added to the portfolio [33], or changes in the shopfloor management, as well as increased worker competencies [61].
- *Extendable open source application:*
  Another advantage compared to the majority of simulations reviewed in Section 3 is that OntologySim is implemented with the basic idea of being an open-source publication and software. Thus, the focus of the implementation is to make the simulation easily adaptable and to provide clearly defined and explained interfaces. This makes it possible to simply extend the simulation with self-implemented agents and access the ontology through standardized wrapper methods.
- *Generalization:*
  The interaction of the ontology and the Python modules enables a good generalization of simulation models. It is possible to model different production systems, be it line production, a workshop production or matrix production, or any combination. The ontology can be extended and customized to meet specific requirements. This includes the possible integration of information about tangible objects, such as products, intangible information, such as production planning, and control organization [33,58], or ever-changing, human-centered Industry 4.0 implementations [61].
- *Step wise going back in simulation:*
  Going back within a single simulation run in a step-by-step manner is another special feature of the OntologySim. Going back enables a better analysis of the simulation and increases the traceability and understanding of complex production systems [60]. This feature is made possible by the fact that, at any point in time, a defined state is available, and the past events are stored. These two properties are sufficient to recreate the past and explicitly analyze its states. By doing so, the analysis of "what-

if" scenarios within a single simulation run can be evaluated without the need to instantiate many different simulations, for instance, for time-constraint adherence predictions [59,60].

- *Digital twin:*
  The OntologySim can ideally serve as a digital twin. It possesses the unique ability to serve as digital twin, digital master, and, to some degree, digital shadow, all at once. Starting from any current state in the real system, the ontology can be created and updated externally, manually, or by connection to data sources, such as MES or ERP systems. Then, an instantiation, or, in other terms, a simulation run, can be started directly within the OntologySim framework. This enables increasing digital twin capabilities for production systems [3].

4.3. Classification of simulation

This section classifies the simulation according to VDI3363 [29], as shown in Figure 8. The OntologySim is a discrete event-based simulation. The event-based programming has the advantage of simple implementation, high execution speed, and flexibility; for example, one event can easily trigger several other events [54]. Furthermore, the OntologySim is a multi-agent-based simulation. Each machine and transport unit represents an agent that can be instantiated as to decide independently.

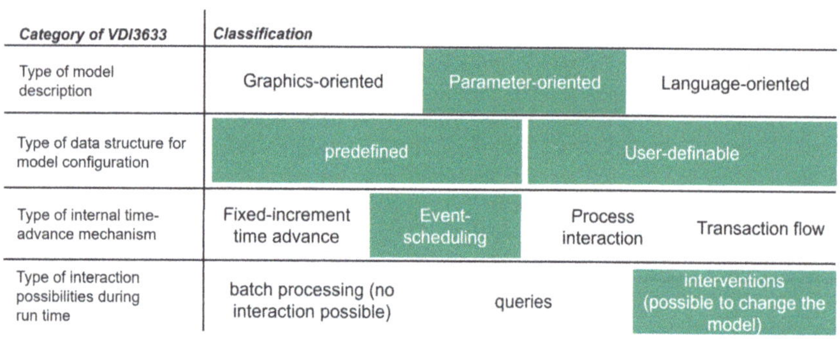

Figure 8. Classification of simulation according to VDI3363 [29].

5. Design Principles for the OntologySim

The OntologySim concept and the structure of the developed ontology presented below is intended to enable mapping diverse production systems, such as line production, workshop production, and matrix production, efficiently, in a modular way. The basic requirement for the OntologySim development is the ability to save the simulation state at any time, without data loss, and the ability to restart the simulation using the saved ontology. From this requirement follows that all entities have to be mapped in this ontology, achieving Ontology Integration level (OIL) 5. Thus, not only machines and transporters are stored in the ontology, but also set-up processes, defects and services, their statistics, the production plan for all products, and relevant future and past events. The storage of any entities in the ontology enables a high degree of flexibility since, on the one hand, relationships from zero to n are possible, and, on the other hand, a high degree of detail can be generated. Furthermore, the information and relationships within the ontology are dynamically adapted. Each executed event or simulation step executes changes to the ontology. Because past events are also stored in the ontology, it is possible to recreate past states from the current state of the ontology. This past information is also seamlessly available to enable machine learning-based production control, for instance, with reinforcement learning [55,58]. The concept and functioning of the ontology are exemplified by the entities Machine, Event, and Product type in the following.

5.1. Machine

The basic components of machines are queues, which are divided into different categories. One is the input and output queues, which serve as a buffer, and the "ProdQueue", in which products are processed. A queue can serve as both an input and output queue, but not as a "ProdQueue" and an input or output queue at the same time. The next level is the position, which serves as the interface to the products. Each position can be reserved in advance by exactly one product. In addition to the queues, the set-up processes and production processes are stored with respective distributions. The machine is also connected to events, past events ("EventLogger"), and various defects and maintenance entities. The exact configuration of a machine is always adapted to the individual use case so that, for example, the number of queues and positions can be flexibly varied. A simplified machine is visualized in Figure 9.

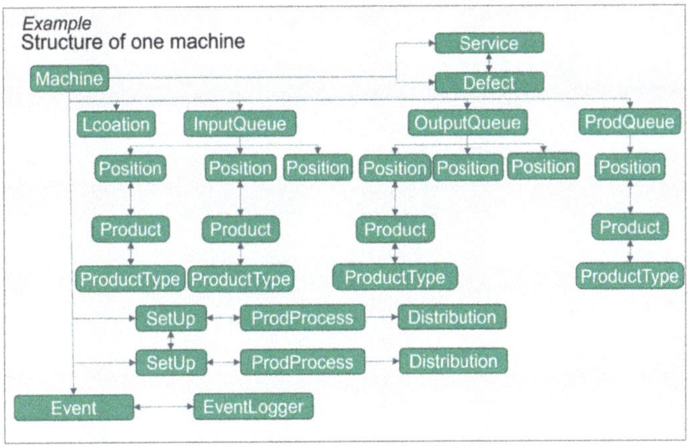

Figure 9. Concept of a machine in OntologySim.

5.2. Event

The states of the simulation are not controlled by state entities and are based solely on events and the information stored within the ontology. The state of an entity, e.g., a machine or transport unit, is defined by the currently valid and connected event. Each event contains information about the start time, duration, and type of event and is connected to the entities, which are influenced by the event. From this information, the state of each entity can be extracted at any time. Events that lie in the future can also be generated. However, the agents and control algorithms do not have access to the future events but only see the current state to preserve comparability in respect to the real world. Creating future events is useful for processing tasks at once in the simulation. For example, if the task is to produce a product on a machine, then two events are created, one to change the machine setup to the new process and the other to feed the product into the machine for processing. This possibility facilitates the implementation of the simulation.

It is always possible to infer in both directions (bidirectionally) in the simulation. This increases flexibility and allows easier access to entities. The disadvantage of this is that more data is stored, which hurts performance. As shown by Reference [17], the performance for compact ontologies is significantly better. For this reason, past events are only made available in one direction, which has the great advantage that, when querying the state and the next steps of a machine, only a few future events need to be queried, and it is not necessary to iterate over all past and future events. Figure 10 illustrates the differences between past and future events and their connectivity to entities.

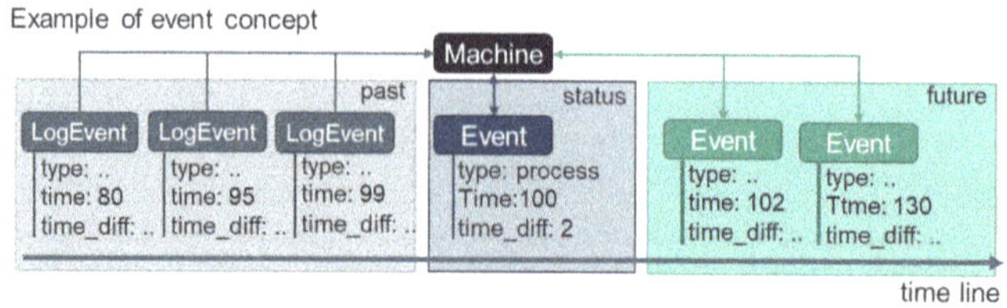

Figure 10. Explanation of the (semi-)bidirectional event-resource mapping.

5.3. Product Type

The mapping of product types and their current state is based on the structure of a Petri net, which is a feasible representation, for instance, in disassembly [62]. The state node in the ontology is equivalent to a place S, and the production process represents transitions T. Each product type has a production plan with start and end nodes. The arrangement and sequence of the process steps can be freely designed, as shown in Figure 11. The individual products always refer to their current state. The Petri net-like structure offers the advantages that different and complex production plans with many potential paths can be stored and iterated through with high performance and that information can be extracted quickly. For example, the calculation of the fastest path, the selection of the process with the shortest machine queue, and the number of production steps still required is possible. The graphic in Figure 11 shows a simple, yet relatively complex, production plan.

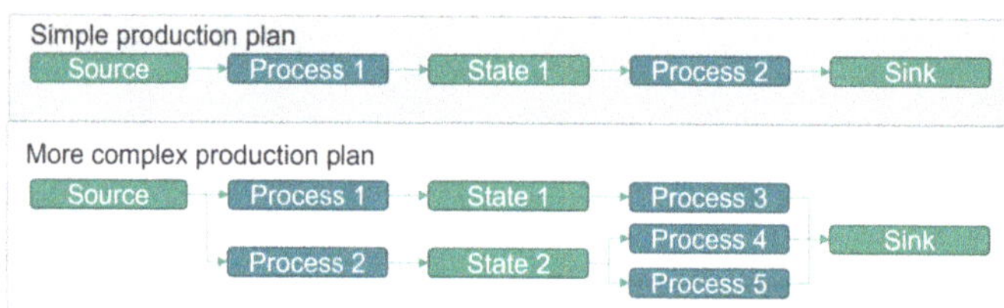

Figure 11. Exemplary concept of product type.

An overview of the classes and relationships and more in-depth information on how the ontology has been implemented in concrete terms are summarized in the ReadTheDocs documentation (https://ontologysim.readthedocs.io, accessed on 30 December 2021) for the OntologySim [63]. The code for the frontend and for the simulation are both published in two seperated github projects (ontologysim_react (https://github.com/larsKiefer/ontologysim_react, accessed on accessed on 27 December 2021) [64], ontologysim (https://github.com/larsKiefer/ontologysim, accessed on 27 December 2021) [65]).

6. Procedure of OntologySim

Based on the simulation steps as shown in the VDI3363 [29] and the adjustments regarding the application of an ontology, the following 4-stage pipeline was designed: Configuration of the simulation, Reasoning/loading of the simulation, Running through the simulation, Logging & storage of KPIs. The individual steps are briefly described below and visualized in Figure 12.

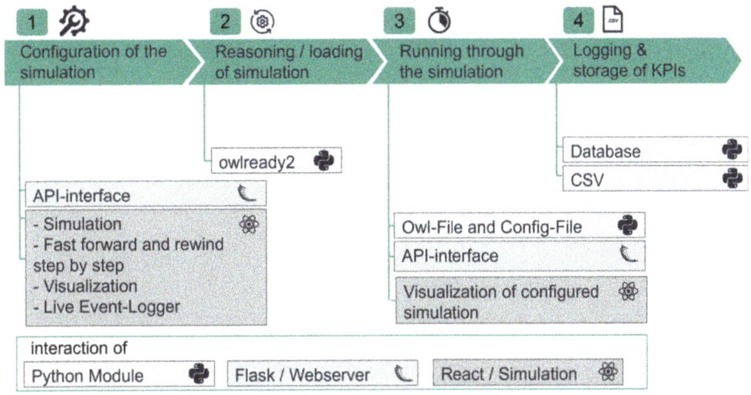

Figure 12. OntologySim pipeline.

6.1. Configuration of the Simulation (1)

The configuration of the simulation is provided via config-files, owl-files, or an API interface. To simplify the configuration from lines to matrix production, different templates and examples can be used. Additionally, the OntologySim provides support for the configuration by visualization means [64].

6.2. Reasoning/Loading of the Simulation (2)

When starting the simulation, the model is created in Owlready2, and the reasoning is started [65]. For performance and resource efficiency reasons, the reasoning is only carried out at the beginning. All changes are made via wrapper methods so that the model remains consistent, and queries are executed directly on top of objects. This leads to a faster simulation procedure. However, the ability to perform reasoning on this ontology during or after simulation runs remains.

6.3. Running through the Simulation (3)

Running through the simulation offers two possibilities. One is the step-by-step run, and the other is the direct run [64], which is used for a fast simulation execution. Running through the simulation step-by-step allows you to go back and forth between simulation, live display of KPIs and events, and visualization of the production. Furthermore, the simulation can also be accessed via an API [63]; thus, the required data can be retrieved at any time. The goal of the step-by-step process is to better understand and analyze the decisions of the agents and algorithms in order to facilitate better algorithm or machine learning model design to enable further studies into the explainability of such systems [55].

6.4. Logging & Storage of KPIs (4)

After the simulation run, all KPIs and events can be obtained either as CSV files or an SQLite database [64,65]. The KPIs are based on the standard defined by Kang et al. [66]. For each run, up to 22 KPIs for machines, 8 KPIs for transporters, 11 KPIs for products, 1 KPI for queues, and 4 KPIs for the general simulation can be stored. There is one summary per KPI and element (machine, transporter, queue) and one per time interval. The configuration allows the time interval to be changed and KPIs to be added and removed.

7. Technical Description

7.1. Basic Building Block of the Ontologysim

The basis of the OntologySim is the Python library Owlready2. The ontology framework is used to store the data and the status of the production. A more detailed description of how Owlready works can be found in Section 2.1 and References [17,26].

The structure of the OntologySim is shown in Figure 13 below. Around the entire ontology, wrapper methods have been designed to standardize access to the ontology. The search queries are realized via unique IDs and via iteration through the connections. This approach has high-speed advantages over SPARQL and SWRL (Semantic Web Rule Language). Building on top of the wrapper methods, simulation, loggers, agents, and KPI modules are implemented. These modules form the basis for ontology-based simulation and contain, in addition to the ontology, the core logic of the simulation. To start the simulation, either config files or owl files are loaded or a request to the webserver is sent. Nevertheless, the wrapper methods are generalized, so that only the config or owl is required to instantiate a simulation run. The agent integration is structured in such a way that standard agents, such as FIFO, Shortest Queue, etc., are pre-implemented, and their programmed agents can be easily created via a predefined interface.

Based on the basic simulation, a web server is integrated, which enables more targeted access to simulation data and lays the foundation for the visualization, as shown in Figure 13. The designed API interface is implemented with Flask, which is a stateful service due to the discrete running simulations. The basic functions of the API are to call KPIs, log data, and create and configure simulations. Together with the simulation module and the web server, this forms the backend of the software structure [64,65]. A frontend Framework, here React, is used for the visualization of the simulation. The data is retrieved from the Flask web server using Ajax calls and stored in the Redux Store. The Redux Store serves as the basis for displaying KPI diagrams, event logger tables, and simulations. The connection between frontend and backend is enabled using AJAX calls.

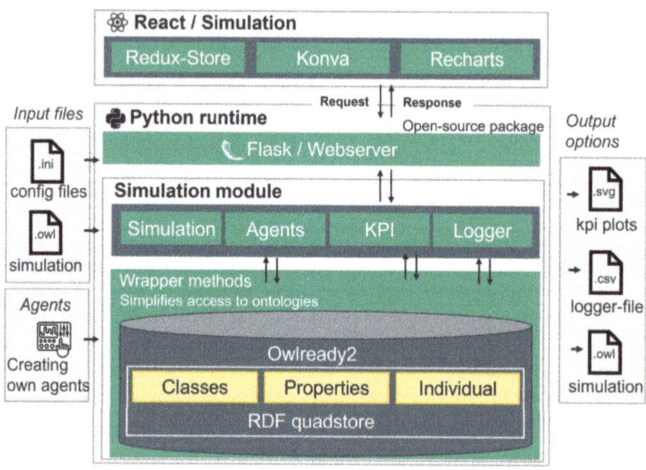

Figure 13. OntologySim structure.

7.2. Visualization

The website (Frontend) visualizes the information available during and after the simulation. In the following, the three most elementary pages are presented in more detail: (1) simulation with back and forward data, hover functions for more information (Figure 14); (2) dashboard and charts with KPIs (Figure 15); and (3) event logging including filtering and sorting of events (Figure 16).

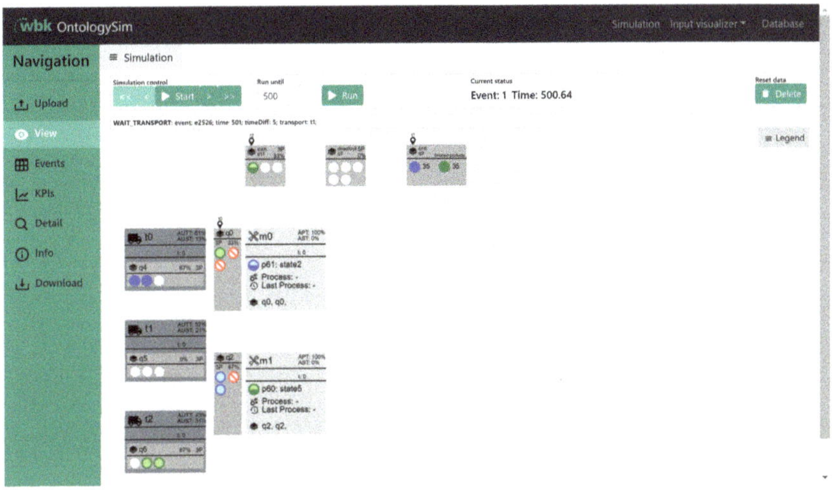

Figure 14. Simulation with back and forward data, hover functions for more information.

The following elements are displayed in the graphic visualization of the simulation: Queue, Process Queue, Machine, and Transporter. In addition, the last event is displayed for the transporter and the machine. To be able to distinguish individual products, various product types are colored differently, and the progress is symbolized by the fill level of the circle and hovering over the product provides further information, such as queue time and the start of production. Furthermore, the visualization makes it possible to go back to steps in the production to better understand the agents' decisions.

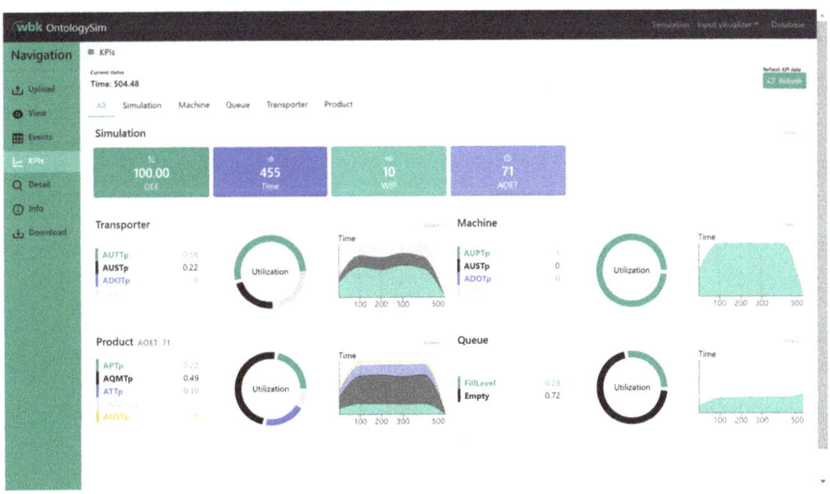

Figure 15. KPI Dashboard of a simulation run.

The KPI overview enables quick analysis of the production. For this purpose, the summarized value and the course of the KPIs are displayed during the simulation. Selected KPIs are visualized graphically, and the remaining KPIs are displayed in tabular form, as shown in Figure 15.

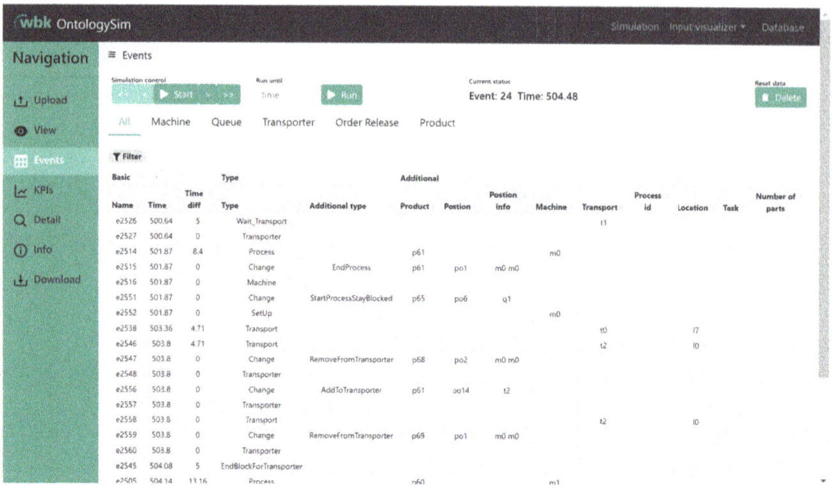

Figure 16. Event logger of simulation.

The event logging page allows the display of all past events and the filtering and sorting of events, as exemplified in Figure 16. For example, this enables to track the path of individual products exactly or to display only one product type on a machine.

All in all, the proposed framework is adaptable, as the strict division of the simulation core and visualization allows simple integration and adaption with existing systems and requirements. Thus, the overview is as presented in Figures 14–16.

8. Discussion

Regarding the aim of "[presenting] a combination of ontology with manufacturing simulation, which is available as an open-source solution to the general public and is distinguished from previous solutions by its flexibility and storability", the proposed OntologySim satisfies all requirements outlined in the literature review in Section 3. As shown in Table 3, the OntologySim uses the underlying ontology (or several ontologies) both as a schema and as the underlying simulation storage and core. Thereby, it provides an application framework with the ability to support multi-agent systems and interventions. The latter is crucial and enabled by the truly ontology-based schema and storage and, thus, enables the OntologySim to achieve Ontology Integration Level 5.

As explained in Table 4, OntologySim can be compared to existing simulation frameworks in terms providing a visualization, changeability via GUI and integration of KPI calculation or flexibility. The latter is based on the ontology simulation core approach and, hence, additionally enables changeability of the simulation and interventions beyond the state-of-the-art. The challenge in developing ontology-based simulations, however, is to achieve high performance in terms of execution speed. The OntologySim is slower compared to SimPy, Anylogic, and other commercially available simulations. Slower, although massive, speed improvements have been achieved by avoiding SPARQL and SWRL rules. Nevertheless, further speed improvements are required to ensure the applicability in much larger systems and, thus, to enable more use cases the benefits of an OntologySim-enabled real digital twin. Furthermore, the implemented web application is not a stateless API because of the state storage in the ontology. In today's web applications, stateless APIs are used to enable multi-user operation. Hence, this multi-user operation should be included and examined in the application for further follow-up studies. Nevertheless, the provided OntologySim as an Ontology Integration Level 5 framework enables the application of truly interlocked digital twin, the analysis of "what if" situations in up-to-the-minute simulations, and the convenience for extendability of an open source software.

Table 3. OntologySim classification according to the literature review.

	Ontology System					Real World			Simulation			
Approach by	**Ontology Schema**	**Multi Agent System**	**Multiple Ontology**	**Ontology as Data Storage**	**Application Framework**	**External Data Source**	**Cyber-Physical-System (CPS)**	**Manufacturing Execution System (MES)**	**Ontology Integration Level**	**Intervention**	**Visualization**	**Main Goal**
Proposed Implementation as Fully ontology-based simulations												
OntologySim [64,65]	●	●	●	●	●	○	○	○	5	●	●	Simulation
Legend: ● regarded ◐ rudimentary regarded ○ not regarded												

Table 4. Comparison of simulation application.

	Applicability				Availability	Modularity		
Simulation	**Visualization**	**GUI Changeability**	**Speed**	**KPI Standards**	**Open Source**	**Intervention**	**Changeability**	**Flexibility**
SimPy [54]	○	◐	●	◐	●	○	○	◐
AnyLogic [56]	●	◐	●	◐	○	○	○	◐
Siemens PLM	●	◐	●	◐	○	○	○	◐
OntologySim	●	◐	◐	●	●	●	●	●
Legend: ● implemented ◐ partly implemented ○ not possible								

Despite the previously described advantages and approaches to circumvent the disadvantages of an ontology-based production simulation, in general, detrimental issues are as follows: Simulation speed is extenuated due to numerous knowledge graph or ontology queries. While queries, for instance, with SPARQL, are more flexible than existing frameworks, they are yet typically more complex to apply than semi-graphical GUIs. The direct integration into existing MES or ERP systems is not (yet) regarded.

9. Summary and Outlook

In order to enable fully autonomous digital twins that interact with real world entities and allow the digital representation of changing and flexible production systems, an ontology-based simulation model, the OntologySim, is presented. The OntologySim is an open-source fully ontology-based event-discrete simulation, which has high flexibility and modularity due to the developed ontology schema and the defined interfaces. Due to the fully ontology-based approach, it is possible to change and save simulations during runtime. The designed user interface allows a detailed analysis of the agents using KPIs, event, and simulation display. Since OntologySim is published as open source (AGPL-3.0 License), we hope to contribute to the growing collaboration and exchange in the production and simulation community.

Follow up research shall continue the development to enable to join and separate products. This would offer new possibilities to better map production processes in the industry and likewise enable portfolio external products and their influences on the production system to be analyzed [33]. Furthermore, work on the OntologySim is ongoing and feedback from the community is being incorporated. To further close the research gap as outlined before, experimental studies with real world use cases shall be continued to be conducted. Last, but not least, the advantages provided by the ontology core can be researched with the integration of autonomous production control, understandable reinforcement learning [55], and novel production planning approaches that make use of the available real-time data and experimentation ability in further research projects.

Author Contributions: Conceptualization, M.C.M., L.K., A.K. and G.L.; methodology, M.C.M., L.K., A.K. and G.L.; software, L.K. and M.C.M.; validation, L.K. and M.C.M.; formal analysis, L.K. and M.C.M.; investigation, L.K. and M.C.M.; resources, M.C.M., A.K. and G.L.; data curation, L.K. and M.C.M.; writing—original draft preparation, L.K. and M.C.M.; writing—review and editing, M.C.M., L.K., A.K. and G.L.; visualization, L.K. and M.C.M.; supervision,A.K. and G.L.; project administration, M.C.M..; funding acquisition, M.C.M., A.K. and G.L. All authors have read and agreed to the published version of the manuscript.

Funding: This research work was undertaken in the context of DIGIMAN4.0 project ("DIGItal MANufacturing Technologies for Zero-defect Industry 4.0 Production", http://www.digiman4-0.mek.dtu.dk/, accessed on 27 December 2021). DIGIMAN4.0 is a European Training Network supported by Horizon 2020, the EU Framework Programme for Research and Innovation (Project ID: 814225).

Conflicts of Interest: The authors declare no conflict of interest. The funders had no role in the design of the study; in the collection, analyses, or interpretation of data; in the writing of the manuscript, or in the decision to publish the results.

References

1. Nee, A.Y.; Ong, S.; Chryssolouris, G.; Mourtzis, D. Augmented reality applications in design and manufacturing. *CIRP Ann.* **2012**, *61*, 657–679. [CrossRef]
2. Duray, R.; Ward, P.T.; Milligan, G.W.; Berry, W.L. Approaches to mass customization: Configurations and empirical validation. *J. Oper. Manag.* **2000**, *18*, 605–625. [CrossRef]
3. May, M.C.; Overbeck, L.; Wurster, M.; Kuhnle, A.; Lanza, G. Foresighted digital twin for situational agent selection in production control. *Procedia CIRP* **2021**, *99*, 27–32. [CrossRef]
4. Mourtzis, D. Simulation in the design and operation of manufacturing systems: State of the art and new trends. *Int. J. Prod. Res.* **2020**, *58*, 1927–1949. [CrossRef]
5. Durán-Muñoz, I.; Bautista-Zambrana, M.R. Applying ontologies to terminology: Advantages and disadvantages. *Hermes-J. Lang. Commun. Bus.* **2013**, 65–77. [CrossRef]
6. Cheng, H.; Zeng, P.; Xue, L.; Shi, Z.; Wang, P.; Yu, H. Manufacturing ontology development based on Industry 4.0 demonstration production line. In Proceedings of the 2016 Third International Conference on Trustworthy Systems and Their Applications (TSA), Wuhan, China, 18–22 September 2016; pp. 42–47.
7. Lamy, J.B. Owlready: Ontology-oriented programming in Python with automatic classification and high level constructs for biomedical ontologies. *Artif. Intell. Med.* **2017**, *80*, 11–28. [CrossRef]
8. Gruber, T.R. Toward principles for the design of ontologies used for knowledge sharing? *Int. J. -Hum.-Comput. Stud.* **1995**, *43*, 907–928. [CrossRef]
9. Shamsfard, M.; Barforoush, A.A. Learning ontologies from natural language texts. *Int. J. -Hum.-Comput. Stud.* **2004**, *60*, 17–63. [CrossRef]
10. Knublauch, H.; Oberle, D.; Tetlow, P.; Wallace, E.; Pan, J.; Uschold, M. A semantic web primer for object-oriented software developers. In *W3c Working Group Note W3C*; 2006. Available online: https://www.w3.org/TR/sw-oosd-primer/ (accessed on 27 December 2021).
11. Wand, Y.; Weber, R. An ontological model of an information system. *IEEE Trans. Softw. Eng.* **1990**, *16*, 1282–1292. [CrossRef]
12. Gruninger, Michael, L.J. Ontology: Applications and design. *Commun. ACM* **2002**, *45*, 39–41.
13. Berners-Lee, T.; Hendler, J.; Lassila, O. The semantic web. *Sci. Am.* **2001**, *284*, 34–43. [CrossRef]
14. McGuinness, D.L.; Van Harmelen, F. OWL web ontology language overview. *W3C Recomm.* **2004**, *10*, 2004.
15. Jurisica, I.; Mylopoulos, J.; Yu, E. Ontologies for knowledge management: An information systems perspective. *Knowl. Inf. Syst.* **2004**, *6*, 380–401. [CrossRef]

16. Pérez, J.; Arenas, M.; Gutierrez, C. Semantics and Complexity of SPARQL. In *The Semantic Web-ISWC 2006*; Springer: Berlin/Heidelberg, Germany, 2006; pp. 30–43.
17. Lamy, J.B. Ontology-oriented programming for biomedical informatics. In *Transforming Healthcare with the Internet of Things*; IOS Press: Amsterdam, The Netherlands 2016; pp. 64–68.
18. Kalyanpur, A.; Pastor, D.J.; Battle, S.; Padget, J.A. Automatic Mapping of OWL Ontologies into Java. In *SEKE*; Citeseer: Princeton, NJ, USA, 2004; Volume 4, pp. 98–103.
19. Goldman, N.M. Ontology-oriented programming: Static typing for the inconsistent programmer. In *The Semantic Web-ISWC 2003*; Springer: Berlin/Heidelberg, Germany, 2003; pp. 850–865.
20. Seaborne, A.; Manjunath, G.; Bizer, C.; Breslin, J.; Das, S.; Davis, I.; Harris, S.; Idehen, K.; Corby, O.; Kjernsmo, K.; et al. SPARQL/Update: A language for updating RDF graphs. *W3c Memb. Submiss.* **2008**, *15*, 1–13.
21. World Wide Web Consortium. OWL 2 Web Ontology Language Document Overview. 2012. Available online: https://www.w3.org/TR/owl2-overview/ (accessed on 27 December 2021).
22. Knublauch, H.; Fergerson, R.W.; Noy, N.F.; Musen, M.A. The Protégé OWL plugin: An open development environment for semantic web applications. In *The Semantic Web-ISWC 2004*; Springer: Berlin/Heidelberg, Germany, 2004; pp. 229–243.
23. Kalyanpur, A.; Parsia, B.; Sirin, E.; Grau, B.C.; Hendler, J. Swoop: A web ontology editing browser. *J. Web Semant.* **2006**, *4*, 144–153. [CrossRef]
24. Haase, P.; Lewen, H.; Studer, R.; Tran, D.T.; Erdmann, M.; d'Aquin, M.; Motta, E. The Neon Ontology Engineering Toolkit. 2008. Available online: http://neon-toolkit.org/wiki/Main_Page.html (accessed on 27 December 2021).
25. Horridge, M.; Bechhofer, S. The owl api: A java api for owl ontologies. *Semant. Web* **2011**, *2*, 11–21. [CrossRef]
26. Jean-Baptiste, L. Constructs, restrictions, and class properties. In *Ontologies with Python*; Springer: Berlin/Heidelberg, Germany, 2021; pp. 135–156.
27. Adel, A.Z.; Zebari, S.; Jacksi, K. Football Ontology Construction using Oriented Programming. *J. Appl. Sci. Technol. Trends* **2020**, *1*, 24–30.
28. Liu, C.; Jiang, P.; Jiang, W. Web-based digital twin modeling and remote control of cyber-physical production systems. *Robot. -Comput.-Integr. Manuf.* **2020**, *64*, 101956. [CrossRef]
29. Ingenieure, V.D. *VDI 3633 Simulation von Logistik-, Materialfluß- und Produktionssytemen—VDI 3633 Entwurf | Begriffsdefinitionen*; Beuth: Düsseldorf, Germany, 1996.
30. HrdliczNa, V. Leitfaden für Simulationsbenutzer in ProduNtion und LogistiN. *ASIM Mitteilungen* **1997**, *58*.
31. May, M.C.; Kiefer, L.; Kuhnle, A.; Stricker, N.; Lanza, G. Decentralized multi-agent production control through economic model bidding for matrix production systems. *Procedia CIRP* **2021**, *96*, 3–8. [CrossRef]
32. Greschke, P. *Matrix-Produktion: Konzept einer Taktunabhängigen Fließfertigung*; BoD–Books on Demand Gmbh: Norderstedt, Germany, 2020.
33. May, M.C.; Schmidt, S.; Kuhnle, A.; Stricker, N.; Lanza, G. Product Generation Module: Automated Production Planning for optimized workload and increased efficiency in Matrix Production Systems. *Procedia CIRP* **2021**, *96*, 45–50. [CrossRef]
34. Panetto, H.; Baïna, S.; Morel, G. Mapping the IEC 62264 models onto the Zachman framework for analysing products information traceability: A case study. *J. Intell. Manuf.* **2007**, *18*, 679–698. [CrossRef]
35. Wolfswinkel, J.F.; Furtmueller, E.; Wilderom, C.P. Using grounded theory as a method for rigorously reviewing literature. *Eur. J. Inf. Syst.* **2013**, *22*, 45–55. [CrossRef]
36. Karageorgos, A.; Mehandjiev, N.; Weichhart, G.; Hämmerle, A. Agent-based optimisation of logistics and production planning. *Eng. Appl. Artif. Intell.* **2003**, *16*, 335–348. [CrossRef]
37. Mönch, L.; Stehli, M. An ontology for production control of semiconductor manufacturing processes. In *Multiagent System Technologies*; MATES 2003. Lecture Notes in Computer Science; Springer: Berlin/Heidelberg, Germany, 2003; pp. 156–167.
38. Scholz, J.; Schabus, S. An indoor navigation ontology for production assets in a production environment. In *Geographic Information Science*; GIScience 2014. Lecture Notes in Computer Science; Springer: Cham, Switzerland, 2014; pp. 204–220.
39. Viljanen, K.; Tuominen, J.; Hyvönen, E. Ontology libraries for production use: The Finnish ontology library service ONKI. In *European Semantic Web Conference*; Springer: Berlin/Heidelberg, Germany, 2009; pp. 781–795.
40. Katti, B. Ontology-Based Approach to Decentralized Production Control in the Context of Cloud Manufacturing Execution Systems. Ph.D. Thesis, Technical University Kaiserslautern, Kaiserslautern, Germany, 2020.
41. Terkaj, W.; Pedrielli, G.; Sacco, M. Virtual factory data model. In Proceedings of the Workshop on Ontology and Semantic Web for Manufacturing, Graz, Austria, 24–27 July 2012; pp. 29–43.
42. Silver, G.A.; Miller, J.A.; Hybinette, M.; Baramidze, G.; York, W.S. An ontology for discrete-event modeling and simulation. *Simulation* **2011**, *87*, 747–773. [CrossRef] [PubMed]
43. Benjamin, P.; Patki, M.; Mayer, R. Using ontologies for simulation modeling. In Proceedings of the 2006 Winter Simulation Conference, Monterey, CA, USA, 3–6 December 2006; pp. 1151–1159.
44. Du, J.; Jing, H.; Choo, K.K.R.; Sugumaran, V.; Castro-Lacouture, D. An ontology and multi-agent based decision support framework for prefabricated component supply chain. *Inf. Syst. Front.* **2020**, *22*, 1467–1485. [CrossRef]
45. Terkaj, W.; Urgo, M. Ontology-based modeling of production systems for design and performance evaluation. In Proceedings of the 2014 12th IEEE International Conference on Industrial Informatics (INDIN), Porto Alegre, Brazil, 27–30 July 2014; pp. 748–753.

46. Terkaj, W.; Tolio, T.; Urgo, M. A virtual factory approach for in situ simulation to support production and maintenance planning. *CIRP Ann.* **2015**, *64*, 451–454. [CrossRef]
47. Warden, T.; Porzel, R.; Gehrke, J.D.; Herzog, O.; Langer, H.; Malaka, R. *Towards Ontology-Based Multiagent Simulations: The Plasma Approach*; ECMS: Kuala, Lumpur, 2010; pp. 50–56.
48. Gurjanov, A.; Zakoldaev, D.; Shukalov, A.; Zharinov, I. The ontology in description of production processes in the Industry 4.0 item designing company. *J. Phys.* **2018**, *1059*, 012010. [CrossRef]
49. Fumagalli, L.; Pala, S.; Garetti, M.; Negri, E. Ontology-based modeling of manufacturing and logistics systems for a new MES architecture. In *IFIP International Conference on Advances in Production Management Systems*; Springer: Berlin/Heidelberg, Germany, 2014; pp. 192–200.
50. Mazzola, L.; Kapahnke, P.; Vujic, M.; Klusch, M. *CDM-Core: A Manufacturing Domain Ontology in OWL2 for Production and Maintenance*; KEOD: Setubal, Portugal, 2016; pp. 136–143.
51. Guizzardi, G.; Wagner, G. Towards an ontological foundation of discrete event simulation. In Proceedings of the 2010 Winter Simulation Conference, Baltimore, MD, USA, 5–8 December 2010; pp. 652–664.
52. Ansari, F.; Khobreh, M.; Seidenberg, U.; Sihn, W. A problem-solving ontology for human-centered cyber physical production systems. *CIRP J. Manuf. Sci. Technol.* **2018**, *22*, 91–106. [CrossRef]
53. Chen, R.S.; Tu, M.A. Development of an agent-based system for manufacturing control and coordination with ontology and RFID technology. *Expert Syst. Appl.* **2009**, *36*, 7581–7593. [CrossRef]
54. Matloff, N. Introduction to discrete-event simulation and the simpy language. *Dept. Comput. Sci. Univ. Calif. Davis* **2008**, *2*, 1–33.
55. Kuhnle, A.; May, M.C.; Schäfer, L.; Lanza, G. Explainable reinforcement learning in production control of job shop manufacturing system. *Int. J. Prod. Res.* **2021**, 1–23. [CrossRef]
56. Borshchev, A.; Brailsford, S.; Churilov, L.; Dangerfield, B. Multi-method modelling: AnyLogic. *Discrete-Event Simulation and System Dynamics for Management Decision Making*; Wiley: West Sussex, UK, 2014; pp. 248–279.
57. *ISO 22400-2*; Automation Systems and Integration—Key Performance Indicators (KPIs) for Manufacturing Operations Management—Part 2: Definitions and Descriptions. International Organization for Standardization: Geneva, Switzeralnd, 2014. [CrossRef]
58. Overbeck, L.; Hugues, A.; May, M.C.; Kuhnle, A.; Lanza, G. Reinforcement Learning Based Production Control of Semi-automated Manufacturing Systems. *Procedia CIRP* **2021**, *103*, 170–175. [CrossRef]
59. May, M.C.; Maucher, S.; Holzer, A.; Kuhnle, A.; Lanza, G. Data analytics for time constraint adherence prediction in a semiconductor manufacturing use-case. *Procedia CIRP* **2021**, *100*, 49–54. [CrossRef]
60. May, M.C.; Behnen, L.; Holzer, A.; Kuhnle, A.; Lanza, G. Multi-variate time-series for time constraint adherence prediction in complex job shops. *Procedia CIRP* **2021**, *103*, 55–60.
61. Kandler, M.; May, M.C.; Kurtz, J.; Kuhnle, A.; Lanza, G. Development of a Human-Centered Implementation Strategy for Industry 4.0 Exemplified by Digital Shopfloor Management. In *Towards Sustainable Customization: Bridging Smart Products and Manufacturing Systems*; Springer: Berlin/Heidelberg, Germany, 2021; pp. 738–745. [CrossRef]
62. Wurster, M.; Michel, M.; May, M.C.; Kuhnle, A.; Stricker, N.; Lanza, G. Modelling and condition-based control of a flexible and hybrid disassembly system with manual and autonomous workstations using reinforcement learning. *J. Intell. Manuf.* **2022**, *33*, 575–591.
63. Kiefer, L.; May, M.C. Read the Docs: OntologySim. 2021. Available online: https://ontologysim.readthedocs.io (accessed on 30 December 2021).
64. Kiefer, L.; May, M.C. OntologySim_react. 2021. Available online: https://github.com/larsKiefer/ontologysim_react (accessed on 27 December 2021).
65. Kiefer, L.; May, M.C. OntologySim. 2021. Available online: https://github.com/larsKiefer/ontologysim (accessed on 27 December 2021). [CrossRef]
66. Kang, N.; Zhao, C.; Li, J.; Horst, J.A. A Hierarchical structure of key performance indicators for operation management and continuous improvement in production systems. *Int. J. Prod. Res.* **2016**, *54*, 6333–6350.

applied sciences

MDPI

Article

Facility Layout Problem with Alternative Facility Variants

Jiří Kubalík *, Lukáš Kurilla † and Petr Kadera †

Czech Institute of Informatics, Robotics and Cybernetics, Czech Technical University in Prague, 16636 Prague, Czech Republic
* Correspondence: jiri.kubalik@cvut.cz
† These authors contributed equally to this work.

Abstract: The facility layout problem is one of the fundamental production system management problems. It has a significant impact on overall system efficiency. This paper introduces a new facility layout problem that allows for choosing from multiple variants of each facility. The need for choosing the most suitable selection from the facility variants while at the same time optimizing other layout quality indicators represents a new optimization challenge. We build on our previous work where single- and multi-objective evolutionary algorithms using indirect representation were proposed to solve the facility layout problem. Here, the evolutionary algorithms are adapted for the problem of facility variants, including the new solution representation and variation operators. Additionally, a cooling schedule, whose role is to control the exploration/exploitation ratio during the course of the optimization process, is proposed. It was inspired by the cooling schedule used in the simulated annealing technique. The extended evolutionary algorithms have been experimentally evaluated on two data sets, with and without the alternative variants of facilities. The obtained results demonstrate the capability of the extended evolutionary algorithms to solve the newly formulated facility layout problem efficiently. It also shows that the cooling schedule improves the convergence of the algorithms.

Keywords: facility layout problem; evolutionary algorithms; numerical modeling; multi-objective optimization; performance analysis

Citation: Kubalík, J.; Kurilla, L.; Kadera, P. Facility Layout Problem with Alternative Facility Variants. *Appl. Sci.* **2023**, *13*, 5032. https://doi.org/10.3390/app13085032

Academic Editors: Guido Tosello, Roque Calvo and José A. Yagüe-Fabra

Received: 30 January 2023
Revised: 5 April 2023
Accepted: 12 April 2023
Published: 17 April 2023

1. Introduction

The facility layout problem (FLP) is one of the most important problems in production management and industrial engineering. It is a nonlinear combinatorial optimization problem defined as finding the most efficient arrangement of non-overlapping facilities on the factory floor with respect to one or more objectives subject to various constraints. Typically, the optimization goals are to maximize the utilization of the area available, minimize material-handling costs, or fulfill the adjacency or distance requirements between facilities. Furthermore, various constraints are imposed on the solution sought, such as the total area available, maximum acquisition or material handling costs, maximum maintenance costs, and other indirect expenses limits. The problem of finding the optimal layout of a set of elements occurs at various levels of a complex manufacturing environment, e.g., when arranging machines in a workshop, production lines in a production hall, or buildings on factory premises [1–3].

The FLP is a challenging task that has a direct impact on production efficiency, as it directly affects manufacturing costs, work in process, lead times, and productivity. A good layout of facilities contributes to the overall efficiency of operations and can reduce total operating costs by 20% to 50% [1]. Additionally, it has been shown that more than 35% of system efficiency is likely to be lost by applying incorrect layouts and location designs [4].

Numerous variants of the FLP problem have been formulated and researched. They can be categorized by different aspects as static or dynamic FLP, single- or multi-objective FLP [5], single- or multi-floor FLP, or FLP with equal or unequal facility areas, etc. [6–9].

When considering facility characteristics, several factors and design issues are addressed in the literature, such as the production variety and volume, the facility shapes and dimensions, the material handling system chosen, and different possible movement types allowed for parts, to name a few. Regarding the shapes and dimensions, there are two types of facilities: facilities with regular rectangular shapes and those with irregular shapes, i.e., generally polygons. In the former case, a facility can be defined by fixed length and width, and this facility is called fixed or rigid block [10]. An important FLP variant with numerous practical uses is the FLP with loosely-defined facility shapes and dimensions. In that case, each facility is associated with its area and a shape constraint, e.g., the maximum aspect ratio allowed (quotient between greater and lesser side measurements) or the minimum side length.

Layout problems are known to be generally NP-hard, which means that no exact algorithm exists that can provide an optimal solution in a reasonable polynomial time. On the one hand, there are many works in the literature that successfully use mathematical programming approaches to solve particular FLP variants. For instance, discrete quadratic assignment problem models were used for FLP with regular-shape equal-area facilities [11,12], and continuous linear and non-linear mixed integer programming models were used for FLP with irregularly-shaped facilities of unequal areas [13–17]. On the other hand, these approaches are applicable only to a rather small number of FLP instances, typically a few dozen facilities at most. Thus, many recent developments in solving the FLP are based on iterative metaheuristic approaches such as Tabu Search [18], Ant System [19,20], Particle Swarm Optimization [21], Variable Neighborhood Search [22], Simulated Annealing [23], and Evolutionary Algorithms [24–31].

We build on our previous work [32] that considers the FLP with workstations to be optimally placed into a production hall. A realistic definition of the hall and workstations was introduced there. The hall is defined as a rectangular area with obstacles (i.e., where no workstation can be placed) and communications (i.e., the areas that can be used only by relevant parts of the workstations). Workstations have a rectangular working area to which one or more handling zones can be attached. Furthermore, input/output points are optionally defined as well. These are used to define dependencies among workstations, such as links for material or semi-product handling. Each workstation can appear in the floor plan in one of six possible orientations. To solve this FLP, single- and multi-objective evolutionary algorithms have been proposed in [32]. The main feature of the evolutionary algorithms is that they use an indirect representation. This means that the exact position and rotation of the workstations are not directly encoded into the solution representation. Instead, the floor plan is represented indirectly using a so-called priority list that encodes the order and how the workstations will be added to the floor plan.

In this paper, we propose a new FLP variant with a new definition of loosely-defined facilities. Here, the facilities are not allowed to freely change their shapes within allowed boundaries. Instead, there are only several possible fixed-shape alternatives defined for each facility. The optimization task is to generate an optimal layout using exactly one variant per facility. Therefore, the optimization algorithm can not morph the shape of facilities arbitrarily. Instead, it has to choose among the feasible alternatives available for each facility. An important aspect of this FLP formulation is that the alternatives for a particular facility differ not only in shape and dimensions but also in other characteristics such as construction costs, operating costs, etc. Thus, a given selection of workstations' alternatives affects various quality and performance indicators of the corresponding layout.

This work was motivated by our research activities carried out in cooperation with the ŠKODA AUTO a.s. company. Our joint project aimed to propose machine learning-based approaches to optimally place multiple assembly lines' workstations within a production hall, given particular objectives and constraints. This is an optimization problem that occurs every time the product portfolio assigned to the respective hall is to be changed. Then, the assembly lines must be re-configured accordingly and placed in the production hall most effectively while respecting the products' assembly workflow, production hall infrastructure, logistics, and other requirements and constraints. An essential part of this optimization process is the re-design of some of the existing workstations and the design of new ones. The idea behind the possible use of multiple workstation variants is inspired by real-life best practices used by the experts when designing feasible layouts of the workstations. Usually, many constraints and restrictions affect the workstation's shape, such as internal dependencies related to the particular technology implemented by the workstation, the workstation's interface with the outside world, etc. Thus, only a few well-defined realizations might be considered for a given workstation.

To solve this FLP, we adopt the approach presented in [32] and further extend it so that multiple alternatives of each workstation can be defined and considered during the optimization process. To improve the exploration capabilities of the evolutionary algorithms, a new variation operator is defined for the extended representation. Additionally, a cooling schedule similar to the one used in simulated annealing [33,34] is introduced. Its role is to control the search mode along the optimization process. It slowly turns the search mode from a global search to a local one in order to restrict variation operators' scope to the tail section of the priority list as the algorithm converges. All in all, the primary goal of this paper is to introduce a new variant of FLP, its properties, and the implications for the solutions one can get, compared to the original FLP formulation with a single alternative defined for each workstation.

The main contributions of this work are:

- A new FLP with alternative variants of workstations was formulated. The use of the alternatives is a new FLP feature and represents a new challenge for optimization methods. To the best of our knowledge, this type of FLP has not yet been solved in the literature.
- Single- and multi-objective evolutionary algorithms to solve the problem were proposed.
- A cooling schedule that controls the search mode during the optimization process was incorporated into the evolutionary algorithms.
- The efficiency of the evolutionary algorithms and the impact of the cooling schedule on their performance at solving the new FLP problem were analyzed.
- Altogether, the new FLP formulation with redundant workstation definitions plus the proposed algorithms for finding high-quality plant layouts make up a truly flexible manufacturing system [35], which further contributes to sustainability within the whole production enterprise.
- A data set for the new FLP formulation (possibly a new benchmark in the field) was created. It will be made publicly available upon publication of this paper.

The paper is organized as follows. The problem solved in this work is defined in Section 2. Section 3 provides a concise description of the most important features of the approach we build on here. The proposed representation and algorithmic extensions that allow for using alternative workstation variants are described in Section 4. Section 5 describes the experiment set-up. Results obtained with the compared methods are presented and discussed in Sections 6 and 7. Finally, Section 8 concludes the paper.

2. Problem Definition

The FLP solved in this work is defined as follows. Givens include a particular specification of the hall, a set of N workstations $\mathcal{W} = \{w_1, \dots, w_N\}$, and a set of U pairwise dependencies between workstations $\mathcal{L} = \{l_1, \dots, l_U\}$. The hall specification defines its geometry, including restricted areas and internal communications; see Figure 1a. Each workstation is defined by its geometry, including the working area, handling areas, and position of its input/output points, and by other characteristics such as its size and cost. At least one of the workstations is defined in two possible variants, which can differ in any of the workstation's characteristics; see Figure 1b. The dependencies $\mathcal{L}$ define a set of source–destination links between workstations. Each link $l_i = (w_j, w_k)$ defines a directed connection between the output point of w_j and the input point of w_k. These are typically used to represent the material flow between workstations. The goal is to find a floor plan containing exactly one variant of each workstation that is optimal with respect to the given optimization objective(s) subject to the no-overlap constraint imposed on all workstation–workstation pairs as well as workstation–hall pairs. See [32] for formal definitions of the constraints.

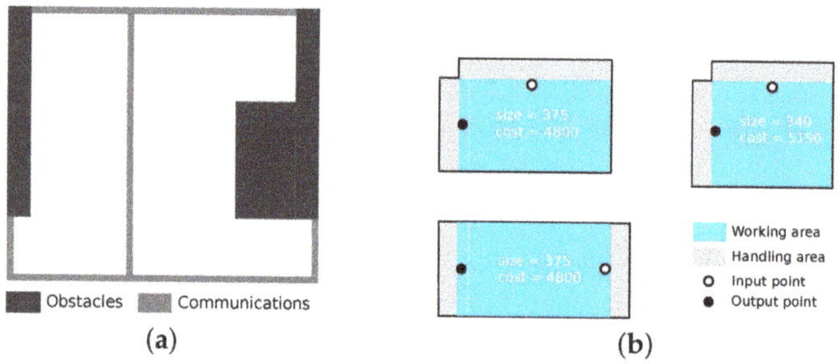

Figure 1. An example of (**a**) the production hall geometry and (**b**) three alternative variants of the same workstation.

In the following, we use the term *FLP with alternatives* for this new FLP formulation and the term *FLP without alternatives* for the original FLP as proposed in [32].

3. Preliminaries

The main feature of the original method introduced in [32] is that the evolutionary algorithms (see the pseudocode in Appendix A) use an indirect representation. This means that the layout of the floor plan, i.e., the exact position and rotation of the workstations, is not directly encoded into the solution representation. Instead, the solution is represented using a so-called *priority list*, which encodes the order in which the workstations will be added to the floor plan using placement heuristics designed for the problem. The priority list is thus a sequence of tuples

$$t_j = (\texttt{id}, \texttt{orientation}, \texttt{heuristic}),$$

where each tuple at the position $j = 1 \ldots N$ in the priority list specifies which workstation (`id`) with what orientation (`orientation`) will be added to the floor plan using what placement heuristic (`heuristic`). Mapping from the priority list to the floor plan is accomplished through an iterative process. It starts with an empty floor plan. Then, the priority list is traversed from left to right, and the workstations are successively added one by one to the developed floor plan according to the 'instructions' given in the corresponding tuple. The mapping process is illustrated in Figure 2. For more details, refer to [32].

An important aspect of this representation is its redundancy. On the one hand, each priority list maps to exactly one floor plan. On the other hand, multiple different priority lists can potentially map to the same floor plan. This is an important feature, as it allows for converging to an optimal solution via different search trajectories.

In [32], a single crossover and several mutation operators were used. The crossover operator is a variant of a standard order-based crossover operator [36]. It takes for its input two parental priority lists and starts generating the offspring by randomly choosing two crossover points. Standard tournament selection is used to select the parents. Then, the head and tail parts are inherited from the first parent. The middle part is filled in with the missing elements in the same order in which they appear in the second parent. The crossover points delimiting the head, middle, and tail parts are found randomly anew for each crossover operation. In the following, we denote this operator as a *2-point crossover*. We also implement a simple *1-point crossover* operator. It generates the offspring by inheriting a randomly-chosen head part from the first parent and then appending missing elements in the same order in which they appear in the second parent.

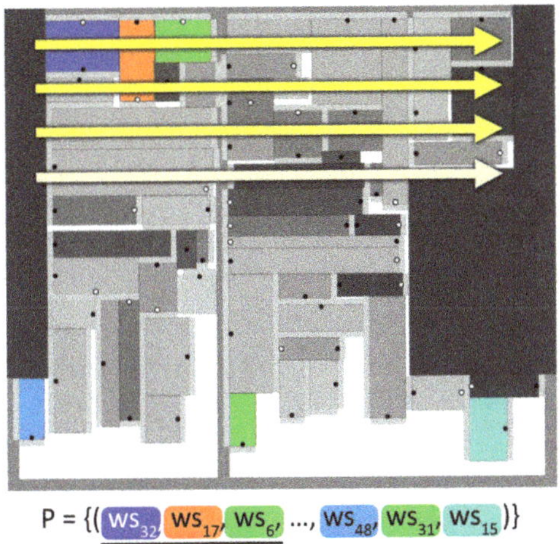

Figure 2. Illustration of the effect of mapping of the priority list P to the floor plan. Each term $\mathtt{ws}_\mathtt{i}$ stands for a tuple representing a particular variant of the workstation *i*. Workstations are placed into the floor plan in a top-down manner as indicated by the yellow arrows of decreasing intensity. The head part of the priority list ($\mathtt{ws}_{32}$, $\mathtt{ws}_{17}$, and $\mathtt{ws}_{6}$) maps to the top region of the floor plan. The tail part of the priority list ($\mathtt{ws}_{48}$, $\mathtt{ws}_{31}$, and $\mathtt{ws}_{15}$) maps to the bottom region of the floor plan.

The mutation operators operate on a single priority list. They are designed to modify a randomly-chosen element of the parental priority list so that it changes either its position within the priority list or the way the corresponding workstation is added to the floor plan, i.e., its `orientation` or `heuristic`.

In the original work, two optimization criteria, `maxFreeSpace` and `minConnDist`, were defined. The first one maximizes the compactness of the floor plan, defined as the maximization of the free space remaining between the lower horizon of the placed workstations and the bottom edge of the hall. This objective ensures that among the maximally-compact floor plans, the ones with the remaining space concentrated towards the bottom edge of the hall are preferred. As its name suggests, this objective drives the optimization towards floor plans that have the greatest potential for placing new workstations. This objective exhibits desired properties, as it drives the search towards maximally-compact floor plans and simultaneously maximizes the utilization of the available area. Contrary to traditional variants of FLP space occupation-related objectives that try to pack the entities within the smallest rectangular envelope, our `maxFreeSpace` objective better optimizes for space utilization. The `minConnDist` objective minimizes the total length of all connections between workstations calculated as the sum of Euclidean distances between the output point of the source workstation and the input point of the destination workstation over all pairs of interconnected workstations in $\mathcal{L}$; see [32] for details.

4. Proposed Method

4.1. Extended Representation

Here, we extend the representation so that it allows for multiple variants of each workstation. The priority list is composed of tuples t_i' of the following form

$$t_j' = (\mathtt{id}, \mathtt{orientation}, \mathtt{heuristic}, \mathtt{variant}),$$

with the new attribute `variant` defined as a tuple (`geometry`, `size`, `cost`), where

- `geometry` is a definition of the workstation's variant geometry, including its shape, position of its handling areas, and the position of its input/output points (see Figure 1b),
- `size` is the size of the working area of this variant, and

- `cost` is the cost of this particular realization of the workstation.

There are n_i alternative variants defined for each workstation w_i. Thus, the size of the priority list is $M = \sum_{i=1}^{N} n_i$. The whole solution, i.e., the floor plan, is represented by an extended priority list, which is a permutation of a set of tuples T that contains all variants of all workstations

$$T = \{t'_{1,1}, \ldots, t'_{1,n_1}, \ldots, t'_{N,1}, \ldots, t'_{N,n_N}\}. \quad (1)$$

The extended priority list is interpreted via an iterative process similar to the one used for the original single-shape representation version. The priority list is traversed from left to right, and the corresponding workstations are added to the floor plan in a first-come-first-served manner. This means that the left-most occurrence of each workstation's variant within the priority list is used in the floor plan; i.e., it is an active variant. The remaining inactive ones are just ignored.

4.2. Mutation

In order to reflect the extended representation and the fact that the search space significantly increases with the introduced workstation alternatives, we have added a new mutation operator to the evolutionary algorithms. This operator swaps two elements of the priority list associated with the same workstation. Given the parental priority list P as a permutation of set T, see (1), the mutation works in the following steps:

1. Randomly choose a position k within the parental priority list P;
2. Find a set of positions S_k in P, excluding the position k, that are associated with the same workstation as the element $P[k]$,
   $S_k = \{l : (P[l].\texttt{id} = P[k].\texttt{id}) \wedge (l \neq k)\}$;
3. Randomly choose one position j from S_k;
4. Swap elements at positions j and k.

If one of those two swapped elements is the left-most occurrence of the workstation in the priority list, then this action most likely leads to the effective modification of the corresponding floor plan. Otherwise, the mutation operation does not affect the floor plan represented by the priority list. This is a so-called silent mutation, and it has been shown to be beneficial for preserving diversity within the population in evolutionary algorithms [37,38].

Note that each workstation variant can be present in the priority list with only 1 out of 12 configurations, given there are 6 possible orientations and 2 placement heuristics. Variants present in the priority list can change their configuration via standard mutations that change the `orientation` and `heuristic`. Importantly, this new mutation operator works with particular workstation variant configurations, as they are stored in the priority list. By executing the effective mutation, a new variant of a certain workstation with its particular configuration becomes active (i.e., will be used in the floor plan). However, the other one, which becomes inactive, stays intact with its configuration in the priority list, thus available for its later re-use.

4.3. Cooling Schedule

Simulated annealing (SA) is a metaheuristic optimization technique. It searches for the optimal solution in an iterative process, where in each iteration, it generates a random neighbor of the current solution and either accepts the neighbor as the new current solution or rejects it. The neighbor is always accepted if it is not worse than the current solution. It may also be accepted with probability p^c, even if it is worse than the current solution. The probability p^c is high at the beginning of the run and gradually decreases through time, approaching zero towards the end of the run. This way, SA gradually changes its search mode from a global exploration to a local one. The probability p^c changes according to the *cooling schedule*, which is either defined by the user or automatically adapted during the course of the run [34].

Here, we adopt a similar approach to control the search mode along the optimization process. The idea behind it is based on the relation between the priority list and the mapping process. The mapping process starts at the upper-left corner of the empty hall. Then, the workstations are added to the floor plan one by one, filling up the available hall space in a left-right and top-down manner. The workstations are served in the order they come in the priority list; see Figure 2. This means that changes made to the head part of the priority list have a greater impact on the resulting floor plan than changes made within the tail part of the priority list. This is reflected by the cooling schedule proposed in this work, as it gradually restricts variation operators' scope to the tail section of the priority list as the algorithm converges. It is realized by preventing the variation operators (i.e., crossover and mutations) from perturbing elements of the priority list whose position is less than or equal to the left-most variation point. In this way, the search mode slowly turns from a global one to a local one.

There are many possible ways to define the cooling schedule. The one used in this work is shown in Figure 3. It sets the left-most variation point to 1 for the first half of the run. Then, it jumps to $|T|/4$, making the leading quarter of the priority list fixed. As the run progresses towards the end, the fixed portion of the priority list linearly increases to $3 * |T|/4$. Thus, in the first half of the run, the algorithm works in a standard global exploration mode. Then, the operation mode changes to the local one. However, even in the final phase of the run, the exploration capabilities stay at a significant level, as the tail quarter of the priority list can still be modified by variation operators.

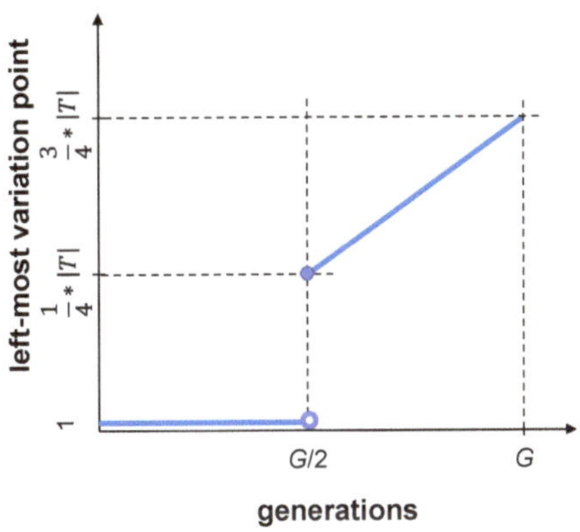

Figure 3. Cooling schedule.

4.4. Optimization Criteria

In this work, we use two optimization criteria defined in [32]—the `maxFreeSpace` and `minConnDist`. The `maxFreeSpace` objective maximizes the free space concentrated towards one side of the hall. The `minConnDist` objective minimizes the sum of Euclidean distances between input and output points of interconnected workstations.

In addition to the two optimization criteria, the `minCost` objective is defined to make use of the newly defined `cost` attribute of the workstation. It minimizes the sum of `cost` values of particular workstation variants used in the given floor plan. Note that this is an optimization objective that explicitly defines quality for any particular selection of the workstations' variants. In the following, we use the term *explicit objective* in connection with the `minCost`. On the contrary, the `maxFreeSpace` and `minConnDist` objectives are *implicit* ones in the sense that they are not directly calculated out of the parameter values of the selected workstations' variants. Instead, their value depends on how well the workstations' variants are placed in the floor plan.

The `size` and `cost` attributes were added to the workstation definition in order to illustrate another possible dimension of the optimization problem. With these new attributes, the optimization task can be formulated as a multi-objective optimization problem that seeks trade-off solutions to optimize both the implicit objective, i.e., `maxFreeSpace`, and the explicit objective, `minCost`.

5. Experiments

In this section, the experimental setup is described. First, the data sets are described. Then, the compared algorithms are listed together with their configurations used in the experiments. Finally, the experiments, tested hypotheses, and the evaluation scheme used to analyze the algorithms' performance are described.

5.1. Data Set

Two data sets are used in the experiment:

- Data set D_A (data set without alternative workstation variants)—a data set derived from the one used in [32]. It contains 58 workstations, each provided in just a single variant. The length of the priority list is $|T| = 58$. New attributes `cost` and `size` have been added to the workstation definitions. The total cost of all workstations is $cost_{static} = 284.200$. The total working area of the workstations is 33.829 m^2.
- Data set D_B (data set with alternative workstation variants)—a data set derived from the data set D_A so that a second variant has been defined for 30 randomly-chosen workstations out of the original 58. Thus, the length of the priority list increases to $|T| = 88$. The two variants for each such workstation differ in both the working area size and cost. The size is inversely proportional to the cost. The total cost of the set of workstations with only the lower-cost variants and the set of workstations with only the higher-cost variants is 238,900 and 329,000, respectively.

5.2. Compared Algorithms

Single-objective (EA) and multi-objective evolutionary algorithms (MOEA) are used in the experiments. We use the notation 'A_V_C' to refer to the particular algorithm's variant, with $A \in \{\text{EA, MOEA}\}$, $V \in \{\text{T, F}\}$ denoting whether the alternative workstations' variants are used (T—true) or not (F—false) and $C \in \{\text{T, F}\}$ denoting whether the cooling scheme is used or not. The evolutionary algorithms were run with the following configuration:

- Population size: 200;
- Maximum number of generations (G): 1000;
- Tournament size: 4;
- Probability of crossover: 0.6;
- Probability of mutation: 0.5.

Note that the algorithms were run with the same parameter setting in all experiments. No parameter tuning was carried out, since one of the goals of the experiments is to demonstrate the viability and robustness of the proposed algorithms, given that no knowledge about the best values of the parameters is available.

5.3. Goals of Empirical Investigation

The following three experiments were carried out to evaluate the performance of the single- and multi-objective evolutionary algorithms on the FLP with alternatives:

- Experiment 1—The goal of this experiment was to analyze and compare the performance of the algorithms on the FLP with and without alternatives. Note that the search space in the case of the FLP with alternative shapes is much bigger than the search space of the single-shape FLP. Thus, the question was whether the algorithms could converge to solutions that were at least as good as the solutions found on the single-shape FLP. Experiments with both the EA and MOEA were carried out. In the former case, the `maxFreeSpace` optimization objective was used. In the latter one, the `maxFreeSpace` and `minConnDist` objectives were used. Note that, in this case, both optimization objectives were of the implicit type, as the workstation cost was not considered here.

- Experiment 2—The goal of this experiment was to investigate the effect of using the cooling scheme on the quality of solutions. Again, experiments with the EA and MOEA were carried out while optimizing the same objectives as in Experiment 1.
- Experiment 3—The goal of this experiment was to investigate the performance of the MOEA, given the optimization objectives were a mixture of explicit and implicit ones. The MOEA was run with the implicit `maxFreeSpace` and the explicit `minCost` objectives. As mentioned in Section 4.4, the `minCost` objective can easily be improved just by choosing the cheaper variant for each workstation. On the contrary, improving the `maxFreeSpace` objective requires finding a better placement for the given selection of the workstation variants. So, it is likely that the algorithm may prematurely converge to the set of solutions composed of low-cost/high-size workstation variants, thus exhibiting low `maxFreeSpace` values. Thus, the question was whether the MOEA could efficiently search for a diverse set of high-quality solutions.

5.4. Evaluation

One hundred independent runs were carried out with each tested algorithm's variant in each experiment. In the case of the single-objective optimization, the median best-of-run values were compared. In the case of multi-objective optimization, there is typically no single best solution produced by the MOEA. Instead, a set of so-called non-dominated solutions is delivered by the MOEA at the end of its run. We used a so-called *hypervolume* performance metric [39], frequently used in the literature, to assess the quality of the set of non-dominated solutions. It calculates the hypervolume of the multi-dimensional region enclosed by the set of non-dominated solutions and a reference point; see Figure 4.

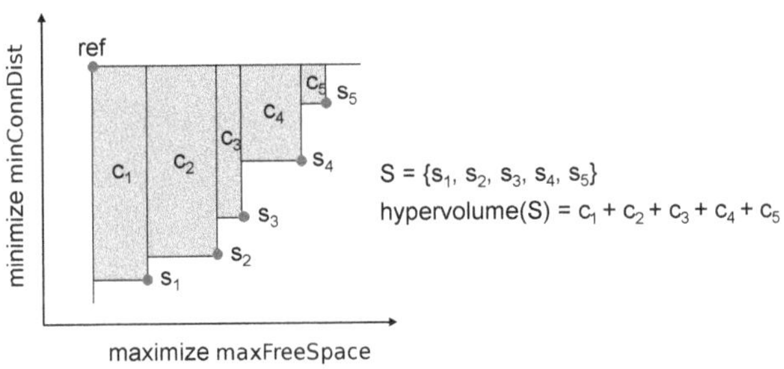

Figure 4. Illustration of the hypervolume calculation for a set of non-dominated solutions S, given the maximization objective `maxFreeSpace` and minimization objective `minConnDist`.

The coordinates of the reference point were calculated as the worst value of the respective objective observed in the final non-dominated fronts over all 100 runs. The hypervolume expresses the size of the region that is dominated by the non-dominated set. Similarly to the single-objective case, we then compared the median of the hypervolume values.

Note that the same random generator seeds were used for all series of 100 independent runs in all experiments. This implies that the runs with the cooling schedule followed the same search trajectory as those without the cooling schedule, up to generation 500 (i.e., $G/2$). Then, the two trajectories departed from each other and went their own way. This allowed us to accurately measure the effect of the cooling schedule.

In order to assess the statistical significance of the differences among the methods, we analyzed them pair-wise using the Wilcoxon rank-sum test, which rejects the null hypothesis that the compared result sets are sampled from continuous distributions with equal medians at the 5% significance level.

6. Results

6.1. Experiment 1

First, we analyzed the performance of the evolutionary algorithms on the extended FLP with alternative workstations' variants. Tables 1 and 2 show the results obtained for the single-objective and multi-objective cases, respectively. In Table 1, the best median `maxFreeSpace` value of 10,501 was observed for the variant EA_F_T (i.e., without alternatives, with cooling). This is significantly better than the median `maxFreeSpace` value achieved by EA_T_T. Also, the results obtained by EA_F_F are slightly better than those achieved by EA_T_F, although not significantly so as indicated by the p-value of 0.37.

Table 1. Results for the single-objective case of Experiment 1 (variants = false vs. variants = true) and Experiment 2 (cooling = false vs. cooling = true). The median of the set of 100 best-of-run values is presented for each method plus the p-values calculated pairwise with the Wilcoxon rank-sum test. The statistically significant differences are highlighted in bold.

	Median	***p*-Value**			
	`maxFreeSpace` [m^2]	**EA_T_T**	**EA_T_F**	**EA_F_T**	**EA_F_F**
EA_T_T	10,176	-	**2.2 × 10^{-8}**	**3.1 × 10^{-2}**	**1.6 × 10^{-6}**
EA_T_F	9879	-	-	**2.9 × 10^{-13}**	0.37
EA_F_T	10,501	-	-	-	**1.6 × 10^{-11}**
EA_F_F	9940	-	-	-	-

Table 2. Results for the multi-objective case of Experiment 1 (variants = false vs. variants = true) and Experiment 2 (cooling = false vs. cooling = true). The median of the set of 100 hypervolume values is presented for each method plus the p-values calculated pairwise with the Wilcoxon rank-sum test. The best values are highlighted in bold.

	Median	***p*-Value**			
	Hypervolume	**MOEA_T_T**	**MOEA_T_F**	**MOEA_F_T**	**MOEA_F_F**
MOEA_T_T	1.61×10^7	-	0.16	**9.76 × 10^{-4}**	7.97×10^{-2}
MOEA_T_F	1.55×10^7	-	-	**1.2 × 10^{-5}**	**2.13 × 10^{-3}**
MOEA_F_T	1.71×10^7	-	-	-	5.96×10^{-2}
MOEA_F_F	1.65×10^7	-	-	-	-

The same trend is also observed for the multi-objective optimization case; see Table 2. Again, the runs not using alternative variants yielded significantly better results than the corresponding runs with alternatives. Here, the difference between results obtained with and without alternatives was even stronger than in the single-objective case.

These observations confirm our hypothesis that the FLP with alternatives is significantly harder than the one without alternatives. The algorithms were not able to converge on the extended data set D_B to produce solutions of the same quality as the ones obtained from the data set D_A, given the same computational resources.

6.2. Experiment 2

In Table 1, a clear trend that the cooling schedule helps to converge to better solutions can be observed. EA_T_T significantly outperformed EA_T_F, and the same held for EA_F_T and EA_F_F.

Using the cooling schedule led to better solutions in the multi-objective optimization case as well, but the differences were not as profound as in the single-objective case. While MOEA_T_T is not significantly better than MOEA_T_F in terms of the median hypervolume, as indicated by the p-value of 0.16, the run using the cooling schedule resulted in a better set of non-dominated solutions compared to the corresponding run not using the cooling schedule, in 70 out of 100 runs. A similar ratio, 73:27, was observed for MOEA_F_T and MOEA_F_F. This indicates that the cooling schedule has a positive effect on performance, even in the multi-objective optimization scenario.

The observed difference in the effect of the cooling schedule in the single- and multi-objective optimization may also point to the fact that in the multi-objective case, it is even more important to keep the population maximally diverse throughout the whole run. This applies particularly to the head parts of individuals' priority lists. As soon as the cooling schedule becomes effective, the head parts get 'frozen' and cannot be modified any more. Suppose a majority of head parts of the priority lists in the population determine floor plans that are good in terms of one particular objective. In that case, there is only a small chance that additional floor plans, good in terms of the other objective, will effectively evolve from then on. In the single-objective case, this is not an issue.

6.3. Experiment 3

Figure 5 clearly demonstrates the ability of the MOEA to efficiently evolve a population of candidate solutions and to converge to a diverse set of high-quality solutions even when the optimization criteria are a mixture of explicit (`minCost`) and implicit (`maxFreeSpace`) ones. It shows all non-dominated solutions from the initial and final populations of all independent runs. One can see that, starting from very poor solutions with the initial population, the MOEA converges to much better solutions in the end. In particular, the final non-dominated fronts have shifted both to the right and downwards, thus significantly improving both objectives. However, the median `maxFreeSpace` value of the final non-dominated solutions still remains well below the `maxFreeSpace` value obtained with a single-objective EA from the data set D_A (i.e., the value of EA_F_T in Table 1). We hypothesize this can be attributed to the MOEA having to deal with a mixture of explicit and implicit objectives, where the explicit ones are much easier to improve than the implicit ones. Thus, the optimization algorithms need to be further enhanced to cope with this situation efficiently. On the contrary, the vast majority of final solutions lie below the $cost_{static}$ value of `minCost`. Many of them are also very good in terms of the `maxFreeSpace` value.

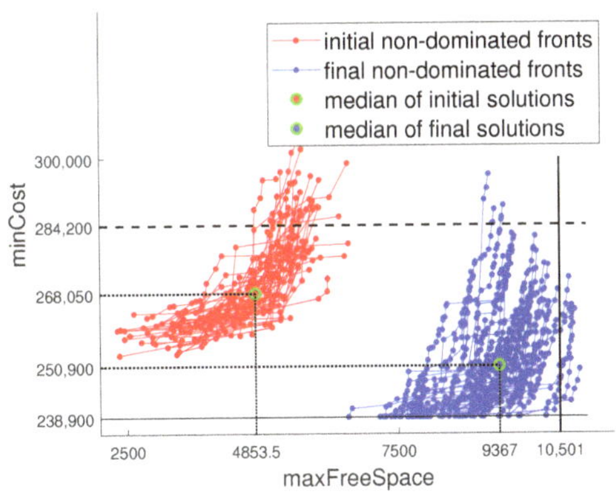

Figure 5. Evolution of non-dominated solutions using MOEA_T_T. Initial (red) and final (blue) sets of non-dominated solutions generated in 100 independent runs are shown. The horizontal dashed line indicates the `minCost` = $cost_{static}$ value of solutions produced for the data set D_A (i.e., without alternative workstations). The vertical solid line indicates the median `maxFreeSpace` value obtained with single-objective EA on data set D_A.

We also experimented with different genetic operators, namely the 1-point and 2-point crossovers. Figure 6 shows the final sets of non-dominated solutions obtained with MOEA_T_T using the 1-point and MOEA_T_T using the 2-point crossover. As for the median `minCost` value, the results are almost equal. A bigger difference, in favor of the 2-point crossover, can be observed with respect to the median `maxFreeSpace` value. The 2-point crossover also outperforms the 1-point one in terms of the hypervolume dominated by the final set of solutions. The median hypervolume dominated by the 2-point crossover solutions is 2.83×10^9, while the median hypervolume of the 1-point crossover solutions is 2.72×10^9. This is a statistically significant difference, as indicated by the p-value of 4.3×10^{-4}. This observation clearly shows the importance of using appropriate sampling operators.

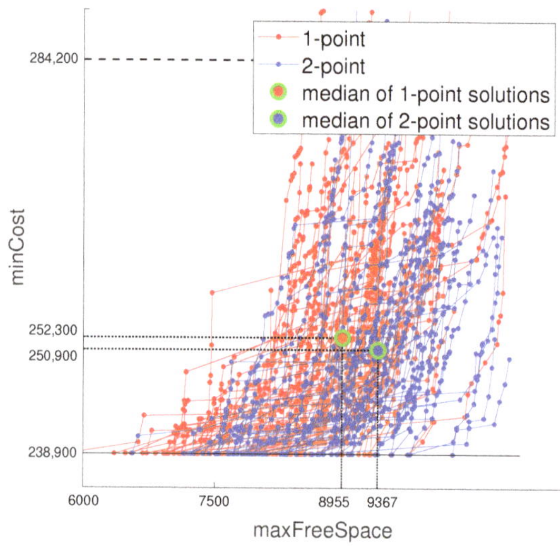

Figure 6. Comparison of MOEA_T_T using 1-point and MOEA_T_T using 2-point crossover. The sets of non-dominated solutions generated in 100 independent runs are shown.

Figure 7 shows illustrative examples of two floor plans evolved in a single run of MOEA_T_T. Several workstations are realized differently in these two floor plans; examples are the three workstations marked 'A', 'B', and 'C'.

Figure 7. Examples of two floor plans using different workstation variants: (**a**) a floor plan with `maxFreeSpace` = 10.010 m^2 and `minCost` = 261.500. (**b**) a floor plan with `maxFreeSpace` = 10.620 m^2 and `minCost` = 243.900. Workstations marked with 'A', 'B', and 'C' are examples of workstations realized differently in the two floor plans.

7. Discussion

7.1. Applicability of the Proposed Approach

The proposed FLP, with the realistic definition of workstations together with the finite set of possible fixed-shape realizations available for each workstation, fills the current gap in the literature. Existing FLP formulations involving loosely-defined facility shapes and dimensions rely on the concept that the facilities can freely vary their shape and dimensions within some limits. However, this is not a general case. In many practical situations, facilities can only be realized by a small number of fixed-shape variants. One example is the automotive industry, with assembly lines composed of dozens of interconnected workstations. Every time a new car model is to be introduced to production, the workstations and assembly lines must be re-configured and reorganized within a production hall accordingly. Importantly, the design of a feasible workstation realization is a highly-constrained optimization problem itself. It involves various constraints related to the specifics of the technological procedure implemented by the workstation and to the workstation's interface with other workstations it interacts with.

7.2. Advantages and Disadvantages of (MO)EA

In general, EAs are optimization techniques with many advantages—they can be used for an arbitrary type of optimization problem (continuous, discrete, multimodal, single/multi-objective, nonlinear, constrained, etc.). Thus, they are particularly well-suited for the FLP problem as well.

From a practical point of view, EAs are easy to implement. They also allow easy incorporation of arbitrary constraints and optimization objectives. This is often much easier than extending the model for a given mathematical programming method.

Perhaps the most important advantage of the proposed approach is its ability to deliver a diverse pool of high-quality solutions. This is demonstrated in Figure 7, which illustrates the ability of the multi-objective algorithm to evolve high-quality non-dominated solutions in a single run while different workstation variants are automatically chosen for different floor plans. In general, this technique generates multiple solutions from which the user can choose the best one according to his/her expert knowledge. We believe that this is the feature that makes the method truly usable in practice. Often, when solving real-world problems, not all aspects of the desired solution can be captured within the formal definition of the optimization problem. Thus, only the most important ones and the ones that can be formally described are used to control the search process. Even then, it is very beneficial if the user is provided with a set of diverse, high-quality solutions from which he/she can further select the best-looking one in the end.

EAs also have a known limitation that they might be computationally intensive because they require processing the entire population of candidate solutions in each generation. On the other hand, when solving many real-world problems, including the FLP, the solver has enough time to generate a high-quality solution. Moreover, the EA can be interrupted at any time and return the current best solution generated thus far.

7.3. Potential for Improvement

While the experimental results show that the proposed evolutionary algorithms are effective and robust, there is still some room for further improvement. As mentioned above, it is absolutely essential that the evolutionary algorithm maintain a diverse set of candidate solutions during the course of the whole run while converging to ever-improving solutions. To attain this, the trade-off between exploration and exploitation has to be ensured. To achieve this, the algorithms have to ensure a balance between exploration and exploitation.

It is also important that efficient genetic operators are used to sample new candidate solutions. It was demonstrated that the performance of the 1-point crossover and the 2-point one differ significantly. This suggests that new types of operators for the indirect representation are worth exploring. Last but not least, the experiments showed that the multi-objective algorithm experiences difficulties when simultaneously optimizing explicit and implicit objectives. The algorithm tends to prioritize the explicit objective over the implicit one. This is not unexpected, as improving the explicit objective value is much easier than improving the implicit one. Techniques for diversity preservation such as novelty search, niching, or fitness sharing might help to remedy these issues.

8. Conclusions

We formulated a new FLP problem that involves alternative variants of facilities. This is a new FLP feature and represents a new challenge for optimization methods. The realistic definition of workstations and their variants suits many practical situations in factory floor optimization and management.

We proposed extensions to the single- and multi-objective evolutionary algorithms to solve the problem. Besides including a new mutation operator, a cooling schedule was proposed to control the search mode during the optimization process. It progressively restricts the scope of the crossover and mutation operators, thus changing the optimization mode from a global one at the early stages of the run to a local one at the end of the run. Experiments showed that the cooling schedule significantly improves the performance of the algorithms.

The efficiency of the extended evolutionary algorithms has been tested and analyzed for the data sets with and without alternative facility variants. It was demonstrated that the algorithms are able to efficiently solve the extended FLP, which involves a much larger search space than the FLP without alternatives. However, there is still room left for improvement of the optimization algorithms.

Finally, the performance of the multi-objective algorithm was analyzed within an optimization scenario that involves both explicit and implicit objectives, where the explicit ones are much easier to improve than the implicit ones. Although the algorithm demonstrated its potential to generate well-fit solutions, it also showed that its limits could be pushed even further. This remains a challenge for our future research.

In our future research, we will investigate different variants of the cooling schedule procedure. This can involve investigation of the linear vs. exponential schedule, static vs. dynamic setting of the schedule parameters, deterministic vs. probabilistic schedule, automatic identification of the trigger points based on the statistics of the current population, etc.

While the current formulation of the proposed FLP assumes only rectangular workstation shapes, it can be straightforwardly adapted to consider free-form, possibly non-convex, rectilinear shapes as well. New constraints and objectives can easily be introduced as well. This is also the topic of our further research.

Author Contributions: Conceptualization, J.K., L.K. and P.K.; Methodology, J.K., L.K. and P.K.; Software, J.K.; Validation, J.K.; Formal analysis, J.K.; Investigation, J.K., L.K. and P.K.; Data curation, J.K.; Writing—original draft, J.K., L.K. and P.K.; Funding acquisition, P.K. All authors have read and agreed to the published version of the manuscript.

Funding: This work was funded by the Ministry of Education, Youth, and Sport of the Czech Republic within the project Cluster 4.0 number CZ.02.1.01/0.0/0.0/16_026/0008432. The publication was created with support from the project "Regeneration of Used Batteries from Electric Vehicles" (Slovak ITMS2014+code 313012BUN5), which is part of the Important Project of Common European Interest (IPCEI) called European Battery Innovation (code OPII-MH/DP/2021/9.5-34), announced as a part of Operational Program Integrated Infrastructure.

Data Availability Statement: The data are available from the corresponding author upon request.

Acknowledgments: This work was funded by the Ministry of Education, Youth, and Sport of the Czech Republic within the project Cluster 4.0 number CZ.02.1.01/0.0/0.0/16_026/0008432. The publication was created with support from the project "Regeneration of Used Batteries from Electric Vehicles" (Slovak ITMS2014+code 313012BUN5), which is part of the Important Project of Common European Interest (IPCEI) called European Battery Innovation (code OPII-MH/DP/2021/9.5-34), announced as a part of Operational Program Integrated Infrastructure.

Conflicts of Interest: The authors declare no conflict of interest.

Nomenclature

The following nomenclature and abbreviations are used in this manuscript:

EA	Evolutionary Algorithm
FLP	Facility Layout Problem
FLP without alternatives	FLP with a single variant for each workstation
FLP with alternatives	FLP with multiple possible variants defined for each workstation
MOEA	Multi-Objective Evolutionary Algorithm
SA	Simulated Annealing
A	Flag indicating the evolutionary algorithm type, $A \in \{\text{EA, MOEA}\}$
C	Flag indicating the usage of the cooling scheme
$cost_{static}$	Total cost of the workstations of the FLP without alternatives
G	Maximum number of generations
$\mathcal{L}$	Set of pairwise dependencies between workstations
M	Size of the priority list for the FLP with alternatives
N	Number of workstations
n_i	Number of variants of the workstation w_i
p^c	Probability of accepting a worsening step in the SA method
P	Priority list
S_k	Set of positions in P, excluding the position k, that are associated with the same workstation as the element $P[k]$
t_j	Tuple representing the j-th element of the priority list, used with the FLP without alternatives
t_i'	Extended version of t_i used with the FLP with alternatives
T	Set of tuples that contains all variants of all workstations
U	Number of pairwise dependencies between workstations
V	Flag indicating the FLP variant
w_i	i-th workstation
$\mathcal{W}$	Set of all workstations to be placed in the floor plan

Appendix A. Pseudocode of the Evolutionary Algorithms

Algorithm A1: Single-objective evolutionary algorithm, EA

Input: $PopSize$ ... population size
G ... maximum number of generations
P_C ... crossover rate
P_M ... mutation rate
$EliteSize$ ... number of elite individuals
Output: Floorplan represented by the best individual

```
init(population)
evaluate(population)                                          # NSGA-II specific
g ← 0
while g < G do
    tempPop ← {}
    while |tempPop| < PopSize do
        par1 ← select(population)                             # NSGA-II specific
        if rand() < P_C then
            par2 ← select(population)                         # NSGA-II specific
            child ← crossover(par1, par2)
            if rand() < P_M then
                mutate(child)
        else
            child ← par1
            mutate(child)
        evaluate(child)                                       # NSGA-II specific
        tempPop ← tempPop + child
    tempPop ← tempPop ∪ elite(population, EliteSize)          # NSGA-II specific
    population ← getUnique(tempPop, PopSize)                  # NSGA-II specific
    g ← g + 1
return bestOf(population)                                     # NSGA-II specific
```

The EA starts with a random initialization and evaluation of the initial population of individuals. Evaluation of an individual means that, first, the individual's priority list is mapped to the actual floor plan, and then the quality measure of the floor plan is calculated. Then, the algorithm iterates through G generations, lines 4–20. In each generation, a new population is created as follows. A population $tempPop$ is created through a standard evolutionary process using selection, crossover, and mutation. When the $tempPop$ of size $PopSize$ has been generated, a number of $EliteSize$ top best unique individuals from the previous generation are added to $tempPop$, line 18. Finally, the best $PopSize$ unique individuals from $tempPop$ are selected to the new $population$, line 19. In the end, the best floor plan of the last population is returned.

Multi-objective evolutionary algorithm, MOEA, implements the NSGA-II algorithm [40]. The pseudocode of the EA algorithm applies to MOEA as well, with only a few modifications listed below:

- Floor plan's quality assessment, line 2 and line 16: contrary to EA, in MOEA, multiple quality measures are calculated for each solution.
- Parental selection, line 7 and line 9: the EA uses a standard tournament selection that chooses the best competitor solution with respect to the single objective. In MOEA, the *crowded-comparison operator* is used within the tournament selection. It guides the selection process towards a uniformly-spread-out Pareto-optimal front by considering both the quality and uniqueness of the solutions; see [40].
- Forming the final $population$ in generation $g + 1$, line 18 and line 19: instead of the actions taken on line 18 and line 19 of the EA, MOEA adopts the procedure that prefers high-quality and unique solutions out of the set $tempPop \cup population$; see Figure 2 in [40].
- Selection of the final solution(s), line 21: in EA, a single best-of-run solution is returned. In MOEA, the whole set of all non-dominated solutions of the last population is returned.

References

1. Tompkins, J.; White, J.; Bozer, Y.; Tanchoco, J. *Facilities Planning*; Wiley: Hoboken, NJ, USA, 2010.
2. Papakostas, N.; O'Connor Moneley, J.; Hargaden, V. Integrated simulation-based facility layout and complex production line design under uncertainty. *CIRP Ann.* **2018**, *67*, 451–454. [CrossRef]
3. YounesSinaki, R.; Sadeghi, A.; Mosadegh, H.; Almasarwah, N.; Suer, G. Cellular manufacturing design 1996–2021: A review and introduction to applications of Industry 4.0. *Int. J. Prod. Res.* **2022**, 1–52.
4. Izadinia, N.; Eshghi, K. A robust mathematical model and ACO solution for multi-floor discrete layout problem with uncertain locations and demands. *Comput. Ind. Eng.* **2016**, *96*, 237–248. [CrossRef]
5. Zhang, Z.; Wu, L.; Wu, Z.; Zhang, W.; Jia, S.; Peng, T. Energy-Saving Oriented Manufacturing Workshop Facility Layout: A Solution Approach Using Multi-Objective Particle Swarm Optimization. *Sustainability* **2022**, *14*, 2788. [CrossRef]
6. Drira, A.; Pierreval, H.; Hajri-Gabouj, S. Facility layout problems: A survey. *Annu. Rev. Control.* **2007**, *31*, 255–267. [CrossRef]
7. Hosseini-Nasab, H.; Fereidouni, S.; Ghomi, S.M.T.F.; Fakhrzad, M.B. Classification of facility layout problems: A review study. *Int. J. Adv. Manuf. Technol.* **2017**, *94*, 957–977. [CrossRef]
8. Pérez-Gosende, P.; Mula, J.; Díaz-Madroñero, M. Facility layout planning. An extended literature review. *Int. J. Prod. Res.* **2021**, *59*, 3777–3816. [CrossRef]
9. Guo, W.; Jiang, P.; Yang, M. Unequal area facility layout problem-solving: A real case study on an air-conditioner production shop floor. *Int. J. Prod. Res.* **2023**, *61*, 1479–1496. [CrossRef]
10. Moatari-Kazerouni, A.; Chinniah, Y.; Agard, B. Integrating occupational health and safety in facility layout planning, part I: Methodology. *Int. J. Prod. Res.* **2015**, *53*, 3243–3259. [CrossRef]
11. Koopmans, T.C.; Beckmann, M.J. Assignment Problems and the Location of Economic Activities. *Econometrica* **1957**, *25*, 53. [CrossRef]
12. Loiola, E.M.; de Abreu, N.M.M.; Boaventura-Netto, P.O.; Hahn, P.; Querido, T. A survey for the quadratic assignment problem. *Eur. J. Oper. Res.* **2007**, *176*, 657–690. [CrossRef]
13. Ejeh, J.O.; Liu, S.; Papageorgiou, L.G. Optimal multi-floor process plant layout with production sections. *Chem. Eng. Res. Des.* **2018**, *137*, 488–501. [CrossRef]
14. Klausnitzer, A.; Lasch, R. Optimal facility layout and material handling network design. *Comput. Oper. Res.* **2019**, *103*, 237–251. [CrossRef]
15. Vázquez-Román, R.; Díaz-Ovalle, C.O.; Jung, S.; Castillo-Borja, F. A reformulated nonlinear model to solve the facility layout problem. *Chem. Eng. Commun.* **2019**, *206*, 476–487. [CrossRef]

16. Gülşen, M.; Murray, C.C.; Smith, A.E. Double-row facility layout with replicate machines and split flows. *Comput. Oper. Res.* **2019**, *108*, 20–32. [CrossRef]
17. Baykasoglu, A.; Subulan, K.; Hamzadayı, A. Capability-based machine layout with a matheuristic-based approach. *Expert Syst. Appl.* **2022**, *198*, 116900. [CrossRef]
18. Scholz, D.; Petrick, A.; Domschke, W. STaTS: A Slicing Tree and Tabu Search based heuristic for the unequal area facility layout problem. *Eur. J. Oper. Res.* **2009**, *197*, 166–178. [CrossRef]
19. Komarudin.; Wong, K.Y. Applying Ant System for solving Unequal Area Facility Layout Problems. *Eur. J. Oper. Res.* **2010**, *202*, 730–746. [CrossRef]
20. Wong, K.Y.; Komarudin. Solving facility layout problems using Flexible Bay Structure representation and Ant System algorithm. *Expert Syst. Appl.* **2010**, *37*, 5523–5527. [CrossRef]
21. Derakhshan Asl, A.; Wong, K.Y. Solving Unequal-area Static and Dynamic Facility Layout Problems Using Modified Particle Swarm Optimization. *J. Intell. Manuf.* **2017**, *28*, 1317–1336. [CrossRef]
22. Guan, J.; Lin, G. Hybridizing variable neighborhood search with ant colony optimization for solving the single row facility layout problem. *Eur. J. Oper. Res.* **2016**, *248*, 899–909. [CrossRef]
23. Saraswat, A.; Venkatadri, U.; Castillo, I. A framework for multi-objective facility layout design. *Comput. Ind. Eng.* **2015**, *90*, 167–176. [CrossRef]
24. Aiello, G.; Scalia, G.L.; Enea, M. A multi objective genetic algorithm for the facility layout problem based upon slicing structure encoding. *Expert Syst. Appl.* **2012**, *39*, 10352–10358. [CrossRef]
25. García-Hernández, L.; Arauzo-Azofra, A.; Salas-Morera, L.; Pierreval, H.; Corchado, E. Facility Layout Design Using a Multi-objective Interactive Genetic Algorithm to Support the DM. *Expert Syst. J. Knowl. Eng.* **2015**, *32*, 94–107. [CrossRef]
26. Gonçalves, J.F.; Resende, M.G. A biased random-key genetic algorithm for the unequal area facility layout problem. *Eur. J. Oper. Res.* **2015**, *246*, 86–107. [CrossRef]
27. Liu, J.; Liu, J. Applying multi-objective ant colony optimization algorithm for solving the unequal area facility layout problems. *Appl. Soft Comput.* **2019**, *74*, 167–189. [CrossRef]
28. Garcia-Hernandez, L.; Garcia-Hernandez, J.; Salas-Morera, L.; Carmona-Muñoz, C.; Alghamdi, N.; de Oliveira, J.V.; Salcedo-Sanz, S. Addressing Unequal Area Facility Layout Problems with the Coral Reef Optimization algorithm with Substrate Layers. *Eng. Appl. Artif. Intell.* **2020**, *93*, 103697. [CrossRef]
29. Krömer, P.; Platoš, J.; Snášel, V. Solving the Single Row Facility Layout Problem by Differential Evolution. In Proceedings of the 2020 Genetic and Evolutionary Computation Conference, GECCO '20, New York, NY, USA, 8–12 July 2020; pp. 210–218. [CrossRef]
30. García-Hernández, L.; Palomo-Romero, J.M.; Salas-Morera, L.; Arauzo-Azofra, A.; Pierreval, H. A novel hybrid evolutionary approach for capturing decision maker knowledge into the unequal area facility layout problem. *Expert Syst. Appl.* **2015**, *42*, 4697–4708. [CrossRef]
31. Reisinger, J.; Zahlbruckner, M.A.; Kovacic, I.; Kán, P.; Wang-Sukalia, X.; Kaufmann, H. Integrated multi-objective evolutionary optimization of production layout scenarios for parametric structural design of flexible industrial buildings. *J. Build. Eng.* **2022**, *46*, 103766. [CrossRef]
32. Kubalík, J.; Kadera, P.; Jirkovský, V.; Kurilla, L.; Prokop, Š. Plant Layout Optimization Using Evolutionary Algorithms. In *Industrial Applications of Holonic and Multi-Agent Systems: 9th International Conference, HoloMAS 2019, Linz, Austria, 26–29 August 2019*; Mařík, V., Kadera, P., Rzevski, G., Zoitl, A., Anderst-Kotsis, G., Tjoa, A.M., Khalil, I., Eds.; Springer International Publishing: Cham, Switzerland, 2019; pp. 173–188.
33. Kirkpatrick, S.; Gelatt, C.D.; Vecchi, M.P. Optimization by Simulated Annealing. *Science* **1983**, *220*, 671–680. [CrossRef]
34. Anand, S.; Saravanasankar, S.; Subbaraj, P. Customized simulated annealing based decision algorithms for combinatorial optimization in VLSI floorplanning problem. *Comput. Optim. Appl.* **2012**, *52*, 667–689. [CrossRef]
35. Yadav, A.; Jayswal, S. Makespan minimization in flexible manufacturing system using genetic algorithm. *Int. J. Appl. Eng. Res.* **2018**, *13*, 9328–9334.
36. Umbarkar, A.; Sheth, P.D. Crossover Operators in Genetic Algorithms: A Review. *ICTACT J. Soft Comput.* **2015**, *6*, 1083–1092
37. Hill, S.; O'Riordan, C. Examining the impact of Neutral theory on Genetic Algorithm population evolution. In Proceedings of the 2015 7th International Joint Conference on Computational Intelligence (IJCCI), Lisbon, Portugal, 12–14 November 2015; Volume 1, pp. 196–203.
38. Galván-López, E.; Poli, R.; Kattan, A.; O'Neill, M.; Brabazon, A. Neutrality in evolutionary algorithms... What do we know? *Evol. Syst.* **2011**, *2*, 145–163. [CrossRef]
39. Knowles, J.; Corne, D. On metrics for comparing nondominated sets. In Proceedings of the 2002 Congress on Evolutionary Computation, CEC'02 (Cat. No.02TH8600), Honolulu, HI, USA, 12–17 May 2002; Volume 1, pp. 711–716. [CrossRef]
40. Deb, K.; Pratap, A.; Agarwal, S.; Meyarivan, T. A fast and elitist multiobjective genetic algorithm: NSGA-II. *IEEE Trans. Evol. Comput.* **2002**, *6*, 182–197. [CrossRef]

applied sciences

MDPI

Article

Machine Learning in Manufacturing towards Industry 4.0: From 'For Now' to 'Four-Know'

Tingting Chen [1,*], Vignesh Sampath [2], Marvin Carl May [3], Shuo Shan [1], Oliver Jonas Jorg [4], Juan José Aguilar Martín [5], Florian Stamer [3], Gualtiero Fantoni [4], Guido Tosello [1] and Matteo Calaon [1]

1 Department of Civil and Mechanical Engineering, Technical University of Denmark, 2800 Kongens Lyngby, Denmark
2 Autonomous and Intelligent Systems Unit, Tekniker, Member of Basque Research and Technology Alliance, 20600 Eibar, Spain
3 wbk Institute of Production Science, Karlsruhe Institute of Technology (KIT), Kaiserstr. 12, 76131 Karlsruhe, Germany
4 Department of Civil and Industrial Engineering, University of Pisa, 56122 Pisa, Italy
5 Department of Design and Manufacturing Engineering, School of Engineering and Architecture, University of Zaragoza, 50009 Zaragoza, Spain
* Correspondence: tchen@dtu.dk; Tel.: +45-5028-1068

Abstract: While attracting increasing research attention in science and technology, Machine Learning (ML) is playing a critical role in the digitalization of manufacturing operations towards Industry 4.0. Recently, ML has been applied in several fields of production engineering to solve a variety of tasks with different levels of complexity and performance. However, in spite of the enormous number of ML use cases, there is no guidance or standard for developing ML solutions from ideation to deployment. This paper aims to address this problem by proposing an ML application roadmap for the manufacturing industry based on the state-of-the-art published research on the topic. First, this paper presents two dimensions for formulating ML tasks, namely, 'Four-Know' (Know-what, Know-why, Know-when, Know-how) and 'Four-Level' (Product, Process, Machine, System). These are used to analyze ML development trends in manufacturing. Then, the paper provides an implementation pipeline starting from the very early stages of ML solution development and summarizes the available ML methods, including supervised learning methods, semi-supervised methods, unsupervised methods, and reinforcement methods, along with their typical applications. Finally, the paper discusses the current challenges during ML applications and provides an outline of possible directions for future developments.

Keywords: machine learning; Industry 4.0; manufacturing; artificial intelligence; smart manufacturing; digitization

Citation: Chen, T.; Sampath, V.; May, M.C.; Shan, S.; Jorg, O.J.; Aguilar Martín, J.J.; Stamer, F.; Fantoni, G.; Tosello, G.; Calaon, M. Machine Learning in Manufacturing towards Industry 4.0: From 'For Now' to 'Four-Know'. *Appl. Sci.* **2023**, *13*, 1903. https://doi.org/10.3390/app13031903

Academic Editors: Alexandre Carvalho and Richard (Chunhui) Yang

Received: 23 November 2022
Revised: 18 January 2023
Accepted: 27 January 2023
Published: 1 February 2023

1. Introduction

Within the fourth industrial revolution, coined as 'Industry 4.0', the way products are manufactured is changing dramatically [1]. Moreover, the way humans and machines interact with one another in manufacturing has seen enormous changes [2], developing towards an 'Industry 5.0' notion [3]. The digitalization of businesses and production companies, the inter-connection of their machines through embedded system and the Internet of Things (IoT) [4], the rise of cobots [5,6], and the use of individual workstations and matrix production [7] are disrupting conventional manufacturing paradigms [1,8]. The demand for individualized and customized products is continuously increasing. Consequently, order numbers are surging while batch sizes diminish, to the extremes of fully decentralized 'batch size one' production. The demand for a high level of variability in production and manufacturing through Mass Customization is inevitable. Mass Customization in turn requires manufacturing systems which are increasingly more flexible and adaptable [7–9].

Machine Learning (ML) is one of the cornerstones for making manufacturing (more) intelligent, and thereby providing it with the needed capabilities towards greater flexibility and adaptability [10]. These advances in ML are shifting the traditional manufacturing era into the smart manufacturing era of Industry 4.0 [11]. Therefore, ML plays an increasingly important role in manufacturing domain together with digital solutions and advanced technologies, including the Industrial Internet of Things (IIoT), additive manufacturing, digital twins, advanced robotics, cloud computing, and augmented/virtual reality [11]. ML refers to a field of Artificial Intelligence (AI) that covers algorithms learning directly from their input data [12]. Despite most researchers focusing on finding a single suitable ML solution for a specific problem, efforts have already been undertaken to reveal the entire scope of ML in manufacturing. Wang et al. presented frequently-used deep learning algorithms along with an assessment of their applications towards making manufacturing "smart" in their 2018 survey [13]. In particular, they discussed four learning models: Convolutional Neural Networks, Restricted Boltzmann Machines, Auto-Encoders, and Recurrent Neural Networks. In their recent literature review on "Machine Learning for Industrial Applications", Bertolini et al. [12] identified, classified, and analyzed 147 papers published during a twenty-year time span from Jan. 2000 to Jan. 2020. In addition, they provided a classification on the basis of application domains in terms of both industrial areas and processes, as well as their respective subareas. Within these domains, the authors analyzed the different trends concerning supervised, unsupervised, and reinforced learning techniques, including the most commonly used algorithms, Neural Networks (NNs), Support Vector Machine (SVM), and Tree-Based (TB) techniques. The goal of another literature review from Dogan and Birant [14] was to provide a sound comprehension of the major approaches and algorithms from the fields of ML and data mining (DM) that have been used to improve manufacturing in the recent past. Similarly, they investigated research articles from the period of the past two decades and grouped the identified articles under four main subjects: scheduling, monitoring, quality, and failure.

While these classifications and trend analyses provide an excellent overview of the extent of ML applications in manufacturing, they mainly focus on introducing ML algorithms; the implementation of ML solution for different tasks in an industrial environment from scratch has not yet been fully discussed. In general, a comprehensive formulation of industrial problems prior to the development of ML solutions seems lacking. Therefore, the issue we aim to address in this paper is how ML can be implemented to improve manufacturing in the transition towards Industry 4.0. From this issue, we derive the following research questions:

- RQ1: How does ML benefit manufacturing, and what are the typical ML application cases?
- RQ2: How are ML-based solutions developed for problems in manufacturing engineering?
- RQ3: What are the challenges and opportunities in applying ML in manufacturing contexts?

To answer these research questions, more than a thousand research articles retrieved from two well-known research databases were systematically identified, screened, and analyzed. Subsequently, the articles were classified within a two-dimensional framework, which takes value-based development stages into account on one axis and manufacturing levels on the other. The development stage concerns visibility, transparency, predictive capacity, and adaptability, whereas the four manufacturing levels are product, process, machine, and system.

The rest of this paper is structured as follows. Section 1 introduces the key concepts, research questions, and motivations. Section 2 proposes the methodology of 'Four-know' and 'Four-level' to establish a two-dimensional framework for helping to formulate industrial problems effectively. Based on the proposed framework, a systematic literature review is carried out and the identified articles are analysed and classified. Section 3 describes a six-step pipeline for the application of ML in manufacturing. Section 4 explains different ML methods, presenting where and how they have been applied in manufacturing according to the prior identified research articles. Section 5 formulates common challenges and potential future directions; finally, the paper concludes in Section 6 with a summary and discussion of the authors' findings.

2. Overview of Machine Learning in Manufacturing

Despite numerous ML studies and their promising performance, it remains very difficult for non-experts working in the manufacturing industry to begin developing ML solutions for their specific problems. The first challenging part of application is to formulate the actual problems to be solved [15]. Therefore, this section aims to overcome this problem by introducing the categories of Four-Know and Four-Level to help formulate ML tasks in manufacturing and describing the benefits of applying ML in manufacturing from ML use cases categorized using the Four-Know and Four-Level concepts (RQ1). Lastly, an overview and developing trends in recent ML studies are provided as formulated by Four-Know and Four-Level.

2.1. *Introduction of Four-Know and Four-Level*

According to the *Acatech* Industrie 4.0 Maturity levels [16], the development towards Industry 4.0 in manufacturing can be structured into the following six successive stages: computerization, connectivity, visibility, transparency, predictive capacity, and adaptability. The first two stages, computerization and connectivity, provide the basis for digitization, while the rest are analytic capabilities required for achieving Industry 4.0. ML, as powerful data analytics tools are normally applied in the last four stages. Inspired by the *Acatech* Industrie 4.0 Maturity levels, ML studies in manufacturing can be categorized into four subjects: Know-what, Know-why, Know-when, and Know-how, which to a degree overlap with visibility, transparency, predictive capacity, and adaptability, respectively. The Four-Know definitions are presented below:

- *Know-what* deals with understanding of the current states of machines, processes, or production systems, which can help in rapid decision-making. It should be noted that Know-what goes beyond visualization of real-time data. Instead, data should be processed, analyzed, and distilled into information which enables decision-making. For instance, typical examples of Know-what in manufacturing are defect detection in quality control [17,18], fault detection in process/machine monitoring [19,20], and soft sensor modelling [21,22].
- *Know-why*, based on the information from Know-what, aims to identify inner patterns from historical data, thereby discovering the reasons for a thing happening. Know-why includes the identification of interactions among different variables [23] and the discovery of cause-effect relationship between an event and other variables [24,25]. On one hand, Know-why can indicate most important factors for understanding Know-what. On the other hand, Know-why is the prerequisite for Know-when, as the reliability of predictions is heavily dependent upon the quality of casual inference.
- *Know-when*, built on Know-why, involves timely predictions of events or prediction of key variables based on historical data, allowing the decision-maker can take actions at early stages. For instance, Know-when in manufacturing includes quality prediction based on relevant variables [26,27], predictive maintenance via detection of incipient anomalies before break-down [28,29], and predicting Remaining Useful Life (RUL) [30,31].
- *Know-how*, on the foundation of Know-when, can recommend decisions that help adapt to expected disturbance and can aid in self-optimization. Examples in manufacturing include prediction-based process control [27,32], scheduling of predictive maintenance tasks [33,34], dynamic scheduling in the flexible production [35,36], and inventory control [34].

The aim of applying ML in manufacturing is to achieve production optimization across four different levels: product, process, machine, and system. Therefore, the use cases for applying ML can be further categorized by these different levels, as shown in Figure 1 and Table 1, which answer RQ1 in terms of ML typical use cases.

Table 1. Typical ML use cases categorized by Four-Level and Four-Know.

Level	Know-What	Know-Why	Know-When	Know-How
Product	Defect detection [37], Product design [38]	Correlation between process and quality [23]	Quality prediction [26]	Quality improvement [39]
Process	Process monitoring [40]	Root cause analysis of process failure [41], Process modelling [42]	Process fault prediction [43], Process characteristics prediction [44]	Self-optimizing process planning [45], Adaptive process control [46]
Machine	Machine tool monitoring [47]	Fault diagnosis [48], Downtime prediction [49]	RUL prediction [50], Tool wear prediction [51]	Adaptive compensation of errors [52,53],
System	Anomaly detection [54]	Root cause analysis of production disturbances or casual-relationship discovery [55]	Production performance prediction [56], Human behavior control [57]	Predictive scheduling [58], Adaptive production control [59]

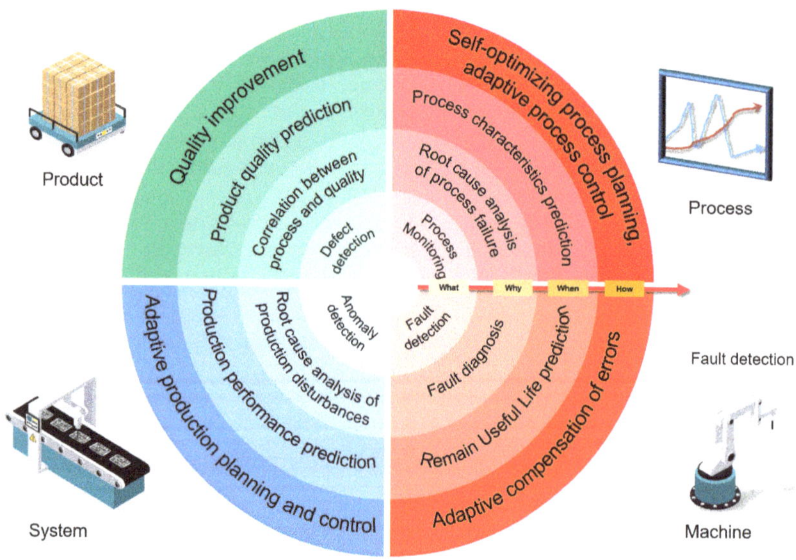

Figure 1. Four-Level and Four-Know categorization of ML applications. The Four-Know categories, from Know-what to Know-how, are respectively demonstrated by the four concentric circles, from the inner circle to the outer circle, with each circle divided into four quarters according to the Four Levels.

2.2. Literature Review Methodology

In order to address the research questions laid out in Section 1, a systematic literature review following the PRISMA methodology [60] was carried out. Two well-known research databases, Scopus (Elsevier) and Web of Science (WoS), were chosen for retrieving documents. The overall literature review process is shown in Figure 2.

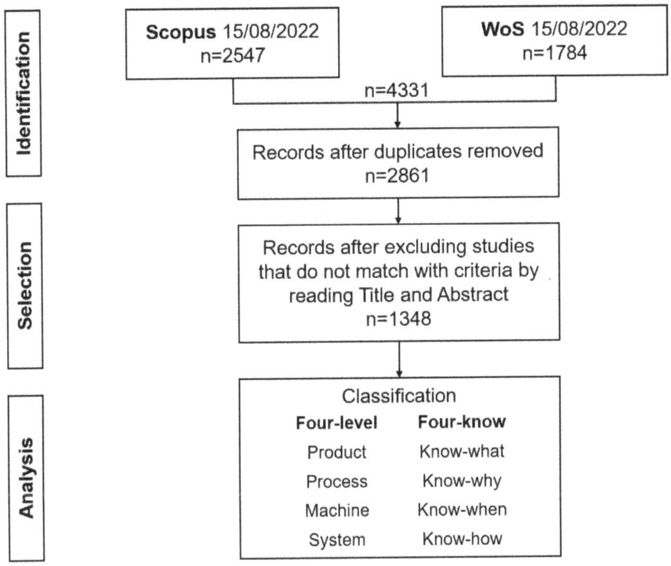

Figure 2. The overall literature review process following PRISMA. All identified documents were screened and assessed for eligibility, then subjected to Four-Level and Four-Know classification.

Table 2 shows the limitations used when performing the document search. It should be noted that the query strings were used for Title, Abstract, and Keywords as well as Keyword Plus (only in WoS).

Table 2. Limitations for document searching.

Item	Description
Query string	("manufacturing" OR "industry 4.0" OR "industrie 4.0") AND ("machine learning" OR "deep learning" OR "supervised learning" OR "semi-supervised learning" OR "unsupervised learning" OR "reinforcement learning")
Year	Published from 2018 to 2022
Language	English
Subject/Research area	Engineering
Document type	Article

Following the document search, 2547 documents were found from Scopus and 1784 from WoS. The identified publications from the two databases were merged and duplicates were removed, resulting in 2861 publications. The documents were then evaluated and selected by reading the Title and Abstract field, and articles that did not meet the following selection criteria were excluded:

- The study dealt with the context of manufacturing;
- The study dealt with ML applications in specific fields.

Therefore, conceptual models, frameworks, and studies that only focused on algorithm development were considered to be out of scope.

Finally, the remaining 1348 documents were analyzed and classified based on the Four-Level and Four-Know categories. Figure 3 shows the trend of ML applications in manufacturing over the past five years from the Four-Level perspective. Figure 4 reveals the detailed distribution of ML applications in Four-Know terms. It should be noted that because the literature review was conducted in August 2022, the actual numbers for the full year 2022 should be higher. As can be seen, there has been a gradual increase in the number of ML publications in manufacturing in all levels over the past five years. Typically, what stands out in this figure is the dominance of the product level. From Figure 4, it can be seen that recent ML applications in product level are mainly focused on Know-what and Know-when. A similar pattern can be found at the machine level. Interestingly, a considerable growth in Know-how is observed at the process and system levels compared to the others. The reason for this may be correlated with higher demand for adaptability with respect to changes on the process and system levels.

The identified documents were analyzed and classified according to their applied ML methods, providing examples for non-experts when dealing with similar tasks.

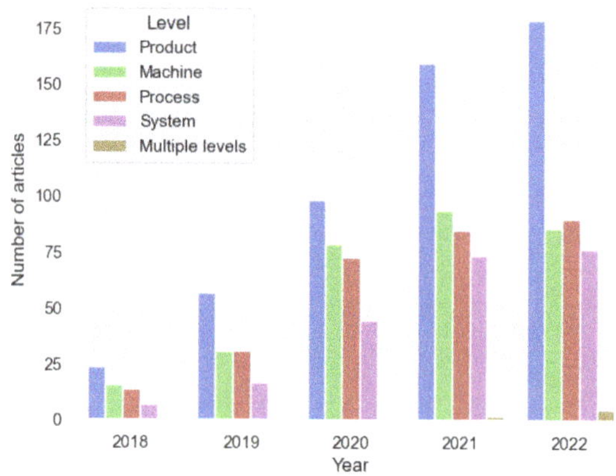

Figure 3. Trends in ML publications in manufacturing in the past five years by Four-Level grouping.

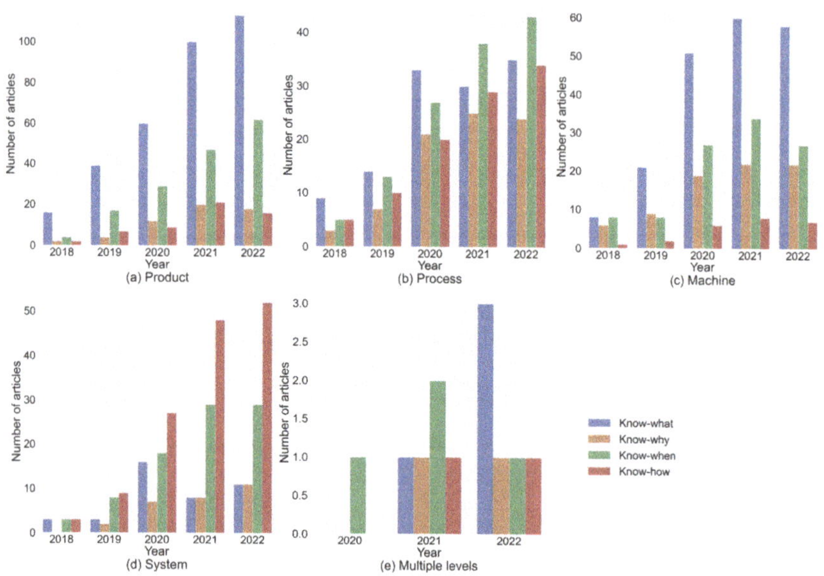

Figure 4. Four-Know development trends for each level over the past five years.

3. Pipeline of Applying Machine Learning in Manufacturing

ML is a technique capable of extracting knowledge from data automatically [12]. Increasing research on ML has shown that it is an appealing solution when tackling complex challenges. In recent years, more and more manufacturing industries have begun to leverage the benefits of ML by developing ML solutions in several industrial fields. However, despite plenty of off-the-shelf ML models, there are challenges when applying ML to real-world problems [61]. In particular, it is harder for small and medium-sized enterprises to develop in-house ML solutions, as commercial ML solutions are normally confidential and inaccessible. Therefore, this section aims to provide a pipeline for applying ML for those who are starting from scratch (RQ2). Applying machine learning in manufacturing normally involves the following six steps: (i) data collection, (ii) data cleaning, (iii) data transformation, (iv) model training, (v) model analysis, and (vi) model push, as shown in Figure 5.

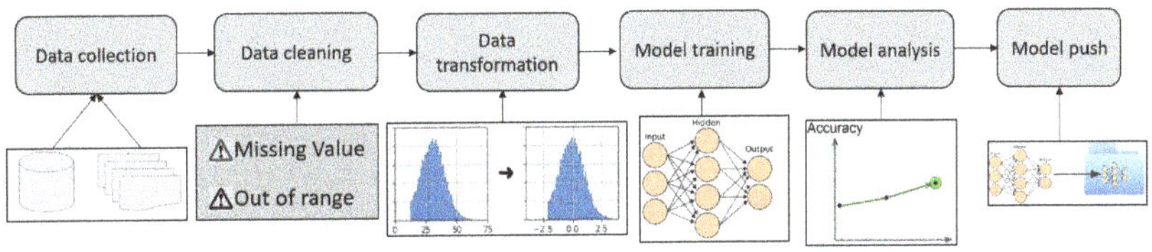

Figure 5. Pipeline of applying machine learning in manufacturing.

3.1. Data Collection

The lifeblood of any machine learning model is data. In order for an ML model to learn, clean data samples must be continuously fed into system throughout the training process. When the collected data are highly imbalanced or otherwise inadequate, the desired task may not be achievable. Data can be collected from different sources, including machines, processes, or production with the aid of sensors or external databases. In terms of data types, the data used in machine learning can be generally categorized as follows:

- *Image data*, matrices of pixels with two or more dimensions, such as gray-scale images or colored images. Image data can acquired by with vision systems, through data transformations such as simple concatenation of several one-dimensional vectors with same length, or by the transformation of images from the spatial domain to the frequency domain.
- *Tabular data* organized in a table, where normally one axis represents attributes and another axis represents observations. Tabular data are typically observed in production data, where the attributes of events of interest are collected. Though tabular data share a similar data structure with image data, the latter are more focused on one-dimensional interaction among attributes, while image data typically stress spatial interactions in both dimensions.
- *Time series data*, sequences of one or more attributes over time, with the former corresponding to univariate time series and the latter multivariate time series. In manufacturing, time series data are normally acquired with sensors whenever there is a need for monitoring time flow changes of data.
- *Text data*, including written documents with words, sentences or paragraphs. Examples of text data in manufacturing include maintenance reports on machines and descriptions of unexpected disturbances or events in production.

3.2. Data Cleaning

Real-world industrial data are highly susceptible to noisy, missing, and inconsistent data due to several factors. Low-quality noisy data can lead to less accurate ML models. Data cleaning [62] is a crucial step when organizing data into a consistent data structure across packages, and can improve the quality of the data, leading to more accurate ML models. It is usually performed as an iterative approach. Methods include filling in missing values, smoothing noisy data, removing outliers, resolving data inconsistencies, etc.

3.3. Data Transformation

Data transformation is the process of transforming unstructured raw data into data better suited for model construction. Data transformation can be broadly classified into mandatory transformations and optional quality transformations. Mandatory transformations must be carried out to convert the data into a usable format and then deliver the transformed data to the destination system. These include transforming non-numerical data into numerical data, resizing data to a fixed size, etc. It should be noted that data transformations are not always straightforward. Indeed, in certain situations data types can be interconvertible by leveraging specific processing techniques, as shown in Figure 6. For instance, univariate time series can be converted into image data using the Gramian Angular Field (GAF) or Markov Transition Field (MTF) [63] methods. Unstructured text data can be converted into tabular data via word embedding [64]. Tabular data can be transformed into image data by projecting data into a 2D space and assigning pixels, as in Deepinsight [65] or Image Generator for Tabular Data (IGTD) [66]. Image data are preferable for data analysis, as they allow the power of Convolutional Neural Networks (CNNs) [67] to be exploited.

In real-world applications, data are normally high-dimensional and redundant. When performing data modelling directly in the original high-dimensional space, the computational efficiency can be very low. Hence, it is necessary to reduce the dimensionality in order to obtain better representation for data modelling. This is achieved by feature selection, which selects the most informative feature subset from raw data, or feature extraction, which generates new lower-dimensional features. After feature engineering, features are either manually designed, so-called "handcrafted features" [68], or automatically learned from data, so-called "automatic features". Handcrafted features are heavily dependent on domain knowledge, and normally have physical meaning. However, these features are highly subjective [69] and inevitably lack implicit key features [70,71].

By contrast, automatic features driven by data require no prior knowledge. Therefore, they have been gaining increasing research attention in recent years. Conventionally, automatic features are obtained by linear transformations such as Principle Component Analysis (PCA) [72] or Independent Component Analysis (ICA) [73]. However, with the development of Artificial Neural Networks (ANNs), direct learning of implicit features has become possible by optimizing the loss function. Thus, neural networks have gradually developed into an end-to-end solution where knowledge is directly learned from raw data without human effort. Typically, CNNs [74] and Recurrent Neural networks (RNNs) [75] are used for image data and time series data, respectively.

A summary of typical features for different data types can be seen in Table 3.

Table 3. Typical features for different data types.

Data Type	Handcrafted Features	Automatic Features
Image data	LBP [76], SIFT [77], HOG [78]	ICA, CNNs
Tabular data	feature selection	PCA, ICA, ANNs
Time series data	Time domain: mean, min, max, etc. Frequency domain: power spectrum [78] Time-frequency domain: DWT [79], STFT [80]	ICA, RNNs
Text data	Bag of Words (BoW) [81]	Word2vec [82]

3.4. Model Training

After selecting the features, it is necessary to form the correct data structure for each individual ML model used in the subsequent steps. Note that different ML algorithms might require different data models for the same task. Furthermore, results can be improved through normalization or standardization. Then, the ML models can be applied in the actual modelling phase. The first step in training a machine learning model typically involves selecting a model type that is appropriate for the nature of the data and the problem at hand. After a model has been chosen, it can be trained by providing it with the training data and using an optimization algorithm to find the set of parameters that provide the best performance on those data. Depending on the task, either unsupervised, semi-supervised, supervised, or reinforcement learning can be applied. These are individually introduced in the following section.

3.5. Model Analysis

Analysis of model performance is an important step in choosing the right model. This stage emphasizes how effective the selected model will perform in the future and helps to make the final decision with regard to model selection. Performance analysis evaluates models using different metrics, e.g., accuracy, precision, recall, and F1-score (the weighted average of precision and recall) for classification tasks and the root mean square error (RMSE) for regression tasks.

3.6. Model Push

Although state-of-the-art ML models improve predictive performance, they contain millions of parameters, and consequently require a large number of operations per inference. Such computationally intensive models make deployment in low-power or resource-constrained devices with strict latency requirements quite difficult. Several methods, including model pruning [83], model quantization [84], and knowledge distillation [85], have been suggested in the literature as ways to compress these dense models.

Overall, In the context of manufacturing applications, data collection, data cleaning, data transformation, model training, model analysis, and model push are key steps in the implementation of utilizing historical data with ML in order to optimize production and improve efficiency, quality, and productivity. For instance, data collection involves gathering data from various sources, such as sensor data, production logs, and quality control records. Data cleaning involves removing any errors, inconsistencies, or irrelevant information from the data. Data transformation involves preparing the data for analysis via formatting in a way that is suitable for the chosen model. Model training involves using the cleaned and transformed data to train a machine learning model. Model analysis involves evaluating the performance of the model and identifying any areas for improvement. Model push involves deploying the model in a production environment and making predictions or decisions based on the model. All of these steps are critical to ensuring that the results from ML models are accurate, reliable, and useful for manufacturing production.

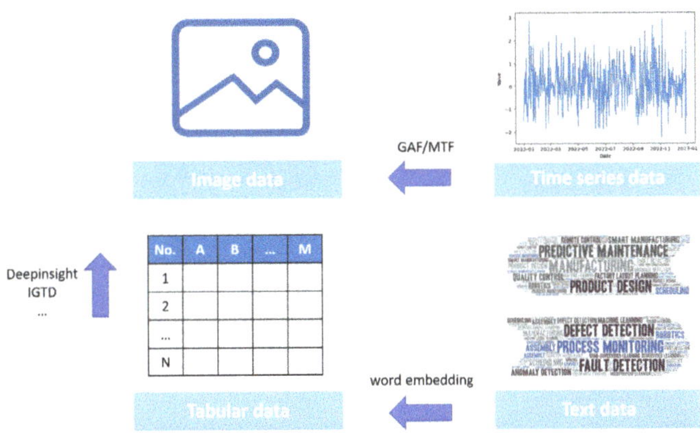

Figure 6. Data types used in ML and their convertibility.

4. Machine Learning Methods and Applications

Model development is the core of ML-based solutions, as the selection of an ML model plays a critical roles in the outcome. Therefore, this section aims to provide a comprehensive overview of ML methods and their potential possibilities in manufacturing applications, including supervised learning methods, semi-supervised learning methods, unsupervised learning methods, and reinforcement learning methods. In addition, example typical applications for each category of ML method are listed to support model selection.

4.1. Supervised Learning Methods

Supervised learning methods aim to learn an approximation function f that can map inputs x to outputs y with the guidance of annotations $(x_1, y_1), (x_2, y_2), \ldots, (x_N, y_N)$. In supervised learning, the algorithm analyzes a labeled dataset and derives an inferred function which can be applied to unseen samples. It should be noted that labeled dataset is a necessity for supervised learning, and as such it requires a large amount of data and high labeling costs. Supervised learning methods are generally used for dealing with two problems, namely, regression and classification. The difference between regression and classification is in the data type of the output variables; regression predicts continuous numeric values ($y \in \mathbb{R}$), while classification predicts categorical values ($y \in \{0, 1\}$). In terms of principles, supervised learning methods can be further categorized into four groups: tree-based methods, probabilistic-based methods, kernel-based methods, and neural network-based methods.

Tree-based methods: Tree-based methods aim at partitioning the feature space into several regions until the datapoints in each region share a similar class or value, as depicted in Figure 7. After space partitioning, a series of if–then rules with a tree-like structure can be obtained and used to determine the target class or value. Compared with the black-box models in other supervised methods, Tree-based methods are easily understandable models that offer better model interpretability. Decision trees [86], in which only a single tree is established, are the most basic of tree-based methods. It is simple and effective to train a decision tree, and the results are intuitively understandable, though this approach is very prone to overfitting. A tree ensemble is an extension of the decision tree concept. Instead of establishing a single tree, multiple trees are established in parallel or in sequence, referred to as bagging [87] and boosting [88], respectively. Commonly used tree ensemble methods include Random Forest [89], Adaptive Boosting (AdaBoost) [88], and Extreme Gradient Boosting (XGBoost) [90].

Thanks to their better model interpretability, tree-based methods can be used to identify the most important factors leading up to events. Their possible applications in manufacturing are mainly in the Know-why and Know-when stages. For instance, examples of Know-why tasks with tree-based methods at the product and machine level include identifying the influencing factors that lead to quality defects [91] or machine failure [92], thereby allowing the manufacturer to diagnose problems effectively. In addition, the identified important factors when using tree-based methods can help in further predicting target values such as product quality [93](Know-when, product level) or events of interest before they happen, such as machine breakdown [31] (Know-when, machine level).

Probabilistic-based methods: For a given input, probabilistic-based methods provide probabilities for each class as the output. Probabilistic models are able to explain the uncertainties inherent to data, and can hierarchically build complex models. Widely used probabilistic-based methods include Bayesian Optimization (BO) [94] and Hidden Markov Models (HMM) [95].

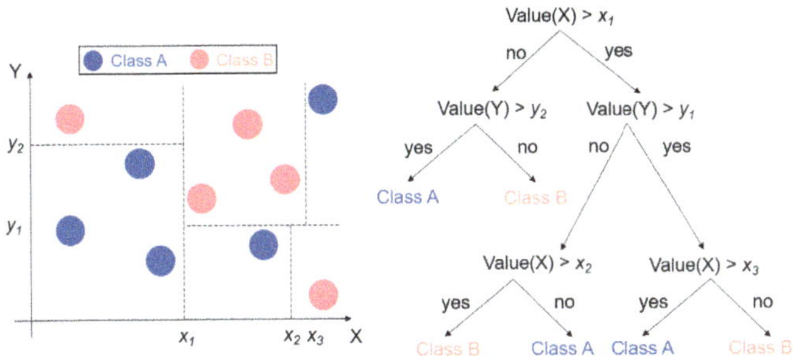

Figure 7. The principle of a decision tree. As shown, the feature space is partitioned into several rectangles in which the input point can find the corresponding class.

The dependencies among different variables can be well captured by Bayesian networks [94], enabling a greater likelihood of predicting the target. This can be potentially beneficial for manufacturing when it comes to Know-what and Know-when tasks, for instance, detection or prediction of events such as quality issues [96] (product level), machine failure [97] (machine level), or dynamic process modelling [98] (process level).

Markov chains [95], on the other hand, are a type of probabilistic model that describe a sequence of possible events in which the probability of each event depends only on the state attained in the previous event. Markov chains can be utilized in manufacturing to model and analyze the behavior of systems (Know-why, system level) such as production lines [99] or supply chains [100]. In addition, the capability of predicting future states with Markov chains enables applications predicting joint maintenance in production systems [101] (Know-when, system level) and optimizing production scheduling [102] (Know-how, system level).

Kernel-based methods: As depicted in Figure 8, kernel-based methods utilize a defined kernel function to map input data into a high-dimensional implicit feature space [103]. Instead of computing the targeted coordinates, kernel-based methods normally compute the inner product between a pair of data points in the feature space. However, kernel-based methods have low efficiency, especially with respect to large-scale input data. Due to the promising capability of kernel-based methods in classification and regression, they can be utilized in the Know-what and Know-when stages in manufacturing, such as defect detection [104] (Know-what, product level), quality prediction [105] (Know-when, product level), and wear prediction in machinery [106] (Know-when, machine level). There are different types of kernel-based methods in supervised learning, such as SVM [107] and Kernel–Fisher discriminant analysis (KFD) [108].

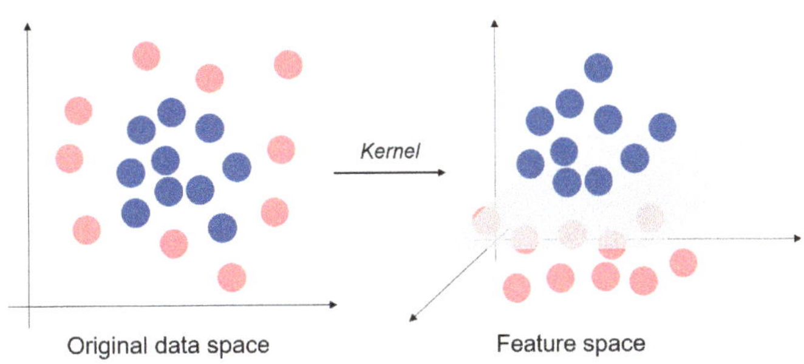

Figure 8. The principle of kernel-based methods. Using a kernel, the linearly inseparable input data are transformed to another feature space in which they become linearly separable.

Neural-network-based methods: Inspired by biological neurons and their ability to communicate with other connected cells, neural network-based methods employ artificial neurons. A typical neural network, such as ANNs, consists of an input layer, hidden layer, and output layer, as illustrated in Figure 9. Common ANNs types include CNNs [109], RNNs [110], and Deep Belief Network (DBN) [111].

Thanks to their powerful feature extraction capability when using matrix-like data, CNNs are widely used for image processing. In terms of possible applications in manufacturing, CNNs can be used in the Know-what stage to perform image-based quality control [112] (Know-what, product level) or image-based process monitoring [113] (Know-what, process level). In addition, by converting time series data from sensors to 2D images [114], CNNs can be used to detect and diagnosis machine failure as well.

RNNs are typically used to process sequential input data such as time series data or sequential images. Therefore, in terms of possible applications in manufacturing, RNNs are well-suited to the Know-when stage for analyzing sensor data or live images from machines, processes, or production systems. For instance, RNNs can enable the real-time performance prediction, such as the remaining useful life of machinery [115] (Know-when, machine level), process behavior prediction [116] (Know-when, process level), or the prediction of production indicators for real-time production scheduling [117] (Know-when, system level).

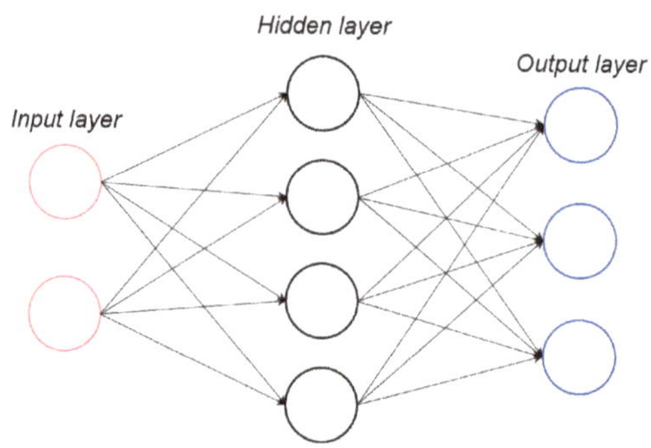

Figure 9. The scheme of an ANN, which normally consists of an input layer, hidden layer and output layer.

The typical supervised learning approaches applied in manufacturing are summarized in Table A1.

4.2. Unsupervised Learning Methods

Unsupervised learning algorithms aim to identify patterns in data sets containing data points that are not labeled. Unsupervised learning eliminates the need for labeled data and manual feature engineering, allowing for more general, flexible, and automated ML methods. As a result, unsupervised learning methods draw patterns and highlight areas of interest, revealing critical insight into the production process and opportunities for improvement. This can allow manufacturers to make better production-focused decisions, driving their business forward. The primary goal of unsupervised learning is to identify hidden and interesting patterns in unlabeled data. In terms of principles, there are three types of unsupervised tasks: Dimension Reduction [118,119], Clustering [120], and Association Rules [121]. Many aspects of unsupervised learning can be beneficial in manufacturing applications. First, clustering algorithms can be used to identify outliers in manufacturing data. Another aspect is to handle high dimensional data, e.g., for manufacturing cost estimation, quality improvement methodologies, production process optimization, better understanding of the customer's data, etc. Usually, a dimensional reduction support algorithm is required to handle data complexity and high dimensionality. Finally, it is challenging to perform root cause analysis in large-scale process execution due to the complexity of services in data centers. Association rule-based learning can be employed to conduct root cause analysis and to identify correlations between variables in a dataset.

Dimensional reduction is the process of converting data from a high-dimensional space to a low-dimensional space while preserving important characteristics of the original data.

Principal component analysis (PCA) [118]: The main idea of PCA is to minimize the number of interrelated variables in a dataset while preserving as much of the dataset's inherent variance as possible. A new set of variables, called principal components (PCs), are generated; these are uncorrelated and sorted such that the first few variables retain the majority of the variance included in all of the original variables. A pictorial representation of PCA is shown in Figure 10.

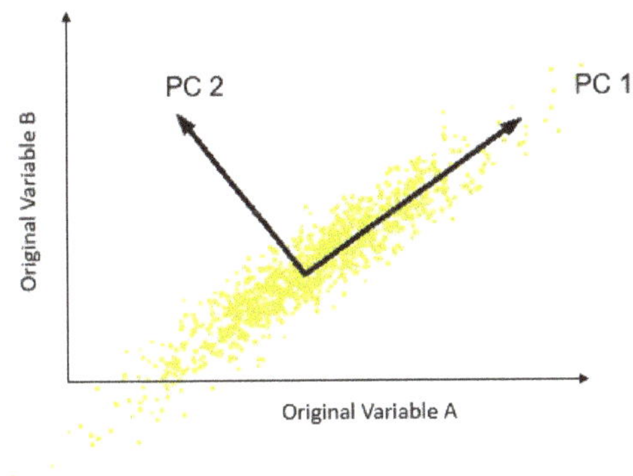

Figure 10. Principal Component Analysis.

The five steps below can be used to condense the entire process of extracting principal components from a raw dataset.

1. Say we wish to condense d features in our data matrix X to k features. The first step is to standardize the input data:
$$z = x - \mu/\sigma$$
where μ is the mean and σ is the standard deviation.
2. Next, it is necessary to find the covariance matrix of the standardized input data. The covariance of variables X and Y can be written as follows:
$$\text{cov}(X,Y) = \frac{1}{n-1}\sum_{i=1}^{n}(Xi - \bar{x})(Yi - \bar{y}). \tag{1}$$
3. The third steps is to find all of the eigenvalues and eigenvectors of the covariance matrix:
$$A\vec{v} = \lambda\vec{v} \tag{2}$$
$$A\vec{v} - \lambda\vec{v} = 0 \tag{3}$$
$$\vec{v}(A - \lambda I) = 0. \tag{4}$$
4. Then, the eigenvector corresponding to the largest eigenvalue is the direction with the maximum variance, the eigenvector corresponding to second-largest eigenvalue is the direction with the second maximum variance, etc.
5. To obtain k features, it is necessary to multiply the original data matrix by the matrix of eigenvectors corresponding to the k largest eigenvalues.

PCA is particularly useful for processing manufacturing data, which typically have a large number of variables, making it difficult to identify patterns and trends. A variety of applications of PCA in manufacturing are listed below:

1. Quality improvement (Know-why, product level): by analyzing the variations of a product's features, PCA can be used to identify the causes of product defects [122].
2. Machine monitoring (Know-why, machine level): by analyzing sensor data from a machine, PCA can be used to detect incipient patterns in the data that indicate potential issues with the machinery, such as wear and tear [123].
3. Process optimization (Know-why, process level): by analyzing variations in the process data, PCA can be used to identify the most important factors that affect the process, allowing the manufacturer to optimize the process and thereby reduce costs [124].

Autoencoder (AE) [119] is another popular method for reducing the dimensionality of high-dimensional data. AE alone does not perform classification; instead, it provides a compressed feature representation of high-dimensional data. The typical structure of AE consists of an input layer, one hidden or encoding layer, one reconstruction or decoding layer, and an output layer. The training strategy of AE includes encoding input data into a latent representation that can reconstruct the input. To learn a compressed feature representation of input data, AE tries to reduce the reconstruction error, that is, to minimize the difference between the input and output data. An illustration of AE is shown in Figure 11.

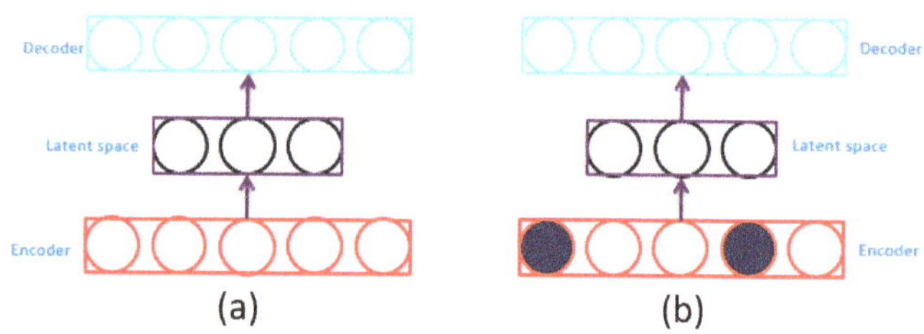

Figure 11. A pictorial representation of (**a**) an Autoencoder and (**b**) a Denoising Autoencoder. An autoencoder is trained to reconstruct its input, while a denoising autoencoder is trained to reconstruct a "clean" version of its input from a corrupted or "noisy" version of the input.

There are different types autoencoders that can be used for high-dimensional data. *Stacked Autoencoder* (SAE) [119] is built by stacking multiple layers of AEs in such a way that the output of one layer serves as the input of the subsequent layer. *Denoising autoencoder* (DAE) [125] is a variant of AE that has a similar structure except for the input data. In DAE, the input is corrupted by adding noise to it; however, the output is the original input signal without noise. Therefore, unlike AE, DAE has the ability to recover the original input from a noisy input signal. *Convolutional autoencoder* [126] is another interesting variant of AE, employing convolutional layers to encode and decode high-dimensional data.

AEs can be used for a variety of applications in manufacturing, such as:

1. Anomaly detection (Know-what): an AE can be trained to reconstruct normal data and detect abnormal data by measuring the reconstruction error, which allows the manufacturer to detect and address issues such as product defects [124] and machinery failure [127].
2. Feature selection (Know-why): an AE can be used to identify the most important features in the data and remove the noise and irrelevant information, which can be used for diagnosis of product defects or to detect events of interests [128].
3. Dimensionality reduction: an AE can be used to reduce the dimensionality of large and complex datasets, making it easier to identify patterns and trends [129].

Furthermore, AEs can be used in conjunction with other techniques, such as clustering or classification, to improve the accuracy of prediction and enhance the interpretability of the results [130]. Additionally, AEs can be used for data visualization. By reducing the dimensionality of the data, AEs allow high-dimensional data to be visualized clearly and interpretably [129] in a way that can be easily understood by non-technical stakeholders.

Clustering: The objective of clustering is to divide the set of datapoints into a number of groups, ensuring that the datapoints within each group are similar to one another and different from the datapoints in the other groups. Clustering methods are powerful tools, allowing manufacturers examine large and complex datasets and gain meaningful insights. There are different clustering methods available, each with their own strengths and weaknesses, and the choice of method depends on the characteristics of the data and the problem to be solved. Among the widely used clustering methods are *Centroid-based Clustering* [120], *Density-based Clustering* [131], *Distribution-based Clustering* [132], and *Hierarchical Clustering* [133]. Clustering algorithms have a wide range of applications in manufacturing. For instance, clustering can be used to group manufactured inventory parts according to different features [134] (Know-what). The obtained clusters can be used as a guideline for warehouse space optimization [135]. Clustering can be used for anomaly detection [136] (Know-what) and process optimization [137] (Know-how), and can be used in conjunction with other techniques to improve the interpretability of results.

Association rule-based learning [121]: Association rule-based learning is an unsupervised data-mining technique that finds important interactions among variables in a dataset. It is capable of identifying hidden correlations in datasets by measuring degrees of similarity. Hence, association rule-based learning is suitable in the Know-why stage in manufacturing. For instance, association rule-based learning can be utilized to accurately depict the relationship between quantifiable shop floor indicators and appropriate causes of action under various conditions of machine utilization (Know-why, system level), which can be used to establish an appropriate management strategy [138].

4.3. Semi-Supervised Learning Methods

Unsupervised learning methods do not have any input guidance during training, which reduces labeling costs; however, their performance is normally less accurate. Therefore, semi-supervised learning methods can be used to take advantage of the accuracy achieved by supervised learning while limiting costs thanks to the reduction in labeling effort. Therefore, researchers have turned to data augmentation [139,140] to enlarge dataset, with the inputs and labels generated massively based on the existing dataset in a controlled way while incurring no extra cost in the labeling phase. Taking an image with its label as an example, it can be enriched by basic transformations such as rotation, translation, flipping, noise injection, etc. It can be enriched by adversarial data augmentation, such as by generating synthetic dataset using generative models, e.g., Generative Adversarial Network (GAN) [141] and Variational AutoEncoder (VAE) [142], thereby obtaining new images for training ML models at low cost. However, the improvements obtainable with data augmentation are limited, and more real data are better than more synthetic data [143]. Therefore, increasing attention is being paid to the combination of supervised learning and unsupervised learning, namely, semi-supervised learning, in which both unlabeled data and labeled data are leveraged during training.

Semi-supervised learning methods can be generally divided into two groups: data augmentation-based methods and semi-supervised mechanism-based methods. An overview of semi-supervised methods is provided in Figure 12.

Data augmentation: through data augmentation, labeled data can be enlarged and augmented by adding model predictions of newly unlabeled data with high confidence as pseudo-labels, as shown in Figure 13. However, the model continues to be run in a fully supervised manner. In addition, the quality of the pseudo-labels can highly affect model performance, and incorrect pseudo-labels with high confidence are inevitable due to their nature. To improve the quality of pseudo-labels, there are hybrid methods combining pseudo-labels and consistency regularization, such as MixMatch [144] and FixMatch [145]. Nevertheless, data augmentation-based methods are simple, and there is no need to carefully design the loss. Therefore, data augmentation-based methods can be potentially useful for non-experts in manufacturing for enlarging labeled dataset when it is easy to collect massive amounts of unlabeled data.

Semi-supervised mechanisms: by contrast, semi-supervised mechanism-based methods are more focused on the mechanism of utilizing both labeled data and unlabeled data. The principle of semi-supervised mechanisms is illustrated in Figure 14, where both labeled data and unlabeled data can be model inputs while their losses are calculated in a different way. Semi-supervised mechanism-based methods can be further categorized into consistency-based methods, graph-based methods, and generative-based methods.

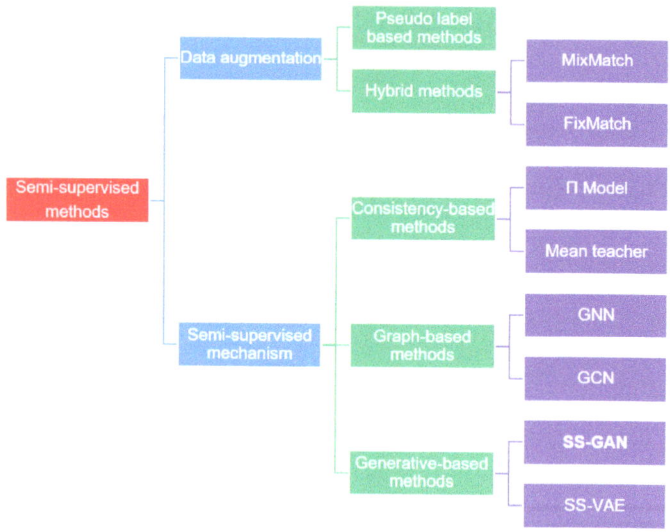

Figure 12. Overview of semi-supervised methods.

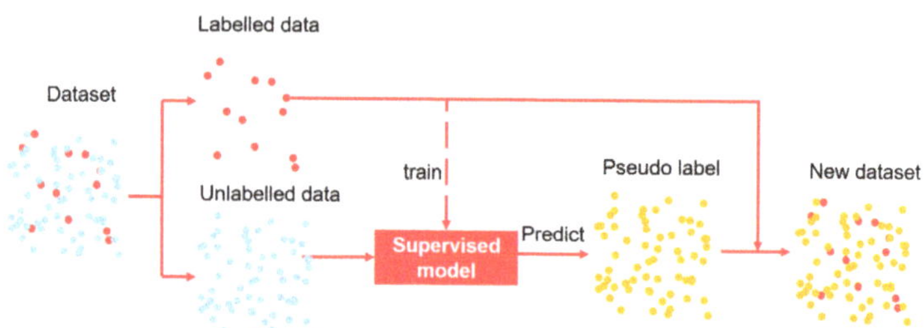

Figure 13. Data augmentation-based methods.

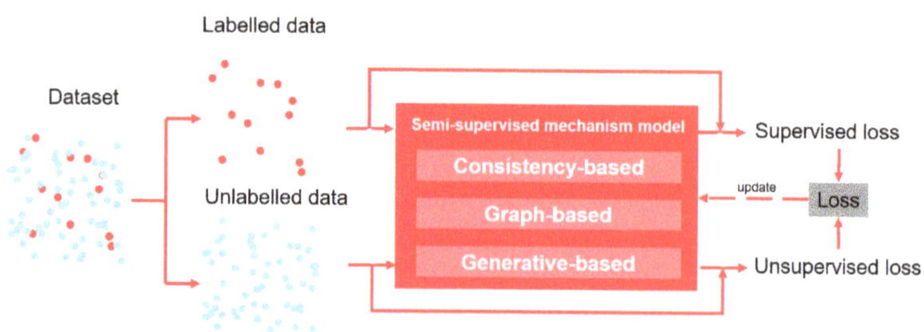

Figure 14. Semi-supervised mechanism-based methods.

Consistency-based methods take advantage of the consistency of model outputs after perturbations [146]; therefore, consistency regularization can be applied for unlabeled data. Consistency constraint can be either imposed between the predictions from perturbed inputs from the same sample, for instance, the π model [147], or between the predictions from two models with the same architecture, such as MeanTeacher [148]. Thanks to the perturbations in consistency-based methods, model generalization can be enhanced [149]. In terms of applications in manufacturing, depending on the output values consistency-based methods can be used in the Know-what and Know-when stages. For instance, consistency-based methods can be utilized in quality monitoring based on images (Know-what, product level).

Graph-based methods aim to establish a graph from a dataset by denoting each data point as a node, with the edge connecting two nodes representing the similarity between them. Label propagation is then performed on the established graph, with the information from labeled data used to infer the labels of the unlabeled data. Graph-based methods result in the connected nodes being closer in the feature space, while disconnected nodes repel each other. Therefore, graph-based methods can be used to address the problem of poor class separation due to intra-class variations and inter-class similarities [18]. Consequently, graph-based methods can be potentially useful for defect classification [18] (Know-what, product level) or machine health state monitoring [150] (Know-what, machine level) where there are problems with insufficient label information or poor class separation. However, it should be noted that graph-based methods are normally transductive methods, meaning that the constructed graph is only valid for the trained data and rebuilding the graph is necessary when it comes to new data. Typical examples of graph-based methods include Graph Neural Networks (GNNs) [151] and Graph Convolution Networks (GCNs) [152].

The main point of generative-based methods is to learn patterns from a dataset and to model data distributions, allowing the model to be used to generate new samples. Then during training, the model can be updated using the combination of the supervised loss (for existing data with labels) and unsupervised loss (for synthetic data). An inherent advantage of generative-based methods is that the labeled data can be enriched by a trained model which has learned the data distribution. Therefore, generative-based methods are well-suited for situations where it is difficult to collect labeled data, such as process fault detection [153] (Know-what, process level) and anomaly detection in machinery [154] (Know-what, machine level). Examples include the semi-supervised GAN series (SS-GANs), such as Categorical Generative Adversarial Network (CatGAN) [155], Improved GAN [156], and semi-supervised VAEs (SS-VAEs) [157].

Table A3 lists semi-supervised applications in manufacturing taken from the selected documents in Section 2.2.

4.4. Reinforcement Learning Methods

Reinforcement Learning (RL) algorithms consist of two elements, namely, an ***agent*** acting within an ***environment*** (see Figure 15). The agent is acting, and is therefore subject to the desired learning process by directly interacting with and manipulating the environment. Based on [158], the procedure of a learning cycle is as follows: first, the agent is presented with an observation of the environment state $s_t \in \mathbb{S}$; then, based on this observation (along with internal decision making), the selection of an action $a_t \in \mathbb{A}$. $\mathbb{S}$ refers to the state space, that is, the set of possible observations that could occur in the environment. The observation has to provide sufficient information on the current environment or system state in order for the agent to select actions in an ideal way to solve the control problem. For selecting the action, $\mathbb{A}$ refers to the action space, that is, the set of possible actions chosen by the agent. After a_t is performed (in a given state s_t), the environment moves to the resulting state s_{t+1} and the agent receives a reward r_{t+1}. Then, the reinforcement learning cycle continues to iterate as shown in Figure 15. The agent aims to maximize the (discounted) long-term cumulative reward by improving the selection of actions towards an optimum. In other words, the RL agent wants to learn an optimal control policy for the environment.

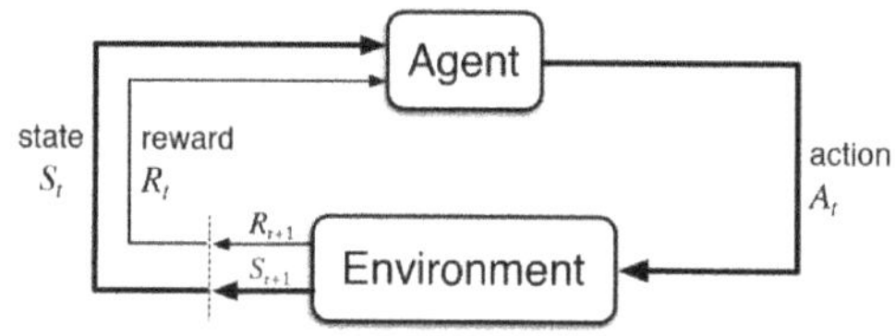

Figure 15. Overview of the Reinforcement Learning approach based on [158].

In general, RL approaches can be split into model-based, i.e., the agent has an internal model of how the environment works, and model-free. The latter is most common thanks to the advent of deep learning, and simplifies application, as feature selection can be applied. Model-free approaches themselves can be divided into short value-based or policy-based approaches by their approach to storing state-action value pairs, which are used to select the action for optimal value return; the latter directly optimize the action selection policy. In contrast to the other machine learning techniques, RL does not require large dataset, only a clearly specified environment. Typically, an RL agent is trained on a simulation or digital twin model [159]; after successful training, it can be implemented on the Know-how level for its original purpose. Otherwise, the agent starts with random non-optimal actions, leading to undesired system behavior.

Considering the aim of achieving the Know-how level for autonomous control in processes, machines, or systems, RL is extremely important for applications in future production. In addition, multi-agent RL is becoming of interest to the research community [33], and can even be applied for controlling products [160]. However, RL remains under-exploited in the industrial area, especially in respect to other machine learning techniques [161].

As of now, applied approaches can be summarized as shown in Table A4. Note that the applications reviewed here are implemented in a simulation or digital twin [159], and features are manually crafted from raw data.

5. Challenges and Future Directions

A large number of ML use cases have shown the great potential for addressing complex manufacturing problems, from knowing what is happening to knowing how employ self-adapting or self-optimizing systems. The data-driven mechanisms in ML enable broader applications in different fields as well as at different levels, from individual products to whole systems. However, in spite of the great potential and advantages offered by ML and numerous off-the-shelf ML models, there are critical challenges to overcome before the successful application of ML in manufacturing can be realized. The following demonstrate typical challenges that manufacturing industries might confront during the application and deployment of ML-based solutions, along with corresponding future directions for tackling these challenges (RQ3).

- *Lack of data*. Preparing the data used for ML is not a simple task, as the scale and the quality of data can greatly affect the performance of ML models. The most common challenge involves preparing a large amount of organized input data, and ensuring high-quality labels if labels are needed. Despite manufacturing data becoming increasingly more accessible due to the development of sensors and the Internet of Things, gathering meaningful data is time-consuming and costly in many cases, for example, fault detection and RUL prediction. This issue might be alleviated by the Synthetic Minority Over-sampling Technique (SMOTE) [162]. However, SMOTE cannot capture complex representative data, as it often relies on interpolation [163]. Data augmentation [139,164] or transfer learning [165] may address this problem. The aim of data augmentation is to enlarge dataset by means of transforming data [139], by transforming both data and labels, as with MixUp [166], or by generating synthetic data using generative models [167,168]. On the contrary, instead of focusing on expanding data, transfer learning aims to leverage knowledge from similar external datasets. A typically used method in transfer learning is parameter transfer, where a pretrained model from a similar dataset is employed for initialization [165]. Another situation involving lack of data is that certain data cannot be shared due to data privacy and security issues. In confronting this problem, Federated Learning (FL) [169] might be a potential opportunity to enable model training across multiple decentralized devices while holding local data privately.

- *Limited computing resources*. The high performance of ML models always comes with high computational complexity. In particular, obtaining high accuracy with a neural network requires on millions or even billions of parameters [170]. However, limited computing resources in industries makes it a challenge to deploy heavy ML models in real-time industrial environments. Possible approaches include model compression via pruning and sharing of model parameters [171] and knowledge distillation [172]. Parameter pruning aims to reduce the number of model parameters by removing redundant parameters without any effect on model performance. By contrast, seeking the same goal, knowledge distillation focuses on distilling knowledge from a cumbersome neural network to a lightweight network to allow it to be deployed more easily with limited computing resources.
- *Changing circumstances*. Most ML applications in manufacturing focus only on model development and verification in off-line environments. However, when deploying these models in running production, their performance may be degraded due to changing circumstances, leading to changes in data distribution, that is, drift [173,174]. Therefore, manual model adjustment over time, which is time-consuming, is usually unavoidable [175]. However, this could be addressed in the future by automatic model adaption [174], in which data drifts are automatically detected and handled with less resources.
- *Interpretability of results*. Many expectations have been placed on ML to overcome all types of problems without the need for prior knowledge. In particular, ML models are expected to directly learn higher level knowledge such as Know-when and Know-how, which is difficult for human beings to obtain in manufacturing. However, without the foundations of early-stage knowledge and an understanding of the data, the results inferred from big data by black-box ML models are meaningless and unreliable. For instance, predictions blindly obtained from all data, including both relevant and irrelevant data, might even degrade performance due to the GIGO (garbage in, garbage out) phenomenon [176]. To overcome this problem, future directions within ML development might include incorporating physical models into ML models [177] or obtaining Four-know knowledge successively.
- *Uncertainty of results*. Related to the challenge of interpretability is the challenge of uncertain results. The success of manufacturing depends heavily on the quality of the resulting products. As every manufacturing process has a degree of variability, almost all industrial manufacturers use statistical process control (SPC) to ensure a stable and defined quality of products [178]. A central element of statistical process control is the determination and handling of statistical uncertainty. The uncertainty of ML results often cannot be quantified reliably and efficiently, even with today's state-of-the-art [179–181]. Furthermore, model complexity and severe non-linearity in ML can hinder the evaluation of uncertainty [182]. Although there are promising approaches, e.g., Gaussian mixture models for NN [183,184] and Probabilistic Neural Network (PNN) [184], or the use of Baysian Networks [180], there are several limitations limiting potential applications, such as high computational cost and simplified assumptions [184]. Therefore, future research needs to make progress on the general theory of integrating uncertainty into ML methods to allow manufacturing in order to ensure high quality and stability in production.

To summarize, while ML is a fairly open tool which can be used to handle a variety of problems in manufacturing, it is necessary to have an understanding of the hidden challenges in ML application in order to provide more realistic and robust outcomes. For instance, early in ML application in manufacturing, one might face the problem of lacking data. During the deployment of ML-based solutions, one might confront challenges around integrating the solution into the industrial environment. After deployment, one might encounter the challenge of evaluating ML results on product and process in terms of interpretability and uncertainty. The future directions pointed out in this review can help to address the above-mentioned challenges and ensure reliable improvements in manufacturing contexts.

6. Conclusions

It is fully recognized that ML is playing an increasingly critical role in the digitization of manufacturing industries towards Industry 4.0, leading to improved quality, productivity, and efficiency. This review has paper aimed to address the issue of how ML can improve manufacturing, posing three research questions related to the above issue in the introduction. To address these research questions, we carried out a literature review assessing the state-of-the-art based on 1348 published scientific articles.

To answer RQ1, we first introduced the concepts of the 'Four-Know' (Know-what, Know-why, Know-when, Know-how) and 'Four-Level' (Product, Process, Machine, System) categories to help formulate ML tasks in manufacturing. By mapping ML use cases into the Four-Know and Four-Level matrix, we provide an understanding of typical ML use cases and their potential benefits for improving manufacturing. To further support RQ1, the identified ML studies were classified using the 'Four-Know' and 'Four-Level' perspective to provide an overview of ML publications in manufacturing. The results showed that current ML applications are mainly focused on the product level, in particular in terms of Know-what and Know-when. In addition, considerable growth in Know-how was observed at the process and system levels, which might be correlated to higher demand for adaptability to changes on these levels.

To fill the gap between academic research and manufacturing industries, we provided an actionable pipeline for the implementation of ML solutions by production engineers from ideation through to deployment, thereby answering RQ2. To further explain the 'model training' step, which is the core stage in the pipeline, a holistic review of ML methods was provided, including supervised, semi-supervised, unsupervised, and reinforcement learning methods along with their typical applications in manufacturing. We hope that this can provide support in method selection for decision-makers considering ML solutions.

Finally, to answer RQ3, we uncovered the current challenges that manufacturing industry is likely to encounter during application and deployment, and provided possible future directions for tackling these challenges as possible developments for ensuring more reliable and robust outcomes in manufacturing.

Author Contributions: Conceptualization, T.C., O.J.J., M.C.M., V.S. and G.F.; methodology, T.C. and S.S.; formal analysis, T.C. and S.S.; writing—original draft preparation, T.C., V.S., S.S., M.C.M., O.J.J. and F.S.; writing—review and editing, T.C., V.S., M.C.M., S.S., O.J.J., M.C., G.F., G.T., J.J.A.M. and F.S.; supervision, M.C., G.F., G.T., J.J.A.M. and F.S.; funding acquisition, G.T. All authors have read and agreed to the published version of the manuscript.

Funding: This research was funded by a European Training Network supported by Horizon 2020, grant number 814225.

Institutional Review Board Statement: Not applicable.

Informed Consent Statement: Not applicable.

Data Availability Statement: Not applicable.

Acknowledgments: This research work was undertaken in the context of the DIGIMAN4.0 project ("DIGItal MANufacturing Technologies for Zero-defect Industry 4.0 Production", https://www.digiman4-0.mek.dtu.dk/, accessed on 1 January 2023). DIGIMAN4.0 is a European Training Network supported by Horizon 2020, the EU Framework Programme for Research and Innovation (Project ID: 814225).

Conflicts of Interest: The authors declare no conflict of interest.

Appendix A

Table A1. Categories of supervised learning applications.

Ref.	Year	Level	Know-What	Know-Why	Know-When	Know-How	Data Type	Method Type	Case	Field
[104]	2018	Product	✓				Image	Kernel	Defect detection	Metallic powder bed fusion
[185]	2018	Product	✓				Tabular	Kernel	Product monitoring	Metal frame process in mobile device manufacturing
[186]	2020	Process	✓	✓	✓		Time series, Image	Kernel	Temperature prediction and potential anomaly detection	Additive manufacturing
[19]	2022	Product, Process	✓	✓			Tabular	Kernel	Fault detection and classification	Semiconductor Etch Equipment
[105]	2022	Product			✓		Tabular	Kernel	Quality prediction	Additive manufacturing
[106]	2022	Machine			✓		Time series	Kernel	Wear prediction	Metal forming
[187]	2022	System		✓	✓		Time series	Kernel	Content prediction	Steel making
[114]	2018	Machine	✓	✓			Time series (Image)	NN	Fault diagnosis	Motor bearning and pump
[188]	2020	System		✓	✓		Image	NN	Cost estimation	
[112]	2020	Product	✓				Image	NN	Defect detection	Battery manufacturing
[189]	2022	Product	✓				Time series	NN	Quality assurance	Fused deposition modeling
[190]	2022	Process	✓				Time series	NN	Process optimization	Wire arc additive manufacturing
[191]	2022	Process			✓	✓	Tabular	NN	Parameter optimization	Laser powder bed fusion
[192]	2022	Process	✓				Image	NN	Object detection	Robotic grasp
[193]	2022	Product	✓				Image	NN, kernel	Defect detection	Roller manufacturing
[194]	2022	Machine			✓		Time series (Image)	NN, kernel	Tool condition monitoring	Machining
[195]	2019	Product			✓		Time series	Tree	Material removal prediction	Robotic grinding
[196]	2022	Product			✓		Image (Tabular)	Tree	Porosity prediction	Powder-bed additive manufacturing
[197]	2019	Product	✓				Image	Probabilistic	Online quality inspection	Powder-bed additive manufacturing
[198]	2018	System				✓	Tabular	Hybrid	Scheduling	Flexible Manufacturing Systems (FMSs)

Table A2. Categories of unsupervised learning applications.

Ref.	Year	Level	Know-What	Know-Why	Know-When	Know-How	Data Type	Method Type	Case	Field
[120]	2021	Machine	✓				Time series	Clustering	Tool Condition clustering	Autonomous manufacturing
[131]	2021	Machine	✓				Time series	Clustering	Tool health monitoring	Machine tool health Monitoring
[132]	2019	Machine	✓	✓			Time series	Clustering	Defect Identification	Manufacturing systems
[199]	2021	Process	✓				Tabular, Time series	Clustering	Condition monitoring	Manufacturing Condition monitoring
[133]	2020	System	✓				Time series	Clustering	Condition monitoring	Manufacturing Condition monitoring
[125]	2018	Product	✓				Image, Text	Autoencoder	Defect Identification	Fabric industry
[119]	2019	Product	✓				Image	Autoencoder	Defect Identification	Automatic Optical Inspection
[200]	2021	Product	✓				Image	Autoencoder	Defect Identification	Printed circuit board manufacturing
[201]	2022	Machine	✓		✓		Tabular, Time series	Autoencoder	Anomaly detection	Steel rolling Process
[126]	2022	Process	✓		✓		Image	Autoencoder	Anomaly detection	Industrial Anomaly detection
[126]	2022	Process	✓		✓		Image, Text	Autoencoder	Anomaly detection	Semi conductor manufacturing
[202]	2022	Machine			✓		Time series	PCA	Predictive maintenance	Fan-motor system
[118]	2022	Machine	✓	✓			Time series	PCA	Anomaly detection	Programmable logic controllers
[121]	2015	Process	✓				Tabular	Association rule	Predictive maintenance	Wooden door manufacturing

Table A3. Categories of semi-supervised learning applications.

Ref.	Year	Level	Know-What	Know-Why	Know-When	Know-How	Data Type	Method Type	Case	Field
[203]	2020	Product	✓				Image	Data augmentation	Quality control	Automated Surface Inspection
[204]	2021	Process	✓				Image	Data augmentation	Measurement in process	Positioning of welding seams
[205]	2019	System			✓		Tabular	Data augmentation	Energy consumption modelling	Steel industry
[206]	2020	Product, System			✓		Time series	Data augmentation	Quality prediction	Continuous-flow manufacturing.
[207]	2021	Machine	✓				Time series	Consistency-based	Predictive quality control	Semiconductor manufacturing
[149]	2020	Product	✓				Image	Consistency-based	Quality monitoring	Metal additive manufacturing
[18]	2021	Product	✓				Image	Graph-based	Quality control	Automated Surface Inspection
[150]	2022	Machine	✓	✓			Time series	Graph-based	Machine health state diagnosis	Manipulator
[208]	2022	Machine		✓	✓		Tabular	Graph-based	Predict tool tip dynamics	Machine tool
[209]	2021	Product			✓		Image	Generative-based	Assessing manufacturability of cellular structures	Direct metal laser sintering process
[210]	2019	Product	✓				Time series	Generative-based	Quality inferred from process	laser powder-bed fusion
[211]	2020	Product	✓				Image	Generative-based	Quality diagnosis	Wafer fabrication
[212]	2021	Product	✓				Image	Generative-based	Quality control	Automated Surface Inspection
[213]	2020	Machine			✓		Time series	Generative-based	Remaining useful life prognostics	Turbofan engine and rolling bearing
[214]	2021	Machine	✓	✓			Tabular	Generative-based	Machine condition monitoring	Vacuum system in styrene petrochemical plant
[153]	2021	Machine	✓	✓			Time series	Generative-based	Anomaly detection for predictive maintenance	Press machine
[154]	2022	Process	✓				Time series (image)	Generative-based	Process fault detection	Die casting process

Table A4. Categories of reinforcement learning applications.

Ref.	Year	Level	Know-What	Know-Why	Know-When	Know-How	Data Type	Method Type	Case	Field
[32]	2021	Process	✓			✓	Tabular	Value-based	Quality control	Statistical Process Control
[215]	2022	System			✓	✓	Tabular	Value-based	Scheduling	Semiconductor fab
[216]	2021	System				✓	Tabular	Value-based	Throughput control	Flow shop
[217]	2021	Machine				✓	Tabular	Value-based	Scheduling & Maintenance	Multi-state single machine
[34]	2020	System				✓	Tabular	Value-based	Quality Control & Maintenance	Production system
[218]	2022	System				✓	Tabular	Value-based	Lead time management	Flow shop
[219]	2020	Process				✓	Tabular	Value-based	Robotic arm control	Soft fabric manufacturing
[220]	2021	System			✓	✓	Tabular	Value-based	Layout planning	Greenfield factories
[221]	2020	Machine				✓	Tabular	Value-based	Maintenance scheduling	Preventive maintenance
[33]	2022	Machine			✓	✓	Tabular	Policy-based	Maintenance scheduling	Parallel machines
[222]	2021	Process				✓	Tabular	Policy-based	Improving efficiency	Automated product disassembly
[223]	2021	System				✓	Tabular	Policy-based	Dispatching	Job shop
[224]	2022	System				✓	Tabular	Policy-based	Scheduling & maintenance	Semiconductor fab
[225]	2022	System				✓	Tabular	Policy-based	Yield optimization	Multi-agent RL
[57]	2022	System				✓	Tabular	Policy-based	Human Worker Control	Flow shop
[59]	2022	System				✓	Tabular	Policy-based	Scheduling & dispatching	Disassembly job shop
[160]	2021	Product				✓	Tabular	Policy-based	Multi-agent production control	Job shop
[226]	2022	Process				✓	Tabular	Both	Parameter optimisation	Manufacturing processes
[227]	2019	Process				✓	Tabular	Both	Online parameter optimisation	Injection molding
[228]	2022	System				✓	Tabular	Both	scheduling	Matrix production system

References

1. Abele, E.; Reinhart, G. *Zukunft der Produktion: Herausforderungen, Forschungsfelder, Chancen*; Hanser: München, Germany, 2011.
2. Zizic, M.C.; Mladineo, M.; Gjeldum, N.; Celent, L. From industry 4.0 towards industry 5.0: A review and analysis of paradigm shift for the people, organization and technology. *Energies* **2022**, *15*, 5221. [CrossRef]
3. Huang, S.; Wang, B.; Li, X.; Zheng, P.; Mourtzis, D.; Wang, L. Industry 5.0 and Society 5.0—Comparison, complementation and co-evolution. *J. Manuf. Syst.* **2022**, *64*, 424–428. [CrossRef]
4. Vukovic, M.; Mazzei, D.; Chessa, S.; Fantoni, G. Digital Twins in Industrial IoT: A survey of the state of the art and of relevant standards. In Proceedings of the 2021 IEEE International Conference on Communications Workshops (ICC Workshops), Montreal, QC, Canada, 14–23 June 2021. [CrossRef]
5. Mourtzis, D.; Fotia, S.; Boli, N.; Vlachou, E. Modelling and quantification of industry 4.0 manufacturing complexity based on information theory: A robotics case study. *Int. J. Prod. Res.* **2019**, *57*, 6908–6921. [CrossRef]
6. Galin, R.; Meshcheryakov, R.; Kamesheva, S.; Samoshina, A. Cobots and the benefits of their implementation in intelligent manufacturing. *IOP Conf. Ser. Mater. Sci. Eng.* **2020**, *862*, 032075. [CrossRef]
7. May, M.C.; Schmidt, S.; Kuhnle, A.; Stricker, N.; Lanza, G. Product Generation Module: Automated Production Planning for optimized workload and increased efficiency in Matrix Production Systems. *Procedia CIRP* **2020**, *96*, 45–50. [CrossRef]
8. Lu, Y. Industry 4.0: A survey on technologies, applications and open research issues. *J. Ind. Inf. Integr.* **2017**, *6*, 1–10. [CrossRef]
9. Miqueo, A.; Torralba, M.; Yagüe-Fabra, J.A. Lean manual assembly 4.0: A systematic review. *Appl. Sci.* **2020**, *10*, 8555. [CrossRef]
10. Wuest, T.; Weimer, D.; Irgens, C.; Thoben, K.D. Machine learning in manufacturing: Advantages, challenges, and applications. *Prod. Manuf. Res.* **2016**, *4*, 23–45. [CrossRef]
11. Rai, R.; Tiwari, M.K.; Ivanov, D.; Dolgui, A. Machine learning in manufacturing and industry 4.0 applications. *Int. J. Prod. Res.* **2021**, *59*, 4773–4778. [CrossRef]
12. Bertolini, M.; Mezzogori, D.; Neroni, M.; Zammori, F. Machine Learning for industrial applications: A comprehensive literature review. *Expert Syst. Appl.* **2021**, *175*, 114820. [CrossRef]
13. Wang, J.; Ma, Y.; Zhang, L.; Gao, R.X.; Wu, D. Deep learning for smart manufacturing : Methods and applications. *J. Manuf. Syst.* **2018**, *48*, 144–156. [CrossRef]
14. Dogan, A.; Birant, D. Machine learning and data mining in manufacturing. *Expert Syst. Appl.* **2021**, *166*, 114060. [CrossRef]
15. Alshangiti, M.; Sapkota, H.; Murukannaiah, P.K.; Liu, X.; Yu, Q. Why is developing machine learning applications challenging? a study on stack overflow posts. In Proceedings of the 2019 ACM/IEEE International Symposium on Empirical Software Engineering and Measurement (ESEM), Porto de Galinhas, Brazil, 19–20 September 2019; pp. 1–11.
16. Zeller, V.; Hocken, C.; Stich, V. Acatech Industrie 4.0 maturity index—A multidimensional maturity model. In Proceedings of the IFIP International Conference on Advances in Production Management Systems, Seoul, Republic of Korea, 26–30 August 2018; Springer: Cham, Switzerland, 2018; pp. 105–113.
17. Yang, L.; Fan, J.; Huo, B.; Li, E.; Liu, Y. A nondestructive automatic defect detection method with pixelwise segmentation. *Knowl.-Based Syst.* **2022**, *242*, 108338. [CrossRef]
18. Wang, Y.; Gao, L.; Gao, Y.; Li, X. A new graph-based semi-supervised method for surface defect classification. *Robot. Comput. Integr. Manuf.* **2021**, *68*, 102083. [CrossRef]
19. Kim, S.H.; Kim, C.Y.; Seol, D.H.; Choi, J.E.; Hong, S.J. Machine Learning-Based Process-Level Fault Detection and Part-Level Fault Classification in Semiconductor Etch Equipment. *IEEE Trans. Semicond. Manuf.* **2022**, *35*, 174–185. [CrossRef]
20. Peng, S.; Feng, Q.M. Reinforcement learning with Gaussian processes for condition-based maintenance. *Comput. Ind. Eng.* **2021**, *158*, 107321. [CrossRef]
21. Zheng, W.; Liu, Y.; Gao, Z.; Yang, J. Just-in-time semi-supervised soft sensor for quality prediction in industrial rubber mixers. *Chemom. Intell. Lab. Syst.* **2018**, *180*, 36–41. [CrossRef]
22. Kang, P.; Kim, D.; Cho, S. Semi-supervised support vector regression based on self-training with label uncertainty: An application to virtual metrology in semiconductor manufacturing. *Expert Syst. Appl.* **2016**, *51*, 85–106. [CrossRef]
23. Srivastava, A.K.; Patra, P.K.; Jha, R. AHSS applications in Industry 4.0: Determination of optimum processing parameters during coiling process through unsupervised machine learning approach. *Mater. Today Commun.* **2022**, *31*, 103625. [CrossRef]
24. Antomarioni, S.; Ciarapica, F.E.; Bevilacqua, M. Association rules and social network analysis for supporting failure mode effects and criticality analysis : Framework development and insights from an onshore platform. *Saf. Sci.* **2022**, *150*, 105711. [CrossRef]
25. Pan, R.; Li, X.; Chakrabarty, K. Semi-Supervised Root-Cause Analysis with Co-Training for Integrated Systems. In Proceedings of the 2022 IEEE 40th VLSI Test Symposium (VTS), San Diego, CA, USA, 25–27 April 2022. [CrossRef]
26. Chen, R.; Lu, Y.; Witherell, P.; Simpson, T.W.; Kumara, S.; Yang, H. Ontology-Driven Learning of Bayesian Network for Causal Inference and Quality Assurance in Additive Manufacturing. *IEEE Robot. Autom. Lett.* **2021**, *6*, 6032–6038. [CrossRef]
27. Sikder, S.; Mukherjee, I.; Panja, S.C. A synergistic Mahalanobis–Taguchi system and support vector regression based predictive multivariate manufacturing process quality control approach. *J. Manuf. Syst.* **2020**, *57*, 323–337. [CrossRef]
28. Cerquitelli, T.; Ventura, F.; Apiletti, D.; Baralis, E.; Macii, E.; Poncino, M. Enhancing manufacturing intelligence through an unsupervised data-driven methodology for cyclic industrial processes. *Expert Syst. Appl.* **2021**, *182*, 115269. [CrossRef]
29. Kolokas, N.; Vafeiadis, T.; Ioannidis, D.; Tzovaras, D. A generic fault prognostics algorithm for manufacturing industries using unsupervised machine learning classifiers. *Simul. Model. Pract. Theory* **2020**, *103*, 102109. [CrossRef]

30. Verstraete, D.; Droguett, E.; Modarres, M. A deep adversarial approach based on multisensor fusion for remaining useful life prognostics. In Proceedings of the 29th European Safety and Reliability Conference (ESREL 2019), Hannover, Germany, 22–26 September 2020; pp. 1072–1077. [CrossRef]
31. Wu, D.; Jennings, C.; Terpenny, J.; Gao, R.X.; Kumara, S. A Comparative Study on Machine Learning Algorithms for Smart Manufacturing: Tool Wear Prediction Using Random Forests. *J. Manuf. Sci. Eng. Trans. ASME* **2017**, *139*, 071018. [CrossRef]
32. Viharos, Z.J.; Jakab, R. Reinforcement Learning for Statistical Process Control in Manufacturing. *Meas. J. Int. Meas. Confed.* **2021**, *182*, 109616. [CrossRef]
33. Luis, M.; Rodríguez, R.; Kubler, S.; Giorgio, A.D.; Cordy, M.; Robert, J.; Le, Y. Multi-agent deep reinforcement learning based Predictive Maintenance on parallel machines. *Robot. Comput. Integr. Manuf.* **2022**, *78*, 102406.
34. Paraschos, P.D.; Koulinas, G.K.; Koulouriotis, D.E. Reinforcement learning for combined production-maintenance and quality control of a manufacturing system with deterioration failures. *J. Manuf. Syst.* **2020**, *56*, 470–483. [CrossRef]
35. Liu, Y.H.; Huang, H.P.; Lin, Y.S. Dynamic scheduling of flexible manufacturing system using support vector machines. In Proceedings of the 2005 IEEE Conference on Automation Science and Engineering, IEEE-CASE 2005, Edmonton, AB, Canada, 1–2 August 2005; Volume 2005, pp. 387–392. [CrossRef]
36. Zhou, G.; Chen, Z.; Zhang, C.; Chang, F. An adaptive ensemble deep forest based dynamic scheduling strategy for low carbon flexible job shop under recessive disturbance. *J. Clean. Prod.* **2022**, *337*, 130541. [CrossRef]
37. de la Rosa, F.L.; Gómez-Sirvent, J.L.; Sánchez-Reolid, R.; Morales, R.; Fernández-Caballero, A. Geometric transformation-based data augmentation on defect classification of segmented images of semiconductor materials using a ResNet50 convolutional neural network. *Expert Syst. Appl.* **2022**, *206*, 117731. [CrossRef]
38. Krahe, C.; Marinov, M.; Schmutz, T.; Hermann, Y.; Bonny, M.; May, M.; Lanza, G. AI based geometric similarity search supporting component reuse in engineering design. *Procedia CIRP* **2022**, *109*, 275–280. [CrossRef]
39. Onler, R.; Koca, A.S.; Kirim, B.; Soylemez, E. Multi-objective optimization of binder jet additive manufacturing of Co-Cr-Mo using machine learning. *Int. J. Adv. Manuf. Technol.* **2022**, *119*, 1091–1108. [CrossRef]
40. Jadidi, A.; Mi, Y.; Sikström, F.; Nilsen, M.; Ancona, A. Beam Offset Detection in Laser Stake Welding of Tee Joints Using Machine Learning and Spectrometer Measurements. *Sensors* **2022**, *22*, 3881. [CrossRef]
41. Sanchez, S.; Rengasamy, D.; Hyde, C.J.; Figueredo, G.P.; Rothwell, B. Machine learning to determine the main factors affecting creep rates in laser powder bed fusion. *J. Intell. Manuf.* **2021**, *32*, 2353–2373. [CrossRef]
42. Verma, S.; Misra, J.P.; Popli, D. Modeling of friction stir welding of aviation grade aluminium alloy using machine learning approaches. *Int. J. Model. Simul.* **2022**, *42*, 1–8. [CrossRef]
43. Gerling, A.; Ziekow, H.; Hess, A.; Schreier, U.; Seiffer, C.; Abdeslam, D.O. Comparison of algorithms for error prediction in manufacturing with automl and a cost-based metric. *J. Intell. Manuf.* **2022**, *33*, 555–573. [CrossRef]
44. Akbari, P.; Ogoke, F.; Kao, N.Y.; Meidani, K.; Yeh, C.Y.; Lee, W.; Farimani, A.B. MeltpoolNet: Melt pool characteristic prediction in Metal Additive Manufacturing using machine learning. *Addit. Manuf.* **2022**, *55*, 102817. [CrossRef]
45. Dittrich, M.A.; Uhlich, F.; Denkena, B. Self-optimizing tool path generation for 5-axis machining processes. *CIRP J. Manuf. Sci. Technol.* **2019**, *24*, 49–54. [CrossRef]
46. Xi, Z. Model predictive control of melt pool size for the laser powder bed fusion process under process uncertainty. *ASCE-ASME J. Risk Uncertain. Eng. Syst. Part B Mech. Eng.* **2022**, *8*, 011103. [CrossRef]
47. Li, X.; Liu, X.; Yue, C.; Liu, S.; Zhang, B.; Li, R.; Liang, S.Y.; Wang, L. A data-driven approach for tool wear recognition and quantitative prediction based on radar map feature fusion. *Measurement* **2021**, *185*, 110072. [CrossRef]
48. Xia, B.; Wang, K.; Xu, A.; Zeng, P.; Yang, N.; Li, B. Intelligent Fault Diagnosis for Bearings of Industrial Robot Joints Under Varying Working Conditions Based on Deep Adversarial Domain Adaptation. *IEEE Trans. Instrum. Meas.* **2022**, *71*, 1–13. [CrossRef]
49. May, M.C.; Neidhöfer, J.; Körner, T.; Schäfer, L.; Lanza, G. Applying Natural Language Processing in Manufacturing. *Procedia CIRP* **2022**, *115*, 184–189. [CrossRef]
50. Xu, X.; Li, X.; Ming, W.; Chen, M. A novel multi-scale CNN and attention mechanism method with multi-sensor signal for remaining useful life prediction. *Comput. Ind. Eng.* **2022**, *169*, 108204. [CrossRef]
51. Shah, M.; Vakharia, V.; Chaudhari, R.; Vora, J.; Pimenov, D.Y.; Giasin, K. Tool wear prediction in face milling of stainless steel using singular generative adversarial network and LSTM deep learning models. *Int. J. Adv. Manuf. Technol.* **2022**, *121*, 723–736. [CrossRef]
52. Verl, A.; Steinle, L. Adaptive compensation of the transmission errors in rack-and-pinion drives. *CIRP Ann.* **2022**, *71*, 345–348. [CrossRef]
53. Frigerio, N.; Cornaggia, C.F.; Matta, A. An adaptive policy for on-line Energy-Efficient Control of machine tools under throughput constraint. *J. Clean. Prod.* **2021**, *287*, 125367. [CrossRef]
54. Bozcan, I.; Korndorfer, C.; Madsen, M.W.; Kayacan, E. Score-Based Anomaly Detection for Smart Manufacturing Systems. *IEEE/ASME Trans. Mechatron.* **2022**, *27*, 5233–5242. [CrossRef]
55. Bokrantz, J.; Skoogh, A.; Nawcki, M.; Ito, A.; Hagstr, M.; Gandhi, K.; Bergsj, D. Improved root cause analysis supporting resilient production systems. *J. Manuf. Syst.* **2022**, *64*, 468–478. [CrossRef]
56. Long, T.; Li, Y.; Chen, J. Productivity prediction in aircraft final assembly lines: Comparisons and insights in different productivity ranges. *J. Manuf. Syst.* **2022**, *62*, 377–389. [CrossRef]

57. Overbeck, L.; Hugues, A.; May, M.C.; Kuhnle, A.; Lanza, G. Reinforcement Learning Based Production Control of Semi-automated Manufacturing Systems. *Procedia CIRP* **2021**, *103*, 170–175. [CrossRef]
58. May, M.C.; Behnen, L.; Holzer, A.; Kuhnle, A.; Lanza, G. Multi-variate time-series for time constraint adherence prediction in complex job shops. *Procedia CIRP* **2021**, *103*, 55–60. [CrossRef]
59. Wurster, M.; Michel, M.; May, M.C.; Kuhnle, A.; Stricker, N.; Lanza, G. Modelling and condition-based control of a flexible and hybrid disassembly system with manual and autonomous workstations using reinforcement learning. *J. Intell. Manuf.* **2022**, *33*, 575–591. [CrossRef]
60. Liberati, A.; Altman, D.G.; Tetzlaff, J.; Mulrow, C.; Gøtzsche, P.C.; Ioannidis, J.P.; Clarke, M.; Devereaux, P.J.; Kleijnen, J.; Moher, D. The PRISMA statement for reporting systematic reviews and meta-analyses of studies that evaluate health care interventions: Explanation and elaboration. *J. Clin. Epidemiol.* **2009**, *62*, e1–e34. [CrossRef]
61. Sampath, V.; Maurtua, I.; Aguilar Martín, J.J.; Gutierrez, A. A survey on generative adversarial networks for imbalance problems in computer vision tasks. *J. Big Data* **2021**, *8*, 27. [CrossRef]
62. Polyzotis, N.; Roy, S.; Whang, S.E.; Zinkevich, M. Data lifecycle challenges in production machine learning: A survey. *ACM Sigmod Rec.* **2018**, *47*, 17–28. [CrossRef]
63. Wang, Z.; Oates, T. Imaging time-series to improve classification and imputation. In Proceedings of the Twenty-Fourth International Joint Conference on Artificial Intelligence, Buenos Aires, Argentina, 25–31 July 2015.
64. Lee, G.; Flowers, M.; Dyer, M. Learning distributed representations of conceptual knowledge. In Proceedings of the International 1989 Joint Conference on Neural Networks, Washington, DC, USA, 18–22 June 1989. [CrossRef]
65. Zhu, Y.; Brettin, T.; Xia, F.; Partin, A.; Shukla, M.; Yoo, H.; Evrard, Y.A.; Doroshow, J.H.; Stevens, R.L. Converting tabular data into images for deep learning with convolutional neural networks. *Sci. Rep.* **2021**, *11*, 11325. [CrossRef]
66. Sharma, A.; Vans, E.; Shigemizu, D.; Boroevich, K.A.; Tsunoda, T. DeepInsight: A methodology to transform a non-image data to an image for convolution neural network architecture. *Sci. Rep.* **2019**, *9*, 11399. [CrossRef]
67. LeCun, Y.; Bottou, L.; Bengio, Y.; Haffner, P. Gradient-based learning applied to document recognition. *Proc. IEEE* **1998**, *86*, 2278–2324. [CrossRef]
68. Nanni, L.; Ghidoni, S.; Brahnam, S. Handcrafted vs. non-handcrafted features for computer vision classification. *Pattern Recognit.* **2017**, *71*, 158–172. [CrossRef]
69. Alkinani, M.H.; Khan, W.Z.; Arshad, Q.; Raza, M. HSDDD: A Hybrid Scheme for the Detection of Distracted Driving through Fusion of Deep Learning and Handcrafted Features. *Sensors* **2022**, *22*, 1864. [CrossRef]
70. Chen, Z.; Zhang, L.; Cao, Z.; Guo, J. Distilling the Knowledge from Handcrafted Features for Human Activity Recognition. *IEEE Trans. Ind. Inform.* **2018**, *14*, 4334–4342. [CrossRef]
71. Albawi, S.; Mohammed, T.A.; Al-Zawi, S. Understanding of a convolutional neural network. In Proceedings of the 2017 International Conference on Engineering and Technology (ICET), Antalya, Turkey, 21–23 August 2017; pp. 1–6.
72. Pearson, K. LIII. On lines and planes of closest fit to systems of points in space. *Lond. Edinb. Dublin Philos. Mag. J. Sci.* **1901**, *2*, 559–572. [CrossRef]
73. Comon, P. Independent component analysis, a new concept? *Signal Process.* **1994**, *36*, 287–314. [CrossRef]
74. LeCun, Y.; Boser, B.; Denker, J.S.; Henderson, D.; Howard, R.E.; Hubbard, W.; Jackel, L.D. Backpropagation applied to handwritten zip code recognition. *Neural Comput.* **1989**, *1*, 541–551. [CrossRef]
75. Mikolov, T.; Karafiát, M.; Burget, L.; Cernocký, J.; Khudanpur, S. Recurrent neural network based language model. In *Interspeech*; Makuhari: Chiba-city, Japan, 2010; Volume 2, pp. 1045–1048.
76. Ojala, T.; Pietikainen, M.; Maenpaa, T. Multiresolution gray-scale and rotation invariant texture classification with local binary patterns. *IEEE Trans. Pattern Anal. Mach. Intell.* **2002**, *24*, 971–987. [CrossRef]
77. Lowe, D.G. Distinctive image features from scale-invariant keypoints. *Int. J. Comput. Vis.* **2004**, *60*, 91–110. [CrossRef]
78. Dalal, N.; Triggs, B. Histograms of oriented gradients for human detection. In Proceedings of the 2005 IEEE Computer Society Conference on Computer Vision and Pattern Recognition (CVPR'05), San Diego, CA, USA, 20–26 June 2005; Volume 1, pp. 886–893.
79. Shensa, M.J. The discrete wavelet transform: Wedding the a trous and Mallat algorithms. *IEEE Trans. Signal Process.* **1992**, *40*, 2464–2482. [CrossRef]
80. Gröchenig, K. The short-time Fourier transform. In *Foundations of Time-Frequency Analysis*; Springer Science & Business Media: Berlin/Heidelberg, Germany, 2001; pp. 37–58.
81. Harris, Z.S. Distributional structure. *Word* **1954**, *10*, 146–162. [CrossRef]
82. Mikolov, T.; Chen, K.; Corrado, G.; Dean, J. Efficient estimation of word representations in vector space. *arXiv* **2013**, arXiv:1301.3781.
83. Liu, D.; Kong, H.; Luo, X.; Liu, W.; Subramaniam, R. Bringing AI to edge: From deep learning's perspective. *Neurocomputing* **2021**, *485*, 297–320. [CrossRef]
84. Gray, R.M.; Neuhoff, D.L. Quantization. *IEEE Trans. Inf. Theory* **1998**, *44*, 2325–2383. [CrossRef]
85. Sampath, V.; Maurtua, I.; Aguilar Martín, J.J.; Iriondo, A.; Lluvia, I.; Rivera, A. Vision Transformer based knowledge distillation for fasteners defect detection. In Proceedings of the 2022 International Conference on Electrical, Computer and Energy Technologies (ICECET), Prague, Czech Republic, 20–22 July 2022; pp. 1–6.
86. Shelden, R. Decision Tree. *Chem. Eng. Prog.* **1970**, *66*, 8.
87. Breiman, L. Bagging predictors. *Mach. Learn.* **1996**, *24*, 123–140. [CrossRef]

88. Freund, Y.; Schapire, R.E.; others. Experiments with a new boosting algorithm.. In Proceedings of the Thirteenth International Conference on International Conference on Machine Learning (ICML'96), Bari, Italy, 3–6 July 1996; Volume 96, pp. 148–156.
89. Ho, T.K. Random decision forests. In Proceedings of the 3rd International Conference on Document Analysis and Recognition, Montreal, QC, Canada, 14–16 August 1995; Volume 1, pp. 278–282.
90. Chen, T.; Guestrin, C. XGBoost. In Proceedings of the 22nd ACM SIGKDD International Conference on Knowledge Discovery and Data Mining, San Francisco, CA, USA, 13–17 August 2016. [CrossRef]
91. Choi, S.; Battulga, L.; Nasridinov, A.; Yoo, K.H. A decision tree approach for identifying defective products in the manufacturing process. *Int. J. Contents* **2017**, *13*, 57–65.
92. Sugumaran, V.; Muralidharan, V.; Ramachandran, K. Feature selection using decision tree and classification through proximal support vector machine for fault diagnostics of roller bearing. *Mech. Syst. Signal Process.* **2007**, *21*, 930–942. [CrossRef]
93. Hung, Y.H. Improved ensemble-learning algorithm for predictive maintenance in the manufacturing process. *Appl. Sci.* **2021**, *11*, 6832. [CrossRef]
94. Močkus, J. On bayesian methods for seeking the extremum. In *Optimization Techniques IFIP Technical Conference Novosibirsk, Novosibirsk, Russia, 1–7 July 1974*; Marchuk, G.I., Ed.; Springer: Berlin/Heidelberg, Germany, 1975; pp. 400–404.
95. Baum, L.E.; Petrie, T. Statistical inference for probabilistic functions of finite state Markov chains. *Ann. Math. Stat.* **1966**, *37*, 1554–1563. [CrossRef]
96. Papananias, M.; McLeay, T.E.; Mahfouf, M.; Kadirkamanathan, V. A Bayesian framework to estimate part quality and associated uncertainties in multistage manufacturing. *Comput. Ind.* **2019**, *105*, 35–47. [CrossRef]
97. Patange, A.D.; Jegadeeshwaran, R. Application of bayesian family classifiers for cutting tool inserts health monitoring on CNC milling. *Int. J. Progn. Health Manag.* **2020**, *11*. [CrossRef]
98. Pandita, P.; Ghosh, S.; Gupta, V.K.; Meshkov, A.; Wang, L. Application of Deep Transfer Learning and Uncertainty Quantification for Process Identification in Powder Bed Fusion. *ASME J. Risk Uncertain. Part B Mech. Eng.* **2022**, *8*, 011106. [CrossRef]
99. Farahani, A.; Tohidi, H.; Shoja, A. An integrated optimization of quality control chart parameters and preventive maintenance using Markov chain. *Adv. Prod. Eng. Manag.* **2019**, *14*, 5–14. [CrossRef]
100. El Haoud, N.; Bachiri, Z. Stochastic artificial intelligence benefits and supply chain management inventory prediction. In Proceedings of the 2019 International Colloquium on Logistics and Supply Chain Management (LOGISTIQUA), Paris, France, 12–14 June 2019; pp. 1–5.
101. Feng, M.; Li, Y. Predictive Maintenance Decision Making Based on Reinforcement Learning in Multistage Production Systems. *IEEE Access* **2022**, *10*, 18910–18921. [CrossRef]
102. Sobaszek, Ł.; Gola, A.; Kozłowski, E. Predictive scheduling with Markov chains and ARIMA models. *Appl. Sci.* **2020**, *10*, 6121. [CrossRef]
103. Hofmann, T.; Schölkopf, B.; Smola, A.J. Kernel methods in machine learning. *Ann. Stat.* **2008**, *36*, 1171–1220. [CrossRef]
104. Gobert, C.; Reutzel, E.W.; Petrich, J.; Nassar, A.R.; Phoha, S. Application of supervised machine learning for defect detection during metallic powder bed fusion additive manufacturing using high resolution imaging. *Addit. Manuf.* **2018**, *21*, 517–528. [CrossRef]
105. McGregor, D.J.; Bimrose, M.V.; Shao, C.; Tawfick, S.; King, W.P. Using machine learning to predict dimensions and qualify diverse part designs across multiple additive machines and materials. *Addit. Manuf.* **2022**, *55*, 102848. [CrossRef]
106. Kubik, C.; Knauer, S.M.; Groche, P. Smart sheet metal forming: Importance of data acquisition, preprocessing and transformation on the performance of a multiclass support vector machine for predicting wear states during blanking. *J. Intell. Manuf.* **2022**, *33*, 259–282. [CrossRef]
107. Cortes, C.; Vapnik, V. Support-vector networks. *Mach. Learn.* **1995**, *20*, 273–297. [CrossRef]
108. Mika, S.; Ratsch, G.; Weston, J.; Scholkopf, B.; Mullers, K. Fisher discriminant analysis with kernels. In Proceedings of the Neural Networks for Signal Processing IX: Proceedings of the 1999 IEEE Signal Processing Society Workshop (Cat. No.98TH8468), Madison, WI, USA, 25 August 1999; pp. 41–48. [CrossRef]
109. Fukushima, K.; Miyake, S. Neocognitron: A self-organizing neural network model for a mechanism of visual pattern recognition. In *Competition and Cooperation in Neural Nets*; Springer: Berlin/Heidelberg, Germany, 1982; pp. 267–285.
110. Rumelhart, D.E.; Hinton, G.E.; Williams, R.J. Learning representations by back-propagating errors. *Nature* **1986**, *323*, 533–536. [CrossRef]
111. Hinton, G.E. Deep belief networks. *Scholarpedia* **2009**, *4*, 5947. [CrossRef]
112. Badmos, O.; Kopp, A.; Bernthaler, T.; Schneider, G. Image-based defect detection in lithium-ion battery electrode using convolutional neural networks. *J. Intell. Manuf.* **2020**, *31*, 885–897. [CrossRef]
113. Ho, S.; Zhang, W.; Young, W.; Buchholz, M.; Al Jufout, S.; Dajani, K.; Bian, L.; Mozumdar, M. DLAM: Deep Learning Based Real-Time Porosity Prediction for Additive Manufacturing Using Thermal Images of the Melt Pool. *IEEE Access* **2021**, *9*, 115100–115114. [CrossRef]
114. Wen, L.; Li, X.; Gao, L.; Zhang, Y. A new convolutional neural network-based data-driven fault diagnosis method. *IEEE Trans. Ind. Electron.* **2017**, *65*, 5990–5998. [CrossRef]
115. Al-Dulaimi, A.; Zabihi, S.; Asif, A.; Mohammadi, A. A multimodal and hybrid deep neural network model for remaining useful life estimation. *Comput. Ind.* **2019**, *108*, 186–196. [CrossRef]

116. Huang, J.; Segura, L.J.; Wang, T.; Zhao, G.; Sun, H.; Zhou, C. Unsupervised learning for the droplet evolution prediction and process dynamics understanding in inkjet printing. *Addit. Manuf.* **2020**, *35*, 101197. [CrossRef]
117. Huang, J.; Chang, Q.; Arinez, J. Product completion time prediction using a hybrid approach combining deep learning and system model. *J. Manuf. Syst.* **2020**, *57*, 311–322. [CrossRef]
118. Cohen, J.; Jiang, B.; Ni, J. Machine Learning for Diagnosis of Event Synchronization Faults in Discrete Manufacturing Systems. *J. Manuf. Sci. Eng.* **2022**, *144*, 071006. [CrossRef]
119. Mujeeb, A.; Dai, W.; Erdt, M.; Sourin, A. One class based feature learning approach for defect detection using deep autoencoders. *Adv. Eng. Inform.* **2019**, *42*, 100933. [CrossRef]
120. Kasim, N.; Nuawi, M.; Ghani, J.; Rizal, M.; Ngatiman, N.; Haron, C. Enhancing Clustering Algorithm with Initial Centroids in Tool Wear Region Recognition. *Int. J. Precis. Eng. Manuf.* **2021**, *22*, 843–863. [CrossRef]
121. Djatna, T.; Alitu, I.M. An application of association rule mining in total productive maintenance strategy: An analysis and modelling in wooden door manufacturing industry. *Procedia Manuf.* **2015**, *4*, 336–343. [CrossRef]
122. Chiang, L.H.; Colegrove, L.F. Industrial implementation of on-line multivariate quality control. *Chemom. Intell. Lab. Syst.* **2007**, *88*, 143–153. [CrossRef]
123. You, D.; Gao, X.; Katayama, S. WPD-PCA-based laser welding process monitoring and defects diagnosis by using FNN and SVM. *IEEE Trans. Ind. Electron.* **2014**, *62*, 628–636. [CrossRef]
124. Moshat, S.; Datta, S.; Bandyopadhyay, A.; Pal, P. Optimization of CNC end milling process parameters using PCA-based Taguchi method. *Int. J. Eng. Sci. Technol.* **2010**, *2*, 95–102. [CrossRef]
125. Mei, S.; Wang, Y.; Wen, G. Automatic fabric defect detection with a multi-scale convolutional denoising autoencoder network model. *Sensors* **2018**, *18*, 1064. [CrossRef] [PubMed]
126. Maggipinto, M.; Beghi, A.; Susto, G.A. A Deep Convolutional Autoencoder-Based Approach for Anomaly Detection With Industrial, Non-Images, 2-Dimensional Data: A Semiconductor Manufacturing Case Study. *IEEE Trans. Autom. Sci. Eng.* **2022**. [CrossRef]
127. Yang, Z.; Gjorgjevikj, D.; Long, J.; Zi, Y.; Zhang, S.; Li, C. Sparse autoencoder-based multi-head deep neural networks for machinery fault diagnostics with detection of novelties. *Chin. J. Mech. Eng.* **2021**, *34*, 54. [CrossRef]
128. Cheng, R.C.; Chen, K.S. Ball bearing multiple failure diagnosis using feature-selected autoencoder model. *Int. J. Adv. Manuf. Technol.* **2022**, *120*, 4803–4819. [CrossRef]
129. Ramamurthy, M.; Robinson, Y.H.; Vimal, S.; Suresh, A. Auto encoder based dimensionality reduction and classification using convolutional neural networks for hyperspectral images. *Microprocess. Microsyst.* **2020**, *79*, 103280. [CrossRef]
130. Angelopoulos, A.; Michailidis, E.T.; Nomikos, N.; Trakadas, P.; Hatziefremidis, A.; Voliotis, S.; Zahariadis, T. Tackling faults in the industry 4.0 era—A survey of machine-learning solutions and key aspects. *Sensors* **2019**, *20*, 109. [CrossRef]
131. de Lima, M.J.; Crovato, C.D.P.; Mejia, R.I.G.; da Rosa Righi, R.; de Oliveira Ramos, G.; da Costa, C.A.; Pesenti, G. HealthMon: An approach for monitoring machines degradation using time-series decomposition, clustering, and metaheuristics. *Comput. Ind. Eng.* **2021**, *162*, 107709. [CrossRef]
132. Song, W.; Wen, L.; Gao, L.; Li, X. Unsupervised fault diagnosis method based on iterative multi-manifold spectral clustering. *IET Collab. Intell. Manuf.* **2019**, *1*, 48–55. [CrossRef]
133. Subramaniyan, M.; Skoogh, A.; Muhammad, A.S.; Bokrantz, J.; Johansson, B.; Roser, C. A generic hierarchical clustering approach for detecting bottlenecks in manufacturing. *J. Manuf. Syst.* **2020**, *55*, 143–158. [CrossRef]
134. Srinivasan, M.; Moon, Y.B. A comprehensive clustering algorithm for strategic analysis of supply chain networks. *Comput. Ind. Eng.* **1999**, *36*, 615–633. [CrossRef]
135. Das, J.N.; Tiwari, M.K.; Sinha, A.K.; Khanzode, V. Integrated warehouse assignment and carton configuration optimization using deep clustering-based evolutionary algorithms. *Expert Syst. Appl.* **2023**, *212*, 118680. [CrossRef]
136. Stojanovic, L.; Dinic, M.; Stojanovic, N.; Stojadinovic, A. Big-data-driven anomaly detection in industry (4.0): An approach and a case study. In Proceedings of the 2016 IEEE International Conference on Big Data (Big Data), Washington, DC, USA, 5–8 December 2016; pp. 1647–1652.
137. Saldivar, A.A.F.; Goh, C.; Li, Y.; Chen, Y.; Yu, H. Identifying smart design attributes for Industry 4.0 customization using a clustering Genetic Algorithm. In Proceedings of the 2016 22nd International Conference on Automation and Computing (ICAC), Colchester, UK, 7–8 September 2016; pp. 408–414.
138. Chen, W.C.; Tseng, S.S.; Wang, C.Y. A novel manufacturing defect detection method using association rule mining techniques. *Expert Syst. Appl.* **2005**, *29*, 807–815. [CrossRef]
139. Shorten, C.; Khoshgoftaar, T.M. A survey on image data augmentation for deep learning. *J. Big Data* **2019**, *6*, 60. [CrossRef]
140. Iwana, B.K.; Uchida, S. An empirical survey of data augmentation for time series classification with neural networks. *PLoS ONE* **2021**, *16*, e0254841. [CrossRef]
141. Goodfellow, I.; Pouget-Abadie, J.; Mirza, M.; Xu, B.; Warde-Farley, D.; Ozair, S.; Courville, A.; Bengio, Y. Generative adversarial networks. *Commun. ACM* **2020**, *63*, 139–144. [CrossRef]
142. Kingma, D.P.; Welling, M. Auto-encoding variational bayes. *arXiv* **2013**, arXiv:1312.6114.
143. Wong, S.C.; Gatt, A.; Stamatescu, V.; McDonnell, M.D. Understanding data augmentation for classification: When to warp? In Proceedings of the 2016 International Conference on Digital Image Computing: Techniques and Applications (DICTA), Gold Coast, Australia, 30 November–2 December 2016; pp. 1–6.

144. Berthelot, D.; Carlini, N.; Goodfellow, I.; Papernot, N.; Oliver, A.; Raffel, C.A. Mixmatch: A holistic approach to semi-supervised learning. *Adv. Neural Inf. Process. Syst.* **2019**, *32*, 5049–5059.
145. Sohn, K.; Berthelot, D.; Carlini, N.; Zhang, Z.; Zhang, H.; Raffel, C.A.; Cubuk, E.D.; Kurakin, A.; Li, C.L. Fixmatch: Simplifying semi-supervised learning with consistency and confidence. *Adv. Neural Inf. Process. Syst.* **2020**, *33*, 596–608.
146. Yang, X.; Song, Z.; King, I.; Xu, Z. A Survey on Deep Semi-supervised Learning. *arXiv* **2021**, arXiv:2103.00550.
147. Sajjadi, M.; Javanmardi, M.; Tasdizen, T. Regularization with stochastic transformations and perturbations for deep semi-supervised learning. *Adv. Neural Inf. Process. Syst.* **2016**, *29*, 1171–1179.
148. Tarvainen, A.; Valpola, H. Mean teachers are better role models: Weight-averaged consistency targets improve semi-supervised deep learning results. *Adv. Neural Inf. Process. Syst.* **2017**, *30*, 1195–1204.
149. Li, X.; Jia, X.; Yang, Q.; Lee, J. Quality analysis in metal additive manufacturing with deep learning. *J. Intell. Manuf.* **2020**, *31*, 2003–2017. [CrossRef]
150. Zhao, B.; Zhang, X.; Zhan, Z.; Wu, Q.; Zhang, H. A Novel Semi-Supervised Graph-Guided Approach for Intelligent Health State Diagnosis of a 3-PRR Planar Parallel Manipulator. *IEEE/ASME Trans. Mechatron.* **2022**, *27*, 4786–4797. [CrossRef]
151. Gilmer, J.; Schoenholz, S.S.; Riley, P.F.; Vinyals, O.; Dahl, G.E. Neural message passing for quantum chemistry. In Proceedings of the International Conference on Machine Learning, Sydney, Australia, 6–11 August 2017; pp. 1263–1272.
152. Kipf, T.N.; Welling, M. Semi-supervised classification with graph convolutional networks. *arXiv* **2016**, arXiv:1609.02907.
153. Serradilla, O.; Zugasti, E.; Ramirez de Okariz, J.; Rodriguez, J.; Zurutuza, U. Adaptable and explainable predictive maintenance: Semi-supervised deep learning for anomaly detection and diagnosis in press machine data. *Appl. Sci.* **2021**, *11*, 7376. [CrossRef]
154. Song, J.; Lee, Y.C.; Lee, J. Deep generative model with time series-image encoding for manufacturing fault detection in die casting process. *J. Intell. Manuf.* **2022**, 1–14. [CrossRef]
155. Springenberg, J.T. Unsupervised and semi-supervised learning with categorical generative adversarial networks. *arXiv* **2015**, arXiv:1511.06390.
156. Salimans, T.; Goodfellow, I.; Zaremba, W.; Cheung, V.; Radford, A.; Chen, X. Improved techniques for training gans. *Adv. Neural Inf. Process. Syst.* **2016**, *29*, 2234–2242.
157. Kingma, D.P.; Mohamed, S.; Jimenez Rezende, D.; Welling, M. Semi-supervised learning with deep generative models. *Adv. Neural Inf. Process. Syst.* **2014**, *27*, 3581–3589.
158. Sutton, R.S.; Barto, A.G. *Reinforcement Learning: An Introduction*; MIT Press: Cambridge, MA, USA, 2018.
159. May, M.C.; Overbeck, L.; Wurster, M.; Kuhnle, A.; Lanza, G. Foresighted digital twin for situational agent selection in production control. *Procedia CIRP* **2021**, *99*, 27–32. [CrossRef]
160. May, M.C.; Kiefer, L.; Kuhnle, A.; Stricker, N.; Lanza, G. Decentralized multi-agent production control through economic model bidding for matrix production systems. *Procedia Cirp* **2021**, *96*, 3–8. [CrossRef]
161. Yao, M. Breakthrough Research In Reinforcement Learning From 2019. 2019. Available online: https://www.topbots.com/top-ai-reinforcement-learning-research-papers-2019 (accessed on 1 September 2022).
162. Gao, R.X.; Wang, L.; Helu, M.; Teti, R. Big data analytics for smart factories of the future. *CIRP Ann.* **2020**, *69*, 668–692. [CrossRef]
163. Kozjek, D.; Vrabič, R.; Kralj, D.; Butala, P. Interpretative identification of the faulty conditions in a cyclic manufacturing process. *J. Manuf. Syst.* **2017**, *43*, 214–224. [CrossRef]
164. Wen, Q.; Sun, L.; Yang, F.; Song, X.; Gao, J.; Wang, X.; Xu, H. Time series data augmentation for deep learning: A survey. *arXiv* **2020**, arXiv:2002.12478.
165. Pan, S.J.; Yang, Q. A survey on transfer learning. *IEEE Trans. Knowl. Data Eng.* **2009**, *22*, 1345–1359. [CrossRef]
166. Zhang, H.; Cisse, M.; Dauphin, Y.N.; Lopez-Paz, D. mixup: Beyond empirical risk minimization. *arXiv* **2017**, arXiv:1710.09412.
167. Bao, J.; Chen, D.; Wen, F.; Li, H.; Hua, G. CVAE-GAN: Fine-grained image generation through asymmetric training. In Proceedings of the IEEE International Conference on Computer Vision, Venice, Italy, 22–29 October 2017; pp. 2745–2754.
168. Yoon, J.; Jarrett, D.; Van der Schaar, M. Time-series generative adversarial networks. *Adv. Neural Inf. Process. Syst.* **2019**, *32*, 5508–5518.
169. McMahan, B.; Moore, E.; Ramage, D.; Hampson, S.; y Arcas, B.A. Communication-efficient learning of deep networks from decentralized data. In Proceedings of the Artificial Intelligence and Statistics, Fort Lauderdale, FL, USA, 20–22 April 2017; pp. 1273–1282.
170. Cheng, Y.; Wang, D.; Zhou, P.; Zhang, T. Model compression and acceleration for deep neural networks: The principles, progress, and challenges. *IEEE Signal Process. Mag.* **2018**, *35*, 126–136. [CrossRef]
171. Gou, J.; Yu, B.; Maybank, S.J.; Tao, D. Knowledge distillation: A survey. *Int. J. Comput. Vis.* **2021**, *129*, 1789–1819. [CrossRef]
172. Hinton, G.; Vinyals, O.; Dean, J. Distilling the knowledge in a neural network. *arXiv* **2015**, arXiv:1503.02531.
173. Schlimmer, J.C.; Granger, R.H. Incremental learning from noisy data. *Mach. Learn.* **1986**, *1*, 317–354. [CrossRef]
174. Gama, J.; Žliobaitė, I.; Bifet, A.; Pechenizkiy, M.; Bouchachia, A. A survey on concept drift adaptation. *ACM Comput. Surv.* **2014**, *46*, 44. [CrossRef]
175. Baier, L.; Jöhren, F.; Seebacher, S. Challenges in the Deployment and Operation of Machine Learning in Practice. In Proceedings of the ECIS 2019 27th European Conference on Information Systems, Stockholm, Sweden, 8–14 June 2019.
176. Canbek, G. Gaining insights in datasets in the shade of "garbage in, garbage out" rationale: Feature space distribution fitting. *Wiley Interdiscip. Rev. Data Min. Knowl. Discov.* **2022**, *12*, e1456. [CrossRef]

177. Moges, T.; Yang, Z.; Jones, K.; Feng, S.; Witherell, P.; Lu, Y. Hybrid modeling approach for melt-pool prediction in laser powder bed fusion additive manufacturing. *J. Comput. Inf. Sci. Eng.* **2021**, *21*, 050902. [CrossRef]
178. Colledani, M., Statistical Process Control. In *CIRP Encyclopedia of Production Engineering*; Laperrière, L., Reinhart, G., Eds.; Springer: Berlin/Heidelberg, Germany, 2014; pp. 1150–1157. [CrossRef]
179. Abdar, M.; Pourpanah, F.; Hussain, S.; Rezazadegan, D.; Liu, L.; Ghavamzadeh, M.; Fieguth, P.; Cao, X.; Khosravi, A.; Acharya, U.R.; et al. A review of uncertainty quantification in deep learning: Techniques, applications and challenges. *Inf. Fusion* **2021**, *76*, 243–297. [CrossRef]
180. Yong, B.X.; Brintrup, A. Multi Agent System for Machine Learning Under Uncertainty in Cyber Physical Manufacturing System. In *Service Oriented, Holonic and Multi-Agent Manufacturing Systems for Industry of the Future*; Borangiu, T., Trentesaux, D., Leitão, P., Giret Boggino, A., Botti, V., Eds.; Studies in Computational Intelligence; Springer International Publishing: Cham, Switzerland, 2020; Volume 853, pp. 244–257. [CrossRef]
181. Tavazza, F.; DeCost, B.; Choudhary, K. Uncertainty Prediction for Machine Learning Models of Material Properties. *ACS Omega* **2021**, *6*, 32431–32440. [CrossRef]
182. Arkov, V. Uncertainty Estimation in Machine Learning. *arXiv* **2022**. [CrossRef]
183. Zhang, B. Data-Driven Uncertainty Analysis in Neural Networks with Applications to Manufacturing Process Monitoring. Ph.D. Thesis, Purdue University Graduate School, West Lafayette, IN, USA, 2021. [CrossRef]
184. Zhang, B.; Shin, Y.C. A probabilistic neural network for uncertainty prediction with applications to manufacturing process monitoring. *Appl. Soft Comput.* **2022**, *124*, 108995. [CrossRef]
185. Lee, S.; Kim, S.B. Time-adaptive support vector data description for nonstationary process monitoring. *Eng. Appl. Artif. Intell.* **2018**, *68*, 18–31. [CrossRef]
186. Gaikwad, A.; Yavari, R.; Montazeri, M.; Cole, K.; Bian, L.; Rao, P. Toward the digital twin of additive manufacturing: Integrating thermal simulations, sensing, and analytics to detect process faults. *IISE Trans.* **2020**, *52*, 1204–1217. [CrossRef]
187. Zhang, C.J.; Zhang, Y.C.; Han, Y. Industrial cyber-physical system driven intelligent prediction model for converter end carbon content in steelmaking plants. *J. Ind. Inf. Integr.* **2022**, *28*, 100356. [CrossRef]
188. Ning, F.; Shi, Y.; Cai, M.; Xu, W.; Zhang, X. Manufacturing cost estimation based on the machining process and deep-learning method. *J. Manuf. Syst.* **2020**, *56*, 11–22. [CrossRef]
189. Westphal, E.; Seitz, H. Machine learning for the intelligent analysis of 3D printing conditions using environmental sensor data to support quality assurance. *Addit. Manuf.* **2022**, *50*, 102535. [CrossRef]
190. Qin, J.; Wang, Y.; Ding, J.; Williams, S. Optimal droplet transfer mode maintenance for wire+ arc additive manufacturing (WAAM) based on deep learning. *J. Intell. Manuf.* **2022**, *33*, 2179–2191. [CrossRef]
191. Lapointe, S.; Guss, G.; Reese, Z.; Strantza, M.; Matthews, M.; Druzgalski, C. Photodiode-based machine learning for optimization of laser powder bed fusion parameters in complex geometries. *Addit. Manuf.* **2022**, *53*, 102687. [CrossRef]
192. Zhang, T.; Zhang, C.; Hu, T. A robotic grasp detection method based on auto-annotated dataset in disordered manufacturing scenarios. *Robot. Comput. Integr. Manuf.* **2022**, *76*, 102329. [CrossRef]
193. Singh, S.A.; Desai, K. Automated surface defect detection framework using machine vision and convolutional neural networks. *J. Intell. Manuf.* **2022**, 1–17. [CrossRef]
194. Duan, J.; Hu, C.; Zhan, X.; Zhou, H.; Liao, G.; Shi, T. MS-SSPCANet: A powerful deep learning framework for tool wear prediction. *Robot. Comput. Integr. Manuf.* **2022**, *78*, 102391. [CrossRef]
195. Gao, K.; Chen, H.; Zhang, X.; Ren, X.; Chen, J.; Chen, X. A novel material removal prediction method based on acoustic sensing and ensemble XGBoost learning algorithm for robotic belt grinding of Inconel 718. *Int. J. Adv. Manuf. Technol.* **2019**, *105*, 217–232. [CrossRef]
196. Gawade, V.; Singh, V.; Guo, W. Leveraging simulated and empirical data-driven insight to supervised-learning for porosity prediction in laser metal deposition. *J. Manuf. Syst.* **2022**, *62*, 875–885. [CrossRef]
197. Aminzadeh, M.; Kurfess, T.R. Online quality inspection using Bayesian classification in powder-bed additive manufacturing from high-resolution visual camera images. *J. Intell. Manuf.* **2019**, *30*, 2505–2523. [CrossRef]
198. Priore, P.; Ponte, B.; Puente, J.; Gómez, A. Learning-based scheduling of flexible manufacturing systems using ensemble methods. *Comput. Ind. Eng.* **2018**, *126*, 282–291. [CrossRef]
199. Guo, S.; Chen, M.; Abolhassani, A.; Kalamdani, R.; Guo, W.G. Identifying manufacturing operational conditions by physics-based feature extraction and ensemble clustering. *J. Manuf. Syst.* **2021**, *60*, 162–175. [CrossRef]
200. Kim, J.; Ko, J.; Choi, H.; Kim, H. Printed circuit board defect detection using deep learning via a skip-connected convolutional autoencoder. *Sensors* **2021**, *21*, 4968. [CrossRef]
201. Jakubowski, J.; Stanisz, P.; Bobek, S.; Nalepa, G.J. Anomaly Detection in Asset Degradation Process Using Variational Autoencoder and Explanations. *Sensors* **2021**, *22*, 291. [CrossRef]
202. Sarita, K.; Devarapalli, R.; Kumar, S.; Malik, H.; Garcia Marquez, F.P.; Rai, P. Principal component analysis technique for early fault detection. *J. Intell. Fuzzy Syst.* **2022**, *42*, 861–872. [CrossRef]
203. Zheng, X.; Wang, H.; Chen, J.; Kong, Y.; Zheng, S. A generic semi-supervised deep learning-based approach for automated surface inspection. *IEEE Access* **2020**, *8*, 114088–114099. [CrossRef]
204. Zhang, W.; Lang, J. Semi-supervised training for positioning of welding seams. *Sensors* **2021**, *21*, 7309. [CrossRef]

205. Chen, C.; Liu, Y.; Kumar, M.; Qin, J.; Ren, Y. Energy consumption modelling using deep learning embedded semi-supervised learning. *Comput. Ind. Eng.* **2019**, *135*, 757–765. [CrossRef]
206. Jun, J.h.; Chang, T.W.; Jun, S. Quality prediction and yield improvement in process manufacturing based on data analytics. *Processes* **2020**, *8*, 1068. [CrossRef]
207. Shim, J.; Cho, S.; Kum, E.; Jeong, S. Adaptive fault detection framework for recipe transition in semiconductor manufacturing. *Comput. Ind. Eng.* **2021**, *161*, 107632. [CrossRef]
208. Qiu, C.; Li, K.; Li, B.; Mao, X.; He, S.; Hao, C.; Yin, L. Semi-supervised graph convolutional network to predict position-and speed-dependent tool tip dynamics with limited labeled data. *Mech. Syst. Signal Process.* **2022**, *164*, 108225. [CrossRef]
209. Guo, Y.; Lu, W.F.; Fuh, J.Y.H. Semi-supervised deep learning based framework for assessing manufacturability of cellular structures in direct metal laser sintering process. *J. Intell. Manuf.* **2021**, *32*, 347–359. [CrossRef]
210. Okaro, I.A.; Jayasinghe, S.; Sutcliffe, C.; Black, K.; Paoletti, P.; Green, P.L. Automatic fault detection for laser powder-bed fusion using semi-supervised machine learning. *Addit. Manuf.* **2019**, *27*, 42–53. [CrossRef]
211. Lee, H.; Kim, H. Semi-supervised multi-label learning for classification of wafer bin maps with mixed-type defect patterns. *IEEE Trans. Semicond. Manuf.* **2020**, *33*, 653–662. [CrossRef]
212. Liu, J.; Song, K.; Feng, M.; Yan, Y.; Tu, Z.; Zhu, L. Semi-supervised anomaly detection with dual prototypes autoencoder for industrial surface inspection. *Opt. Lasers Eng.* **2021**, *136*, 106324. [CrossRef]
213. Verstraete, D.; Droguett, E.; Modarres, M. A deep adversarial approach based on multi-sensor fusion for semi-supervised remaining useful life prognostics. *Sensors* **2019**, *20*, 176. [CrossRef]
214. Souza, M.L.H.; da Costa, C.A.; de Oliveira Ramos, G.; da Rosa Righi, R. A feature identification method to explain anomalies in condition monitoring. *Comput. Ind.* **2021**, *133*, 103528. [CrossRef]
215. Lee, Y.H.; Lee, S. Deep reinforcement learning based scheduling within production plan in semiconductor fabrication. *Expert Syst. Appl.* **2022**, *191*, 116222. [CrossRef]
216. Marchesano, M.G.; Guizzi, G.; Santillo, L.C.; Vespoli, S. A deep reinforcement learning approach for the throughput control of a flow-shop production system. *IFAC-PapersOnLine* **2021**, *54*, 61–66. [CrossRef]
217. Yang, H.; Li, W.; Wang, B. Joint optimization of preventive maintenance and production scheduling for multi-state production systems based on reinforcement learning. *Reliab. Eng. Syst. Saf.* **2021**, *214*, 107713. [CrossRef]
218. Schneckenreither, M.; Haeussler, S.; Peiró, J. Average reward adjusted deep reinforcement learning for order release planning in manufacturing. *Knowl.-Based Syst.* **2022**, *247*, 108765. [CrossRef]
219. Tsai, Y.T.; Lee, C.H.; Liu, T.Y.; Chang, T.J.; Wang, C.S.; Pawar, S.J.; Huang, P.H.; Huang, J.H. Utilization of a reinforcement learning algorithm for the accurate alignment of a robotic arm in a complete soft fabric shoe tongues automation process. *J. Manuf. Syst.* **2020**, *56*, 501–513. [CrossRef]
220. Klar, M.; Glatt, M.; Aurich, J.C. An implementation of a reinforcement learning based algorithm for factory layout planning. *Manuf. Lett.* **2021**, *30*, 1–4. [CrossRef]
221. Huang, J.; Chang, Q.; Arinez, J. Deep reinforcement learning based preventive maintenance policy for serial production lines. *Expert Syst. Appl.* **2020**, *160*, 113701. [CrossRef]
222. Zhang, H.; Peng, Q.; Zhang, J.; Gu, P. Planning for automatic product assembly using reinforcement learning. *Comput. Ind.* **2021**, *130*, 103471. [CrossRef]
223. Kuhnle, A.; May, M.C.; Schaefer, L.; Lanza, G. Explainable reinforcement learning in production control of job shop manufacturing system. *Int. J. Prod. Res.* **2021**, *60*, 5812–5834. [CrossRef]
224. Valet, A.; Altenmüller, T.; Waschneck, B.; May, M.C.; Kuhnle, A.; Lanza, G. Opportunistic maintenance scheduling with deep reinforcement learning. *J. Manuf. Syst.* **2022**, *64*, 518–534. [CrossRef]
225. Huang, J.; Su, J.; Chang, Q. Graph neural network and multi-agent reinforcement learning for machine-process-system integrated control to optimize production yield. *J. Manuf. Syst.* **2022**, *64*, 81–93. [CrossRef]
226. Zimmerling, C.; Poppe, C.; Stein, O.; Kärger, L. Optimisation of manufacturing process parameters for variable component geometries using reinforcement learning. *Mater. Des.* **2022**, *214*, 110423. [CrossRef]
227. Guo, F.; Zhou, X.; Liu, J.; Zhang, Y.; Li, D.; Zhou, H. A reinforcement learning decision model for online process parameters optimization from offline data in injection molding. *Appl. Soft Comput. J.* **2019**, *85*, 105828. [CrossRef]
228. Hofmann, C.; Liu, X.; May, M.; Lanza, G. Hybrid Monte Carlo tree search based multi-objective scheduling. *Prod. Eng.* **2022**, *17*, 133–144. [CrossRef]

applied sciences

MDPI

Article

Application of Machine Learning for Prediction and Process Optimization—Case Study of Blush Defect in Plastic Injection Molding

Alireza Mollaei Ardestani [1], Ghasem Azamirad [2,*], Yasin Shokrollahi [3], Matteo Calaon [1], Jesper Henri Hattel [1], Murat Kulahci [4], Roya Soltani [5] and Guido Tosello [1]

1 Department of Civil and Mechanical Engineering, Technical University of Denmark, 2800 Kgs. Lyngby, Denmark
2 Department of Mechanical Engineering, Yazd University, Yazd P.O. Box 8915818411, Iran
3 Department of Mechanical Engineering, Amirkabir University of Technology, Tehran P.O. Box 15875-4413, Iran
4 Department of Applied Mathematics and Computer Science, Technical University of Denmark, 2800 Kgs. Lyngby, Denmark
5 Yazd Poolika Co., Yazd Industrial Zone, Yazd P.O. Box 8947183963, Iran
* Correspondence: azamirad@yazd.ac.ir

Abstract: Injection molding is one of the most important processes for the mass production of plastic parts. In recent years, many researchers have focused on predicting the occurrence and intensity of defects in injected molded parts, as well as the optimization of process parameters to avoid such defects. One of the most frequent defects of manufactured parts is blush, which usually occurs around the gate location. In this study, to identify the effective parameters on blush formation, eight design parameters with effect probability on the influence of this defect have been investigated. Using a combination of design of experiments (DOE), finite element analysis (FEA), and ANOVA, the most significant parameters have been identified (runner diameter, holding pressure, flow rate, and melt temperature). Furthermore, to provide an efficient predictive model, machine learning methods such as basic artificial neural networks, their combination with genetic algorithms, and particle swarm optimization have been applied and their performance analyzed. It was found that the basic artificial neural network (ANN), with an average accuracy error of 1.3%, provides the closest predictions to the FEA results. Additionally, the process parameters were optimized using ANOVA and a genetic algorithm, which resulted in a significant reduction in the blush defect area.

Keywords: plastic injection molding; design of experiments; machine learning; digital twin; process optimization

Citation: Mollaei Ardestani, A.; Azamirad, G.; Shokrollahi, Y.; Calaon, M.; Hattel, J.H.; Kulahci, M.; Soltani, R.; Tosello, G. Application of Machine Learning for Prediction and Process Optimization—Case Study of Blush Defect in Plastic Injection Molding. *Appl. Sci.* **2023**, *13*, 2617. https://doi.org/10.3390/app13042617

Academic Editor: Joamin Gonzalez-Gutierrez

Received: 20 January 2023
Revised: 13 February 2023
Accepted: 14 February 2023
Published: 17 February 2023

1. Introduction

With the widespread use of algorithms in the engineering sciences, lots of additional costs associated with time-consuming and expensive tests have been eliminated from the product design and production development cycle. Today's modeling, prediction, and optimization methods have extremely reduced the need to use traditional experimental trials and measurements for product and process improvement. These techniques include statistical methods such as Analysis of Variances (ANOVA), machine learning methods such as Artificial Neural Networks (ANNs), as well as optimization methods using meta-heuristic algorithms. Nowadays, the use of Finite Element Analysis (FEA) methods combined with modern optimization methods has helped manufacturers select the optimal levels of input parameters and achieve the highest quality products [1,2]. Due to the complex behavior of polymers, particularly when processed by injection molding, many parameters can affect the product quality. Hence, monitoring and controlling each of these parameters as well as their interactions is vital to prevent injection defects. For example, a fundamental quality of injection molded parts is based on the reduction of warpage. Among statistical methods, Genetic Algorithm (GA) is known as a popular technique for reducing warpage [3], as well as other defects. Additionally, ANNs have been widely used to predict the shrinkage values of the examined parts [4]. Wang et al. [5], by taking an ANN approach, found that the shrinkage of the manufactured part has an inverse relationship with process parameters such as packing time, injection pressure, holding pressure, and melt temperature, and a direct relationship with cooling time and mold temperature. Altan et al. [6] conducted a study to reduce the shrinkage of a manufactured part. They studied parts composed of polypropylene and polystyrene. After the design of experiments based on the Taguchi method, an ANOVA was performed. As a result, optimal levels for design parameters were obtained, and by employing these optimal parameters in the simulation, the shrinkage of the parts was reduced significantly. An ANN was also trained to predict response values, and it was found that this kind of network has a good ability for this purpose. In a study conducted by Chen et al. [7], it was found that the parameters of holding pressure and melt temperature are the most effective factors in the warpage of the parts. A model was also presented that demonstrated the ability to predict parts warpage in both simulations and experimental tests with relatively high accuracy. It is also possible to use several optimization methods, such as Taguchi method, ANN, GA, Response Surface Methodology (RSM), Particle Swarm Optimization (PSO), etc. simultaneously to optimize the parameters involved in the part's warpage. Finally, according to the prediction error and the number of iterations performed, the best optimization method can be selected [8–11]. The Taguchi optimization method has been used to reduce the warpage of the parts and improve their strength. Consequently, the holding pressure was found to be the most significant factor in the warpage of the parts [12]. Similar research has been conducted using the Taguchi, Non-dominated Sorting Genetic Algorithm (NSGA-II), and RSM methods to optimize warpage, shrinkage, and residual stress by Li et al. [13]. Reduction of warpage by considering mold filling as a constraint using the RBF (Radial Basis Function) method was achieved in [14]. The RBF method was also used to perform a multi-objective optimization to reduce production time, clamping force, and warpage in the study by Kitayama et al. [15]. ANNs and GA were used to study shrinkage. Researchers found a relationship between design variables and the target parameter, which showed a great ability of ANN to predict the results [16]. In two separate studies, Kitayama et al. arranged two multi-objective optimizations that both used the RBF method to improve the weld lines. In one of these studies, the weld lines were optimized while reducing the clamping force. In a second study, the goal of the research was to reduce cycle time and weld lines [17,18]. Other than the determination of the optimal levels for the factors related to the process parameters, meta-heuristic algorithms have been also used to determine the optimal values for the composition ratio of polymers to reduce the maximum shrinkage in the molded products [19]. Xu et al. used the Taguchi method, ANN, multi-objective PSO algorithm, and a combination of ANN with PSO to attain the optimal level for design parameters to reduce the part weight, shrinkage, and flash size [20]. Along with the shrinkage of the produced part, other parameters such as clamping force have been inspected simultaneously in a multi-objective optimization using the RBF method [21]. To reduce the part shrinkage, the cycle time, and injection time at the same time, a multi-objective optimization using an ANN-based program was carried out [22]. Some studies have examined other types of defects. For example, the research by Tabi et al. [23] aimed at improving the needle-shaped defects around the gate location. The role of residual stress in the occurrence of cracks in parts [24], and the use of RSM for cutting down the average shear rate around the gate [25] were also investigated. Furthermore, improving the optical properties of polymer optics as a result of reducing residual stress in injection molded plastic lenses was the subject of another study [26]. Li et al. employed Kriging and NSGA-II methods to investigate the effect of runner diameter and process parameters on the quality, cost, and production efficiency [27]. The Taguchi method has been used to investigate the effect of process parameters on the mechanical properties of parts produced with recycled plastic. It was revealed that injection time with

40.5% and material temperature with 43.3% of impact are the most important factors in warpage and yield stress of the produced part, respectively [28]. Additionally, Martowibowo et al. [29] used a GA, with a prediction error of approximately 1%, and could find the optimal values of process parameters in such a way that the production time of the part was significantly reduced. In other studies, FEM and conventional optimization methods were used to decrease the cycle time to a minimum by providing an optimal design for the cooling system [15,30]. Eladl et al. [31] studied the effect of process parameters on the formation of flash defect. During the study, polypropylene (PP) and acrylonitrile butadiene styrene (ABS) were investigated. The outputs of the study were part mass, flow length, and flash formation. After employing DOE and statistical analysis, it was shown the injection speed and packing pressure were the most influential factors on the flash area for both materials. Regi et al. [32] used a different technique to investigate the flow propagation in the molds. They used a transparent window on one of the walls of the mold to visually observe the flow of the material. The objective of the study was to compare FEA simulation results with experiments with focus on flow hesitation. A high-speed camera was used to record the mold filling phase with two different materials of PP and ABS. After employing DOE and ANOVA, the results demonstrated that flow progression and hesitation are dependent on wall thickness, injection velocity, and material type. Loaldi et al. [33], using the same techniques (DOE and FEA) focused on the experimental validation of the injection molding simulation of microparts, also with a focus on flash formation. Results confirmed that higher values of holding pressure, injection speed, mold temperature, melt temperature generate larger flash areas. The trends were correctly predicted by the FEA flow simulations. Chen et al. [34] employed the Taguchi, ANOVA, backpropagation ANN, GA, and Davidson-Fletcher-Powell methods to improve the product quality. As a result of this optimization, as well as increasing demands in terms of products quality, issues such as waste, number of defective parts, need for inspection during production, need for recycling, and production time were also decreased. By using machine learning methods, Finkeldey et al. [35] were able to accurately predict the weight and thickness of the manufactured part. Mehat et al. [36] used process parameters optimization instead of adding additives to improve the mechanical properties in a part produced with recycled plastic. Clearly, considerable research efforts have been focused recently to optimize injection molding parts characteristics and minimize several types of defects. However, very limited research has investigated the factors affecting the blush defect. At least, one study can be indicated to have addressed the influence of process parameters on blush: results by Llado et al. [37] indicated that injection flow rate and melt temperature are the most effective parameters. Blush defect not only deteriorates the appearance of the part but also reduces the lifetime and strength of the product. Therefore, due to its importance and the limited research performed so far, there is a clear need for further investigation on the parameters affecting the incidence and exacerbation of this defect. Despite the importance of preventing the incidence of this defect in plastic injection, so far, only a few studies have investigated its causes and impacts.

The present research has inspected the effects of eight injection molding factors (flow rate, melt temperature, holding pressure, mold temperature, runner diameter, gate diameter, gate angle, included angle) on the blush. The novelty of the present study is the relatively number of investigated parameters, including both process and design factors. By examining a large variety of parameters, the study results in a comprehensive view of the causes of this defect. What is more, contrary to previous research, the interaction effects of design parameters on the incidence of the defect have also been studied in the current study. In addition, for the first time, different types of ANNs have been used to predict the values of blush defect, and a GA is utilized to optimize the levels of effective factors to achieve the lowest probable blush defect.

The research method of this study began with the creation of a CAD model of the injection molded component (plastic bushing). Then, as screening step, the fractional factorial DOE with two levels and the eight design parameters has been conducted. After performing FEA, the results were included in an ANOVA routine to identify effective and ineffective factors. Thereafter, for a more detailed study of the effect of the most relevant parameters, a second design of experiment of CCD (Central Composite Design) type has been performed with five levels on the effective factors.

2. Materials and Methods

2.1. Modeling

Blush is a visual defect that occurs as white halos, usually around the gate location. The part under study was a size bushing produced with an injection mold equipped with two cavities. Bushings are a kind a fitting in piping that can be used to connect two pipes together, change the pipeline and flow direction, derive new pipe branches, and to blind a branch. Polyvinyl chloride (PVC) fittings are commonly used in industry due to their ease of use, durability, and cost effectiveness. The geometry of the bushing and injection mold runners is shown in Figure 1. The gate is of round type, and the length of the sprue is 76 mm with a conic angle of 3°. All the geometrical dimensions to reproduce the CAD model are included in Figure 1.

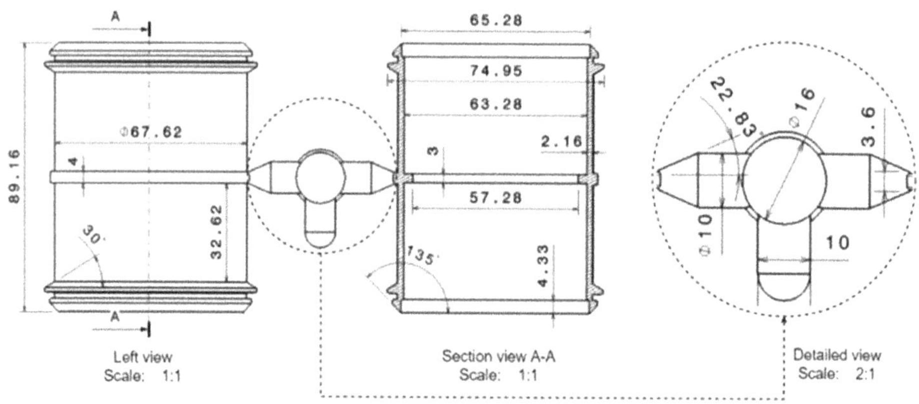

Figure 1. Bushing geometry and dimensions of runner system of the original mold.

The material used in the study is a commercially available grade of polyvinyl chloride (PVC) produced by Solvay ET CIE (Brussels, Belgium) under the commercial name Benvic IR705. This type of PVC is widely used in the production of pipes and fittings. Some of its mechanical, thermal, and rheological properties are presented in Table 1 [34]. Figure 2 shows the PVT data and viscosity characteristics (pressure, specific volume, and temperature) of the material based on the Autodesk Moldflow software database.

Table 1. Properties of Benvic IR705 [34].

Parameter	Value
Maximum allowable shear stress	0.2 MPa
Glass transition temp	79–80 °C
Specific heat	1767 J/kg °C
Conductivity	0.13 W/m °C
density	1.3253 kg/dm^3
Shrinkage	0.60%

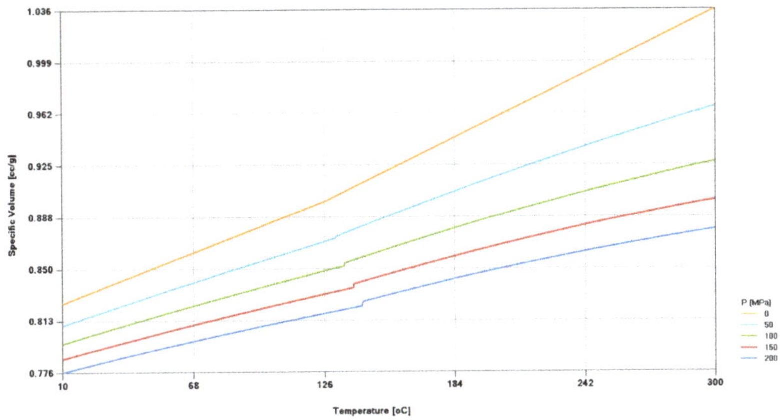

(**a**)

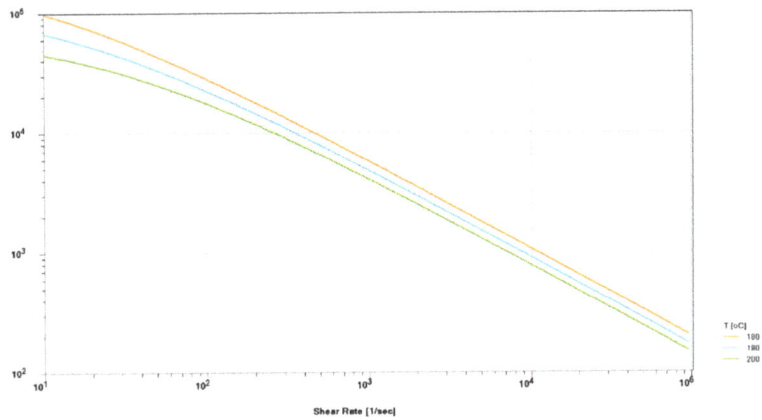

(**b**)

Figure 2. (**a**) pvT data and (**b**) viscosity curves of the Benvic IR705 PVC material.

As the first step, the CAD model of the bushing was created. This model was imported into the Autodesk Moldflow software for mesh generation and creating the runner system. Concurrently, the bushing has been masked using 459,096 3D tetragonal elements for both mold cavities. The overall element size of the bushing is 3.5 mm, and due to the greater sensitivity of the gate location area, the element size of this area is set to 2 mm.

His section may be divided by subheadings. It should provide a concise and precise description of the experimental results, their interpretation, as well as the experimental conclusions that can be drawn.

2.2. *Measuring the Simulation Results*

A visualization of blush simulation is shown in Figure 3a. As an assumption, the shape of the defect was assumed, due to the visual similarity and to avoid adding unnecessary complexity to the problem, assumed to be similar to an ellipse. Therefore, the formula for calculating the area of the ellipse has been applied to determine the area of the defect. Figure 3b represents the measures needed to calculate of the defect area.

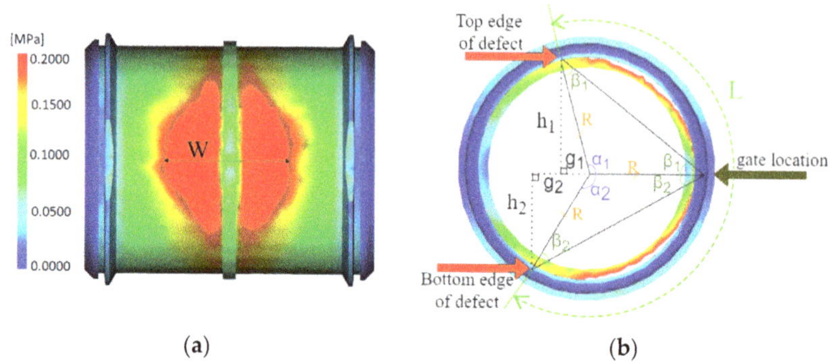

Figure 3. Width of defect on a simulated bushing (**a**) and geometrical measures for calculation of L (**b**).

The area of the ellipse can be calculated by Equation (1):

$$Area = \pi \times {}^{L}\!/_{2} \times {}^{W}\!/_{2} \tag{1}$$

where L and W represent the large and the small diameter of the ellipse, respectively. To calculate the area of the defect, a direct measurement of the value of W, the value of L was calculated by measuring the length of the arc from top edge to bottom edge of the defect, as shown in Figure 3b, via Equations (2)–(7).

$$\beta_1 = \tan^{-1}({}^{h_1}\!/_{g_1}) \tag{2}$$

$$\alpha_1 = 180 - (2 \times \beta_1) \tag{3}$$

Having Equations (2) and (3), the angle of the top arc of defect (α_1) can be calculated. Additionally, Equations (3) and (4) calculate the same parameter (α_2) for the bottom part of the defect.

$$\beta_2 = \tan^{-1}({}^{h_2}\!/_{g_2}) \tag{4}$$

$$\alpha_2 = 180 - (2 \times \beta_2) \tag{5}$$

Equation (6) sums up the amounts of top and bottom angle of the defect's arc. Thus, α_T represents the total angle of the defect's arc.

$$\alpha_T = \alpha_1 + \alpha_2 \tag{6}$$

$$L = \frac{\alpha_T}{360} \times 2\pi R \tag{7}$$

According to Equation (7), the value of L can be calculated. Having the values of L and W, the area of the defect was calculated in each case through Equation (1).

2.3. Experimental Results and Measurements

To validate the simulations, experimental tests were carried out according to the procedure shown in Figure 4. In the first step, the PVC compound with the properties given in Table 1 was fed into the injection molding machine. Next, process parameters were set in the digital process setting panel. After that, the process started with closing the mold, and after injection, the bushing was manufactured. As the next step, accurate and high-resolution imaging was employed to obtain a high-quality representation of the bushing. Eventually, the images were processed by a computer software to better visualize the blush defect. For experimental measurement of the blush defect area, the length and the width of the defect were measured. These measurements enabled us to calculate the defect's area using the ellipse area measurement formula (Equation (1)).

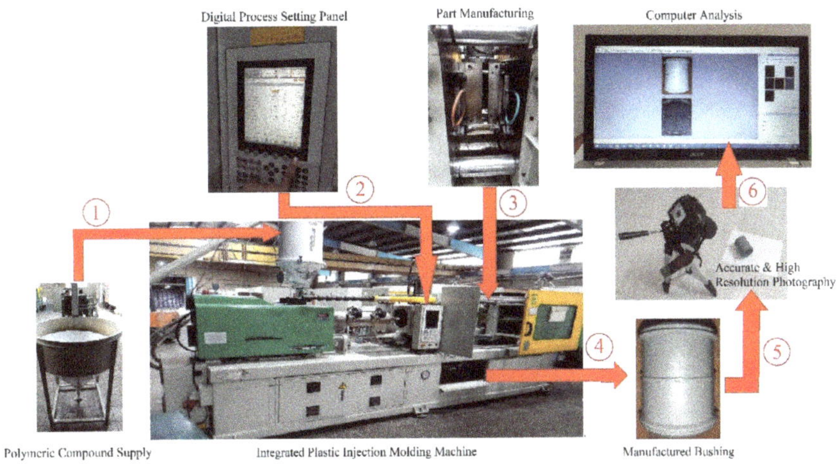

Figure 4. Experimental test procedure.

2.4. Function Approximation by ANOVA

Using Minitab software, a fractional factorial design was first created as the screening step, based on the eight previously mentioned parameters with a ratio of an 1/8. This ratio means that the number of experiments performed in the fractional DOE is 1/8 of the full factorial design. The number of full factorial DOE trials is 256, but with an 1/8 ratio, the fractional DOE is reduced to 32 experiments. The input parameters have been considered to all vary between two levels (see Table 2). After identifying the most effective parameters, a CCD design of the experiments will be conducted to clarify the impact of each effective parameter. After that, the results of ANOVA will represent the effective parameters and eventually a regression predictive model presented by the software.

Table 2. Levels of parameters required for performing simulation analyses.

	Flow Rate (cm^3/s)	Melt Temperature (°C)	Mold Temperature (°C)	Holding Pressure (MPa)	Gate Diameter (mm)	Runner Diameter (mm)	Gate Angle (°)	Included Angle (°)
Lower bound	15	185	25	60	2.5	7	0	25
Higher bound	25	195	35	80	4.5	10	45	45

2.5. Basic and Hybrid Machine Learning Algorithms

2.5.1. ANN

To provide a model for predicting the area of blush, several machine learning algorithms were considered and employed. The algorithms used in this research include basic ANN, combination of ANN and PSO, and combination of ANN and GA. The most straightforward ANN involves an input layer, a hidden layer, and an output layer. Each layer can contain one or more neurons. According to Equation (8), the value of each neuron is multiplied by the specified weight assigned to each link and added to the same value for all the other neurons in the same layer, and eventually enters the neurons of the next layer [7].

$$S_j = \sum_{j=0}^{N} x_i \times w_{ij} \tag{8}$$

where x_i is the output of the *ith* neuron of the previous layer, and w_{ij} is the weight of the link between the *ith* neuron of the previous layer and the *jth* neuron of the present layer. S_j represents the sum of the previous layer's outputs multiplied by the connection weights, which is the net input entering the jth neuron, and N represents the number of inputs to the *jth* neuron in the hidden layer. Each neuron in the network produces its output (O_j) by entering S_j that is a Tansig activator function similar to one indicated in Equation (9) [9]:

$$O_j = F(S_j) = \frac{1 - e^{-S_j}}{1 + e^{-S_j}} \tag{9}$$

2.5.2. Training ANN

- Basic ANN

In this research, the training was conducted with 70% of the available data provided in the DOE phase. Then, 15% of the available data was used as a validation dataset and the rest, 15% as a test dataset. In the first step, the weights were imported to the network randomly. Applying the gradient descent method causes these weights to be continuously updated during successive iterations. To train the network, the mean square error (*MSE*) of the predicted values should move towards minimization. Equations (10)–(12) represent this method [9,20]:

$$E = MSE = {}^{1}\!/_{N} \sum_{i=1}^{N} (y_p - y_t)^2 \tag{10}$$

$$\Delta w_{ij} = -\eta \frac{\partial E}{\partial w_{ij}} \times O_j \tag{11}$$

$$w_{ij}^{m} = w_{ij}^{m-1} + \Delta w_{ij} \tag{12}$$

where η is the learning rate and controls the network convergence (a number between 0 and 1), E represents the *MSE*, N for the number of inputs, y_t desired output, predicted output by the ANN, and m indicates the iterations counter.

- ANN + GA

This method is similar to the basic ANN method, except that instead of using the gradient descend function used in basic ANN, the appropriateness of the weights assigned to each link is determined through the GA process, which is the selection of the best, the crossover, and the mutation, so that the MSE is minimized. In this algorithm, the probability of selecting each parent is assessed through Equations (13) and (14) [9]:

$$f_i = {}^{k}\!/_{m_i} \tag{13}$$

$$p_i = {}^{f_i}\!\Big/_{\sum_{j=1}^{N} f_i} \tag{14}$$

where k is a coefficient, m_i symbolizes the fitness of i_{th} input, N substitutes for the number of generation population, and j indicates the number of generations. Equation (15) is applied to combine the two chromosomes of C_i and C_h to produce the next generation [9].

$$\begin{cases} C_{ij} = C_{ij(1-b)} + C_{hjb} \\ C_{hj} = C_{hj(1-b)} + C_{ijb} \end{cases} \tag{15}$$

where b substitutes for a random number between 0 and 1 and indicates the chromosome's intersection point. In addition, the Equations (16) and (17) are used to model the mutation in this algorithm [9].

$$C_{mn} = \begin{cases} C_{mn} + (C_{mn} - C_{\max}) \times f(g), r > 0.5 \\ C_{mn} + (C_{\min} - C_{mn}) \times f(g), r \leq 0.5 \end{cases} \tag{16}$$

$$f(g) = r_2(1 - g/G_{\max}) \tag{17}$$

where C_{mn} represents the gene C_{mn}, $C_{\min}$ and $C_{\max}$ stands for the higher and lower bounds for genes, and r and r_2 are random numbers while r is between 0 and 1, g indicates the number of present generation population and G_{max} is the maximum generations considered to iterate. So, the algorithm finally reaches a generation with perfectly fitting responses. A flowchart of the ANN + GA is shown in Figure 5 and its pseudocode is reported in Appendix A.

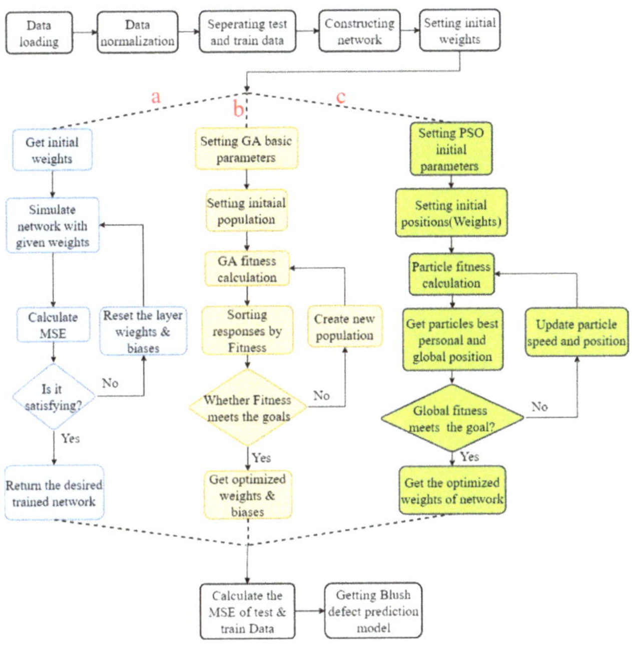

Figure 5. Training flowchart of: (a): basic ANN, (b): ANN + GA, (c): ANN + PSO.

- ANN + PSO

The PSO algorithm is a method established on swarm intelligence inspired by the behavior of a bird flock. The behavior of the particles, their speed, and the direction of their movement are influenced by the best experience of the particle itself in matching the goal and the best experience among all members of the population. The objective of this algorithm is to reduce the average fitness of all members of the population. In Equations (18)–(20), the best personal position of the particle is presented as *Pbest*, and the best position of the whole particle swarm is symbolized by *Gbest*. If the particles are located in an N-dimensional space, the vector $X_i = (X_{i1}, X_{i2}, \ldots, X_{iN})$ and the vector $V_i = (V_{i1}, V_{i2}, \ldots, V_{iN})$ represent the position and velocity of the particle i, respectively. In each generation, the values of these vectors must be updated and compared to the previous generation. The Equations (19)–(21)are used to update these values.

$$w(t) = w_{\max} - \frac{t(w_{\max} - w_{\min})}{t_{\max}} \tag{18}$$

$$v_{i,n}^{t+1} = wv_{i,n}^t + c_1 r_{1,n}(Pbest_{i,n}^t - x_{i,n}^t) + c_2 r_{2,n}(Gbest_n^t - x_{i,n}^t) \tag{19}$$

$$x_{i,n}^{t+1} = x_{i,n}^t + v_{i,n}^{t+1} \tag{20}$$

where t is counter index for generations, c_1 and c_2 stand for two positive coefficients for acceleration while $r_{1,n}$ and $r_{2,n}$ are two random coefficients with uniform distribution in *Nth* dimension of the space, n shows particle number, and m represents the inertia weight employed [7]. Figure 5 depicts a flowchart of steps of optimizing layer weights and training the network using basic ANN, ANN + GA, and ANN + PSO. The pseudocodes for each of the three algorithms are also given in the Appendix A.

In Figure 5, the (a) route shows the basic ANN's flowchart, the (b) route shows the steps of ANN + GA, and the (c) represents the flowchart of ANN + PSO. To calculate the training accuracy of the networks, the Equations (21)–(23) are employed.

$$P_E = \frac{B_P - B_M}{B_M} \tag{21}$$

$$M_E = \frac{\sum_{1}^{n} \left| P_{E_n} \right|}{n} \tag{22}$$

$$Accuracy_{Tr} = (1 - M_E) \times 100 \tag{23}$$

where P_E indicates the prediction error for each data group, B_P and B_M represent the predicted value of the blush defect and the measured value of the defect, respectively. n indicates the number of data sets, M_E represents the average error of the whole data set, and $Accuracy_{Tr}$ shows the accuracy of training.

After training the networks, an optimization was conducted. Additionally, after training an accurate ANN, the prediction accuracy of the network has been measured, and the most accurate network has been employed as the fitness function of a GA to optimize the levels of design parameters. Then, the levels optimized by ANOVA and the GA will be compared together. In the flowchart shown in Figure 6, the research method has been shown. In this figure, the simulation and DOE sections indicate all the steps taken through the creation of digital twin, getting the results of it, and employing statistical analysis on the data. The ANN section demonstrates the steps of using different machine learning methods to analyze the data. Then, as the last step, the GA optimization section reveals the work that has been conducted to optimize the parameters' levels to reach the robust process setting with the lowest blush defect area.

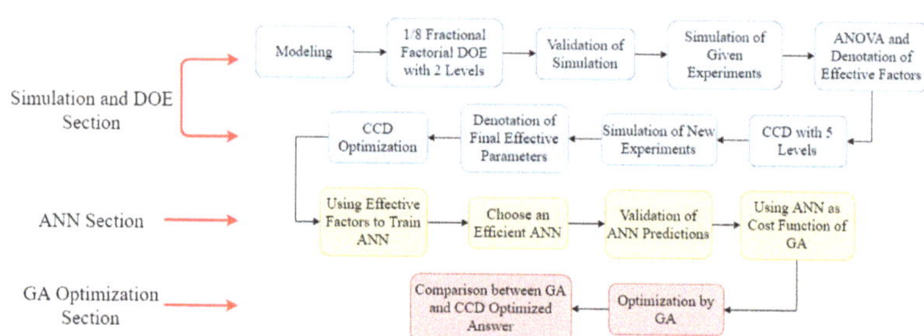

Figure 6. Flowchart of the research method.

3. Results and Discussion

3.1. FEA Validation

To ensure the accuracy of the FEA, experimental tests were performed and compared with the simulation results. Table 3 shows the validation results. As can be observed from this table, the results of FEA are in relatively good agreement with the results of the experimental tests, and the absolute value of the average deviation error is 5.6%. The parameters of flow rate, melt temperature, and holding pressure were varied to verify the results. Other parameters remained constant due to practical limitations in changing the geometry of the mold.

Table 3. Validation of FEA.

	Experimental Test No.	1	2	3	4
Inputs	Flow rate (cm^3/s)	15	15	25	25
	Melt temperature (°C)	185	195	185	195
	Mold temperature (°C)	35	35	35	35
	Holding pressure (MPa)	60	80	60	80
	Runner diameter (mm)	10	10	10	10
	Gate diameter (mm)	3.5	3.5	3.5	3.5
	Gate angle (°)	0	0	0	0
	Included angle (°)	45	45	45	45
Outputs	Experimental defect area (mm^2)	2108	1621	2483	1696
	Simulation defect area (mm^2)	2212	1553	2777	1674
	Deviation error	4.9%	−4.2%	11.8%	−1.3%

According to the shear stress heat map, the area that exceeded the maximum allowable shear stress (0.2 MPa) appeared as the blush [34]. As can be concluded from Table 3, lower amounts of error were observed for simulations associated with process configurations, resulting in a smaller area of defect. Since the objective of the study is to decrease the area of defect, the closer the study gets to its goal, the lower the error amount. Figure 7 shows a comparison between the simulation results for maximum shear stress and the blush defect in an experimental test performed with the same parameter settings.

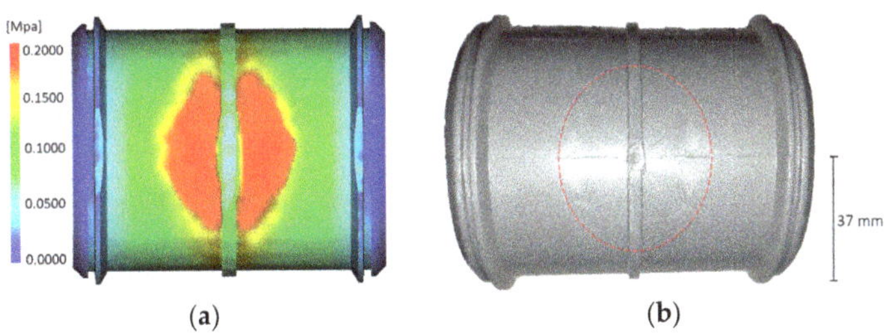

Figure 7. Simulation result (**a**) and enclosed area of defect in the experimental test (**b**) obtained with the same injection molding process settings (melt temperature = 185 °C, mold temperature = 35 °C, flow rate = 25 cm^3/s, holding pressure = 80 MPa, runner diameter = 10 mm, gate diameter = 3.5 mm, gate angle = 0°, and included angle = 45°).

3.2. Statistical Analysis

To determine the effective parameters, a screening step was implemented using the fractional factorial design. The ANOVA results are given in Table 4. These results indicate that the parameters of flow rate, melt temperature, holding pressure, and runner diameter were effective due to the lower *p*-value of 0.05. The parameters of mold temperature, gate diameter, gate angle, and included angle only slightly impacted the defect area. Furthermore, the meaning of squares, which is obtained by dividing the treatment sum of squares by the degrees of freedom, can be used to determine which factors are significant. The higher the mean square, the more of an effect it has on the results [38,39].

Table 4. ANOVA results of the screening step.

No.	Source	Mean of Squares	p-Value
1	Model	3,424,612	0.011
2	Flow rate	4,978,362	0.011
3	Melt temperature	19,348,900	0.001
4	Mold temperature	870,044	0.138
5	Holding pressure	6,600,269	0.007
6	Runner diameter	32,488,905	0.000
7	Gate diameter	458,198	0.250
8	Gate angle	1946	0.934
9	Included angle	96,108	0.571

The DOE intended for the screening step to be executed with all parameters varying between two levels. In the second phase, by performing CCD with an alpha coefficient of 1.4, the number of parameter levels was increased to five (shown in Table 5). This could lead to a better understanding of the parameters' effects on the process outputs.

Table 5. The levels of CCD.

No.	Flow Rate (cm^3/s)	Melt Temperature (°C)	Holding Pressure (MPa)	Runner Diameter (mm)
1	12	185.2	54.6	6.4
2	14	187	59	7
3	19	191	70	8.5
4	24	196	81	10
5	26	197.8	85.4	10.6

To predict the area of the defect with a regression equation, an ANOVA was implemented on the results of the simulations suggested by the CCD. According to Table 6 and considering the p-value of the model equal to 0.000, it can be concluded that the model presented by ANOVA has good accuracy in the prediction of the defect area. The effect of all four parameters known to be effective in the previous step has been reaffirmed by the design of experiments carried out by considering more levels. Diagrams related to effective parameters can be seen in Figure 8.

Table 6. Results of the second ANOVA (CCD).

No.	Source	Mean of Squares	p-Value
1	Model	1,463,290	0
2	**Linear parameters**	**4,473,581**	**0**
3	Flow rate	1,209,310	0.039
4	Melt temperature	1,369,223	0.049
5	Holding pressure	460,154	0.027
6	Runner diameter	16,845,635	0
7	**Squares**	**533,923**	**0.008**
8	Flow rate × Flow rate	6013	0.815
9	Melt temperature × Melt temperature	53,785	0.486
10	Holding pressure × Holding pressure	527	0.945
11	Runner diameter × Runner diameter	1,474,471	0.002
12	**Interactions**	**76,008**	**0.641**
13	Melt temperature × Flow rate	4789	0.834
14	Melt temperature × Holding pressure	9658	0.767
15	Melt temperature × Runner diameter	50,917	0.498
16	Flow rate × Holding pressure	126	0.973
17	Flow rate × Runner diameter	175,496	0.216
18	Holding pressure × Runner diameter	215,063	0.173

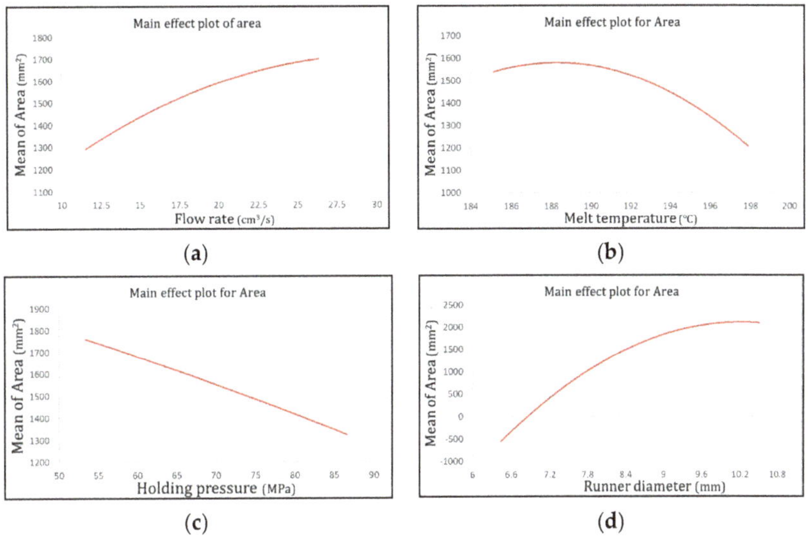

Figure 8. The effect of: (**a**) flow rate, (**b**) melt temperature, (**c**) holding pressure, and (**d**) runner diameter on the defect area.

Figure 8 illustrates the manner of the impact of each parameter on the results. According to the *p*-value obtained from the effects of ANOVA, the runner diameter has the most significant impact on the area of the blush defect. After that, the parameters of holding pressure, flow rate, and melt temperature have the highest impact on the defect area, respectively. Utilizing ANOVA, a regression equation to predict the area of the defect has been worked out using all input parameters presented in Equation (24).

$$Area = -12367 + (25.7 \times A) - (25.4 \times B) - (13.82 \times C) + (3898 \times D) - (193.2 \times D \times D) \tag{24}$$

where *A* indicates the flow rate, *B* the melt temperature, *C* the holding pressure, and *D* represents the runner diameter. The standardized residual and histogram plot shown in Figure 9 reveal the accuracy of Equation (24). As can be seen from the Versus Order plot, the predicted data are in general equally scattered around the zero line. Additionally, the histogram plot in this figure shows that the amount of data with a standard residual equal to or near zero is much higher than the amount of data with higher residuals. Thus, it can be concluded that the defect area amounts predicted by the regression model are in good agreement with the FEA results.

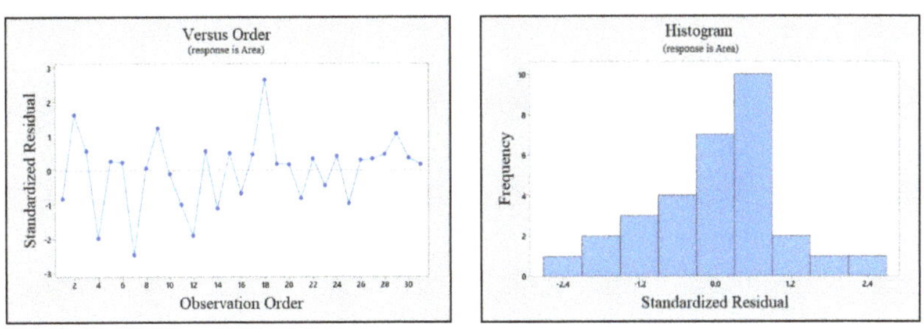

Figure 9. Standardized residual and histogram graph.

3.3. ANN Validation and Comparison with ANOVA

After training the ANN, the network should be verified with the data intended for testing. The test data in this study included 15% of the total data that had not been used in the ANN training phase. Table 7 also compares the prediction accuracy of different types of ANN with that presented by ANOVA.

Table 7. Comparison of response prediction methods.

No.	Predictor	Neurons in Hidden Layer	Number of Particles	Population Size	Iterations	$Accuracy_{Tr}$	Training Time (s)
1	Basic ANN	4			178	99.96%	2
2		6			134	99.97%	2
3		8			91	99.99%	1
4		10			102	99.98%	1
5	ANN + PSO		10		245	97.88%	59
6			30		198	98.23%	178
7			50		179	98.78%	264
8			70		171	99.26%	408
9	ANN + GA			10	276	98.76%	51
10				30	251	98.61%	159
11				50	237	98.33%	256
12				70	209	98.17%	357
13	ANOVA					86.57%	-

According to Table 7, the best networks of each type have been chosen to compare the prediction accuracy. For basic ANN, the trained network in row 3 is considered as the selected basic ANN with the best performance. The type of activator function is "TanSig" for the first layer and "Linear" for the second layer. For ANN + PSO, the trained network in row 7 is regarded as the most suitable algorithm of this type. The trained network in row 11 has also been chosen as the most suitable ANN + GA combination. The mutation rate in this algorithm is 15%, and the cross-over rate is 65%. Additionally, Figure 10 exhibits a comparison between the predicted values via the regression formula provided by ANOVA, selected neural networks of all three types (basic ANN, ANN + GA, and ANN + PSO) with the value obtained from the FEA. Furthermore, Table 8 presents the values predicted by each algorithm for all the data sets compared with ANOVA predicted values and FEA results.

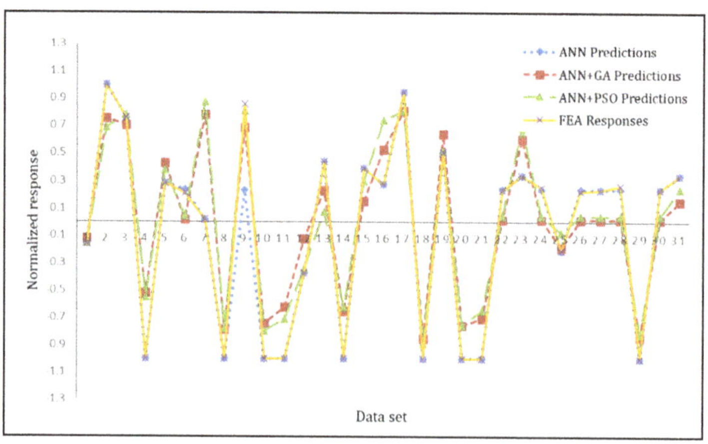

Figure 10. Comparison of normalized neural network and ANOVA prediction values with FEA results.

Table 8. Comparison of normalized values predicted by ANOVA, ANN Algorithms, and FEA.

No.	FEA	ANOVA	Basic ANN	ANN + PSO	ANN + GA
1	−0.161	0.298	−0.161	−0.157	−0.119
2	1	0.849	1	0.688	0.754
3	0.762	0.83	0.762	0.779	0.704
4	−1	−0.219	−1	−0.551	−0.52
5	0.284	0.477	0.284	0.38	0.43
6	0.21	0.417	0.237	0.044	0.016
7	0.015	0.706	0.015	0.876	0.783
8	−1	−0.571	−1	−0.781	−0.786
9	0.859	0.799	0.233	0.811	0.686
10	−1	−0.54	−1	−0.797	−0.745
11	−1	−0.389	−1	−0.71	−0.625
12	−0.376	0.311	−0.376	−0.342	−0.124
13	0.444	0.558	0.444	0.074	0.231
14	−1	−0.37	−1	−0.623	−0.65
15	0.393	0.523	0.393	0.296	0.152
16	0.274	0.629	0.274	0.742	0.531
17	0.945	1	0.945	0.809	0.81
18	−1	−1	−1	−0.886	−0.854
19	0.507	0.679	0.507	0.517	0.644
20	−1	−0.59	−1	−0.757	−0.755
21	−1	−0.42	−1	−0.648	−0.705
22	0.233	0.417	0.237	0.044	0.016
23	0.34	0.647	0.34	0.693	0.606
24	0.252	0.417	0.237	0.044	0.016
25	−0.214	0.276	−0.214	−0.081	−0.191
26	0.225	0.417	0.237	0.044	0.016
27	0.233	0.417	0.237	0.044	0.016
28	0.263	0.417	0.237	0.044	0.016
29	−1	−0.741	−1	−0.82	−0.837
30	0.24	0.417	0.237	0.044	0.016
31	0.336	0.536	0.336	0.241	0.151

As it is clear from Figure 10 and Table 8, ANOVA with 86.5% accuracy and basic ANN with 99.99% accuracy provide the farthest and the closest predictions to the FEA response, respectively. Additionally, according to these data, the basic ANN presented in row 3 of Table 7 (with eight neurons in the hidden layer) has been chosen as the most appropriate prediction model. The graph presented in Figure 11 indicates the training procedure of the basic ANN. During 91 epochs, the basic ANN converged. The MSE trend of training data can be seen alongside this trend for validation and test data. In addition, this figure represents the regression analysis of the basic ANN model for training, validation, and test data. Additionally, the regression model for the whole data set can be seen in the bottom right corner of Figure 11.

3.4. Optimization Using GA

To optimize the parameter levels, a trained ANN with the specifications listed in the third row of Table 4 is used as the cost function. The parameters of melt temperature, flow rate, holding pressure, and runner diameter are considered as the inputs of the GA, and the area of the blush defect is regarded as the only output. The allowable range for each of these parameters is specified in Table 3. Additionally, for the GA, values of 300 were taken into account as the initial population, 300 as the maximum number of generations, 60% as the crossover rate, and 40% as the mutation rate. During the optimization process, the value of the cost function of the GA is constantly decreasing, which indicates the proper performance of the algorithm in finding the optimal values. At the end of the process, the algorithm had reached the optimal response after approximately 170 iterations, after which the response showed little change. The algorithm introduces the values of 197.1,

25.6, 84.9, and 6.4 as the optimal levels for the parameters of melt temperature, flow rate, holding pressure, and runner diameter, respectively. Figure 12 shows the optimization process by the GA. Using experimental validation, it is possible to compare the results of the experimental tests, GA, and ANOVA. This comparison can be seen in Table 5.

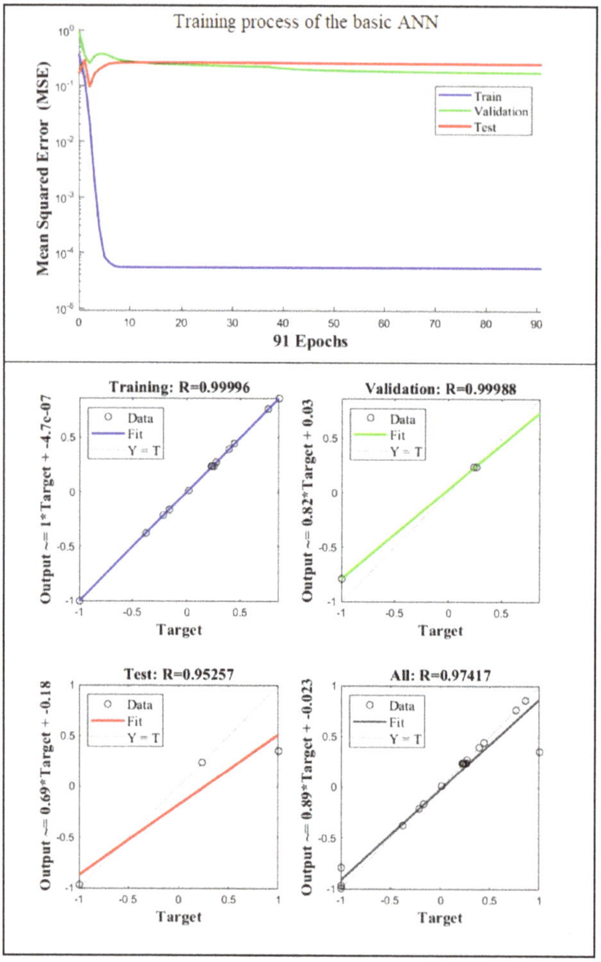

Figure 11. Regression analysis of the basic ANN model.

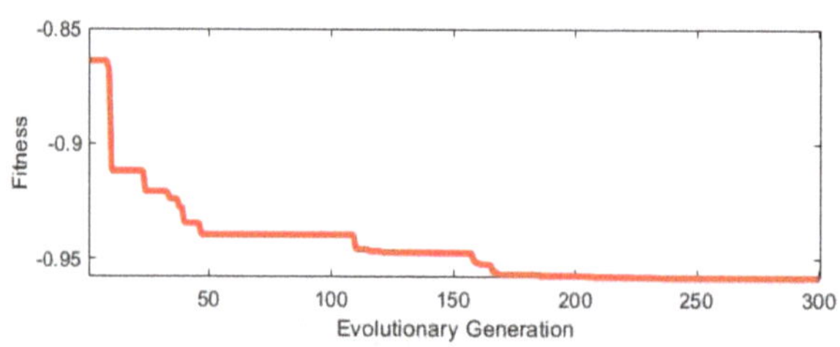

Figure 12. Optimization process by GA.

From the results obtained from Table 9, it can be seen that, on account of optimization using GA, the amount of blush defect area has been reduced by 81.7%. Additionally, the optimization performed using the CCD method has reduced the defect area by 74.0%. Figure 13 shows a visual comparison between the blush area, before and after optimization in the injection molding simulation, alongside the image of the defect in the bushing produced with the optimized parameters.

Table 9. Comparison of the defect area before and after optimization.

Title	Melt Temperature (°C)	Flow Rate (cm^3/s)	Holding Pressure (mm)	Runner Diameter (mm)	Defec tArea (mm^2)
Initial bushing	197.0	12.0	35.0	10.0	1978
Optimal bushing suggested by CCD optimization	197.8	26.0	56.4	6.7	517
Optimal bushing suggested by GA	197.1	25.6	84.9	6.4	362
Experimental test on the dataset suggested by GA	197.1	25.6	84.9	6.4	366

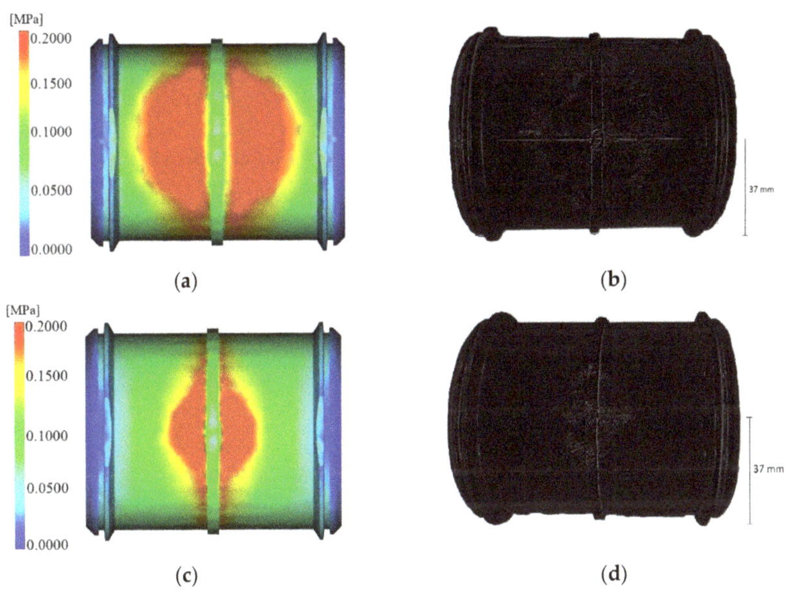

Figure 13. The initial defected bushing area is in simulation (**a**) and experimental test (**b**); and the bushing is after optimization in simulation (**c**) and experimental test (**d**).

4. Conclusions

This study aimed at the determination of the effect of eight process parameters (flow rate, melt temperature, mold temperature, holding pressure, runner diameter, gate diameter, gate angle, and included angle) on the blush defect in PVC bushings produced by injection molding. Some prediction models have been created to estimate the area of the blush defect using ANOVA and ANNs. Among the prediction models used, the basic ANN method with a training accuracy of 99.99% has shown the best performance compared to other prediction methods for predicting the FEA results.

Results showed that flow rate, melt temperature, and runner diameter have a particularly strong effect on the blush defect and can be related to the viscosity of the molten material. The viscosity can increase with rapid cooling of the material, making the material's shear stress exceed the allowable limit and, in turn, promoting blush.

Holding pressure has also affected the blush defect. With the increase in holding pressure, the blush defect decreases. Lower values of holding pressure can cause some semi-cooled material (which has a high viscosity) to enter the mold cavity. Reflowing the material in these high viscosity conditions can cause high amounts of shear rate, which is the underlying reason of blush defect occurrence. The parameters of mold temperature, gate diameter, gate angle, and included angle have a negligible effect on the result.

The key results of the research can be summarized as follows:

- After optimization using the GA through the trained ANN as the cost function, values of 197.1 °C, 25.6 cm^3/s, 84.9 MPa, and 6.4 mm were introduced as the optimal levels for the parameters of melt temperature, injection flow rate, holding pressure, and runner diameter, respectively.
- After applying the optimized parameter levels by GA in a new FEA, the defect area was reduced by 81.7% compared to the pre-optimization defect area.

Author Contributions: Methodology, A.M.A. and G.T.; Software, A.M.A.; Validation, A.M.A., R.S. and G.T.; Formal analysis, R.S.; Investigation, A.M.A.; Resources, G.A.; Writing—original draft, A.M.A.; Writing—review & editing, G.A., Y.S., M.C., J.H.H., M.K. and G.T.; Visualization, A.M.A.; Supervision, G.A. and G.T.; Project administration, G.A. All authors have read and agreed to the published version of the manuscript.

Funding: This research work was partially supported by the DIGIMAN4.0 project ("DIGItal MANufacturing Technologies for Zero-defect Industry 4.0 Production", https://www.digiman4-0.mek.dtu.dk/, accessed on 13 February 2023). DIGIMAN4.0 is a European Training Network supported by Horizon 2020, the EU Framework Programme for Research and Innovation (Project ID: 814225).

Institutional Review Board Statement: Not applicable.

Informed Consent Statement: Not applicable.

Data Availability Statement: Not applicable.

Acknowledgments: This research work was undertaken in the context of DIGIMAN4.0 project ("DIGItal MANufacturing Technologies for Zero-defect Industry 4.0 Production", http://www.digiman4-0.mek.dtu.dk/). DIGIMAN4.0 is a European Training Network supported by Horizon 2020, the EU Framework Programme for Research and Innovation (Project ID: 814225). Yazd Poolica Co. (Yazd, Iran) is thanked for all their support during the experimental steps of the study.

Conflicts of Interest: The authors declare no conflict of interest.

Appendix A. Pseudocodes of the Algorithms

- ANN

    1. Start of program
    2. Train percent = tr
    3. Test percent = ts
    4. Initialize data
    5. Separate input and output data
    6. Normalize all data
    7. Initialize a network structure
    8. Set a random matrix for initial layer weights and biases
    9. repeat
    10. Use tr percent of data for training network
    11. Use Equation (10) to Evaluate trained network with ts percent of data
    12. Use Equations (11) and (12) to reset the weights and biases
    13. Until termination criteria
    14. Simulate the trained network with input data
    15. Calculate and report the MSE

For the ANN + PSO and ANN + GA algorithms, all steps are similar to the above pseudocode. Only the steps for optimizing layer weights (lines 8–12) are overwritten. In other words, the following pseudocodes are replaced with the iterative loop in the upper pseudocode.

- ANN + PSO
    1. Start of weight optimization
    2. for each particle(each neural network)
    3. Initialize particle position(Network weights) and velocity vectors
    4. end for
    5. Fitness = f(X)i
    6. Personal best position = Xpb
    7. Global best position = Xgb
    8. repeat
    9. for particlei i = 1 to Nparticles
    10. Fitness = calculate the Fitness of particlei
    11. if f(Xbp)i < f(Xbp)
    12. Xbp = Xi, Pbest = Fitnessi
    13. end if
    14. if f(Xgb)i > f(Xgb)
    15. Xgb = Xi, Gbest = Fitnessi
    16. end if
    17. update the velocity of the particle using Equations (18 and (19)
    18. update the position of the particle using Equation (20)
    19. end for
    20. Until termination criteria
- ANN + GA
    1. Start of weight optimization
    2. N = number of network weights
    3. G = Number of maximum generations
    4. RecomPercent = r/100
    5. CrossPercent = c/100
    6. MutatPercent = 1 − RecomPercent − CrossPercent
    7. Initialize genes randomly(initial Network weights)
    8. Calculate fitness = f(X)i
    9. Sort Chromosomes according to fitness
    10. for i = 1 to G
    11. Create new population(RecomPercent × N + CrossPercent × N + MutatPercent × N)
    12. Calculate fitness
    13. Sort Chromosomes according to fitness
    14. end for

References

1. Oktem, H.; Erzurumlu, T.; Uzman, I. Application of Taguchi optimization technique in determining plastic injection molding process parameters for a thin-shell part. *Mater. Des.* **2007**, *28*, 1271–1278. [CrossRef]
2. Oliaei, E.; Heidari, B.S.; Davachi, S.M.; Bahrami, M.; Davoodi, S.; Hejazi, I.; Seyfi, J. Warpage and shrinkage optimization of injection-molded plastic spoon parts for biodegradable polymers using Taguchi, ANOVA and artificial neural network methods. *J. Mater. Sci. Technol.* **2016**, *32*, 710–720. [CrossRef]
3. Azad, R.; Shahrajabian, H. Experimental study of warpage and shrinkage in injection molding of HDPE/rPET/wood composites with multiobjective optimization. *Mater. Manuf. Process.* **2019**, *34*, 274–282. [CrossRef]
4. Yin, F.; Mao, H.; Hua, L.; Guo, W.; Shu, M. Back propagation neural network modeling for warpage prediction and optimization of plastic products during injection molding. *Mater. Des.* **2011**, *32*, 1844–1850. [CrossRef]
5. Wang, R.; Zeng, J.; Feng, X.; Xia, Y. Evaluation of effect of plastic injection molding process parameters on shrinkage based on neural network simulation. *J. Macromol. Sci.* **2013**, *52 Pt B*, 206–221. [CrossRef]

6. Altan, M. Reducing shrinkage in injection moldings via the Taguchi, ANOVA and neural network methods. *Mater. Des.* **2010**, *31*, 599–604. [CrossRef]
7. Chen, C.-P.; Chuang, M.-T.; Hsiao, Y.-H.; Yang, Y.-K.; Tsai, C.-H. Simulation and experimental study in determining injection molding process parameters for thin-shell plastic parts via design of experiments analysis. *Expert Syst. Appl.* **2009**, *36*, 10752–10759. [CrossRef]
8. Chen, W.-C.; Liou, P.-H.; Chou, S.-C. An integrated parameter optimization system for MIMO plastic injection molding using soft computing. *Int. J. Adv. Manuf. Technol.* **2014**, *73*, 1465–1474. [CrossRef]
9. Guo, W.; Deng, F.; Meng, Z.; Hua, L.; Mao, H.; Su, J. A hybrid back-propagation neural network and intelligent algorithm combined algorithm for optimizing microcellular foaming injection molding process parameters. *J. Manuf. Process.* **2020**, *50*, 528–538. [CrossRef]
10. Ozcelik, B.; Erzurumlu, T. Comparison of the warpage optimization in the plastic injection molding using ANOVA, neural network model and genetic algorithm. *J. Mater. Process. Technol.* **2006**, *171*, 437–445. [CrossRef]
11. Chen, C.-C.; Su, P.-L.; Chiou, C.-B.; Chiang, K.-T. Experimental investigation of designed parameters on dimension shrinkage of injection molded thin-wall part by integrated response surface methodology and genetic algorithm: A case study. *Mater. Manuf. Process.* **2011**, *26*, 534–540. [CrossRef]
12. Ozcelik, B.; Sonat, I. Warpage and structural analysis of thin shell plastic in the plastic injection molding. *Mater. Des.* **2009**, *30*, 367–375. [CrossRef]
13. Kim, B.-Y.; Nam, G.-J.; Ryu, H.-S.; Lee, J.-W. Optimization of filling process in RTM using genetic algorithm. *Korea-Aust. Rheol. J.* **2000**, *12*, 83–92.
14. Kitayama, S.; Onuki, R.; Yamazaki, K. Warpage reduction with variable pressure profile in plastic injection molding via sequential approximate optimization. *Int. J. Adv. Manuf. Technol.* **2014**, *72*, 827–838. [CrossRef]
15. Kitayama, S.; Yamazaki, Y.; Takano, M.; Aiba, S. Numerical and experimental investigation of process parameters optimization in plastic injection molding using multi-criteria decision making. *Simul. Model. Pract. Theory* **2018**, *85*, 95–105. [CrossRef]
16. Shen, C.; Wang, L.; Li, Q. Optimization of injection molding process parameters using combination of artificial neural network and genetic algorithm method. *J. Mater. Process. Technol.* **2007**, *183*, 412–418. [CrossRef]
17. Kitayama, S.; Hashimoto, S.; Takano, M.; Yamazaki, Y.; Kubo, Y.; Aiba, S. Multi-objective optimization for minimizing weldline and cycle time using variable injection velocity and variable pressure profile in plastic injection molding. *Int. J. Adv. Manuf. Technol.* **2020**, *107*, 3351–3361. [CrossRef]
18. Kitayama, S.; Tamada, K.; Takano, M.; Aiba, S. Numerical optimization of process parameters in plastic injection molding for minimizing weldlines and clamping force using conformal cooling channel. *J. Manuf. Process.* **2018**, *32*, 782–790. [CrossRef]
19. Sin, L.T.; Rahman, W.A.W.A.; Rahmat, A.R.; Tee, T.-T.; Bee, S.T.; Chong-Yu, L. Computer aided injection moulding process analysis of polyvinyl alcohol–starch green biodegradable polymer compound. *J. Manuf. Process.* **2012**, *14*, 8–19. [CrossRef]
20. Xu, G.; Yang, Z.-T.; Long, G.-D. Multi-objective optimization of MIMO plastic injection molding process conditions based on particle swarm optimization. *Int. J. Adv. Manuf. Technol.* **2012**, *58*, 521–531. [CrossRef]
21. Kitayama, S.; Natsume, S. Multi-objective optimization of volume shrinkage and clamping force for plastic injection molding via sequential approximate optimization. *Simul. Model. Pract. Theory* **2014**, *48*, 35–44. [CrossRef]
22. Özek, C.; Çelık, Y.H. Calculating molding parameters in plastic injection molds with ANN and developing software. *Mater. Manuf. Process.* **2012**, *27*, 160–168. [CrossRef]
23. Tábi, T.; Suplicz, A.; Szabó, F.; Kovács, N.K.; Zink, B.; Hargitai, H.; Kovács, J.G. The analysis of injection molding defects caused by gate vestiges. *Express Polym. Lett.* **2015**, *9*, 394–400. [CrossRef]
24. Shin, H.; Park, E.-S. Analysis of crack phenomenon for injection-molded screw using moldflow simulation. *J. Appl. Polym. Sci.* **2009**, *113*, 2702–2708. [CrossRef]
25. Juraeva, M.; Ryu, K.J.; Song, D.J. Gate shape optimization using design of experiment to reduce the shear rate around the gate. *Int. J. Automot. Technol.* **2013**, *14*, 659–666. [CrossRef]
26. Weng, C.; Lee, W.B.; To, S.; Jiang, B.-Y. Numerical simulation of residual stress and birefringence in the precision injection molding of plastic microlens arrays. *Int. Commun. Heat Mass Transf.* **2009**, *36*, 213–219. [CrossRef]
27. Li, S.; Fan, X.; Huang, H.; Cao, Y. Multi-objective optimization of injection molding parameters, based on the Gkriging-NSGA-vague method. *J. Appl. Polym. Sci.* **2020**, *137*, 48659. [CrossRef]
28. Mehat, N.M.; Kamaruddin, S. Investigating the effects of injection molding parameters on the mechanical properties of recycled plastic parts using the Taguchi method. *Mater. Manuf. Process.* **2011**, *26*, 202–209. [CrossRef]
29. Martowibowo, S.Y.; Kaswadi, A. Optimization and simulation of plastic injection process using genetic algorithm and moldflow. *Chin. J. Mech. Eng.* **2017**, *30*, 398–406. [CrossRef]
30. Dimla, D.E.; Camilotto, M.; Miani, F.J.J. Design and optimisation of conformal cooling channels in injection moulding tools. *J. Mater. Process. Technol.* **2005**, *164*, 1294–1300. [CrossRef]
31. Eladl, A.; Mostafa, R.; Islam, A.; Loaldi, D.; Soltan, H.; Hansen, H.N.; Tosello, G. Effect of process parameters on flow length and flash formation in injection moulding of high aspect ratio polymeric micro features. *Micromachines* **2018**, *9*, 58. [CrossRef] [PubMed]

32. Regi, F.; Guerrier, P.; Zhang, Y.; Tosello, G. Experimental Characterization and Simulation of Thermoplastic Polymer Flow Hesitation in Thin-Wall Injection Molding Using Direct In-Mold Visualization Technique. *Micromachines* **2020**, *11*, 428. [CrossRef] [PubMed]
33. Loaldi, D.; Regi, F.; Baruffi, F.; Calaon, M.; Quagliotti, D.; Zhang, Y.; Tosello, G. Experimental validation of injection molding simulations of 3D microparts and microstructured components using virtual design of experiments and multi-scale modeling. *Micromachines* **2020**, *11*, 614. [CrossRef] [PubMed]
34. Chen, W.-C.; Wang, M.-W.; Chen, C.-T.; Fu, G.-L. An integrated parameter optimization system for MISO plastic injection molding. *Int. J. Adv. Manuf. Technol.* **2009**, *44*, 501–511. [CrossRef]
35. Finkeldey, F.; Volke, J.; Zarges, J.-C.; Heim, H.-P.; Wiederkehr, P. Learning quality characteristics for plastic injection molding processes using a combination of simulated and measured data. *J. Manuf. Process.* **2020**, *60*, 134–143. [CrossRef]
36. Mehat, N.M.; Kamaruddin, S. Optimization of mechanical properties of recycled plastic products via optimal processing parameters using the Taguchi method. *J. Mater. Process. Technol.* **2011**, *211*, 1989–1994. [CrossRef]
37. Lladó, J.; Sánchez, B. Influence of injection parameters on the formation of blush in injection moulding of PVC. *J. Mater. Process. Technol.* **2008**, *204*, 1–7. [CrossRef]
38. Sibiya, N.P.; Amo-Duodu, G.; Tetteh, E.K.; Rathilal, S. Model prediction of coagulation by magnetised rice starch for wastewater treatment using response surface methodology (RSM) with artificial neural network (ANN). *Sci. Afr.* **2022**, *17*, e01282. [CrossRef]
39. Haftkhani, A.R.; Abdoli, F.; Sepehr, A.; Mohebby, B. Regression and ANN models for predicting MOR and MOE of heat-treated fir wood. *J. Build. Eng.* **2021**, *42*, 102788. [CrossRef]

applied sciences

MDPI

Article

Study and Simulation of an Under-Actuated Smart Surface for Material Flow Handling

Edoardo Bianchi [1,*], Oliver Jonas Jorg [2], Gualtiero Fantoni [2], Francisco Javier Brosed Dueso [1] and José A. Yagüe-Fabra [1]

1 Department of Design and Manufacturing Engineering, University of Zaragoza, C/Maria de Luna 3, 50018 Zaragoza, Spain
2 Department of Civil and Industrial Engineering, University of Pisa, Largo Lucio Lazzarino, 56122 Pisa, Italy
* Correspondence: ebianchi@unizar.es

Abstract: Smart surfaces are becoming more and more popular in the field of intralogistics, as they combine great flexibility with easy reprogrammability. Pursuing this trend, the following article proposes a modular surface to perform handling tasks, such as sorting, stopping, or slowing down material flows. Differently from the current technology, the surface used is under-actuated, thus, it exploits the speed, already possessed by the object, or the gravity to perform, with a simplified hardware, for the aforementioned tasks. In practice, these handling actions are completed using an array of rotors, of which only the direction of the rotation axis is controlled. Moreover, the axis can only assume certain discrete orientations in the plane, further simplifying the design. Thus, what is created is a controllable and under-actuated friction field, which, in contrast with similar existing systems, does not require active driving forces to manipulate the material flow. In the article, the analytic model of the surface is described, and a software simulation environment is introduced to demonstrate its functioning. In addition, examples of sorting, slowing down, and stopping operations and a validation of the simulation itself are presented.

Keywords: smart surface; friction force field; under-actuation; feeding; simulation; material flow handling

Citation: Bianchi, E.; Jorg, O.J.; Fantoni, G.; Brosed Dueso, F.J.; Yagüe-Fabra, J.A. Study and Simulation of an Under-Actuated Smart Surface for Material Flow Handling. *Appl. Sci.* **2023**, *13*, 1937. https://doi.org/10.3390/app13031937

Academic Editor: Muhammad Junaid Munir

Received: 19 December 2022
Revised: 10 January 2023
Accepted: 31 January 2023
Published: 2 February 2023

1. Introduction

The material transport inside factories and warehouses is a very studied topic. However, companies are always searching for new, cheaper solutions while keeping an eye on efficiency and flexibility [1–4]. Starting with the most basic conveyors, many alternative systems [5–12] have been developed in order to move, sort, orient, and handle the material flow. More recently, the focus of investigation has shifted towards modular devices and surfaces that allow high handling capability, adaptability, and reconfigurability [3,4,13]. This happens because the market increasingly demands a flexible industry, and this is reflected in all production levels, even the internal transportation. These devices are often called smart surfaces because they are controllable and reprogrammable with computers. Thanks to these characteristics, these new systems permit achieving their goals without structural changes [4,13]. In addition, the new devices are capable of identifying a part with sensors and acting in diverse ways, according to their programming. As a result, supervision by an operator is no longer required for the decision process and the system becomes autonomous, promising to reduce errors and costs [14].

The most relevant solutions among the smart surfaces, according to the current state of the art and classified by the operating principle, are: micro electro-mechanical systems, vibrating surfaces, ciliary motion, variable morphology surfaces, pneumatic surfaces, surfaces with rotors, and mobile platforms. Micro electro-mechanical systems (MEMS) [6,15,16] are an array of microscopic cantilevers or tilting planes, actuated electrically, that generate forces to transport the object in contact. The second category of the list are vibrating surfaces. These consist of vibrating plates on which a sequence of supply frequencies is applied to generate a two-dimensional force field for the handling tasks [5,17]. In [12,18–20], ciliary motion is proposed to move and manipulate objects, taking advantage of an array of controllable cilia for contact conveyance. Another class are variable morphology surfaces [9,21–23]. For these devices, gravity together with inclined planes and vertical actuators at different altitudes are used to create a preferential path for the movement and rotation of an object. Additionally, pneumatic surfaces [7,24–30] were introduced to handle materials without direct contact. The working principle for the latter is to have modules with nozzles to move parts by directing the air flow below them. On the other hand, surfaces with rotors take advantage of contact forces made by actively driven wheels to manipulate the material flow [3,11,13,14,31–35]. Finally, mobile platforms [8,36–39] consist of mobile pallets, not connected to fixed axes and free to move on a plane, transporting objects on top. Despite the categorization, not all of these are practically used for macroscopic intralogistics purposes (object size and displacement > cm). In fact, as summarized in Figure 1, some of them, such as MEMS, together with some mobile platforms [8,37–39] and ciliary devices [12,18,19], are more suitable for microscopic transport (object size and displacement < mm). In contrast, pneumatic systems, vibrating surfaces, surfaces with rotors, variable morphology surfaces, some mobile platforms [36], and even tilted brushes (Cilia) [20] can be used to move macroscopic and mesoscopic (object size and displacement > mm, <cm) objects.

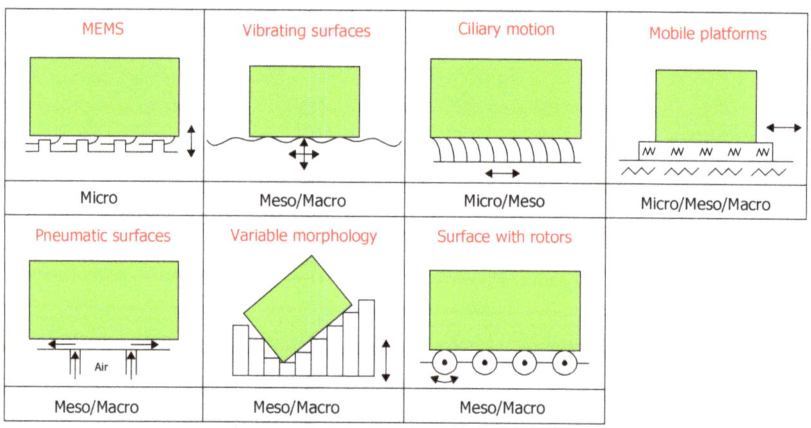

Figure 1. Smart surfaces classification and field of application.

Referring to the previous classification, the system proposed in this paper can be included in the surfaces with rotors category. The literature review conducted on this category identified two main types of modular smart surfaces: systems where each module has only one degree of freedom [3,11,13,35] and systems where each module has two [14,34]. The former, which are the most recent in the literature, exploit multiple omnidirectional and motorized wheels called "omniwheels" positioned in the module with their axes fixed. The layout and number of wheels is not the same in every study, however; as a reference, three or more wheels are usually employed and their spinning axes are positioned out of alignment to create controllable forces in both directions of the plane. As first example of modules with more than three wheels, study [13] alternates units with seven wheels—one large in the center and six small on the sides, to units with five—two large on the sides and three small in the center. For both layouts, the axes of the large wheels are perpendicular to the axes of the small ones to ensure driving forces in the plane. In contrast, in [3,11,35], each module of the device studied, called "Celluveyor", contains three omnidirectional wheels with axes arranged at 120° to one another. Each wheel can rotate at different speeds to control the magnitude of the force exchanged with the transported object, and all three forces together are used to handle the body motion. On the other hand, the second group of systems, i.e., those with modules with two degrees of freedom, consist of units composed of one or more motorized and swiveling wheels. In this case, therefore, as presented in [14,34], the basic element of the module is a simple wheel driven by a motor mounted on its axis of rotation. This wheel, however, unlike the one degree of freedom systems, is also mounted on a vertical axis, which is itself motorized and therefore swiveling. Summarizing, these devices just described are fully implemented and contain several motors per module, specifically, three or more for the first class and at least two for the second. They can actively drive, sort, and manipulate the material flow, creating a totally actuated surface with significant handling skills. However, the use of several motors and their control necessarily involves a greater effort in the management of the system, as well as increasing costs and complexity. According to the current state of the art, a simpler under-actuated device without motors is missing for the same intralogistic tasks. For that reason, seeking cost reduction and simplification of both control and design, without losing the flexibility of this class of systems, in this paper the authors propose an under-actuated surface composed of modules. Unlike existing concepts, each module contains an idle rotor instead of a motorized wheel. The axis of rotation of the mentioned rotor is not driven continuously by a motor but can be oriented in a limited number of directions in the plane of the surface, creating a directionable friction force for object handling. In order to compensate for the under-actuation, the material is moved by exploiting gravity or an initial velocity of the object provided by another system (e.g., a previous conveyor belt). Therefore, compared to the existing systems [3,11,13,14,31–35], the novelty that distinguishes this active surface from the literature technology lies in its simplicity. In fact, the two main characteristics of the system proposed by the authors, compared to those in the same category, are the under-actuation and the limited directions of the rotor axis, both of which are reductions to a minimal form of concept technology, but which save components and make the most of sources already present in the application environment, such as gravity and the speed of the objects in the transport line. Against potential assumptions of performance losses, the authors proved through simulations that the same goals and efficiency [3,4,35] of sorting and handling can be achieved with a minimal design architecture, consisting of few components, while decreasing costs and saving energy.

The focus of this article lies on the study of the surface, the description of the working principle, and its simulation with the purpose of providing an initial proof of concept for possible applications and future developments. Furthermore, the simulation environment allowed numerical results to be obtained for typical intralogistics tasks, giving the opportunity for an initial comparison with the corresponding current technology. The simulations presented in this article are carried out with a customized code developed by the authors using the software MATLAB (Version R2022a). In addition, a validation of the simulation results was obtained with the well-known program for multi-body dynamic simulations Hexagon D&E Adams MSC (Version 2022.1). The authors decided to develop their own simulation environment because of its significant lower computing time. Thus, their software is eligible to control the actual physical system in real time in the future.

The remainder of this paper is organized as follows: Section 2 . describes the concept of the surface. Section 3 explains the analytic model used to describe the working principle. Section 4 describes the simulations of the system for different tasks. Finally, Section 5 reports the results, Section 6 consists in the validation of the latter, and Section 7 describes the conclusions.

2. Concept Description

This section provides a qualitative description of the functioning of the surface and the modules of which it is composed. The operating principle of the surface is based on an array of modules with an idle rolling element and an orientable axis of rotation. These axes are used to generate directionable friction forces on the body in contact with the surface. The origin of the friction forces lies in the very working principle of a rotor. In fact, these are constructed and used (e.g., conveyor rollers) to facilitate motion in one direction (perpendicular to their axis of rotation). This leads to a difference in the friction forces exchanged with an overlying object.

In practice, there will be a smaller component, perpendicular to the axis, caused mainly by rotational friction, and a larger one, parallel to the axis, due to linear friction (F_{perp} and F_{par}, respectively, in Figure 2c). Therefore, the main friction component will be predominantly along the axis, so by orienting the latter, the main force will also be directed. The choices of the axis directions are limited to a fixed number. For example, the simplest configuration is with two orientations (0, 90°), whereas more complex designs are with four (0°, −45°, 45°, 90°) (Figure 2b) or more. In Figure 2b, the four orientations are shown, with the rotors displayed like cylinders (the projection is a rectangle) and the axes like dashed lines. For applications such as slowing or stopping the material flow, two orientations may be enough, whereas for sorting, four orientations should be taken into account. Therefore, modules with four directions will be considered in the following pages. For its intended applications, the module must be used together with others to create the surface (Figure 2a), so the sum of the forces exerted by every unit will control the objects for sorting and feeding purposes. In Figure 2a, the surface setup is summarized, showing the grid of modules with the object on top, and the arrows inside each cell represent directions of motion promoted by them. As noted in the introduction section, the system is under-actuated, so the surface can only slow down the object while it completes its task of sorting or feeding. If, for example, a body is simply placed on the surface without external actuation, it will not move. To overcome this situation, the surface plane can be tilted to use the gravity effect as the missing actuation, or an initial velocity can be given to the part before crossing the surface. Depending on the application and the task of the system, one solution may be preferred over the other, or a combination of both may be chosen. For instance, if the extension of the surface is big, because the objective is to orient a body, the tilted solution could be better, whereas, if the goal is to sort parts in a conveying line, the initial velocity of the object, given by the conveyors before the sorting area, is enough.

In addition, similar to other active surfaces that use rotors for intralogistic purposes, the application is limited to parts with at least one planar surface, in order to have simultaneous contact with three or more rotors. Another limit will be the maximum load per module, but this is related to the resistance of the device, which is not studied in this article. However, the study conducted is done considering the dimensions of the modules and their resistance comparable with the similar actuated existing systems. To provide a reference, the base of a transported object has the minimum dimensions of $b \times h = [20 \text{ cm} \times 20 \text{ cm}]$, and the weight bearable by a single module is 20 kg.

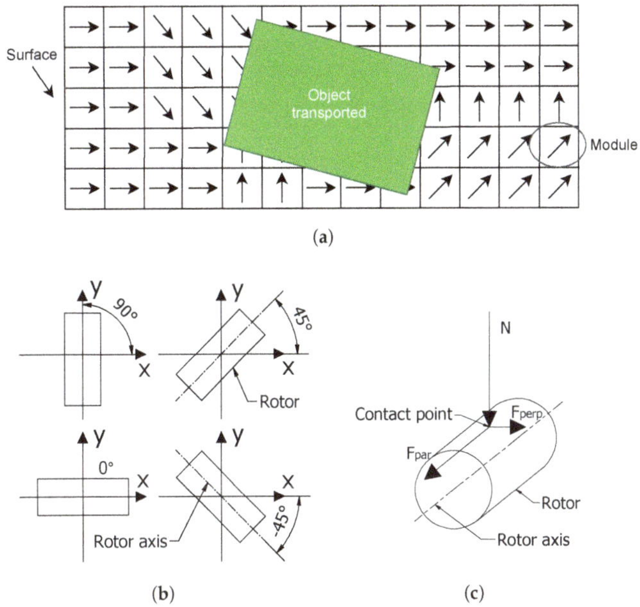

Figure 2. Illustration of: (**a**) the modular active surface, (**b**) the four possible orientations of the rotors, and (**c**) the three contact forces between the object and the rotor.

3. Analytical Model

Continuing with the description of the functioning of the surface, this section reports the hypothesis, the forces involved to move the objects, and the model applied to determine these, respectively, in the subsections *Assumptions*, *Friction Forces Model*, and *Equilibrium Equations*. As stated in Section 2, the working principle of the module is based on the contact forces exchanged between the rotor and the transported object. With this in mind, the analysis starts from the actions on a single module and then expands to the whole surface and to the equations describing the motion of an object.

Figure 3 introduces the notation used in the next pages. In particular, $[x_g, y_g, z_g]$ are the coordinates of the object's center of mass, G, according to the reference frame $\{O, x, y, z\}$, and $\widetilde{G}$ is the projection of G on the object base along the Z axis (Figure 3a).

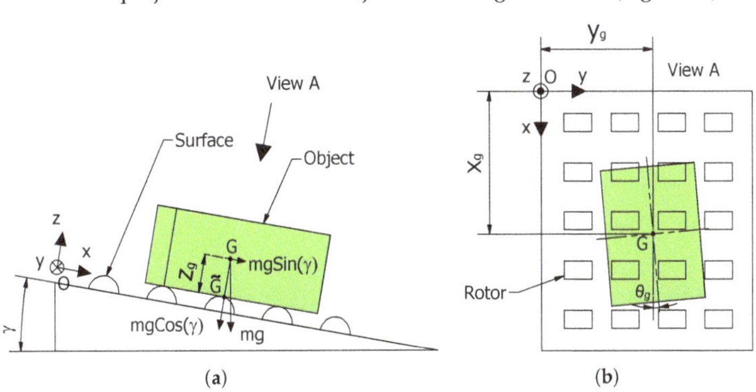

Figure 3. Schematic illustration of the object on the surface: (**a**) lateral view, (**b**) upper view.

3.1. Assumptions

Listed here are the fundamental assumptions that allowed the definition of the analytical model describing the motion of an object on the surface.

1. The first assumption states that there are only the three forces for each interaction rotor-object: two components in the plane of the device and one normal to this (Figure 2c). Considering that the contact area of the module below the object is small compared to the object itself, the previous assertion is plausible and it is translated into the possibility of neglecting the exchanged torques.
2. The second assumption is about the normal component N. In fact, considering that the contact in most of the cases is on more than three rotors, the constraint is hyperstatic and the equilibrium equations of the body are not enough to determine the normal reactions. In particular, for determining the distribution of the forces N_i, an optimization algorithm described in Equation (1) is used.

$$min_{\underline{X}}\, f(\underline{X})\ such\ that \begin{cases} A_{eq} \cdot \underline{X} = \underline{b_{eq}} \\ \underline{l_b} \leq \underline{X} \leq \underline{u_b} \end{cases} \tag{1}$$

where the first vector equation represents the equilibrium in the vertical direction and around the two axes of rotation in the plane, and the second vector equation states the boundary conditions. The function to be minimized $f(\underline{X})$ is the Euclidean norm of the N_i array. The single terms are:

$$A_{eq}(3 \times n) = \begin{bmatrix} 1 & 1 & \dots & 1 \\ y_{\tilde{G}P_1} & y_{\tilde{G}P_2} & \dots & y_{\tilde{G}P_n} \\ -x_{\tilde{G}P_1} & -x_{\tilde{G}P_2} & \dots & -x_{\tilde{G}P_n} \end{bmatrix}, \quad \underline{b_{eq}}(3 \times 1) = \begin{bmatrix} mg\cos(\gamma) \\ -m\ddot{y}z_g \\ m\ddot{x}z_g - mgz_g\sin(\gamma) \end{bmatrix},$$

$$\underline{X}(n \times 1) = \begin{bmatrix} N_1 \\ \dots \\ N_i \\ \dots \\ N_n \end{bmatrix}, \quad \underline{l_b}(n \times 1) = \begin{bmatrix} 0 \\ 0 \\ \dots \\ 0 \end{bmatrix}, \quad \underline{u_b}(n \times 1) = \begin{bmatrix} mg\cos(\gamma) \\ mg\cos(\gamma) \\ \dots \\ mg\cos(\gamma) \end{bmatrix}, \quad f(\underline{X}) = \|\underline{X}\|$$

where n is the number of rotors under the object, m is the mass of the object, g is the gravity acceleration, and γ is the inclination of the surface compared to the ground (Figure 3a). In addition, $x_{\tilde{G}P_i}$ and $y_{\tilde{G}P_i}$ are the x and y components of the vector that connect $\tilde{G}$ to the ith contact point P_i.

3. The third assumption asserts that the inertia of the rotors is negligible, as they are small and made of lightweight material. Instead, the rotational friction given by the spinning of the rotor around its axis is included in the calculation of contact forces.
4. The fourth assumption states that, for an object in motion on the rotors, the friction conditions at the contact points (rotors–object) are as follows: dynamic friction (sliding) in the direction of the axis of the rotors and pure rolling in the perpendicular direction. This is plausible, as the center of instantaneous rotation (CIR) of the rigid body (object) is very unlikely to be located at the center of an underlying rotor. This would guarantee a different static friction and force modeling (F_{par}, F_{perp}). However, with only a few discrete points as supports, it is unlikely that the CIR is located on these. In practice, for the axis direction, this assumption justifies the choice of a smooth Coulomb friction model, which is computationally stable and reliable, as the sliding condition prevails ([40]). The same model is considered for the friction force, due to the rolling, in the perpendicular direction. This value depends on the rotational friction of the rotor axis (assumption 3), so it can also be expressed using the same Coulomb formulation. More complex rotational friction models have been

avoided to reduce complexity and the computational cost (for the simulation) without adding relevant contribution, as this friction component is very small and almost negligible compared to the axis direction component. In conclusion, this assumption, introducing the friction model, results in the boundary condition $F = \mu N$ for the calculation of friction forces.

3.2. Friction Forces Model

In this subsection, considering the previous assumptions regarding sliding and pure rolling and taking into account that the friction forces oppose motion, the directions of the two forces (F_{par}, F_{perp}) and their final formulation are determined. Thanks to the third assumption, these directions are related only to the object velocity at the contact point with the rotor, $\overrightarrow{V_p} = [\dot{x}_p, \dot{y}_p]$; an example is shown in Figure 4a.

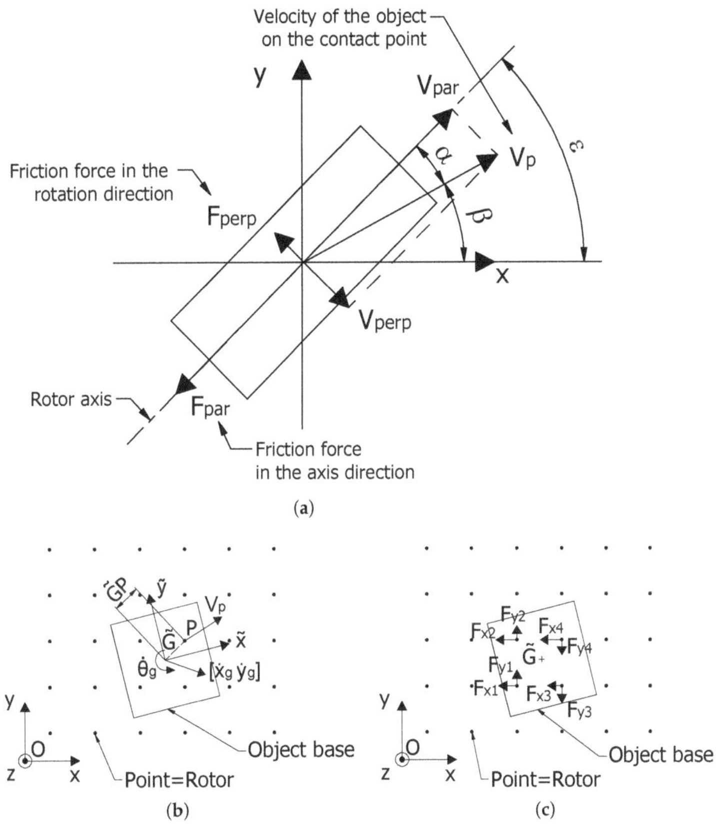

Figure 4. Schematic representations of: (**a**) velocities and forces exchanged in a contact point, (**b**) the calculation of the velocity in the contact point, (**c**) the resulting forces from the contacts.

In particular, the velocity $\overrightarrow{V_p}$ can be split into a perpendicular and a parallel component (Figure 4a), and the two forces F_{par} and F_{perp} are in the opposite directions of these velocities. In Figure 4a, three more angles are introduced: ε, β, and α (all angles are defined positive in the anti-clockwise direction from the x axis):

- ε is the angle between the rotor axis and the x axis;
- β is the angle between the velocity of the object's contact point and the x axis;
- α is obtained by the difference between the previous two ($\alpha = \varepsilon - \beta$), so it is the angle between the velocity of the body's contact point and the rotor axis.

The velocity of the object's contact point is determined using the formula of the kinematics of a rigid body: $\overrightarrow{V_p} = [\dot{x}_g\hat{i}, \dot{y}_g\hat{j}] + \dot{\theta}_g\hat{k} \times \overrightarrow{\widetilde{G}P}$, where $[\dot{x}_g, \dot{y}_g]$ and $\dot{\theta}_g$ are the linear and angular velocity of the object, and $\overrightarrow{\widetilde{G}P}$ is the vector that connects $\widetilde{G}$ to the contact point P. The velocity identification is also reported in Figure 4b, where the piece is schematized with the shape of the base, and the rotors with a grid of points. This scheme presented in Figure 4b is also useful in visualizing how the rotor–object contact points are determined. In fact, knowing the values $[x_g, y_g, \theta_g]$ and the mathematical formulation of the base area with respect to these, it is possible to know the set of coordinates of the points inside the base perimeter. More specifically, given a planar reference frame $\{\widetilde{G}, \widetilde{x}, \widetilde{y}\}$ with coordinates $\widetilde{x}, \widetilde{y}$, centered in $[x_g, y_g]$ ($\widetilde{G}$) (Figure 4b) and oriented by θ_g, let Q be a set of geometrical parameters that depends on the shape of the object base (e.g., radius for a circle or base and height for a rectangle) and let $GD \in Q$. Let Λ be a set of indexes. For every $\lambda \in \Lambda$, let $F_\lambda : \mathbb{R}^2 \times Q \to \mathbb{R}$ be functions such that $W = \{(x, y) : F_\lambda(x, y, GD) = 0, \forall\lambda \in \Lambda\}$ is the perimeter of the object base. The area of the object base, D, is defined as $D = \{(\widetilde{x}, \widetilde{y}) : F_\lambda(\widetilde{x}, \widetilde{y}, GD) \leq 0, \forall\lambda \in \Lambda\}$. Therefore, given the x_R, y_R coordinate pairs that define the position of the rotors in the array (which are fixed), it is verified whether or not they belong to the set, i.e., if $\begin{bmatrix} x_R \\ y_R \end{bmatrix} = \begin{bmatrix} x_g \\ y_g \end{bmatrix} + R_{\theta_g}\begin{bmatrix} \widetilde{x} \\ \widetilde{y} \end{bmatrix}$ has solution for a pair of $\widetilde{x}, \widetilde{y} \in D$, where $R_{\theta_g} = \begin{bmatrix} \cos\theta_g & -\sin\theta_g \\ \sin\theta_g & \cos\theta_g \end{bmatrix}$ is the rotation matrix of the reference frame with coordinates $\widetilde{x}, \widetilde{y}$ and centered in $\widetilde{G}$. Now all the elements to describe the two forces in the plane of the surface (F_{par}, F_{perp}) have been introduced and the result, according to Figure 4a, is summarized in Equation (2a,b).

$$F_{par_i} = \mu_{max} N * \tanh(k|V_p|) * sgn(\cos(\alpha)) \tag{2a}$$

$$F_{perp_i} = \mu_{min} N * \tanh(k|V_p|) * sgn(\sin(\alpha)) \tag{2b}$$

where the two friction coefficients are, respectively:

- μ_{max} is the friction coefficient between the object and the rotor;
- μ_{min} is the friction coefficient that represents the rolling resistance of the rotor.

The *sgn* function models the direction of the forces, according with the orientation of the object velocity in the contact point, whereas *tanh* and the coefficient k (selection of this parameter explained in Section 6) smooth the discontinuities ([41]). The perpendicular force would be ideally 0, however, it is generally small as inertia forces are neglected and only the rotational friction of the rotor around its axis is considered. The parallel friction force can reach significantly higher values and is opposed (in the axis direction) to the object velocity. Therefore, it is the dominant driver to manipulate the object.

3.3. Equilibrium Equations

This subsection shows how the equations of motion of the object on top of the surface are obtained from the contact force components. In practice, once these components are calculated for each contact point, they can be added together to obtain the total reactions on the part and its equilibrium equations. To achieve this, first, the friction forces in the plane at the various contact points are decomposed along the x and y directions, knowing the arrangements of the rotors (ε_i):

$$F_{x_i} = -F_{par_i}\cos(\varepsilon_i) - F_{perp_i}\sin(\varepsilon_i) \tag{3a}$$

$$F_{y_i} = -F_{par_i}\sin(\varepsilon_i) + F_{perp_i}\cos(\varepsilon_i) \tag{3b}$$

The signs introduced for these forces are in accordance with Figure 4a,c. The resulting equilibrium equations are summarized in Equation (4).

$$\begin{cases} \ddot{x}_g = \frac{1}{m}((\sum_{i=1}^{n_i} F_{x_i}) + mg\sin(\gamma)) \\ \ddot{y}_g = \frac{1}{m}(\sum_{i=1}^{n_i} F_{y_i}) \\ \ddot{\theta}_g = \frac{1}{J}(\sum_{i=1}^{n_i}((x_{P_i} - x_g)F_{y_i} - (y_{P_i} - y_g)F_{x_i})) \end{cases} \tag{4}$$

where $\ddot{x}_g$, $\ddot{y}_g$, and $\ddot{\theta}_g$ are, respectively, the two linear and the angular accelerations of the object; x_{p_i} and y_{p_i} are the coordinates of the ith point of contact; and J is the moment of inertia of the object around the axis perpendicular to the surface plane. Considering that also J can be expressed as a function of the mass, it is possible to notice that the inertia affects the system only from a structural point of view. In fact, as it is visible from the Equations (1), (2a,b), and (4), m can be simplified.

4. Simulations

In this section, the simulation process carried out with the analytic model implemented in MATLAB is shown. As a first step, this simulation environment permitted verifying in a general way that moving objects with the proposed theory and concept is possible. Second, it allowed several tests to be carried out in order to prove the usability of the system for some common intralogistics applications, such as sorting, orienting, stopping, and slowing down material flows.

The simulation takes advantage of an iterative loop (shown in Figure 5a) that involves the following steps: initialization, rotor identification, stopping criteria, contact forces and equilibrium computation, and two steps of integration. The initialization of the problem is made at the beginning by providing data from the object and the surface (example: object mass, base shape, initial position and velocity, friction coefficients, module pattern, inclination of the surface, rotor orientations, etc.). After that, the iterative loop can start. For each step (e.g., the sth step), the rotors below the object are identified (thus the contact points as well), knowing the position of the object from the previous iteration ($(s-1)$th) and the disposition of the modules (as introduced in Section 3). Once the rotors below the object and their orientation are determined, the object velocity from the previous iteration ($(s-1)$th) together with the inclination of the surface are evaluated by the stopping criteria: if "$\dot{x}_g \leq 0$ m/s and $\gamma = 0°$", the object is stopped because of negative velocity and null inclination and the loop ends. This is true assuming, in general, that the initial conditions always provide an input towards positive x ($\gamma \geq 0°$ or $\dot{x}_{g0} \geq 0$). When the stopping condition is not reached, the loop continues and the forces in the contact points can be computed (still considering the object velocity from the previous iteration ($(s-1)$th)). With these forces, the equilibrium of the body and the accelerations are calculated according to Section 3. Finally, in order to obtain the velocity and the position, two integration steps of the acceleration vector are implemented. The derived values are the input of the next iteration ($(s+1)$th) of the loop.

So far, the model seems to represent the operation of the surface adequately when the body velocity is greater than zero in the x direction and for the stopping condition. However, when, as an example, the object is about to start from a standstill with rotors and surface inclined, an undesirable phenomena such as reversal of motion (negative x) can occur. This happens because, in the friction assumptions, the static condition is not initially considered (Equation (2a,b)). However, reverse motion is obviously not possible in reality, then, in practice, when the condition of $\dot{x}_g \leq 0$ m/s is achieved ($\gamma > 0°$, otherwise the object is stopped as described before), the process needs to be adjusted. The logic instructions to solve this problem and at the same time implement the analytical model of Section 3 are summarized in the block diagram of Figure 5b, which in practice is executed inside "Friction forces & Object equilibrium" block of Figure 5a. In detail, the process works as follows: first, the N_i terms are calculated as explained in Section 3, then, since the velocity along x is known from the previous iteration, the condition "$\dot{x}_g \leq 0$ m/s" is verified. If it is false and "$\dot{x}_g \geq 0$ m/s", there is no problem of standstill or stopping and the procedure continues as described in Section 3, thus, friction forces calculation (Equation (2a,b)) and equilibrium ((Equations (3a,b) and (4)). In contrast, if the velocity is less or equal to zero ("$\dot{x}_g \leq 0$ m/s" is true), an initial positive speed is assigned to the object $\dot{x}_g = 0.0001$ m/s, the friction model is applied, and it is verified if the gravity effect is stronger than the friction forces in the x direction ("$\sum_{i=1}^{n_i} F_{x_i} + mg \sin\gamma > 0$"). At this point, if gravity wins, the object is moving according to the process defined before (Equations (3a,b) and (4)); in contrast, if gravity is not enough, the friction forces in the x direction will be of the same magnitude of the gravity effect ("$\sum_{i=1}^{n_i} F_{x_i} + mg \sin\gamma = 0$"), and the displacement will be in the y direction (always according to Equations (3a,b) and (4)). To clarify the diagram of Figure 5b, the condition "$\sum_{i=1}^{n_i} F_{x_i} + mg \sin\gamma = 0$" permits the calculation of the F_{par_i} terms, while F_{perp} is the same as the previous case. For instance, this procedure permits us to obtain the motion of the object when it is placed without an initial velocity on the inclined surface, with the rotors oriented at 45°. In fact, with a sequence of displacement in the y and x directions, the movement is achieved.

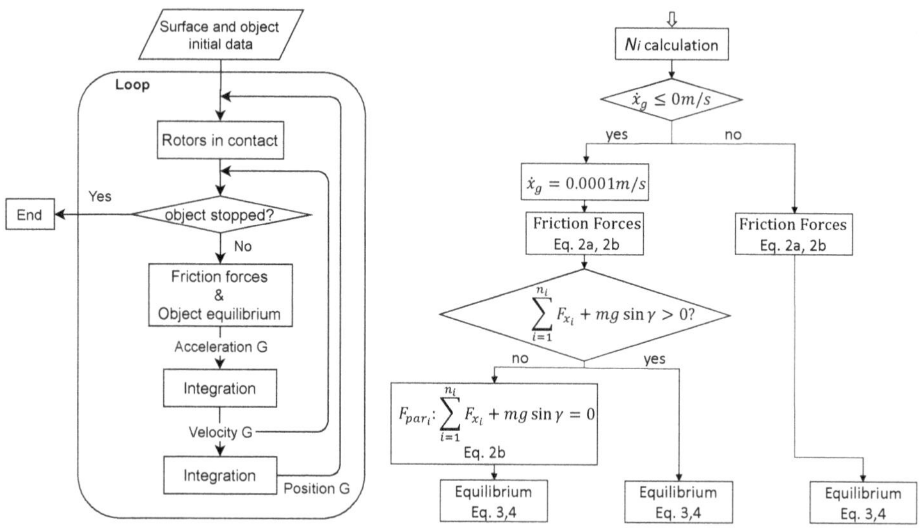

(a) (b)

Figure 5. (**a**) Iterative loop scheme for the simulation. (**b**) Block diagram of the logic process behind the model to avoid the reverse motion and to calculate friction forces (Equation (2a,b)) and equilibrium equations (Equations (3a,b) and (4)).

The simulations conducted to test the surface are divided into three main categories, according to the application of the system:

1. Sorting of material flows on a conveying line;
2. Slowing of material flows on a conveying line;
3. Stopping of material flows on a conveying line.

The first category of simulations concerns sorting. The objective of the setup and the sorting itself is to divert an object from the transport line. The layout considered for the simulation involves an array of modules, of which the subset performing the task has the rotors with the rotation axes inclined at ±45° (in Figure 6a inclined at +45°). The area outside the line, where the objects are directed, is simulated as an array of low-friction (μ_{exsurf}) support points, distributed as the sorting modules. In this zone, the friction is totally opposite to the velocity of the object, without considering any rotor inclination. The sorting is assumed achieved when the body moving on the line is deflected in such a way that it is only in contact with the elements outside the line.

The second simulated application is the slowing activity. The rotor arrangement consists of a few modules within a line to create a controllable friction on the object to modulate its speed, without pushing it to the sides. The layout in this case can use rotors oriented with their axes in the direction of the flow ($\varepsilon_i = 0°$) (Figure 6b) or inclined at $+45°$ in a row and $-45°$ in the other (Figure 6c).

The last application concerns stopping the motion of an object. This task is the extreme application with respect to the previous application of slowing down. In fact, the modules are still placed with their axes in the flow direction (Figure 6b) or oriented at $\pm 45°$ (Figure 6c), but aim to stop the object.

These three types of simulations do not yet include either real-time control of the rotors or position or trajectory tracking for the object, because, as already indicated, the authors' objective in this paper is to verify the mechanical functioning of the surface. However, it is possible to imagine that by having sensors that recognize the object arriving at the surface, the modules can pre-arrange the rotors, as shown in Figure 6, depending on the application or sorting direction. This lays the foundation for using the system as a smart surface.

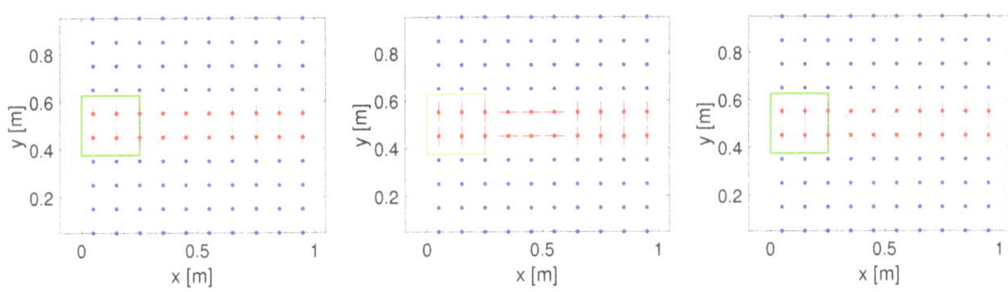

(a) (b) (c)

Figure 6. Examples of simulation layouts for: (**a**) sorting, (**b**) slowing or stopping with $\varepsilon_i = 0°$, and (**c**) slowing or stopping with $\varepsilon_i = \pm 45°$.

Figure 6 introduces the symbols used in the following sections to display the layouts of the simulations implemented in MATLAB: the red elements correspond to the transport line, the blue dots correspond to the area (array of support points) out of the line, and the green square contour corresponds to the object base. Regarding the transport line, dots represent rotors centers, and lines represent the rotors axes. The red lines before the active area (the green square) are oriented perpendicularly to the flow line and their effect is to reduce in a minimum way the motion of the box in the flow direction, and the inclined lines ($\varepsilon_i \neq 90°$) simulate the sorting, the slowing, or the stopping surface.

The fixed initial parameters, which are used in all the simulations, are shown in Table 1. In particular, the first four parameters are about the dimension and the inertia of the object, whereas the following three are the friction coefficients. These last values have been chosen by making the following considerations: μ_{max} must be a medium-high friction value (assumed $\mu = 0.5$, because is similar to the kinetic friction coefficients between paperboard and the conveyor belt in [42,43]), as it models the sliding of the object on the rotor in the direction of the axis; μ_{min} must be a low value ($\mu \leq 0.1$), as it models the rotational friction of the rotors (assumed $\mu = 0.01$, similar to a rolling friction coefficient); and, μ_{exsurf} has to be a low value as well ($\mu = 0.005$ is selected), as it models the area outside the line where one can imagine having load-bearing spheres supporting the material (according to the *Omnitrack* catalog [44], $\mu = 0.005 \div 0.03$). Actually, the coefficients described would depend on the materials in contact, which have not yet been defined. However, the exact values are not relevant for the purpose of proving the surface capabilities; the important thing is that

$\mu_{max} > \mu_{min}, \mu_{exsurf}$ is maintained. Finally, the last parameters are the distances in the x and y directions between two rotors centers and the object initial position and acceleration. Exceptions to these starting conditions are indicated with the results for each particular case, together with the missing parameters such as initial velocity of the object $[\dot{x}_{g0}, \dot{y}_{g0}, \dot{\theta}_{g0}]$ and the inclination of the surface γ.

Table 1. Initial fixed parameters for the MATLAB simulations.

Parameter	Description	Value
$b \times h$	object base dim.	$0.25\ \mathrm{m} \times 0.25\ \mathrm{m}$
z_g	height of G	0.1 m
m	object mass	10 kg
J	moment of inertia	$(1/12)m(h^2 + b^2)$
μ_{max}	max. friction coef.	0.5
μ_{min}	min. friction coef.	0.01
μ_{exsurf}	friction coef. out of the line.	0.005
dx_m	x dist. between 2 rotors	0.1 m
dy_m	y dist. between 2 rotors	0.1 m
$[x_g, y_g, \theta_g]$	obj. initial pos. and orientation	$[b/2, 1\ \mathrm{m}, 0\ \mathrm{rad}]$
$[\ddot{x}_g, \ddot{y}_g, \ddot{\theta}_g]$	obj. initial acc.	$[0\ \mathrm{m/s^2}, 0\ \mathrm{m/s^2}, 0\ \mathrm{rad/s^2}]$

5. Results

This section presents the simulation results for all three categories: sorting, slowing, and stopping. For each category, the authors illustrate their results for relevant parameter sets and indicate a practical use of the program for the real setup design.

5.1. Sorting

This subsection reports the results of the first of the three categories, sorting. Initially, graphical examples of the function under varying input conditions are shown, then an application for the design of the actual system is presented, and finally a comparison with existing sorting systems is given. Figure 7a,b show the sorting towards the two directions, $y > 0$ and $y < 0$, respectively, with an initial object velocity $\dot{x}_{g0} = 1.5\ \mathrm{m/s}$ and without surface inclination. Figure 7c, on the other hand, shows the sorting at different tilting angles of the surface and without an initial velocity. Each plot displays the trajectory of G (object's center of mass) during the sorting and the position and orientation of the object at the end of the simulation time.

In Figure 7a, the green square and its trajectory show that, with rotors inclined at $-45°$, sorting towards $y > 0$ can be achieved, as the object is shifted completely off the line.

The opposite sorting condition was tested as well; the only difference for the initial data was the direction of the rotors ($+45°$). The result is shown in Figure 7b and the trajectory is mirrored to the first one. In fact, as expected in reality, the ending displacements of the object are equal in modulus and with opposite sign for the y and θ displacements (Table 2).

Figure 7a,b show the capability of the surface to deflect the object trajectory for sorting purposes. Furthermore, coherent results were achieved for both sorting directions, for a reasonable initial velocity, and within one second of simulation time. From now on, since the mirroring of the results has been demonstrated for a symmetric rotor arrangement, the outputs are shown only for one sorting direction.

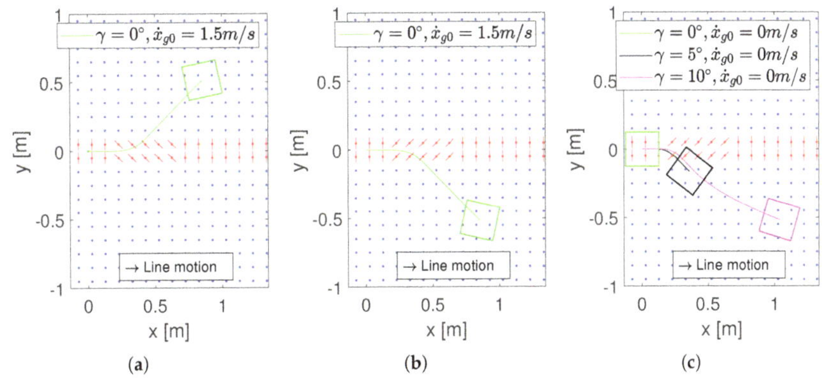

Figure 7. Illustration of the trajectories and the final position and orientation of the object during: (**a**) the sorting towards $y > 0$ (up) task, (**b**) the sorting towards $y < 0$ (down) task, (**c**) the sorting towards $y < 0$ (down) with different γ.

After this first presentation of the sorting capability for the under-actuated system, instead of using an initial object velocity on a horizontal surface, the sorting task was studied with an initially stationary object on an inclined surface with the identical rotor arrangement (Figure 7c).

In Figure 7c, the first case (green) proves that, without any external actuation, the object is not moving in the simulation, as in the real word. In the second case (black square, Figure 7c), with the surface inclination of $\gamma = 5°$, the body moves, but for the chosen simulation time ($t = 1.5$ s) it has not yet completed its sorting. This can be seen from the fact that the black square contour still contains the center of a sorting rotor (red). Finally, the third case (magenta square), where the surface inclination is $\gamma = 10°$, reports a completed sorting activity. Summarizing the black and magenta bodies in Figure 7c demonstrate that it is possible to transport and sort parts by inclining the surface. In particular, the more it is tilted, the faster the object moves, and the quicker the sorting is achieved. In Figure 7c, the values of the position and orientation reached at the end of the simulation ($t_{black} = 1.5$ s) for the $\gamma = 5°$ case (black line) are $[x_g, y_g, \theta_g]_{black} = [0.369 \text{ m}, -0.171 \text{ m}, -36.05°]$ (Table 2), while those of the magenta line for the same value of $y_g = -0.171$ m are $[x_g, y_g, \theta_g]_{mag} = [0.402 \text{ m}, -0.171 \text{ m}, -28.62°]$ and the simulation time for which they are obtained is $t_{mag} = 0.917$ s. Comparing these values, it can be seen that when the surface is more inclined (magenta line), it requires more space in the x direction to perform the sorting $x_{g,mag} > x_{g,black}$, but the time required is less $t_{black} - t_{mag} = 0.583$ s. In addition, different γ also produce different final orientations, in this case smaller for larger γ: $\theta_{g,mag} < \theta_{g,black}$.

Table 2 summarizes the results and the initial data about the previous sorting simulations.

Table 2. Specific initial data and results for the sorting simulations.

$[\dot{x}_{g0}, \dot{y}_{g0}, \dot{\theta}_{g0}]$	γ	**Sim. Time**	**Final** $[x_g, y_g, \theta_g]$	**Figure**
[1.5 m/s,0 m/s,0 rad/s]	0°	1 s	[0.859 m,0.532 m,14.83°]	Figure 7a
[1.5 m/s,0 m/s,0 rad/s]	0°	1 s	[0.859 m,−0.532 m,−14.83°]	Figure 7b
[0 m/s,0 m/s,0 rad/s]	0°	1.5 s	[0 m,0 m,0°]	Figure 7c green
[0 m/s,0 m/s,0 rad/s]	5°	1.5 s	[0.369 m,−0.171 m,−36.05°]	Figure 7c black
[0 m/s,0 m/s,0 rad/s]	10°	1.5 s	[1.057 m,−0.515 m,−12.16°]	Figure 7c magenta

A practical use of the simulation is to determine the minimum number of modules necessary for sorting before building the real system. In general, as it was already presented, this result is affected by various initial conditions. However, it can be interesting to know how many columns of rotor units are necessary to successfully complete the different tasks for different object sizes and velocity ranges. In order to derive this relationship, a set of simulations can be performed. Table 3 reports the results of the discussed analysis for two object sizes: $b \times h = 0.25$ m $\times 0.25$ m, called "S-box", and $b \times h = 0.45$ m $\times$ 0.45 m, called "B-box". "B-box" and "S-box" have the same mass ($m = 10$ kg) and height ($z_g = 0.1$ m).

The counting of the number of columns for the sorting is done starting from the first column with the inclined rotors to the last one touched by the object before being out of the line. As is visible from Table 3, the sorting with low initial velocities is not always possible: with $\dot{x}_{g0}$ less than 0.8 m/s for "S-box" and less than 1.1 m/s for "B-box", the boxes come to a stop within the transport line. Despite this, the speed values considered in the simulations are within the most common operating ranges of conveyors; however, any speed can be implemented in the software. In general, the missed sorting occurs because the deflecting forces also tend to slow the object, which may stop before reaching the target. In addition, the displacement to achieve to be out of the line is related to the dimension of the object, so the bigger it is, the more absolute transversal displacement is required to achieve the sorting. Summarizing, thanks to the simulation described, it is possible to realize such tables depending on the specific application, considering that the results are influenced by the layout of the modules and the initial position. In particular, for Table 3 the "S-box" layout is the same of the sorting simulations of Figure 7b, but with more columns of sorting rotors in order to test a bigger range of velocities. Instead, for the "B-box" there are two more rows of rotors in the sorting line, always with the rotors oriented for sorting towards $y < 0$ ($\varepsilon_i = +45°$), making a total of four instead of two lines.

The data from the simulations carried out to produce Table 3, coupled with further simulations with, as input, only the inclination of the surface, allowed for a comparison of performance with similar existing sorting systems [45]. Figure 8 presents these results. The red dots represent the values of the sorting rate when only the initial velocity is set as input, whereas the blue dots are for when only the inclination of the surface is exploited. Figure 8 shows the most common sorting capacity ranges [45] and, as can be seen, the system proposed by the authors guarantees a medium rate for most input values, with few exceptions in both directions, i.e., high capacity and low capacity. The rate is defined as objects sorted per hour (pcs/h) and can be easily calculated, considering the time needed to perform the sorting and a successful sorting condition. In practice, the condition of successful sorting is when $y_g \geq y_{g,limit}$ (for the rectangular shape: $y_{g,limit} = h/2 + \sqrt{(b/2)^2 + (h/2)^2}$), i.e., when the displacement in the direction y is such that, whatever the orientation of the object, the base is no longer in contact with the sorting rotors. The input velocity values presented in Figure 8 are within the common ranges of use for sorting systems. In particular, Ref. [45] for devices similar to the surface proposed by the authors, but fully actuated and called "Torsional discs", defines input velocities between 0.5 and 1.5 m/s. "Torsional discs" have sorting capacities ranging between 1600 and 4500 pcs/h when the size and weight of the transported objects are comparable to those considered in the authors' simulations ("S-box" and "B-box" characteristics). The system developed in this paper, although it is under-actuated, also allows sorting rates within the same range, providing good performance and in line with the current technology. Obviously, given the under-actuation, there are minimum input speed thresholds, as shown in Table 3, however, they do not seem to limit the application ranges of the device. Tilting the surface and starting from zero initial speed requires longer sorting times and thus provides lower rates than those with velocity as input. However, according to Figure 8, the inclination also provides a predominantly medium sorting capacity, without the need for a prior conveyor. Therefore, the results obtained from the comparison of sorting performances showed that the surface introduced by the authors, with the advantage of the simplicity of the under-actuation, still guarantees sorting capabilities at the same level as current technology, both using initial speed and surface inclination as inputs.

Table 3. Minimum number of rotor columns required to achieve sorting with $\gamma = 0°$, but changing $\dot{x}_{g0}$.

$\dot{x}_{g0}$ [m/s]	n° Columns S-Box	n° Columns B-Box
0.8	3	/
0.9	3	/
1.0	4	/
1.1	4	7
1.2	4	7
1.3	4	7
1.4	4	7
1.5	4	8
1.6	5	8
1.7	5	8
1.8	5	8
1.9	6	9
2.0	6	9
2.1	6	9
2.2	7	10
2.3	7	10
2.4	7	11
2.5	8	11
2.6	8	12
2.7	8	12

To sum up, the results of this subsection show that the surface is capable of sorting with the two different types of input (Figure 7 and Table 2). In addition, thanks to the developed simulation environment, it was possible to obtain the performance of the system proposed by the authors and compare it with the current technology. This showed that the surface provides a medium sorting capacity (Figure 8), in the same range as existing systems. Finally, for certain objects, the initial input speed was associated with the number of rotors required for sorting (Table 3). This was proposed as a further demonstration of the usefulness of the simulation environment for the design and control of the real system.

5.2. Slowing

The second set of results, here presented, is derived from the slowing simulations. The purpose of these simulations is to simplify and speed up the determination of the number of modules required for the slowing task, which, as with sorting, is linked to the initial parameters. Similar to the sorting results, graphical examples of the functioning and an application for the design of the actual system are given in this subsection.

In order to show the results of the slowing task, in this case the evolution of the object speed over time ($[\dot{x}_g, \dot{y}_g, \dot{\theta}_g]$) is reported. Different from the sorting, here the object is not pushed away from the line, so the trajectory does not represent the course of the activity as simply as speed does. Considering the layout of the system, the only velocity component different from zero is in the x direction, which is the one plotted. Four different setups shown in Figure 9 were simulated to demonstrate the concept's operation. The graphs in Figure 10a,c show the results of the slowing simulations for the "S-box" and "B-box" cases, respectively.

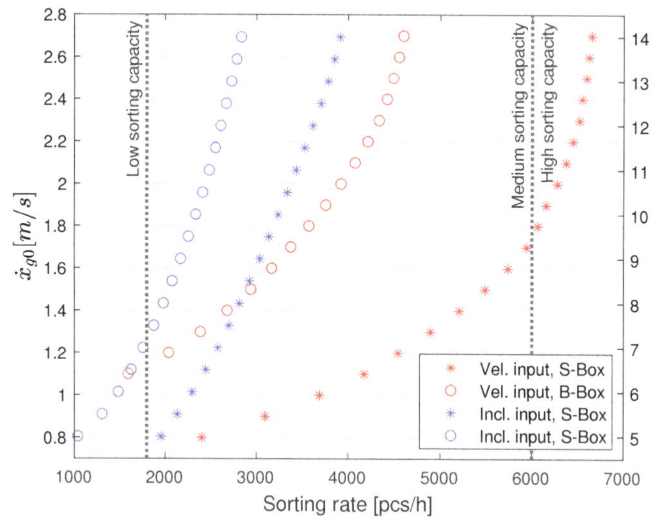

Figure 8. Sorting rates obtained with MATLAB simulations for the "S-box" and "B-box" objects, with, as input, only initial velocity (red) and only inclination of the surface (blue).

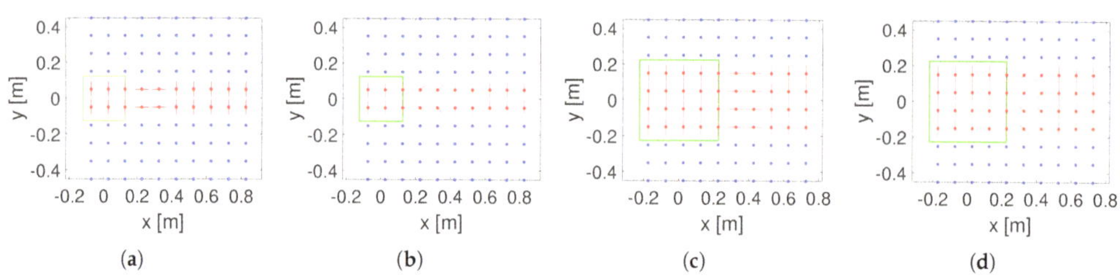

Figure 9. Layouts of the simulations for the slowing and the stopping tasks with: (**a**) "S-box" $\varepsilon_i = 0°$, (**b**) "S-box" $\varepsilon_i = \pm 45°$, (**c**) "B-box" $\varepsilon_i = 0°$, (**d**) "B-box" $\varepsilon_i = \pm 45°$.

The initial data for this analysis are the same as in Table 1, with additionally $\gamma = 0°$ and $[\dot{x}_{g0}, \dot{y}_{g0}, \dot{\theta}_{g0}] = [1.5\ \text{m/s}, 0\ \text{m/s}, 0\ \text{rad/s}]$. As visible from Figure 10a,c, both layouts are capable of reducing the speed of the object, fulfilling the slowing task. However, the speed drop is different with the orientation of the rotors. When the rotors are with $\varepsilon_i = 0°$, all the F_{par} forces in the x direction cause a stronger slowing, whereas with $\varepsilon_i = \pm 45°$ layout, where the F_{par} are inclined like the rotor axes, the slowing is less. For these initial parameters and rotor dispositions, the ratios between the speed before and after the slowing rotors ($r = \dot{x}_{gout}/\dot{x}_{g0}$) are: $r_0^{S-box} = 0.23$, $r_0^{B-box} = 0.18$ and $r_{\pm 45}^{S-box} = 0.53$, $r_{\pm 45}^{B-box} = 0.52$. Additionally, Figure 10 shows the effect of F_{perp}, which is slightly reducing the velocity when the rotors are with $\varepsilon_i = 90°$. In fact, the plots in Figure 10a,c show that the lines are a bit inclined before and after the slowing area, which is included between the "Start slowing" and the "End slowing" dotted lines. It represents the contribution of rotational friction to the deceleration.

As for sorting, a parameters study was conducted within the simulation environment to derive the required number of rotor columns to achieve the slow down task of the objects. The results are displayed in Table 4. Additionally in this case, different initial speed and the two different object sizes were tested ("S-box", "B-box"). The reference layouts for the simulations are the ones in Figure 9. In Table 4, the counting of the columns number begins with the first column with inclination: $\varepsilon_i = 0°$ or $\varepsilon_i = \pm 45°$, that the object encounters starting from the left. For example, considering Figure 9a and assuming that slowing is verified, the result will be 2 columns (with $\varepsilon_i = 0°$). The target condition was set to at least halving the speed of the object while remaining a positive value after the slowing process ($0.5\dot{x}_{g0} \geq \dot{x}_{gout} > 0$). The two configurations ($\varepsilon_i = 0°$ and $\varepsilon_i = \pm 45°$) were possible for the rotor layout, so the configuration with the smallest number of columns was chosen case by case to satisfy the slowing condition. As shown in Table 4 for both layouts and for all speeds analyzed, the number of rotors was identified. The introduction of the $\varepsilon_i = \pm 45°$ configuration was necessary, as for some speeds the variant with $\varepsilon_i = 0°$ generated a total stop of the conveyed object (e.g., for the first two values in Table 4). The $\varepsilon_i = \pm 45°$ positioning also seems promising for the self-alignment of the conveyed material in the center of the transport line. In conclusion, in this subsection, the operation of the surface for slowing tasks was demonstrated (Figure 10). Additionally, as with sorting, certain input speed values of the object were associated with the number of rotors required for the task (Table 4). This provides data for the realization of the physical system and proves the usefulness of simulations for this objective as well.

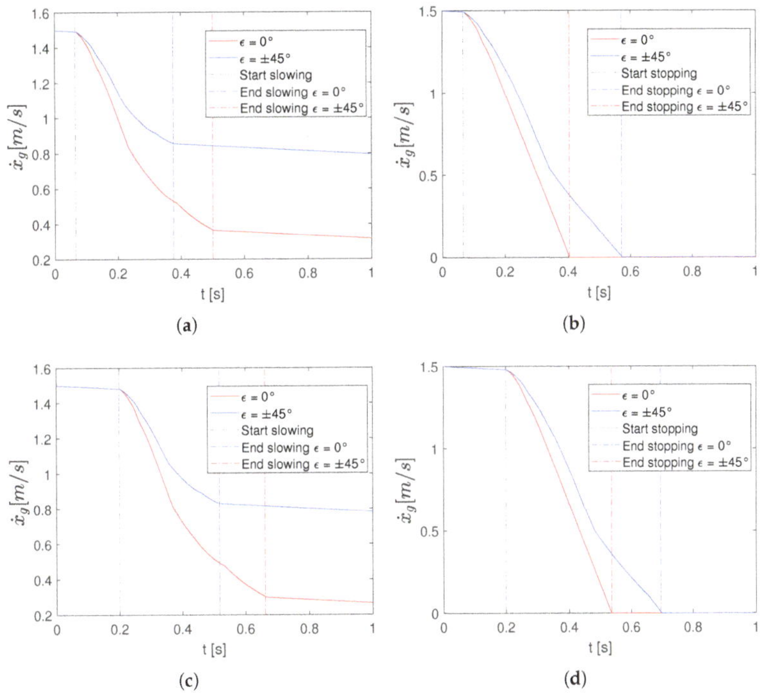

Figure 10. Velocity trends for "S-box" during (**a**) the slowing and (**b**) the stopping, and (**c**) the "B-box" during the slowing and (**d**) the stopping with: $\varepsilon_i = 0°$ (red), $\varepsilon_i = \pm 45°$ (blue).

5.3. Stopping

The last results, as before, consist of graphical examples of the function and an application for the design of the actual system; however, in this subsection, they are related to the simulations of the stopping condition. In this case, the initial parameters and layouts implemented in the software are the same as before (Figure 9), and only the number of columns of the working rotors changes. Starting from the slowing modules from Table 4 and adding to the right an extra column of rotors with $\varepsilon_i = 0°$, the motion of the object can be stopped. The addition of a column proved to be an effective method for simulated speed values; it is not certain that for different values one column is sufficient. Therefore, considering this, Table 4 can also be used for the stopping layout planning. Similar to the slowing, an example (Figure 10b,c) is presented, and only the $\dot{x}_{g0}$ of the object is reported.

However, in this case, the deceleration produced by the rotors of the slowing, plus the extra column, is capable of stopping the object even if the initial velocity simulated is the same as for slowing ($[\dot{x}_{g0}, \dot{y}_{g0}, \dot{\theta}_{g0}] = [1.5\ \text{m/s}, 0\ \text{m/s}, 0\ \text{rad/s}]$). In general, according to Table 4 and Figure 10, the size of the object for the slowing and the stopping tasks has less influence than for the sorting (Table 3). This can be seen from the fact that, compared to Table 3, in Table 4, the numbers of columns to slow (then also to stop) the "S-box" and the "B-box" are only different in one case ($\dot{x}_{g0} = 2.6$ m/s). In the end, also for this subsection, the functioning of the surface for the stopping task was demonstrated (Figure 10b,c) and the table with input speeds and number of rotors was explained. Therefore, as final consideration for the whole section, the simulations enable the determination of the layout of the surface, including the module arrangement and the surface inclination for specific applications, while taking the initial conditions of the object into account and which tasks have to be performed. The same MATLAB environment allowed us to obtain the results here presented, proving graphically and numerically the capabilities of the system proposed in this paper.

Table 4. Minimum number of rotor columns and orientations of their axes (ε_i) required to achieve slowing ($0.5\dot{x}_{g0} \geq \dot{x}_{gout} > 0$) with $\gamma = 0°$, but changing $\dot{x}_{g0}$.

$\dot{x}_{g0}$ [m/s]	n° Columns S-Box	n° Columns B-Box
0.9	$1 \Rightarrow \varepsilon_i = \pm 45°$	$1 \Rightarrow \varepsilon_i = \pm 45°$
1.0	$1 \Rightarrow \varepsilon_i = \pm 45°$	$1 \Rightarrow \epsilon_i = \pm 45°$
1.1	$1 \Rightarrow \varepsilon_i = 0°$	$1 \Rightarrow \varepsilon_i = 0°$
1.2	$1 \Rightarrow \varepsilon_i = 0°$	$1 \Rightarrow \varepsilon_i = 0°$
1.3	$2 \Rightarrow \varepsilon_i = \pm 45°$	$2 \Rightarrow \varepsilon_i = \pm 45°$
1.4	$2 \Rightarrow \varepsilon_i = \pm 45°$	$2 \Rightarrow \varepsilon_i = \pm 45°$
1.5	$2 \Rightarrow \varepsilon_i = 0°$	$2 \Rightarrow \varepsilon_i = 0°$
1.6	$2 \Rightarrow \varepsilon_i = 0°$	$2 \Rightarrow \varepsilon_i = 0°$
1.7	$3 \Rightarrow \varepsilon_i = \pm 45°$	$3 \Rightarrow \varepsilon_i = \pm 45°$
1.8	$3 \Rightarrow \varepsilon_i = 0°$	$3 \Rightarrow \varepsilon_i = 0°$
1.9	$3 \Rightarrow \varepsilon_i = 0°$	$3 \Rightarrow \varepsilon_i = 0°$
2.0	$3 \Rightarrow \varepsilon_i = 0°$	$3 \Rightarrow \varepsilon_i = 0°$
2.1	$4 \Rightarrow \varepsilon_i = 0°$	$4 \Rightarrow \varepsilon_i = 0°$
2.2	$4 \Rightarrow \varepsilon_i = 0°$	$4 \Rightarrow \varepsilon_i = 0°$
2.3	$4 \Rightarrow \varepsilon_i = 0°$	$4 \Rightarrow \varepsilon_i = 0°$
2.4	$5 \Rightarrow \varepsilon_i = 0°$	$5 \Rightarrow \varepsilon_i = 0°$
2.5	$5 \Rightarrow \varepsilon_i = 0°$	$5 \Rightarrow \varepsilon_i = 0°$
2.6	$6 \Rightarrow \varepsilon_i = 0°$	$5 \Rightarrow \varepsilon_i = 0°$
2.7	$6 \Rightarrow \varepsilon_i = 0°$	$6 \Rightarrow \varepsilon_i = 0°$

6. Validation

In order to validate the previous results, this section presents a comparison between the MATLAB environment realized by the authors and a commercial multi-body dynamics software. In the first part of the section, the problem and details on how the validation was carried out are explained, and the second part reports the results.

6.1. Introduction to the Validation

Focusing on the setup of the validation, this subsection is devoted to presenting the objectives, the data, and the ideas behind the comparison. As previously introduced, the sorting system was also modeled in a commercial software for multi-body dynamic simulations: Hexagon D&E Adams MSC. The multi-body dynamic model, called hereinafter Adams for brevity, has two main objectives:

1. Providing further demonstration of the functioning of the under-actuated system introduced in this paper;
2. Comparing the results with the ones from MATLAB and SIMULINK to have a first validation of the analytical model and the self-developed simulation environment.

Simulations with the Adams software require longer computational time compared to the MATLAB simulations ($time_{Adams} \approx 60 \div 300$ s, $time_{MATLAB} \approx 0.1 \div 2$ s, with a HP ProDesk 400 G7). For this reason, Adams was only used for the validation part and not as a standard simulation environment. In fact, using a software as a planning and control tool within the physical system in real time requires very short calculation times, of which the authors' software is capable.

Figure 11a shows one surface configuration that was previously used in the MATLAB environment for sorting. The identical surface configuration was replicated in Adams, shown in Figure 11b. As it is visible in Figure 11b, the sorting setup was designed in Adams using three different types of bodies: a box (orange wire frame) for the object, cylinders (colorful cylinders) for the rotors, and spheres (pink spheres) for the area out of the line. The chosen inertial and dimensional parameters are shown in Table 5. It should be noted that, for the validation, the values of some inertial parameters are different from previous simulations because the mass properties in Adams were introduced by assigning the material of the bodies. The constraints used for the model are: a fixed joint for the spheres and rotational joints for the cylinders, and the box has contact constraints with the other bodies. The parameters of the contact and the joints are illustrated in Table 6. In particular, for the first five values of this table, the selection was done according to the suggestion of [46,47]: St and Pd are the default values for the contact, whereas $Dp = 8$ kg/s instead of 10 kg/s, Stv and Ftv are smaller than the default (respectively, 100 mm/s and 1000 mm/s) to better simulate the stiction at low speeds. On the same line, the MATLAB parameter, k, of the smooth Coulomb model, is set to $k = 5000$ to have a similar friction curve. The friction coefficients are the same values implemented in MATLAB, whereas the joint friction parameters are selected to simulate the rotational friction of the joint, similar to the analytic model. To conclude the description, in Adams, the inclination of the surface is simulated changing the gravity vector direction, and the initial velocity is assigned as an intrinsic input condition to the object. Finally, the simulation time considered was $t = 1$ s and the step size in the range $[1 \times 10^{-4}s \div 0.5 \times 10^{-4}s]$.

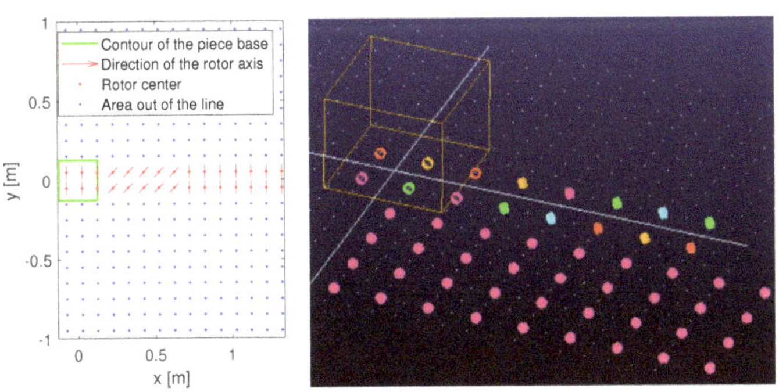

(a) (b)

Figure 11. Model of under-actuated active surface configured for sorting in: (**a**) MATLAB, (**b**) Adams.

Table 5. Inertial and geometrical parameters implemented in Adams for the simulation.

	Objects Parameters	
m	box mass	5.48 kg
I_{xx}	box m. inertia x axis	4.68×10^{-2} kgm^2
I_{yy}	box m. inertia y axis	4.68×10^{-2} kgm^2
I_{zz}	box m. inertia z axis	5.70×10^{-2} kgm^2
$b \times h \times z_g$	object dim.	0.25 m×0.25 m×0.1 m
m_r	rotor mass	8.61×10^{-3} kg
I_{rxx}	rot. m. inertia x axis	4.30×10^{-7} kgm^2
I_{ryy}	rot. m. inertia y axis	2.87×10^{-2} kgm^2
I_{rzz}	rot. m. inertia z axis	2.87×10^{-7} kgm^2
$r_r \times l$	rot. dim.	0.01 m×0.01 m
γ	inclin. of the surf.	$[5 \div 15]^\circ$

Table 6. Contact and friction parameters implemented in Adams for the simulation.

	Contact Parameters	
St	stiffness cont.	10^5 MPa
Dp	damping cont.	8 kg/s
Pd	penetration depth cont.	0.1 mm
Stv	stiction transition vel.	0.1 mm/s
Ftv	friction transition vel.	1 mm/s
μ_{max}	friction coef.	0.5
μ_{exsurf}	friction coef. out the line.	0.005
	Joint friction parameters	
μ_{min}	min friction coef.	0.01
Tv	transition vel.	0.001
Friction	option chosen	*sliding only*

The MATLAB setup (Figure 11a) corresponds to the previous sorting setup in Section 5, Figure 7, but in this case it uses the same geometric and inertial data as Adams to simulate the same sorting process. In this way, the results from the two software differ only because of their intrinsic analytic modeling.

After this brief introduction to the Adams model, the simulations performed to meet the two objectives are now described. The configuration with a tilted surface and no initial object velocity was simulated for different angles of inclination ranging from $\gamma = 5^\circ$ to $\gamma = 14^\circ$ with an increment of 0.25°. Additionally, the horizontal configuration ($\gamma = 0^\circ$) has been simulated with an initial velocity $\dot{x}_{g0} = 0.9$ m/s.

6.2. Results of the Validation

This section reports the results of the validation, which are distinguished according to the two objectives of the Adams model. Concerning the first objective, i.e., the proof of functioning of the under-actuated system, Figure 12 graphically reports the results of some Adams simulations. Considering initially only Figure 12a,b, in Figure 12a, sorting is simulated given an initial velocity of zero ($\dot{x}_{g0} = 0$ m/s) and an inclination of $\gamma = 10^\circ$; in Figure 12b, the initial velocity is $\dot{x}_{g0} = 0.9$ m/s and the inclination is zero ($\gamma = 0^\circ$). In Figure 12a,b, it is possible to see the trajectory taken by the G of the object and its final position and orientation given the simulation time of $t = 1$ s. The object at the end of the simulation is for both images (Figure 12a,b) outside the sorting zone (zone with cylinders), so the activity is considered completed.

(**a**) (**b**) (**c**)

Figure 12. Object trajectories from Adams with: (**a**) $\gamma = 10°$, $\dot{x}_{g0} = 0$ m/s, and $t = 1$ s of simulation, (**b**) $\gamma = 0°$, $\dot{x}_{g0} = 0.9$ m/s, and $t = 1$ s of simulation, and (**c**) $\gamma = 5°$, $\dot{x}_{g0} = 0$ m/s, and $t = 1.8$ s of simulation.

In addition to the case illustrated in Figure 12a, among all the thirty-seven simulations made by varying the inclination of the surface, the ones with $\gamma \geq 9.5°$ (Figure 13a,b) complete the task. For the rest, the target is not really missed, but, because of the limited simulation time ($t = 1$ s), it is not achieved yet. This time was selected because it represents a normal operating condition, for example: there is 1 m of space available in the line for sorting and the conveyor has a speed of 1 m/s. However, as a general rule, when the surface is inclined and with the rotors as in Figure 12, the motion does not self-stop, if $(\sqrt{2}/2)\mu_{min} < \sin(\gamma)$. As proof of this, Figure 12c displays the trajectory of the object with $\gamma = 5°$ but $t = 1.8$ s of simulation time, instead of $t = 1$ s, and the sorting is clearly obtained. Therefore, the sorting is achievable with this under-actuated system, but it takes time and space. Instead, if the surface is not inclined and there is just the initial speed, the time could be not the only cause of the missed sorting. In this case, the object velocity is inevitably reduced by the friction until the part is stopped, because the gravity is not helping the motion. The same result was highlighted before in the MATLAB environment, which confirms that the simulation environment can be used to predict these conditions.

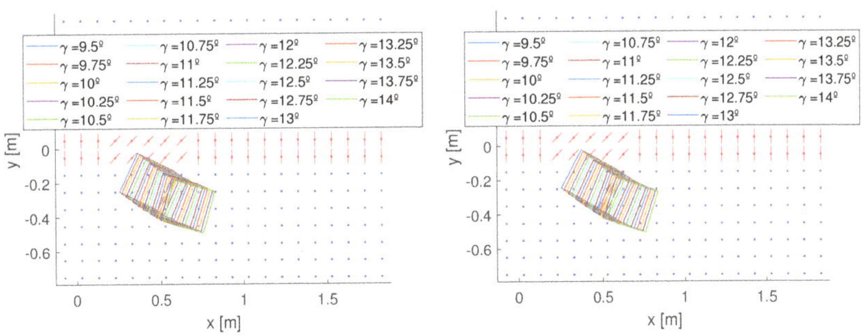

(**a**) (**b**)

Figure 13. Final position with $\gamma > 8.5°$ in: (**a**) Adams MSC, (**b**) MATLAB.

Figure 13 plots the final positions when the sorting is considered achieved for the two software packages (Figure 13a Adams, Figure 13b MATLAB). Having completed the verification of the first objective of the Adams model, it is possible to move on to the second one. Summarizing briefly, objective number two is to exploit the Adams model to show the proximity with the MATLAB results, considering that the same initial data are provided. Figure 14 shows a graphical comparison between the two models referring to the two trajectories of Figure 12a,b. In this case, the trajectories are transposed into a graph (Figure 14) with the corresponding result from MATLAB.

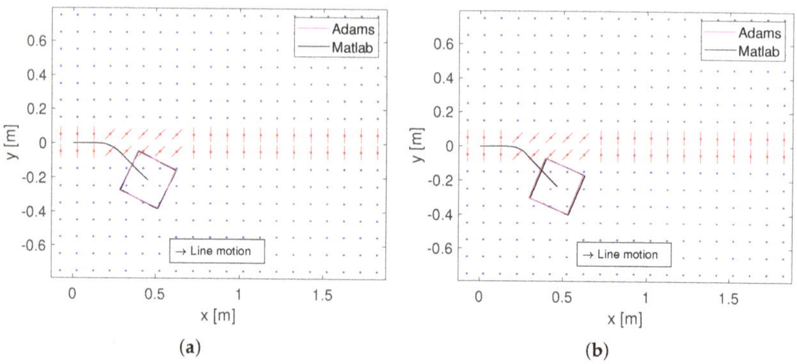

Figure 14. Comparison of the trajectories with: (**a**) $\gamma = 10°$, $\dot{x}_{g0} = 0$ m/s, (**b**) $\gamma = 0°$, $\dot{x}_{g0} = 0.9$ m/s.

As is visible from Figure 14a,b, the two objects follow very close trajectories and end up almost in the same position and with the same orientation. The absolute and the relative percentage errors obtained for these two simulations are: $[|e_x|, |e_y|, |e_\theta|] = [0.0011\ \text{m}, 0.0001\ \text{m}, 1.735°]$, $[|e_{x\%}|, |e_{y\%}|, |e_{\theta\%}|] = [0.24, 0.05, 7.06]$ % when the input is $\gamma = 10°$, $\dot{x}_{g0} = 0$ m/s and $[|e_x|, |e_y|, |e_\theta|] = [0.003\ \text{m}, 0.0037\ \text{m}, 1.56°]$, $[|e_{x\%}|, |e_{y\%}|, |e_{\theta\%}|] = [0.65, 1.59, 6.97]$ % when the input is $\gamma = 0°$, $\dot{x}_{g0} = 0.9$ m/s. The errors are calculated as shown in Equation (5a,b).

$$[e_x, e_y, e_\theta] = \text{FINAL DISP.}_{Adams} - \text{FINAL DISP.}_{Matlab} \tag{5a}$$

$$[e_{x\%}, e_{y\%}, e_{\theta\%}] = \frac{\text{FINAL DISP.}_{Adams} - \text{FINAL DISP.}_{Matlab}}{\text{FINAL DISP.}_{Adams}} \times 100 \tag{5b}$$

The two examples graphically visualize the similarities and the differences between the models. However, to show and summarize all the results for the thirty-seven setups ($\gamma = 5° \div 14°$), the final displacements of the Adams and MATLAB simulations together with the errors between them are reported, respectively, in Figures 15 and 16. Considering the x and y results (Figures 15 and 16), the two models (Adams, MATLAB) are not so different, and the relative percentage errors are below $e_{x\%} < 1$ % and $e_{y\%} < 8$ %. The trends of the percentage errors seem to have a slight tendency to decrease, and the higher error values occur for small γ. In contrast, for high γ values, both coordinates have minimum errors, $e_{x\%} < 0.5$ % and $e_{y\%} < 3$ %. This is because small inclinations generate reduced displacements, especially along y, and the sorting is not completed in the fixed simulation time ($t = 1$ s). This implies that, in the calculation of the percentage error, there are initially very small denominators, and then they become an order of magnitude larger. On the other hand, the absolute errors along γ values keep oscillating, with a bigger frequency for x and smaller for y.

For the orientation, higher $e_{\theta\%}$ values are reached ($e_{\theta\%} \leq 12\%$) compared to those of x and y. In this case, the phenomenon of the larger percentage error for smaller rotations is not noticed, as in a single simulation there can be counter-rotations, particularly when γ is big enough to reach the completed sorting area. Thus, the final orientation represents the sum of clockwise and counterclockwise rotations, and its value has not undergone monotonic growth like for the x and y displacements. In general, larger errors of rotations can be explained by comparing how the calculation of normal forces occurs between the two simulations. Adams uses a penetration model to calculate the contact forces, whereas the optimization algorithm (Equation (1)) is implemented in MATLAB. The distribution of normal forces (and therefore also of friction) mainly affects the calculation of the moment, as the position with respect to the pole is important. Whereas, for the x and y displacements, the over mentioned distribution does not have influence, as the resultant force does not change. In addition, an explanation for others slight deviations in trends is the use of different step sizes in some Adams simulations. This was necessary because in a few cases the specific step size chosen was creating a numerical problem, preventing the completion of the simulation. In conclusion, however, the errors are limited and the two models behave in a similar way, thus giving value to the results provided by MATLAB.

To quantify how different MATLAB and Adams values are, the mean values and standard deviations (SD) of the two types of error are listed in Table 7 for the different displacements.

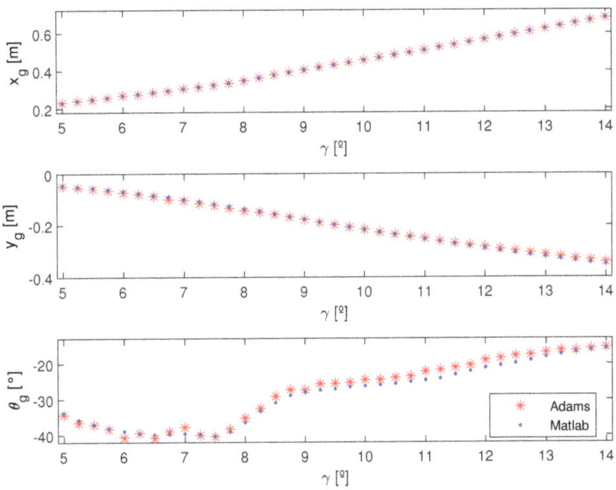

Figure 15. Final displacements obtained with the two software packages: Adams (red) and MATLAB (blue) .

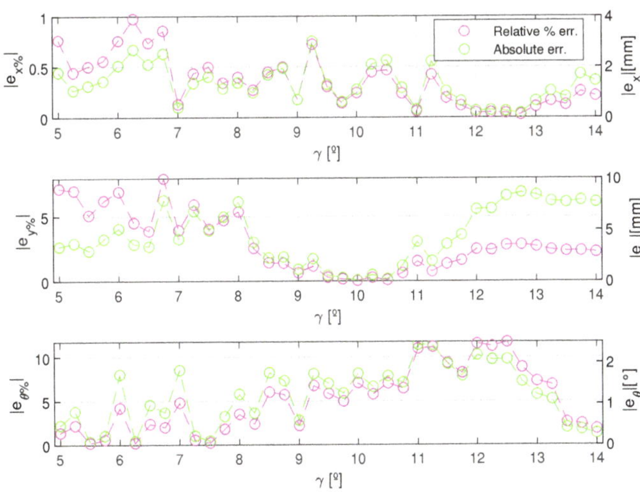

Figure 16. Absolute (green) and relative percentage (magenta) errors between the final displacements resulting from the two software packages.

Table 7. Mean and standard deviation of the relative percentage and absolute errors obtained from the difference between Adams and MATLAB simulations.

	$e_{x\%}$	$e_{y\%}$	$e_{\theta\%}$	e_x [mm]	e_y [mm]	e_θ [°]
Mean	0.02	1.41	−4.47	0.12	0.25	0.96
SD	0.44	3.48	4.36	1.50	5.20	1.02

As results from Table 7 show, the errors of x and y displacements are very limited. This guarantees that the placing made by the real system can be foreseen in advance with an accuracy appropriated to the applications, so when the sorting line is designed, the simulation can be used to define the layout. For the rotation, the errors are bigger, but, despite this, for a simple sorting operation the final orientation is not interesting, so the MATLAB model can be considered reliable. Furthermore, if in a hypothetical application, the errors allowed are the in the range of those shown in Table 7, the model provides usable results. Concerning that, in many cases it is simply important that the object is not rotated by 180° or 90° with respect to the desired condition.

7. Conclusions

In this paper, the authors present a new modular surface for several intra-logistical tasks. In contrast to similar existing systems, their surface is under-actuated, in particular, composed of idle instead of actuated rotors, whose axes can be fixed in defined, discrete positions within the surface plane. The surface can be used in a horizontal orientation by exploiting an object's initial velocity, in a tilted orientation by exploiting the gravitational force on the object, or with a combination of both. The authors derived an analytic model and implemented a programmable simulation environment with the software MATLAB for this modular surface. As result, the functioning of the surface concept for the sorting, stopping, and slowing activities was demonstrated, together with the capabilities of the simulation environment. In particular, the MATLAB code showed its potential for predicting, with very short calculation times ($\approx$ or $<$ 1 s), the number of modules required for the three handling tasks and how the transported object will behave by simply changing the initial conditions. The same code also made it possible to obtain numerical results of sorting performance and thus have a comparison with current technology, showing that the system proposed by the authors guarantees a medium sorting capacity. These results and examples highlight the usefulness of this environment for real system planning and design. In addition, a validation of the concept and of the simulation environment was conducted with the software Adams. As Adams is a highly sophisticated and well accepted commercial software for dynamic simulations frequently used by engineers to simulate and predict the physical interaction of different components adequately, it is a reasonable tool to be considered as a first reference for the comparison.

In conclusion, the simplifications introduced in the surface, such as the under-actuation and the discrete number of orientations for the rotors, are not limiting the handling capabilities and the performances, but rather they are minimizing the number of constructive components requested and, thus, the costs. In fact, the same goals can be achieved with a reduced design and using in a convenient manner the external actuation, for example, gravity or previous conveyors, already in the line. Additionally, the validation with Adams also showed the accuracy of the main simulation. The differences between the two models are limited and the errors acceptable for many intra-logistics applications. This may open the way to other possible tasks for the surface integrated with the MATLAB environment, such as position and trajectory tracking. In these cases, each time the external environment requires a new position or trajectory of the object, the software calculates the orientation to be given to the rotors. The physical system must be integrated with sensors to adjust in real-time the rotors and achieve the tracking objectives. The implementation of sensors and control strategies for trajectory and position tracking greatly increases the adaptability and flexibility of the system.

No reference is made in the article to construction details in order to keep the validity of the work presented as general as possible. At this point, in order to provide some practical elements and above all to highlight the feasibility of the concept, some schematic solutions are proposed in Figure 17.

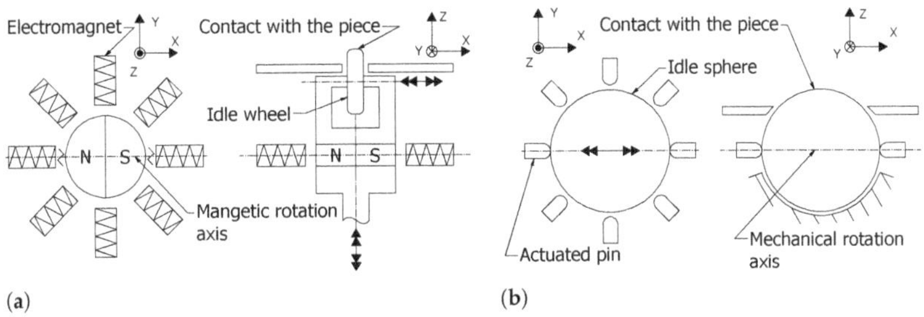

Figure 17. Schematic concepts for the under-actuated module: (**a**) idle wheel concept, (**b**) spherical concept.

Figure 17a shows an idle wheel mounted on a vertical axis of rotation. The operating principle is similar to the functioning of a stepper motor, which could be used for this purpose. In contrast, the concept in Figure 17b depicts a spherical rotor whose axis of rotation is locked using mechanically or electro-mechanically driven pins. New tracking objectives, an accurate design, and the control law for the modules are ongoing research topics.

Author Contributions: Conceptualization, E.B., O.J.J. and G.F.; methodology, E.B.; software, E.B.; validation, E.B.; formal analysis, E.B.; investigation, E.B.; writing—original draft preparation, E.B.; writing—review and editing, E.B., O.J.J., G.F., F.J.B.D. and J.A.Y.-F.; visualization, E.B.; supervision, O.J.J., G.F., F.J.B.D. and J.A.Y.-F.; project administration, F.J.B.D. and J.A.Y.-F. All authors have read and agreed to the published version of the manuscript.

Funding: This research received no external funding.

Institutional Review Board Statement: Not applicable.

Informed Consent Statement: Not applicable.

Data Availability Statement: Not applicable.

Acknowledgments: This research work was undertaken in the context of DIGIMAN4.0 project ("Digital Manufacturing Technologies for Zero-defect", https://www.digiman4-0.mek.dtu.dk/, accessed on 30 January 2023). DIGIMAN4.0 is a European Training Network supported by Horizon 2020, the EU Framework Pro-gramme for Research and Innovation (Project ID: 814225).

Conflicts of Interest: The authors declare no conflict of interest.

Nomenclature

The following nomenclature and abbreviations are used in this manuscript:

CIR	Center of instantaneous rotation
SD	Standard deviation
MEMS	Micro electro-mechanical system
B-Box	Big box
S-Box	Small box
$\{O, x, y, z\}$	Absolute reference frame
$\{\widetilde{G}, \widetilde{x}, \widetilde{y}\}$	Mobile planar reference frame
$\hat{i}, \hat{j}, \hat{k}$	Absolute reference frame versors

g	Acceleration of gravity
m	Object mass, [kg]
J	Object moment of inertia, [kg × m^2]
b, h	Object base width and length, [m, m]
G	Object center of mass
$\widetilde{G}$	Object center of mass projection on the base
P	Generic contact point between the object and the surface
z_g	G vertical coordinate, [m]
$\overrightarrow{\widetilde{G}P}$	Vector connecting $\widetilde{G}$ to P
n	Number of contact points
i	Sub-index to indicate the ith contact point
t	Time, [s]
s	sth step of the loop
x_g, y_g, θ_g	G coordinates and object orientation, [m, m, rad]
$\dot{x}_g, \dot{y}_g, \dot{\theta}_g$	Object velocity components, [m/s, m/s, rad/s]
$\ddot{x}_g, \ddot{y}_g, \ddot{\theta}_g$	Object acceleration components, [m/s^2, m/s^2, rad/s^2]
$x_{g0}, y_{g0}, \theta_{g0}$	G initial ($t = 0$ s) coordinates and object orientation, [m, m, rad]
$\dot{x}_{g0}, \dot{y}_{g0}, \dot{\theta}_{g0}$	Object initial ($t = 0$ s) velocity components, [m/s, m/s, rad/s]
$\ddot{x}_{g0}, \ddot{y}_{g0}, \ddot{\theta}_{g0}$	Object initial ($t = 0$ s) acceleration components, [m/s^2, m/s^2, rad/s^2]
$\overrightarrow{V_p}$	Object velocity in P, [m/s, m/s]
V_{par}	Object velocity in P along the axis direction, [m/s]
V_{perp}	Object velocity in P along the direction perpendicular to the axis, [m/s]
$\dot{x}_p, \dot{y}_p$	Components of V_p, [m/s, m/s]
$\dot{x}_{gout}$	Object velocity along x after the slowing rotors
r	Speed ratio between the velocities after and before the slowing area
F	Force, [N]
μ	Friction coefficient
N	Vertical component of the contact force between the object and a rotor, [N]
ε	Angle between the x axis and the rotor axis, [°]
β	Angle between the x axis and V_p, [°]
α	Difference between ε and β, [°]
F_{par}, F_{perp}	Friction force components of the object-rotor contact, [N, N]
μ_{max}, μ_{min}	Friction coefficient in the directions parallel and perpendicular to the axis
F_x, F_y	Friction force components of the object–rotor contact in the frame directions, [N, N]
μ_{exsurf}	Friction coefficient outside the sorting area
$\underline{X}$	Array of variables
$f(\underline{X})$	Function to be minimized
A_{eq}	Linear equality matrix
$\underline{b_{eq}}$	Linear equality vector
$\underline{l_b}$	Lower bounds vector
$\underline{u_b}$	Upper bounds vector
e_x, e_y, e_θ	Absolute error components, [mm, mm, °]
$e_{x\%}, e_{y\%}, e_{\theta\%}$	Relative percentage error components
Λ, λ	Set of indexes and a one of its generic elements
Q, GD	Set of geometrical parameters and one of its generic elements
F_λ	λth function that is used to define a perimeter of the objet base
W, D	Sets of coordinates that represents the perimeter and the area of the object base

References

1. Fantoni, G.; Santochi, M.; Dini, G.; Tracht, K.; Scholz-Reiter, B.; Fleischer, J.; Kristoffer Lien, T.; Seliger, G.; Reinhart, G.; Franke, J.; et al. Grasping devices and methods in automated production processes. *CIRP Ann.* **2014**, *63*, 679–701. [CrossRef]
2. Kamagaew, A.; Stenzel, J.; Nettsträter, A.; ten Hompel, M. Concept of Cellular Transport Systems in facility logistics. In Proceedings of the 5th International Conference on Automation, Robotics and Applications, Wellington, New Zealand, 6–8 December 2011; pp. 40–45. [CrossRef]

3. Rizescu, C.I.; Pleşea, A.I.; Rizescu, D. Modular Transport and Sorting System with Omnidirectional Wheels. In Proceedings of the 10th International Conference on Manufacturing Science and Education—MSE 2021, Sibiu, Romania, 2–4 June 2021. [CrossRef]
4. Uriarte, C.; Asphandiar, A.; Thamer, H.; Thamer, H.; Benggolo, A.Y.; Freitag, M. Control strategies for small-scaled conveyor modules enabling highly flexible material flow systems. In Proceedings of the ScienceDirect, Procedia CIRP Conference on Intelligent Computation in Manufacturing Engineering, Gulf of Naples, Italy 18–20 July 2018. [CrossRef]
5. Böhringer, K.F.; Bhatt, V.; Donald, B.R.; Goldberg, K. Algorithms for Sensorless Manipulation Using a Vibrating Surface. *Algorithmica* **2000**, *26*, 389–429. [CrossRef]
6. Böhringer, K.F. Programmable Force Fields for Distributed Manipulation, and Their Implementation Using Micro-Fabricated Actuator Arrays. Ph.D. Thesis, Cornell University, Ithaca, NY, USA, 1997.
7. Konishi, S.; Fujita, H. A conveyance system using air flow based on the concept of distributed micro motion systems. *J. Microelectromechan. Syst.* **1994**, *3*, 54–58. [CrossRef]
8. Kritikou, G.; Aspragathos, N. Activation Algorithms for the Micro-Manipulation & Assembly of Hexagonal Microparts on a Programmable Platform. In *Advances in Service and Industrial Robotics: Proceedings of the 27th International Conference on Robotics in Alpe-Adria Danube Region (RAAD 2018), Patras, Greece, 6–8 June 2018*; Springer International Publishing: Berlin/Heidelberg, Germany, 2018. [CrossRef]
9. Oyobe, H.; Hori, Y. Object Conveyance System " Magic Carpet" Consisting of 64 Linear Actuators—Object Position Feedback Control with Object Position Estimation. In Proceedings of the 2001 IEEE/ASME international Conference on Advanced Intelligent Mechatronics Proceedings, Como, Italy, 8–12 July 2001. [CrossRef]
10. Piatkowski, T. Analysis of translational positioning of unit loads by directionally-oriented friction force fields. *Mech. Mach. Theory* **2011**, *46*, 201–217. [CrossRef]
11. Tao, S.; Youpan, Z.; Hui, Z.; Pei, W.; Yongguo, Z.; Guangliang, L. Three-wheel Driven Omnidirectional Reconfigurable Conveyor Belt Design. In Proceedings of the 2019 Chinese Automation Congress (CAC), Hangzhou, China, 22–24 November 2019; pp. 101–105. [CrossRef]
12. Whiting, J.; Mayne, R.; Melhuish, C.; Adamatzky, A. A Cilia-inspired Closed-loop Sensor-actuator Array. *J. Bionic Eng.* **2018**, *15*, 526–532. [CrossRef]
13. Keek, J.S.; Loh, S.L.; Chong, S.H. Design and Control System Setup of an E-Pattern OmniwheeledCellular Conveyor. *Machines* **2021**, *9*, 43. [CrossRef]
14. Krühn, T.; Falkenberg, S.; Overmeyer, L. Decentralized control for small-scaled conveyor modules with cellular automata. In Proceedings of the 2010 IEEE International Conference on Automation and Logistics, Hong Kong, China, 16–20 August 2010; pp. 237–242. [CrossRef]
15. Böhringer, K.F.; Donald, B.R.; MacDonald, N.C. Programmable Force Fields for Distributed Manipulation, with Applications to MEMS Actuator Arrays and Vibratory Parts Feeders. *Int. J. Robot. Res.* **1999**, *18*, 168–200. [CrossRef]
16. Will, P.; Liu, W. *Parts Manipulation on a MEMS Intelligent Motion Surface*; ISI/RR-94-391; University of Southern California Marina Del Rey Information Sciences Inst: Los Angeles, CA, USA, 1994.
17. Frei, P.U.; Wiesendanger, M.; Büchi, R.; Ruf, L. *Simultaneous Planar Transport of Multiple Objects on Individual Trajectories Using Friction Forces*; Kluwer Academic Publishers: Amsterdam, The Netherlands, 2000. [CrossRef]
18. Ataka, M.; Legrand, B.; Buchaillot, L.; Collard, D.; Fujita, H. Design, Fabrication, and Operation of Two-Dimensional Conveyance System With Ciliary Actuator Arrays. *IEEE/ASME Trans. Mechatron.* **2009**, *14*, 119–125. [CrossRef]
19. Becker, K.P.; Chen, Y.; Wood, R.J. Mechanically Programmable Dip Molding of High Aspect Ratio Soft Actuator Arrays. *Adv. Funct. Mater.* **2020**, *30*, 1908919. [CrossRef]
20. Fantoni, G.; Santochi, M. Development and testing of a brush feeder. *CIRP Ann.* **2010**, *59*, 17–20. [CrossRef]
21. Chen, Z.; Deng, Z.; Dhupia, J.S.; Stommel, M.; Xu, W. Motion Modeling and Trajectory Tracking Control for a Soft Robotic Table. *IEEE/ASME Trans. Mechatron.* **2021**, *27*, 2667–2677. [CrossRef]
22. Raptis, I.A.; Hansen, C.; Sinclair, M.A. Design, modeling, and constraint-compliant control of an autonomous morphing surface for omnidirectional object conveyance. *Robotica* **2022**, *40*, 213–233. [CrossRef]
23. Robertson, M.A.; Murakami, M.; Felt, W.; Paik, J. A Compact Modular Soft Surface With Reconfigurable Shape and Stiffness. *IEEE/ASME Trans. Mechatron.* **2019**, *24*, 16–24. [CrossRef]
24. Boutoustous, K.; Laurent, G.J.; Dedu, E.; Matignon, L.; Bourgeois, J.; Le-Fort-Piat, N. Distributed control architecture for smart surfaces. In Proceedings of the The 2010 IEEE/RSJ International Conference on Intelligent Robots and Systems, Taipei, Taiwan, 18–22 October 2010. [CrossRef]
25. Chen, X.; Zhong, W.; Li, C.; Fang, J.; Liu, F. Development of a Contactless Air Conveyor System for Transporting and Positioning Planar Objects. *Micromachines* **2018**, *9*, 487. [CrossRef]
26. Ku, P.J.; Winther, K.; Stephanou, H.; Safaric, R. Distributed control system for an active surface device. In Proceedings of the Proceedings 2001 ICRA. IEEE International Conference on Robotics and Automation (Cat. No.01CH37164), Seoul, Republic of Korea, 21–26 May 2001; Volume 4, pp. 3417–3422. [CrossRef]
27. Mita, Y.; Arai, M.; Tixier, A.; Fujita, H. Bulk Micromachined Durable Air-Flow Microactuator Array for Robust Conveyance Systems. In *Proceedings of the Transducers '01 Eurosensors XV*; Obermeier, E., Ed.; Springer: Berlin/Heidelberg, Germany, 2001; pp. 718–721. [CrossRef]

28. Mobes, S.; Laurent, G.J.; Clevy, C.; Fort-Piat, N.L.; Piranda, B.; Bourgeois, J. Toward a 2D Modular and Self-Reconfigurable Robot for Conveying Microparts. In Proceedings of the 2012 Second Workshop on Design, Control and Software Implementation for Distributed MEMS, Besancon, France, 2–3 April 2012; pp. 7–13. [CrossRef]
29. Turitto, M.; Ratchev, S.; Xue, X.; Hughes, M.; Bailey, C. Pneumatic contactless microfeeder design refinement through CFDsimulation. In Proceedings of the Proceedings of 4M 2007, the Third International Conference on Multi-Material Micro Manufacture, Borovets, Bulgaria, 3–5 October 2007.
30. Zeggari, R.; Yahiaoui, R.; Malapert, J.; Manceau, J.F. Design and fabrication of a new two-dimensional pneumatic micro-conveyor. *Sensors Actuators A-Phys.* **2010**, *164*, 125–130. [CrossRef]
31. Costanzo, M.; Myers, D.H. Diverting Conveyor with Bidirectional Assist Roller. U.S. Patent 9,156,629, 13 October 2015.
32. Mayer, S.H. *Development of a Completely Decentralized Control System for Modular Continuous Conveyors*; KIT Scientific Publishing: Karlsruhe, Germany, 2009; Volume 73.
33. Kleczewski, L. Weighing and Sorting Roller Belt Conveyor and Associated Method. U.S. Patent App. 16/640,445, 18 June 2020.
34. Overmeyer, L.; Ventz, K.; Tobias Krühn, S.F. Interfaced multidirectional small-scaled modules for intralogistics operations. *Logist. Res.* **2010**, *2*, 123–133. [CrossRef]
35. Sanghai, R.A.; Saundalkar, P.P.; Mallick, J.A.; Shah, B. Sort X Consignment Sorter using an Omnidirectional Wheel Array for the Logistics Industry. In Proceedings of the 2020 International Conference on Convergence to Digital World—Quo Vadis (ICCDW), Mumbai, India, 18–20 February 2020. [CrossRef]
36. Bhattacharjee, A.; Lu, Y.; T. Becker, A.; Kim, M. Magnetically Controlled Modular Cubes With Reconfigurable Self-Assembly and Disassembly. *IEEE Trans. Robot.* **2021**, *38*, 1793–1805. [CrossRef]
37. Dang, T.A.T.; Bosch-Mauchand, M.; Arora, N.; Prelle, C.; Daaboul, J. Electromagnetic modular Smart Surface architecture and control in a microfactory context. *Comput. Ind.* **2016**, *81*, 152–170. [CrossRef]
38. Huayu, W.; Shunsuke, Y.; Shuji, T. Fabrication of Multi-Axis Moving Coil Type Electromagnetic Micro-Actuator Using Parylene Beams for Pure In-Plane Motion. In Proceedings of the 2021 21st International Conference on Solid-State Sensors, Actuators and Microsystems (Transducers), Online, 20–25 June 2021; pp. 667–670. [CrossRef]
39. Tisnés, S.D.; Petit, L.; Prelle, C.; Lamarque, F. Modeling and Experimental Validation of a Planar Microconveyor Based on a 2 × 2 Array of Digital Electromagnetic Actuators. *IEEE/ASME Trans. Mechatron.* **2021**, *26*, 1422–1432. [CrossRef]
40. Pennestrì, E.; Rossi, V.; Salvini, P.; Valentini, P.P. Review and comparison of dry friction force models. *Nonlinear Dyn.* **2016**, *83*, 1785–1801. [CrossRef]
41. Green, P.; Worden, K.; Sims, N. On the identification and modelling of friction in a randomly excited energy harvester. *J. Sound Vib.* **2013**, *332*, 4696–4708. [CrossRef]
42. Piatkowski, T.; Wolski, M.; Dylag, K. Angular positioning of the objects by the system of two oblique friction force fields. *Mech. Mach. Theory* **2019**, *140*, 668–685. [CrossRef]
43. Piatkowski, T.; Wolski, M. Model of positioning objects by the system of oblique friction force fields on horizontal and vertically offset planes. *Mech. Mach. Theory* **2021**, *156*, 104155. [CrossRef]
44. Omnitrack. Omnitrack Catalogue. Technical Report. 2022. Available online: https://www.omnitrack.com/wp-content/uploads/2022/11/Catalogue-UK.pdf (accessed on 30 January 2023)
45. Piątkowski, T.; Szempruch, J. Ranges of application of sorting manipulators. *J. Pol. CIMAC* **2011**, *6*, 235–244.
46. Verheul, C. ADAMS Methodology Contact Modeling. Benelux ADAMS User Meet. 2012. Available online: https://www.insumma.nl/wp-content/uploads/SayField_Verheul_ADAMS_Contacts.pdf (accessed on 30 January 2023)
47. Giesbers, J. Contact Mechanics in MSC Adams a Technical Evaluation of the Contact Models in Multibody Dynamics Software MSC Adams. Bachelor's Thesis, Faculty of Engineering Technology Applied Mechanics, London, UK . 2012.

applied sciences

MDPI

Article

Simulation Model for Robotic Pick-Point Evaluation for 2-F Robotic Gripper

Primož Bencak [1], Darko Hercog [2] and Tone Lerher [1,3,*]

[1] Faculty of Logistics, University of Maribor, Slomškov trg 15, 2000 Maribor, Slovenia
[2] Faculty of Electrical Engineering and Computer Science, University of Maribor, Slomškov trg 15, 2000 Maribor, Slovenia
[3] Faculty of Mechanical Engineering, University of Maribor, Slomškov trg 15, 2000 Maribor, Slovenia
* Correspondence: tone.lerher@um.si

Featured Application: This paper presents a simulation model based on cosimulation between ADAMS and MATLAB/Simulink, designed to evaluate pick-points for an arbitrary object gripped with a two-fingered (2-F) robotic gripper.

Abstract: Robotic bin-picking performance has been gaining attention in recent years with the development of increasingly advanced camera and machine vision systems, collaborative and industrial robots, and sophisticated robotic grippers. In the random bin-picking process, the wide variety of objects in terms of shape, weight, and surface require complex solutions for the objects to be reliably picked. The challenging part of robotic bin-picking is to determine object pick-points correctly. This paper presents a simulation model based on ADAMS/MATLAB cosimulation for robotic pick-point evaluation for a 2-F robotic gripper. It consists of a mechanical model constructed in ADAMS/View, MATLAB/Simulink force controller, several support functions, and the graphical user interface developed in MATLAB/App Designer. Its functionality can serve three different applications, such as: (1) determining the optimal pick-points of the object due to object complexity, (2) selecting the most appropriate robotic gripper, and (3) improving the existing configuration of the robotic gripper (finger width, depth, shape, stroke width, etc.). Additionally, based on this analysis, new variants of robotic grippers can be proposed. The simulation model has been verified on a selected object on a sample 2-F parallel robotic gripper, showing promising results, where up to 75% of pick-points were correctly determined in the initial testing phase.

Keywords: intralogistics; robotic bin-picking; simulation model; ADAMS; pick-point determination; MATLAB/Simulink; 2-F robotic gripper; performance analysis

Citation: Bencak, P.; Hercog, D.; Lerher, T. Simulation Model for Robotic Pick-Point Evaluation for 2-F Robotic Gripper. *Appl. Sci.* **2023**, *13*, 2599. https://doi.org/10.3390/app13042599

Academic Editors: Guido Tosello, Roque Calvo and José A. Yagüe-Fabra

Received: 28 December 2022
Revised: 13 February 2023
Accepted: 15 February 2023
Published: 17 February 2023

1. Introduction

The term Industry 4.0 (I4.0) represents the fourth industrial revolution. Its beginning dates to 2010, and the term Industry 4.0 is often used interchangeably with the term "smart industry." The concept's core idea lies in that today's industry needs answers to economic, sociological, and political changes generated by the end-users [1]. However, the industry usually requires some time to adapt to new concepts and adjusts its business accordingly. Therefore, the novelties in the industry need some time to be implemented, as is the case with I4.0. While there is no single definition of what precisely the term I4.0 represents, we can generally divide the I4.0 concept into three major categories, such as Cyber-Physical Systems (CPS), the Internet of Things (IoT), and cloud computing [2]. The first concept describes the ability to couple the physical properties of a system with advanced computational algorithms, such as in the field of predictive maintenance [3]. The IoT relates to the interconnection of various devices that rely on sensors, communication, networking, and information processing technologies [2]. Sensor networks can gather data regarding the manufacturing process and then later use that data to optimize the process. Cloud computing refers to sharing documents, collaboration, distributed production, and resource optimization [4]. The main advantage of cloud computing is scalability, which means extra computing power is ensured if additional demand arises [5]. We can conclude that the I4.0 concepts were developed to improve industrial processes by collecting and using the data for optimization.

Since the industry depends heavily on logistics services, the logistics processes must be adapted accordingly to cope with the increasing demands. Hence, the Logistics 4.0 (L4.0) term has been coined based on the Industry 4.0 term [6,7]. Logistics 4.0 describes advanced usage of various technological advancements such as those in Industry 4.0, such as intelligent devices, IoT, and other CPS, which aim to reorganize some of the basic concepts of logistics [7]. This holds especially true due to the rising trends of E-commerce [8] and the concerning trends of workforce deficiency and the aging population [9]. Logistics' global market size accounted for around EUR 8.43 trillion in the year 2021 [10], while the E-commerce market accounted for over EUR 440 billion in the year 2021 [11].

One of the most critical aspects of logistics is intralogistics, which focuses on the processes carried out inside warehouses and distribution centers and ensures the correct flow of materials and information [12]. One of the essential processes in the warehouse is order-picking of goods according to the customer's order. It is estimated that order-picking accounts for around 55% of the warehouse operation costs [13]. Therefore, automatization in warehouses is necessary to achieve high throughput rates, cope with the increasing trends of "batch size one," and provide the shortest delivery times [8]. In warehouses, the most labor- and time-intensive process is the order-picking process, which requires a lot of (human) intervention and does not add expressive increased value to the product. Several solutions have already been developed to reduce the operators' workload and reduce the non-value-adding transport of the product within the warehouse. Mobile robots are used to transport goods along the operators [14], and mobile robots are equipped with robotic arms to pick the objects directly from the shelving units [15] to the most advanced automated storage and retrieval systems coupled with the robotic bin-pickers [16].

Robotic bin-picking systems aim to reduce or eliminate human intervention during the bin-picking processes for the bulk objects (items) with the help of a (collaborative) robotic arm, a vision system, robotic grippers, and control systems. The most demanding task of robotic bin-picking systems is to automatically and autonomously pick items from the bin that are various by shape, color, and weight and have different mechanical properties. The essential steps in robotic bin-picking can be roughly categorized as (1) object detection, (2) object localization, (3) grasp detection, (4) path/trajectory planning, and (5) object manipulation [17]. The first two steps are in the domain of the vision system, which provides the information of the location and orientation of the object to be bin-picked, while the third step means that to manipulate the object successfully, one must first correctly determine object pick-points. The fourth and fifth steps are in the domain of the robotic subsystem, which determines robotic arm dynamics.

Determining object pick-points can happen before or during the bin-picking operation, directly before the object manipulation [18]. Since the vision system cannot give complete information regarding the objects being observed, the researchers aim to extract possible grasp candidates from partial observation based on RGB-D images or point-cloud, coupled with various machine-learning techniques [19,20]. Lately, the most used are deep-learning methods, where multiple types of neural networks are used (e.g., CNN, DCNN, etc. [21]). Those types of determining-grasp candidates can achieve very high accuracy even on previously unseen (novel) objects [22,23], while usually requiring pretraining and setting the correct hyperparameters, which can also be very time intensive [24]. The second most notable option is to perform mechanical simulations before the bin-picking process in computer simulations or on the physical system, the latter being the most time- and labor-intensive. Since performing experiments on the physical system is not feasible for many objects, computer simulations using multibody physics simulators [25] are practical.

A more significant number of correctly determined object pick-points (or grasp candidates) correspond to a higher possibility of correctly picking an item from a bin, given that the vision system provided the correct object location and pose.

While rigid multibody simulators are highly capable of performing the grasping-point evaluation, it is usually difficult to implement the numerical algorithms which would enable accurate reproduction of the actual (robotic) system [26]. Numerous open-source simulation engines (ODE [27], BULLET [28], DART [29], SIMBODY [30]) exist to date. However, they usually require extensive programming knowledge and are subject to community maintenance. Additionally, they typically suffer from various failure modes, such as interpenetration or other physically unrealistic interactions between the robotic gripper and the objects, which are more analyzed in detail in [26]. For more information regarding the performance of various multibody physics simulators, the authors point out a study by Erez, Tassa, and Todorov [31], in which several different multibody physics engines used in robotics were evaluated. Several other tools have been developed for planning/evaluating robotic gripper grasps, such as GraspIt! [32] and SynGrasp [33]. The latter is oriented more toward multiple DOF robotic grippers, namely, robotic hands. Usually, pick-point evaluation regards static conditions. However, some researchers also evaluate grasp quality according to the robot's trajectory [34].

A GPU-based physics engine, Isaac Gym, is now also being used in pick-point evaluation [35,36]. With the introduction of the Incremental Potential Contact Model (IPC) [37], Kim et al. [38] note that the common issue of most physics simulators can now be more successfully contained. The IPC avoids interpenetration artifacts, which include false sticking to the object or intersecting behavior. In a practical example by [38], the engine is used to simulate parallel gripper grasping with soft fingertips. However, our approach does not focus on a state-of-the-art IPC model but on improving existing approaches, showing possible simulation capabilities.

Therefore, ADAMS 2022.1 simulation software was chosen as the primary physics simulation tool, coupled with the advanced control functionalities provided by MATLAB 2020b and MATLAB/Simulink. Bonilla et al. [39] noted that the ADAMS system software is closely related to simulators, such as GraspIt!, Open-Rave, and OpenGRASP. To provide a robotic bin-picking integrator with a tool, which would enable one to systematically determine object pick-points for various objects in this paper, we explore the below research questions:

RQ1: Can the ADAMS/MATLAB cosimulation be used to determine the systematic grasp quality for the selected 2-F parallel robotic gripper?

RQ2: Can the simulation model also be used to evaluate 2-F robotic grippers or even develop new variants of the 2-F grippers?

RQ3: How does a combination of the analytical and empirical approach of grasp quality evaluation compared to "deep-learning" model-based approaches influence the overall system performance?

A novel simulation model for determining pick-points for a 2-F robotic gripper is proposed and developed to answer the above research questions. This model aims to systematically determine optimal pick-points of the gripping object sent to the collaborative robot for performing bin-picking actions. The model accurately considers the mechanical parameters of the 2-F robotic gripper and an object to be gripped, which results in an accurate bin-picking application. The proposed model is scalable, meaning it can perform simulations on any object. However, several parameters must be determined before pick-point evaluation.

The main contributions of this paper are as follows: (1) The proposed ADAMS/MATLAB cosimulation model can be used to develop the robotic bin-picking setup, where object pick-points must be determined systematically due to object complexity. The proposed simulation model can be integrated with the module to select object pick-points. Furthermore, the proposed simulation model can be used with the robotic bin-picking software to substitute manual determination of pick-points, i.e., via ROS or a similar supported robotic interface.

(2) In case of replacement of the 2-F robotic gripper with another one, the proposed model can be used to determine which of the robotic grippers (or robotic gripper variants) is the most appropriate for the bin-picking process. For example, if the selected robotic gripper achieves a higher overall grasp-quality score on a set of pick-points, it can be considered the most appropriate. (3) The proposed simulation model can also be used to improve the existing configuration of the robotic gripper (finger width, finger depth, finger shape, stroke width, etc.) according to the bin-picking score of the specific configuration. Additionally, based on our analysis, new variants of robotic grippers can be proposed. While the first application requires little or no modification to the existing model, the second and the third applications proposed would require moderate modification to the simulation model to provide the user with the model's potential benefits for pick-point evaluation.

It must be emphasized that the most advanced pick-point determination procedures operate in real-time for various objects based on data gathered with machine vision systems. Nevertheless, many bin-picking applications still focus on bin-picking products of a single type (i.e., electrical outlets). In this case, a study of the object pick-points can be determined prior to the bin-picking application. Since the number of pick-points corresponds to the overall bin-picking success, the object pick-points should be determined systematically and in the most significant number possible. This paper aims to provide an alternative approach to pick-point determination for industrial bin-picking applications. Additionally, the performance in assembly tasks of such robotic bin-picking applications must be known beforehand, and every object and part of the application must be thoroughly tested. In this case, many mechanical parameters and equipment are known in advance, such as grasping force, the exact type of the robotic gripper, gripper closing speed, available cycle time, 3D model of the objects, etc.

The paper is organized as follows. In Section 2, the related works are presented and analyzed. Section 3 presents the simulation model for determining pick-points in detail. The mechanical parameters of the used robotic gripper are presented, and the contact parameters are explained in detail. Next, the ADAMS/MATLAB cosimulation interface is presented, along with a robotic force controller and generation of object pick-points. A Graphical User Interface (GUI) was developed to provide a user-friendly interface in MATLAB/App Designer. Further, several improvements to original simulations are presented, designed to reduce the time needed for more performed simulations. In Section 4, results for a selected test object are presented. The simulation results were validated on a selected robotic bin-picking setup. Lastly, the results are discussed, along with a suggestion for future research directions.

2. Related Works

Many research groups in the robotic research community are working on the problem of successful grasp (or pick-point) determination for various robotic grippers for rigid or nonrigid objects (i.e., cloths, chips packaging, etc.). In a survey by Du, Wang, Lian, and Zhao [18], the authors, in detail, analyze papers dealing with the grasp estimation problems as well as the processes that must happen before the grasping estimation (e.g., object localization, object pose estimation, motion planning). It is evident that the grasp estimation problem mainly relates to the machine-vision challenges regarding when the information about the grasp estimation is directly derived from the RGB-D image/point cloud. The authors categorize the grasp estimation problem into 2D planar grasps (the grasp contact points can uniquely define the gripper's grasp pose) and 6 DoF grasps (the gripper can grasp the object from various angles; 6D gripper pose is essential to conduct the grasp). Since our test object is subjected to the simulation model limitations, we will focus on the 2D grasps in our research model. The authors also divide the methods dealing with the problem into two categories, namely, (i) traditional (e.g., machine learning methods to train classifiers based on manually selected 2D descriptors) and (ii) deep-learning methods [40,41] (e.g., Multilayer Perceptron (MLP), Fully Convolutional Network (FCN)-based methods, and Capsule-based methods). Based on the literature review by

Du, Wang, Lian, and Zhao [18], grasp estimation problems have been gaining attention in recent years.

Wang and Li [42] noted that robot grasping, detection and planning can be roughly divided into three types of problem solving methods: (1) empirical methods (grasp positions on known models are evaluated through physical analysis or virtual environment simulation and stored as a database); (2) analytical methods (first, candidate grasp positions are obtained according to geometric analysis, mechanical analysis, or model reasoning; then, a deep neural network model is established to extract the features of the grasp position, which scores the grasp reliability; and finally, the candidate with the highest score is selected); and (3) detection-based methods that use the neural network as a fitter to directly estimate the parameters of the grasp position from the point cloud or image. According to Wang and Li [42], this method is relatively simple and easy to implement. Therefore, it has gradually become the most used method in grasp detection in the last few years. Since our problem-solving falls into the first two categories, namely, a combination of empirical and analytical methods, we analyze related works. Lastly, we briefly examine papers containing different "deep-learning approaches" as the current state-of-the-art, which show very high accuracy in the grasp-detection area.

Perhaps the closest related work is by Bonilla, Farnioli, Piazza, Catalano, Grioli, Garabini, Gabiccini, and Bicchi [39]. The authors also used the system software ADAMS for conducting grasp configuration determination for a Pisa/IIT SoftHand. They created a batch simulation setup to evaluate grasp affordances on kitchenware objects (a cup, a colander, and a plate). Using MATLAB, they modified simulation parameters so that during each pick-point evaluation, the robotic hand links moved appropriately to form the grasping motion. They performed the grasping experiments on an actual KUKA robot and Pisa/IIT SoftHand prototype by accurately replicating hand/object configuration. The ADAMS simulations served as a means for determining possible grasp configurations that would result in the successful grasping of the object. We further developed our model to execute simulations in parallel compared to their approach. Our model can sweep mechanical parameters in a graphical user interface to find the most appropriate contact and force controller parameters settings. Additionally, a more descriptive results analysis was added. Furthermore, the gripper closing is conducted with the force controller realized in MATLAB/Simulink, which also contains functions for optimizing the required time needed for the execution of a single simulation. Lastly, we provided several measures, based on which we proposed our own set of grasp success metrics. Those are used for the selection of best pick-points and further evaluation.

Taylor, Drumwright, and Hsu [26] discussed possible grasping failures associated with simulating the grasping of rigid objects of four open-source physics engines in the GAZEBO simulator. They exposed that while rigid body simulators are highly capable, it is tricky to implement the numerical algorithms that would enable accurate reproduction of the actual (robotic) system. Usually, this is reflected by the fact that the objects start to slip after some time from the simulated robot's grasp. They identified several failure modes, which may occur during the simulation of quasirigid objects with rigid robots, namely: (1) slip associated with too-low grasping force or friction coefficient, (2) iterative method nonconvergence, (3) rounding errors, (4) regularization errors, (5) constraint stabilization, (6) imprecise contact information, and (7) tangential drift. Of course, these problems can be contained to some extent, however, usually at the expense of increased simulation time. Additionally, they found that while some simulation engines (e.g., SIMBODY) proved more resistant to some types of errors, the simulation time drastically increased, rendering the simulations unusable for evaluating a higher number of pick-points.

Vahrenkamp et al. [43] presented a part-based grasp-planning approach that can generate grasps, which are also applicable to similar, novel objects. They first segmented multiple object meshes according to their shape and volumetric information. Second, they labeled specific parts (segments) of the object according to the robot task the gripper is supposed to perform. Lastly, they used the "Simox grasp planning toolbox" to generate stable grasps by aligning an initially specified hand approach direction with a randomly chosen object surface normal. Finally, they tested their method using the humanoid robot ARMAR-III, equipped with a robotic hand.

Further, Tian et al. [44] presented an approach to transfer grasp configurations from prior examples to novel objects, assuming that the novel and original objects have the same topology and use similar shapes. First, they performed 3D segmentation on the sample objects using geometric and semantic shape characteristics. Next, a grasp space was calculated for each part, and the corresponding grasps were calculated for novel objects using bijective contact mapping.

In the case of detection-based methods, Nechyporenko et al. [45] presented a practical solution to the problem of robot picking in an online shopping warehouse by employing the centroid normal approach method (CNA) on a cost-effective dual-arm robotic system with two grippers. Scene point clouds are matched with the grasping techniques and grippers (2-F and vacuum) by performing visible surface analysis. It does not employ any mechanical model of the object in question since it derives all the needed information for picking from the actual scene. In the Amazon Challenge 2017, for a given object set, they were successful, from 69% to 77.5%, in grasping previously unknown objects. Xu et al. [46] presented an AdaGrasp policy, which is designed to select the most appropriate robotic gripper and its pose for various objects based on cross-convolution between the shape encodings of the gripper and the scene. The policy matches the scene and the gripper geometry under different grasp poses, where a good overlap of the gripper geometry to a 3D geometry of the grasped object will lead to a successful grasp. It uses the "Pybullet" simulation environment to evaluate the grasp quality. The authors also replicated the results on their physical counterparts, where they achieved between an 86% and 80.5% successful grasp rate for single and multiple objects compared to the algorithm performance in simulation. Mahler, Matl, Satish, Danielczuk, DeRose, McKinley, and Goldberg [22] developed a "Dexterity Network" (Dex-Net) 4.0, which was designed to learn the grasping policy for a given set of robotic grippers (2-F and vacuum) by training on synthetic datasets using domain randomization with analytic models of physics and geometry. The policy assumes the following: (i) quasistatic physics (e.g., inertial terms are negligible) with Coulomb friction, (ii) objects are rigid and made of nonporous material, (iii) the robot has a single overhead depth sensor with known performance characteristics, and (iv) the robot has two end effectors with known geometry. They combine the Grasp Quality Convolutional Neural Networks (GQ-CNNs) for each gripper to plan grasps for objects in a given point cloud. On their physical counterpart, they achieved 95% reliability on several of 25 novel objects. Wu, Akinola, Gupta, Xu, Varley, Watkins-Valls, and Allen [23] developed a framework for high-DOF multifingered grasping in clutter that can also be used in various parallel-jaw and multifingered robot hands. It uses simulation-based learning that uses depth and geometry alone (i.e., no texture) to allow accurate domain transfer to real scenes. It does not require any database of grasp examples. Instead, it uses a policy gradient formulation and a learned attention mechanism to generate full 6-DOF grasp poses and all finger joint angles to pick up objects in dense clutter given a single-depth image. They achieved up to 96.7% of grasp success in single-seen objects for two robotic grippers. Laili et al. [47] developed a method that can predict the appropriate grasping point-pair of an unknown object for a specific task with a much lower training cost. They implemented a two-stage predictor, where, firstly, using a Sobel operator on RGB-D data, they proposed robust grasp candidates. Secondly, they used a region-based predictor trained by semisupervised learning. Experimental results demonstrated that the proposed region-based grasping detection method can find an accurate grasp configuration of a new emergent object and achieve an average success rate of 91.5% by using fewer than 100 training samples. Wang and Li [42] proposed a novel two-stage grasp detection method based on visual rotation object detection and point cloud spatial feature scoring. They transformed the depth image to point cloud and proposed a grasp detection method to be made on the point cloud, rather than on RGB image. Since this approach uses the neural network, it requires pretraining prior to operation. They tested their approach on various household and 3D printed objects of different shapes, with the minimum score of 87.3% grasp attempts. Cheng et al. [48] developed a grasp detector based on a Feature Pyramid Network (FPN) for a 2-F parallel gripper. The input for the grasp detector is the RGB-D image obtained from the camera. Their grasp detector works in two stages: first, the detector generates horizontal candidate grasp areas, whilst in the second stage, it refines those poses to predicted rotated grasp poses. The grasp detector achieves up to 93.3% accuracy on a selected object set.

While the most recent approaches rely primarily on "deep-learning" approaches, our simulation model adapts the combination of an analytical and empirical approach for the systematic pick-point determination. This could be considered the slower and more time-consuming approach in many cases. Still, compared to "black-box" methods, the simulation model transparency remains high throughout the pick-point determination process. Additionally, most of the reviewed works from the literature focus on 2-F parallel gripper configurations, while the adaptive grippers are discussed less commonly or rarely. In our simulation model, the system's mechanics (2-F gripper and the gripped objects) can be precisely modeled to account for the possible variants of the 2-F robotic grippers.

Additionally, with the ADAMS/MATLAB cosimulation, complex robotic gripper behavior can be modeled, which is not the subject of point-cloud-based deep-learning methods. Furthermore, the kinematics and the dynamics of the collaborative robot movement can be considered to evaluate the fast movements of the robotic arm during the bin-picking process. Especially in industrial and logistics environments, this remains a significant advantage since those processes usually require extensive verification before the operation. False or poor bin-picking performance could lead to stopping the production line, which is associated with high cost and low-throughput performance.

3. Simulation Model for Object Pick-Point Evaluation

Determining pick-points with simulations is a complex task that requires an accurate description of the robotic order-picking system (a robotic gripper and a gripped object) and all possible physical relationships between the objects (contact forces, friction, etc.). It is necessary to evaluate the influencing parameters and ensure that the system response is as close as possible in simulations compared to a physical system.

The developed simulation model for pick-point evaluation consists of three main parts: (1) a model of a two-fingered robotic gripper, (2) the model of an object to be gripped, and (3) control functions that permit gripper translation and rotation and vertical movement of the gripper and enable the force control of the gripper closing. For this purpose, it was necessary to select a simulation tool that would allow the automatic analysis of arbitrary pick-points, considering the kinematic and dynamic properties of the objects, and would also allow the graphical presentation of the results through animations. Therefore, the MSC ADAMS (Automatic Dynamic Analysis of Mechanical Systems) software package, designed for multibody analysis, was selected for our research study. ADAMS was used to simulate various mechanical systems, enabling verification of forces generated by the interactions between elements while allowing the results to be presented graphically in the form of an animation. The software was selected due to its ability to integrate with the MATLAB/Simulink simulation environment.

Figure 1 schematically presents the simulation model components. The ADAMS/View contains the model of the robotic gripper, gripped object, and various measures for determining forces and other influential parameters. It also serves as a tool for exporting the model for the ADAMS/MATLAB cosimulation. MATLAB/Simulink contains a force controller for closing the robotic gripper fingers and several support and analysis functions. Additional scripts are written in MATLAB, which work as various support functions and enable simulations to be executed parallelly. Lastly, MATLAB/App Designer was used to provide the user with a graphical interface, which enables various parameters settings and analysis of selected pick-points.

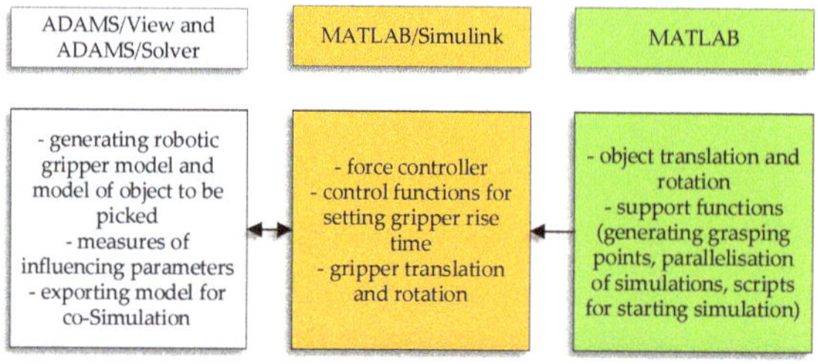

Figure 1. Simulation model components. Source: own.

A few assumptions were made to reduce the overall complexity of the simulation model: (1) the object pick height is always half of the object depth, (2) rotation of the object is permitted only around the (*Y*) axis (in the negative direction of gravitational force), (3) only rigid objects and rigid robotic fingers were considered.

Despite those assumptions, the model still qualifies as applicable to a number of real-world bin-picking problems. The user must first input a 2D top-down image of the analyzed object to evaluate selected object pick-points. The best practice is to use a snapshot of the 3D model with a clear distinction between the background and the model since several calculations (generation of pick-points, collision checks, etc.) rely on the correct image segmentation process. Further, the user must specify the input parameters of the selected 2-F robotic gripper (finger width, finger depth, stroke width, etc.) for the model to function correctly. Parameters for the P force controller must be set beforehand, although the user can fine-tune them if needed. Next, the number of pick-points must be assigned if the "Normal" simulation mode is selected. Otherwise, the maximum number of points is set if the simulation mode is set to "Optimization." In this mode, the user must also specify the successful grasp quality metric (see Section 4.2. for more details) by which the pick-points are considered valid. Pick-points are then generated dynamically, and simulations are run until the selected number of successful pick-points is generated.

3.1. Modeling Robotic Gripper

By importing the 3D model of the robotic gripper from the manufacturer into the ADAMS/View, we can be sure that the dimensions are correct and that there will be no deviations due to the geometry of the 3D model. Next, joint types were assigned to each part (joint) of the robotic gripper, where two or more elements come into contact (Figure 2). Since all joints on the robotic gripper can only move rotationally around their axis, a revolute type of joint was used. This joint type has only one degree of freedom (DoF). Friction in the joints was neglected.

The robotic gripper is a five-link mechanism, meaning it must be actuated in two joints simultaneously to allow the fingertips of the robotic gripper to move linearly. The mechanism is internally linked without intermediate gears with different gear ratios, which means that the set velocity in the first and second joints must be the same to achieve linear fingertip movement.

Lastly, the mass properties of the robot gripper or its individual elements were determined by selecting the appropriate materials in ADAMS/View. The moments of inertia and mass are calculated through the model's geometry. Additionally, a body consisting of various materials can be modeled. However, parts with the same material properties must first be divided in the CAD modeling software and then imported to ADAMS piece by piece. Steel was selected for all components of the robotic gripper except the robotic fingers. For the fingers of the robotic gripper, a nitrile-butadiene rubber (NBR) with a density of 1.5 g/cm^3 was selected according to the robotic gripper manufacturer's specifications. The ADAMS/View interface is shown in Figure 3.

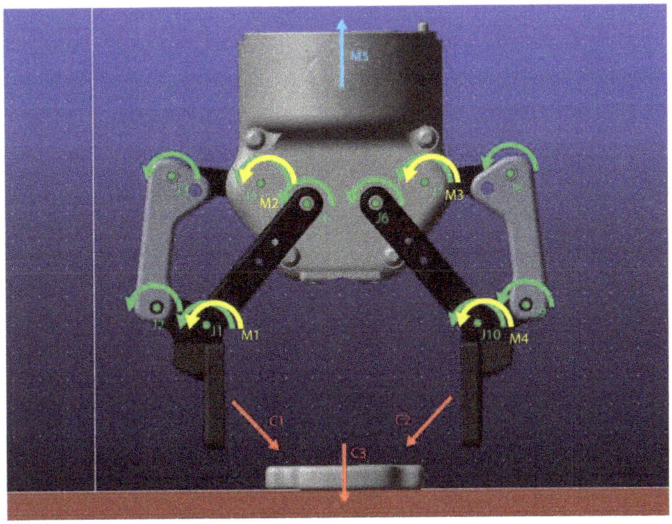

Figure 2. Model of robotic gripper and the object to be gripped in ADAMS/View. Green arrows (J1–J10) indicate revolute joints, yellow arrows (M1–M4) show rotational motions, cyan arrow (M5) indicates general motion (translation and orientation of the gripper), and red arrows indicate three different contacts used (C1–C3). Source: own.

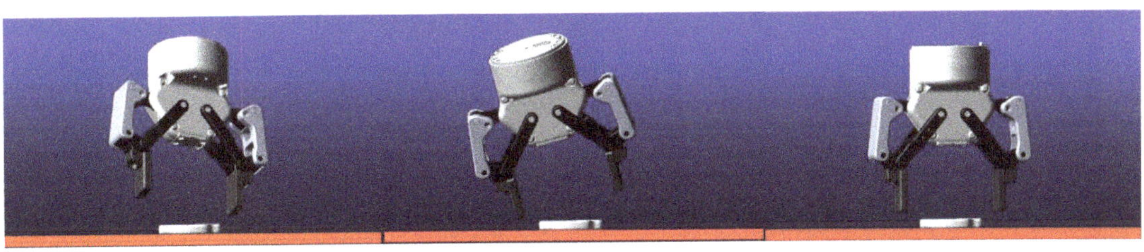

Figure 3. Model of robotic gripper and the object to be gripped in ADAMS/View with different rotation settings. Source: Own.

3.2. Modeling Gripped Object

Similarly, to modelling of the robotic gripper, the 3D model of the object to be gripped was also imported into ADAMS. The object was 3D scanned and postprocessed in a way that can be imported into the ADAMS/View software. Contact forces between the object and the surface and the static and dynamic friction were defined. We also defined the contacts between the robotic gripper's left finger and the object gripped along with the right finger. The contact theory used in the simulations in ADAMS is based on Hertzian Contact theory. Figure 4 shows a friction grip model, where each robotic gripper's fingers supply half of the required contact force.

The minimal required grasping force, $F_{G_{min}}$, applied by the gripper to the load can be calculated via the following equation:

$$F_{G_{min}} = \frac{m \cdot g \cdot S_f}{2 \cdot \mu_s} \tag{1}$$

where m = mass of the object, g = gravitational acceleration (9.81 m/s^2), S_f = safety factor, and μ_s = static friction coefficient.

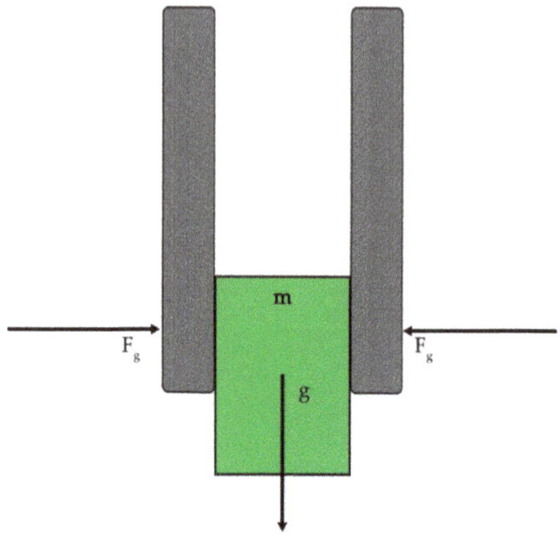

Figure 4. Friction grip model according to Coulomb's friction model. Source: own.

Of course, the minimal required grasping force increases as soon as the robotic gripper starts moving. The most significant factor in calculating the required grasping force is correctly determining the friction coefficient.

Tables of friction coefficients between the individual elements were found in ADAMS Help section of the program, and the other parameters were set according to the real-time response of the simulations. Once the initial model and contact forces were established, measures were added to the model to check the influencing parameters (e.g., friction, contact forces, etc.). In addition, a marker was attached to the fingertip of the two-finger gripper to observe the object's distance from the surface. Furthermore, the object's rotation around the Y-axis was measured. Angular velocity and contact force measures were attached to the left and right robotic gripper fingers. Those variables served as the input of the force controller of the gripper closing. In our model, we defined the following input variables, which are managed via MATLAB/Simulink: (a) the closing speed of the left and right robotic gripper, (b) the final height of the robot gripper, (c) the start and end time of the robotic gripper, (d) initial speed of the robotic gripper, and (e) the initial translation and the rotation of the robotic gripper. The special feature of these variables is that they can be modified at run-time in MATLAB/Simulink. The initial location and orientation of the object and the robotic gripper can be set before the simulation starts by using the "General Point Motion" on the specific part(s) (Figure 5). However, we found that the model works best if the gripper's location and orientation are set using the MATLAB/Simulink variables, while the object's location and orientation are set manually by modifying the batch script (.adm file) prior to the simulation. Therefore, ADAMS models are changed using the MATLAB script by modifying ADAMS model files before running each simulation.

While it is possible to obtain the coefficients of friction between the materials from the existing coefficient of friction tables, it is difficult to determine them accurately in a physical system. In addition, ADAMS models the transition from static to dynamic friction modes continuously rather than as it happens in reality (by jumping from one friction mode to another).

By correctly setting the stiction transition velocity (the velocity at which the static friction is enabled) and friction transition (the velocity at which the static friction turns into dynamic friction mode) velocity parameters, we can obtain a good approximation of this phenomenon.

Table 1 shows model contact parameters, where $C_{LF\text{-}O}$ indicates the contact force between the left robotic finger and the object, $C_{RF\text{-}O}$ is the contact force between the right robotic finger and the object, and $C_{O\text{-}G}$ is the contact force between the object and the surface.

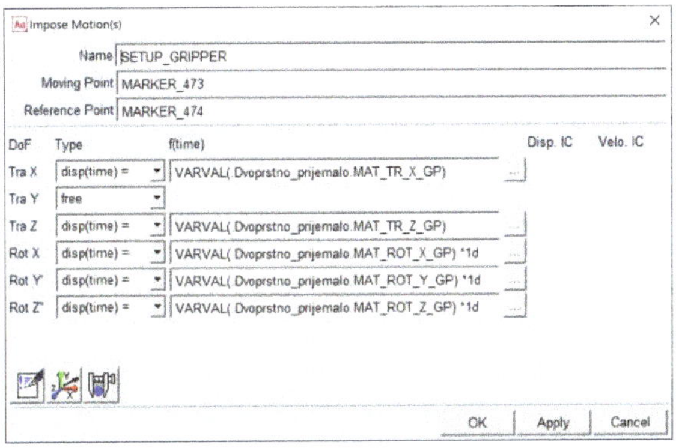

Figure 5. Translation and rotation settings of the robotic gripper in ADAMS/View. Source: own.

Table 1. Model contact parameters. Source: own.

Parameter/Contact	$C_{LF\text{-}O}$	$C_{RF\text{-}O}$	$C_{O\text{-}G}$
Normal force	Impact	Impact	Impact
Stiffness (N/mm)	10^5	10^5	10^5
Force exponent	1.2	1.2	2.2
Damping (Ns/mm)	5.5	5.5	10
Penetration depth (mm)	0.001	0.001	0.01
Friction force type	Coulomb	Coulomb	Coulomb
Coulomb friction	On	On	On
Static coefficient	0.86	0.86	0.4
Dynamic coefficient	0.86	0.86	0.2
Stiction transition velocity (mm/s)	100	100	100
Friction transition velocity (mm/s)	1000	1000	1000

3.3. ADAMS/MATLAB Cosimulation

Preparing the model for cosimulation requires assigning input and variables for exporting created during the modeling of the robotic gripper and gripped object. Once satisfied with the model, it was exported via ADAMS/Controls extension in the form of two ADAMS models and a MATLAB script required for generating the Simulink model. The first one has a ".adm" extension, which allows us to run the model simulations in ADAMS/Solver (batch mode). Animation is not shown during the simulation process, as simulation results are written in several results files for later visualization and analysis. It uses considerably fewer computer resources, meaning the simulation time is significantly lower than those generated in ADAMS/View. The second model has a ".cmd" extension and supports an interactive simulation mode, allowing real-time animation of the model's operation to be shown. Via MATLAB script, an ADAMS system model (written in the form of Level 2 S-Function) is generated, which is then transferred to the simulation model generated by the user.

3.4. Robotic Gripper Force Controller

To approximate the movement of a 2-F robotic gripper closing, the movement of both fingers is controlled by a P force controller. The reference value of the force controller is the desired contact force between the finger of the robotic gripper and an object to be gripped, expressed in Newtons (N). The output of the force controller, which is limited by $[-lim, lim]$, is the robotic gripper closing speed, given in $(°/s)$ since the joints are rotational. The output of the force controller is connected to the input ports of the ADAMS model. Both fingers of the robotic gripper are controlled simultaneously, multiplying the amount of output speed by (-1) for the right finger due to the clockwise direction of the joint's rotation. The control scheme is shown in Figure 6, and the full Simulink model is shown in Figure 7.

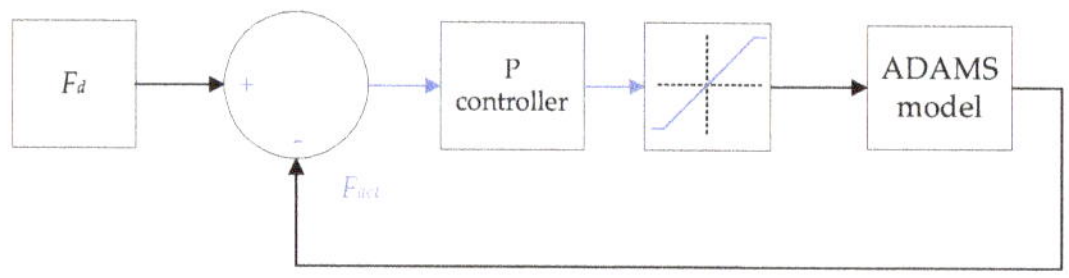

Figure 6. Force P-controller used to control the closing of robotic grippers' fingers. F_d is the desired value of contact force between the robotic grippers finger and the object, and F_{act} is the actual contact force.

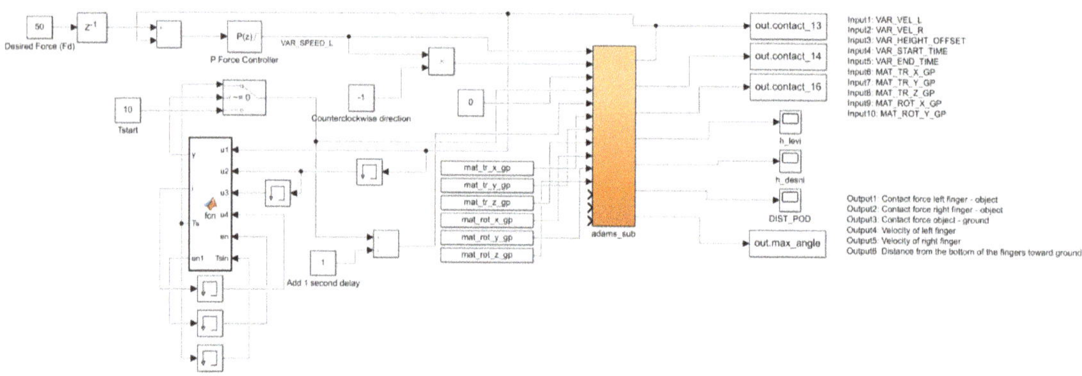

Figure 7. MATLAB/Simulink control diagram. Source: own.

For tuning the force controller, the Ziegler–Nichols method was used. Due to the settings of the controller (at the beginning, the P part must be large enough for the system to react quickly and, at the end, small enough not to cause any overshoot), a compromise was sought between the two values, which resulted in a slow closing of the gripper, but avoided the subsequent rapid response of the controller. Still, the closing velocity remained in the boundary of the actual gripper closing velocity. The parameters of the force controller are presented in Table 2. The problem with higher values of the P part of the controller occurs mainly after a longer grasping time, as the minimal change in velocity causes the object to slip out from under the robotic gripper. As noted by the Taylor, Drumwright, and Hsu [26], the slipping problem may stem from the simulation tool itself. The issue can be contained to a certain extent but cannot be eliminated. An adaptive controller would allow changing the controller coefficients dynamically according to the current response, further minimizing the slipping of the object.

Table 2. Force controller parameters. Source: own.

	P Force Controller
Gain	0.08
Output limit	50/−0.01

In addition, because of the slip problem mentioned above, we set the condition that as soon as the controller reaches the set contact force, the lifting of the object is started. Otherwise, the object's slipping begins before it reaches its final height. Figure 8 shows the close-up of the contact force between the left robotic finger and the selected object.

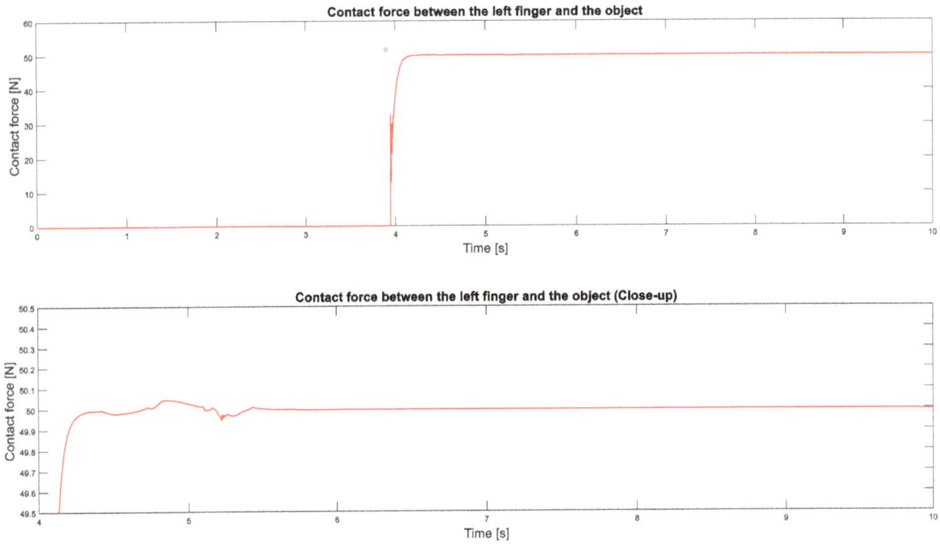

Figure 8. Selected contact force response (red line) between the left robotic finger and the object to be picked for a sample pick-point, F_d = 50 N. Source: own.

3.5. Generating Pick-Points

With each new simulation, a new pick-point point must be defined, which is entered into the ADAMS model (.adm) using a MATLAB script so that it replaces the part of the code of the ADAMS model that describes the position or orientation of the object. The pick-point described in the following text can be categorized as the pick-point pairs due to the parallel 2-F gripper used. However, for simplicity, the term pick-point will be further used throughout the paper. The pick-point generator generates pick-points pseudorandomly. The points are generated according to the size of the object. The object is initially centered under the two-fingered robotic gripper and can be offset by half its length in two directions (X, Z) in both positive and negative (e.g., $+X/-X$) and rotated around the (Y) axis. The maximum rotation around the (Y) axis is between 0° and 360°. Since the combination of the three parameters (the height of the gripper is fixed) can set the object out of its feasible gripping range, three checks are performed to ensure that pick-points are viable and valid. This is conducted by importing the top-down view of the object into the MATLAB script, which performs image segmentation and model detection (Figure 9).

First, the clearance check is performed, ensuring that the object does not lay directly under each of the two fingertips of the robotic gripper. Second, the portion of the object between the fingertips of the robotic gripper is checked. This is conducted by counting the pixels that are positioned between both fingertips. If the number of pixels covers less than the portion of the entire area set during the setup process, the pick-point is discarded, and the process repeats at the first step. Lastly, the shortest distance between both fingertips of the robotic gripper and the object is calculated. By default, the distance between the fingertips of the robotic gripper is the maximum gripper stroke width, which means that the time required to approach the object varies greatly with the position and orientation of the object. Since the output of the force controller is limited, the gripper takes a long time to close fully. Therefore, we aimed to reduce the needed simulation time so that the fingertips would start close to the object to be picked at the beginning of the simulation.

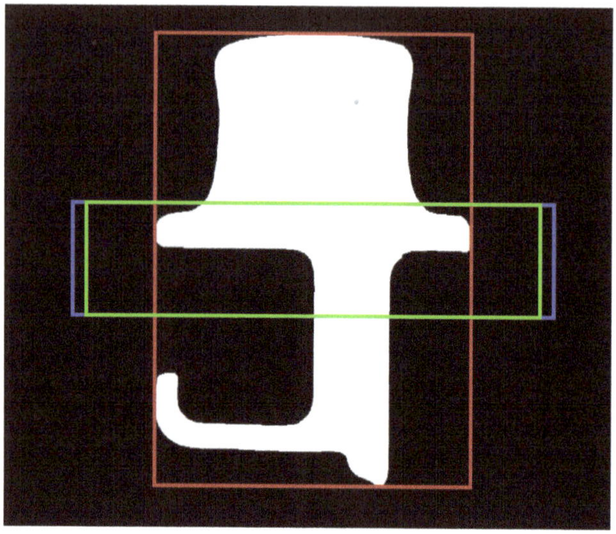

Figure 9. The test object's initial position. The blue rectangle shows robotic fingers, the green rectangle is a possible grasping area, and the red rectangle is the object's bounding box. Source: own.

Since the fingertips of the robotic gripper do not close exactly parallelly (the fingertips of the robotic gripper travel a bit in the (*Y*) axis direction), we compensated the initial gripper height to ensure that the object is always picked on half of the object height.

3.6. Graphical User Interface

A graphical user interface (GUI) was developed in MATLAB/App Designer (Figure 10), enabling a user-friendly simulation model parameter setup. The application is divided into four tabs, each containing a window for setting various model parameters. If simulations are performed for another object, the model must be reimported into ADAMS/View and re-exported. That is a minor drawback since the 3D model of the object is easily switched in ADAMS/View with the new model. However, the friction coefficients must also be altered based on the object material, yet that is possible to change from within the model GUI. Once the appropriate changes are made, a model of the object must first be imported into the GUI. Since the 3D model cannot be directly imported for further analysis, a top-down picture of the object is imported into the GUI. Using the "Image Processing Toolbox," image segmentation and object detection are performed. Second, the "Set Gripper Parameters" tab lets the user specify two-fingered gripper stroke width, finger width, and other gripper-related parameters. Additionally, the user can perform test simulations in interactive mode to ensure everything is set up correctly before performing parallel simulations in batch mode. Third, the "Generate input data" tab generates input data points based on the input parameters and starts parallel simulations. In addition, the user can visualize the generated pick-point to ensure correct input parameters. Lastly, the "Analyze results" tab enables visualization of the pick-point and the graph of (contact) forces between the robotic gripper and the object in question. For each pick-point, three metrics are calculated, which correspond to the grasp quality. Since validation of the model is critical to ensure correct model operation, the user can export a selected number of (n) best and (n) worst pick-points in the form of an object positioning template for testing purposes. The template is also valid in the event of rotating the robotic gripper.

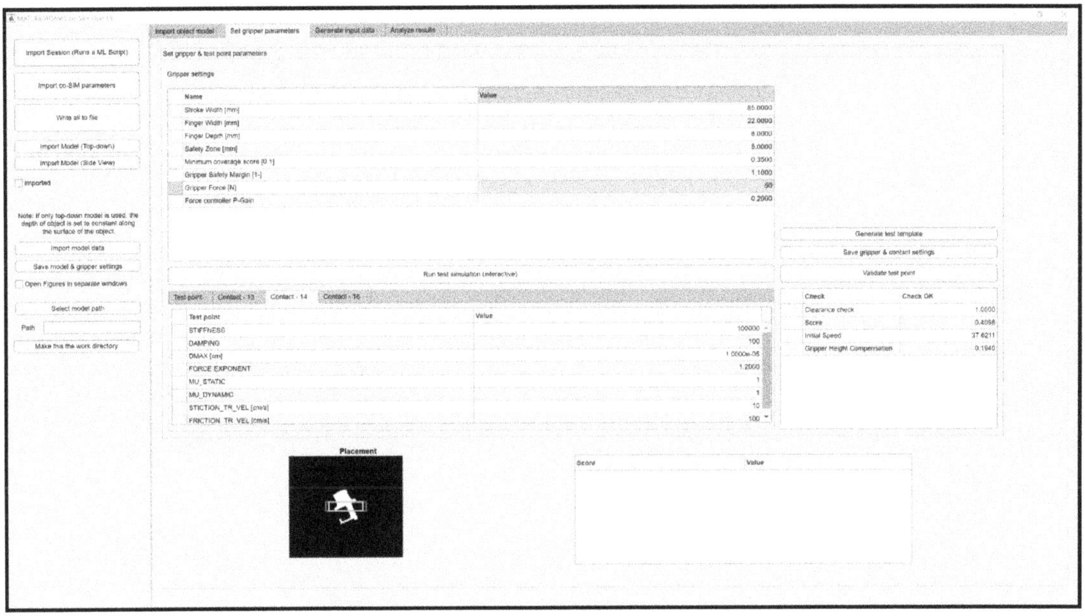

Figure 10. Graphical User Interface (GUI). Source: own.

Since animations cannot be accessed directly from the model GUI, they can be imported into ADAMS/View. First, a file containing model information (.cmd or .amd) must be loaded into ADAMS/View. Second, in ADAMS PostProcessor, the user must import the graphics file generated during the simulation. From there, it is possible to visualize the entire simulation (animation) of the pick-point evaluation.

3.7. Paralleling Simulations

The simulation time for running the analysis of a single pick-point in ADAMS/MATLAB cosimulation (batch mode) was initially approximately one minute, meaning that a maximum of 1440 simulations could be performed per day. Although the simulations reflect the actual situation reasonably well, the model would not be beneficial if left at this stage, as the simulation time is significantly too long. To this end, we tried to optimize the execution performance of the simulations. All three individual components, the model in ADAMS, the execution of the simulation in MATLAB/Simulink, and the execution of the support scripts in MATLAB, impact the simulation run-time. Since support functions are only run once presimulation, the latter takes a negligible amount of time. Therefore, we focused on optimizing the ADAMS and MATLAB/Simulink models. The simulation time is affected by several parameters: sampling time, numeric integration methods, tolerances on the accuracy of the calculations, etc., which were left at default values. We found that a sampling time of 5 milliseconds is sufficient to run the simulations "quasi-continuously" and that there are no excessive discontinuities between simulation points. We tried to improve the performance of the simulation by employing multiple cores simultaneously. Still, we found that the mechanical ADAMS model works best in a single-core mode when the processor runs at maximum clock speed. In addition, when multiple cores are used, the processing time may increase due to processor context switching. The impact of other parameters in ADAMS has not been studied yet and will be the subject of further analysis.

The run-time dependency of MATLAB/Simulink simulations depends on many factors. Among the most influential are: (a) the run time of the simulations, (b) the integration step, (c) the integration method, and (d) the selected simulation mode. MATLAB/Simulink has a built-in tool for checking the execution time of individual blocks, and it was used to identify blocks that can be replaced or simplified with more time-efficient ones. Using the Model Advisor tool, MATLAB/Simulink can check the optimal integration method and suggest other improvements, including settings for running simulations.

In the following, we found it possible to parallelize the simulations using the "parsim" function. With this function, cosimulations can be run simultaneously depending on the available cores of the computer. MATLAB thus assigns a worker to each kernel to run the simulations. With the help of additional scripts that change the position and orientation of the object in each iteration, each simulation can implement completely different pick-points (with different desired forces, etc.). Parallel simulations are carried out as follows: (1) worker simulation team (pool) is initialized, (2) individual ADAMS models for each of the pick-points are generated, (3) simulation parameters for each model are set, (4) simulations are run in parallel (depending on the number of available workers), (5) the entire team of workers is stopped, and (6) the results are stored and prepared for further analysis.

While using the "parsim" function, a single simulation execution time cannot be shortened; however, running multiple simulations in parallel within a much shorter timeframe is possible. In addition, the input pick-points could be generated on a single computer and then distributed to a cluster of computers if necessary.

3.8. Setting the Correct Mechanical Parameters

Since the quality of the simulation model pick-point success prediction relies on the correct mechanical parameter settings, it is crucial to model those parameters correctly. Several mechanical handbooks [49] and webpages offer approximate friction coefficient settings for various materials, which provide a good starting point. Determining friction coefficients along with the stiffness, force exponent, damping, and penetration depth would require extensive test equipment and much time. Therefore, the graphical user interface for the cosimulation model features a tab from where it is possible to run parallel simulations with different parameters (i.e., parameter sweep). The user specifies the number of simulations, start and end values, and the selected step size. From there, it is advisable that the user prints several paper templates and compares the results from the simulation at specific parameters to the existing system. Usually, the first clue to the correct or inappropriate parameter settings is in the model behavior, visible from the simulation-generated animation. According to the experience gained by the modeling, it is first advisable to change the model stiffness parameter by a decade, followed by a damping setting modification. If those two parameters are set up incorrectly, the model may introduce unwanted jolting or unexpected behavior, such as object disappearing. Lastly, if the contact forces are out of the expected values, it is usually a sign of incorrectly set up gain parameters of the P-force controller. Lower gain values cause the robotic gripper's fingers to close more slowly but prevent the generation of excessive force.

4. Results

The simulation model for robotic pick-point evaluation for the 2-F robotic gripper was evaluated with two-fold assessments: (a) simulation time evaluation and (b) grasping performance evaluation. The objective of the first test is to determine the execution time of the simulation model in single and parallel modes. In contrast, the second test directly evaluates the performance of the simulation model compared to its physical experiment with the collaborative robot UR5e, 3D Pickit vision system, and 2-F robotic gripper.

4.1. Simulation Time Evaluation

Several tests were conducted to evaluate the time required for the simulations. We assessed the simulations running in a single batch mode and interactive mode by varying simulation time using Parallel Processing Toolbox. Next, 25 simulations were run in parallel, where all the input models (location and orientation of the object) were the same, and lastly, 25 different models were considered. The number of $N_{\text{sim}} = 25$ was selected due for the time required for subsequent simulations. The simulation time evaluation was performed based on the "hygienic door opener" model, described in the Section 4.2.

Since there is no guarantee that MATLAB will use all the selected workers of the pool in a parallel cluster due to available system resources or possible model errors, we performed five runs for the specified simulation time. However, the average required time for the selected simulation time considers only the three shortest simulation time runs. MATLAB started with the default 12 workers, while this number usually dropped to 9 after executing 25 parallel simulations. The first test runs were performed on a workstation with the following setup: Intel i9-9900K processor, 64 GB DDR4 RAM, Gigabyte nVIDIA GTX 3070 OC, and 1 TB nVMe SSD.

The results of these tests are presented graphically in Figure 11.

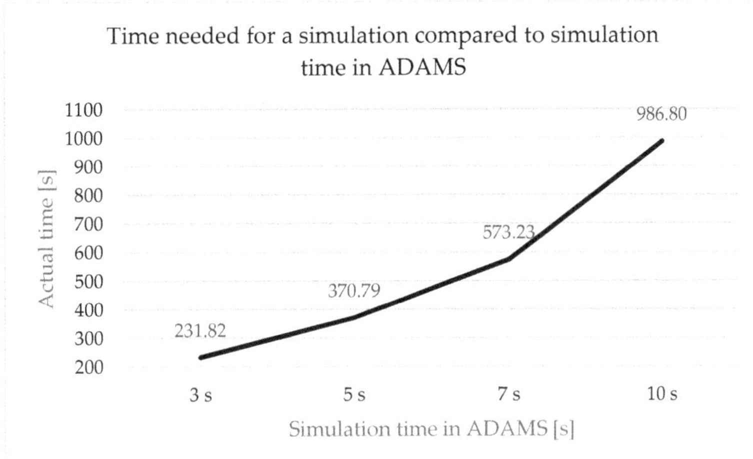

Figure 11. Time required for running 25 parallel simulations compared to simulation time (3–10 s) with the default number of workers (12). Source: own.

Increasing the number of simulations from 25 to 250 does not scale linearly with the time required for the simulations. As seen from Figure 12, in the event of t_{sim} = 5 s, the average time needed to complete 25 simulations is t = 14.61 s. However, when N_{sim} = 50 and N_{sim} = 100 simulations, the time required for a single simulation is t = 13.62 s and t = 12.24 s, respectively. However, it should be noted that the number of workers can drop significantly if more simulations are pending due to possible model errors. On the other hand, the time required for a single simulation also drops significantly with the increase in the total number of simulations.

The time to run a single simulation (in the initial position and orientation) in both modes, normal simulation mode and "ode3" numerical integration method with simulation time t_{sim} = 3 s and integration step T_s = 5 milliseconds, is about t = 53 s in batch mode and about t = 97 s in interactive mode. It must be emphasized that the simulation may take different amounts of time for different pick-points due to the number of (contact) recalculations required.

Under the same simulation conditions, we could run around 600 simulations in 45 min in a batch mode (around 100 simulations for interactive mode) using this function, which equates to around t = 4.5 s per simulation. This would allow us to run around 19,200 simulations in 24 h, almost 13 times the initial simulations. At the same time, we significantly exceeded our target for run-time per simulation.

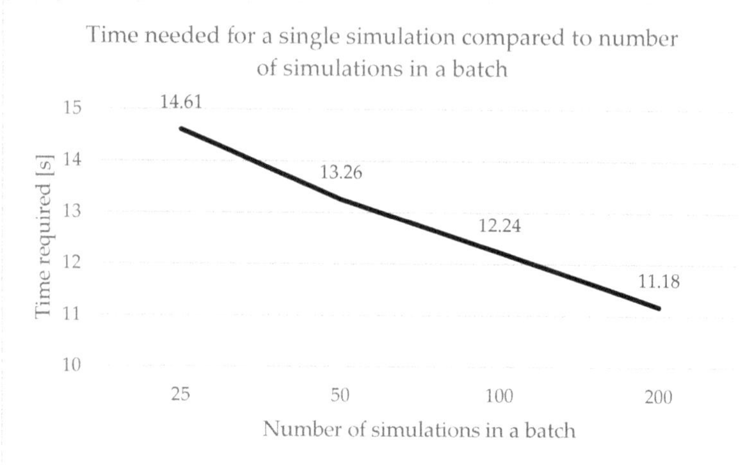

Figure 12. The time needed for a single simulation compared to the total number in a batch. Source: own.

4.2. Performance Evaluation

To elaborate on the accuracy of our simulation model, we selected a "hygienic door opener" object, on which we performed pick-point analyses for the selected robotic gripper. We conducted 50 pick-point evaluations for the chosen object with our simulation model by varying the object's translation and rotation. For each pick-point, we elaborated on grasp quality, a metric derived from three pointers: (a) normalized grasp duration (t_{grasp}), (b) rotation around the main (Y) axis φ_{max}, and (c) the time of absence of contact force between the object and the surface during a successful grasp $t_{grasp-NC}$. Normalized grasp duration (t_{grasp}) is a metric derived from the time contact force between when the gripper fingers and when the object reaches the desired value and then drops back to zero. The rotation (between 0 and 180°) around the main axis (Y) indicates that the set and the actual pick-point are not equal, meaning that it is best to avoid such pick-points since we cannot be sure if the pick-point will be reached.

The total grasp quality score S_{grasp} for the selected pick-point for a single object can be expressed as:

$$S_{grasp} = 0.33 \cdot t_{grasp} + 0.33 \cdot \left(\frac{180^\circ - \varphi_{max}}{180} \right) + 0.33 \cdot t_{grasp-NC}, \tag{2}$$

where S_{grasp} is the grasp quality (score) for the selected pick-point, t_{grasp} is normalized grasp duration, φ_{max} is rotation around the main (Y) axis, and $t_{grasp-NC}$ is the time of absence of contact force between the object and the surface during successful grasp.

After the simulations were completed, we selected ten pick-points which had the highest grasp quality (highest score) and ten with the lowest grasp quality (lowest score). To validate the simulation model, we performed the picking of the object in those pick-points and compared the simulation results with a physical robotic cell workstation. Initial verification of the simulations was carried out by comparing the visual response of the actual bin-picking robotic cell workstation with the simulation model. Still, additional measurement equipment will be needed for accurate verification.

The object had to be placed in the same exact location as in the simulation model, which was achieved with a preprinted template. The template was exported from the GUI and placed directly underneath the robotic cell workstation's robotic gripper. This ensured that the pick-points were set identically in the simulation and physical setup, preventing false input data.

4.3. Test Procedure

A collaborative robot UR5e with a Robotiq 2-F gripper FT-85 was used to elaborate on the simulation data. Additionally, a 3D-printed template holder was mounted on the robot table to ensure that the object stayed in place during the pick-point evaluation procedure (Figure 13).

Figure 13. The test setup consists of UR5e collaborative robot, a Robotiq FT-85 2-F gripper, a 3D-printed template holder, and a paper-printed position template. Source: own.

The Robotiq FT-85 2-F gripper allows the grasping force to be set between 20 and 235 N. The contact force could not be verified this time since we do not possess the force/torque sensor. The grasping force and gripper closing velocity can be set programmatically. Therefore, we decided to set the gripper to the lowest velocity and around 25% of the maximum force. According to manufacturer specification, this would mean that the gripper will grasp the object with approximately 50 N grasping force with a closing velocity of 20 mm/s.

For each of the selected pick-points, the robot arm's initial height had to be modified manually to account for the robotic gripper fingers moving downwards during robot fingers closing.

First, the robotic gripper was positioned h = 10 cm above the object. After the gripper approached the object with a linear motion, the gripper attempted to grasp it. If the grasping is successful, the object is lifted for h = 10 cm above the robot table and then dropped toward the table. This procedure was repeated five times to account for the possible (minimal) offset between the actual and the set object location.

4.4. Test Object: Hygienic Door Opener

The initially tested object is a hygienic door opener, which has the same height across the entire object base surface. It can be characterized as semicomplex with several distinct features, making it appropriate for the initial testing. Additionally, its similar shape can be easily replicated with 3D printing technology for testing purposes. Table 3 shows the object along with several other parameters. The authors can provide the results of the extensive simulation and verification tests, along with the supplementary data (videos, animations), upon request to encourage additional testing of the object.

Table 3. The test object. Source: own.

Selected Object	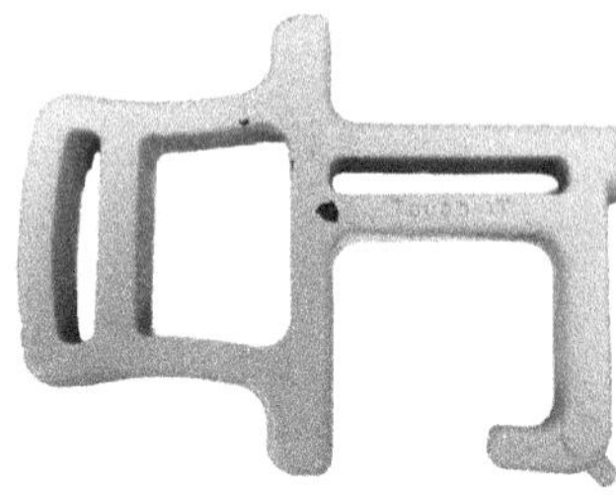
Size ($H \times W \times D$)	$81.9 \times 60.8 \times 9.1$ [mm]
Weight	12.1 g
Friction coefficients (Static/Dynamic) Robotic fingers–object	0.86/0.86
Friction coefficients (Static/Dynamic) Object–ground	0.4/0.2

Out of the 20 selected pick-points (from 50 simulations), the initial testing results show that the simulation model correctly determined if the pick-point would be successful in 75% (15 pick-points) of the cases. This indicates that the mechanical parameters of the model are set accordingly. Table 4 shows the results in more detail.

Table 4. Simulation results. Source: own.

Total pick-points/Selected pick-points	50/20
Correctly determined pick-points	15
Falsely determined pick-points	5
Highest grasp quality score	3.19
Lowest grasp quality score	0.30

To further elaborate on the grasp quality, three sample pick-points were selected to be directly compared in various stages of grasping in simulation and its physical counterpart. Table 5 shows selected pick-points and their displacement parameters–translation around the (X) and the (Z) axis along with rotation around the (Y) axis. The initial position is shown in Figure 9.

Table 5. Test pick-point location, orientation, and score. Source: own.

Pick-Point (N)	X (mm)	Z (mm)	$RotY$ (°)	Grasp Quality Score (S_{grasp})
1	−19.39	6.51	65.90	0.32
2	−3.38	−9.79	−150.05	1.97
3	6.72	−7.43	−167.91	1.61

4.4.1. Analysis of Pick-Point N_1

Figure 14 shows four steps in picking the selected object in the specific location and orientation: (1) prepick, (2) establishing initial contact with the robotic fingers, (3) reaching final contact, and (4) lifting of the object. In the event of the first pick-point, it can be seen that the gripper cannot pick the object successfully in simulation and on the physical test setup. The grasp score for the first pick-point is also low compared to two successful pick-points, indicating that the success metrics are calculated accordingly. Only in the last step are there minor differences between the simulated and the actual response due to the immediate lifting after establishing contact with the physical system. Therefore, the two figures do not match completely. Figure 15 shows contact forces at the described events for pick-point N_1.

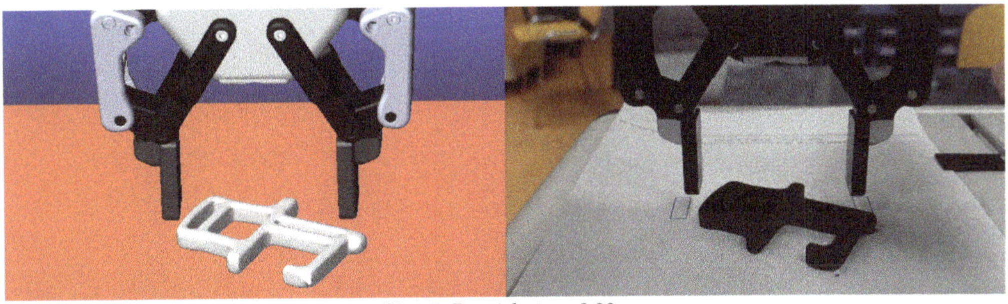

Step 1: Prepick. $t_{sim} = 0.00$ s

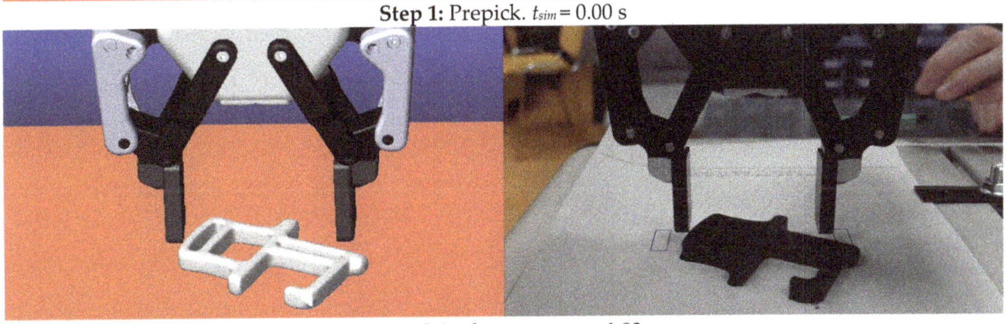

Step 2: Initial contact. $t_{sim} = 1.83$ s

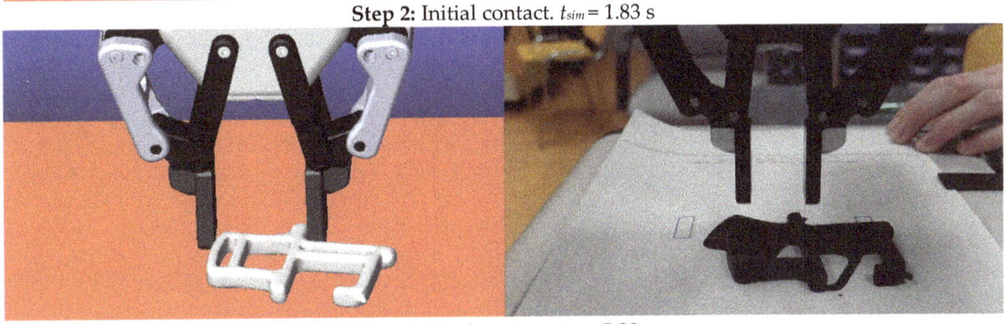

Step 3: Final contact. $t_{sim} = 5.39$ s

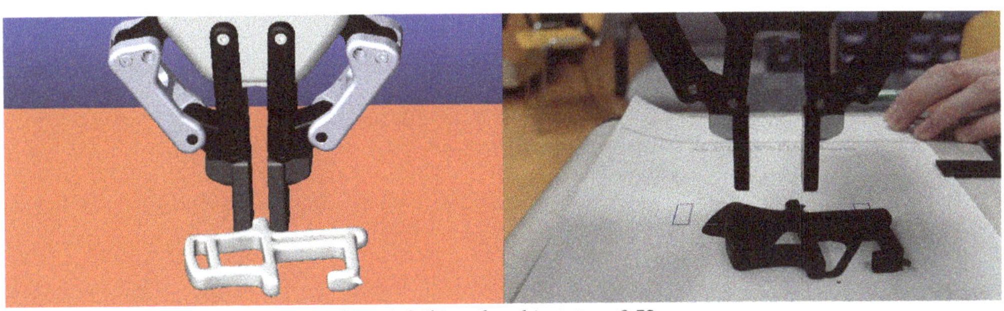

Step 4: Lifting the object. $t_{sim} = 9.52$ s

Figure 14. Graphical analysis of the first pick-point. Source: own.

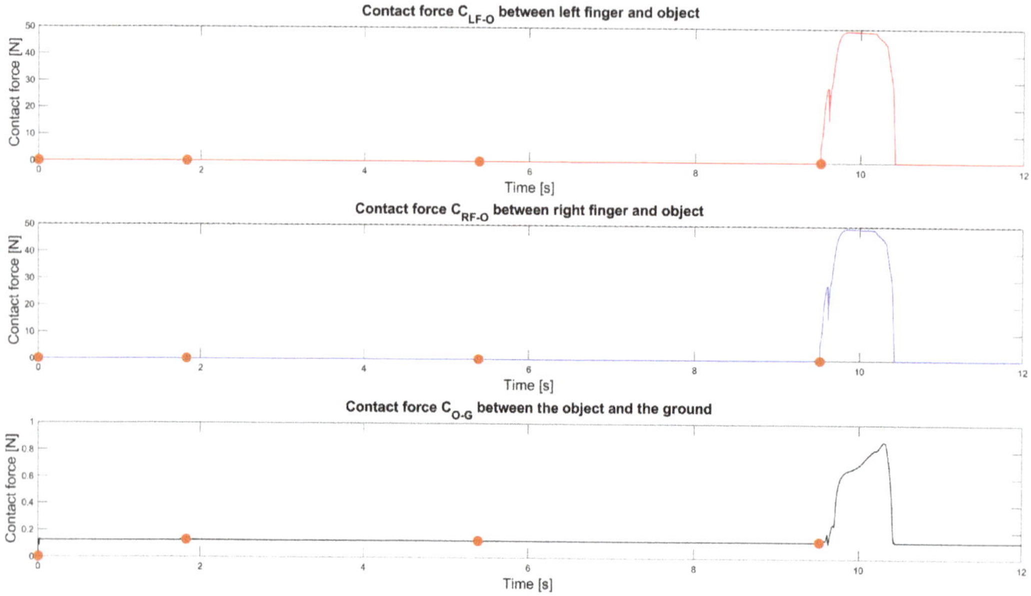

Figure 15. Contact forces between the left and right fingers and the object (red and blue line), and the contact force between the object and the ground (black line) for the first pick-point. The orange circles indicate the events on Figure 14. The dashed line indicates the desired contact force. Source: own.

As can be seen from Figure 14, the robotic gripper starts grasping the object at around t_{sim} = 5.39 s, and the contact force increases to a small value of around $C_{LF\text{-}O}$ = 0.05 N. At around t_{sim} = 9.52 s, the initial contact force is increased to around 49 N since the two bodies (the left gripper finger and the object) come into contact. In the next few milliseconds, the force controller controls the contact force and maintains it for another second. However, when lifting the object starts, the object begins to slip from the gripper fingers; therefore, the contact force $C_{LF\text{-}O}$ value suddenly drops to zero.

4.4.2. Analysis of Pick-Point N_2

From Figure 16, it can be seen that the gripper picked the object successfully in simulation and on the physical test setup. The grasp quality score for the first pick-point is high compared to the first pick-point. The object in the simulation and its physical test setup behaves nearly identically. In the last step, the object starts slipping from the robotic gripper, as explained, due to the P-force controller used. Figure 17 shows the contact forces at the described events for pick-point N_2.

As can be seen from Figure 16, the robotic gripper starts grasping the object at around t_{sim} = 4.60 s. The contact force $C_{LF\text{-}O}$ and the $C_{RF\text{-}O}$ increases to around 50 N. The object is successfully grasped between t_{sim} = 4.60 s and t_{sim} = 6.20 s, as indicated by the nonzero contact force between the object and the surface. Around t_{sim} = 5.70 s and t_{sim} = 6.20 s, the contact force $C_{o\text{-}g}$ becomes zero since the object is lifted from the surface. However, due to the slipping problem mentioned above, the force $C_{o\text{-}g}$ increases back from zero to the object's force of gravity. In the physical test setup, the object is successfully grasped, yet in the simulation, it drops only a few seconds after being grasped.

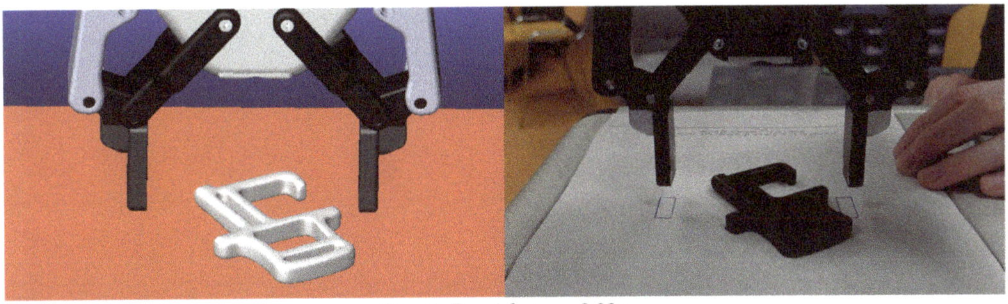

Step 1: Prepick. $t_{sim} = 0.00$ s

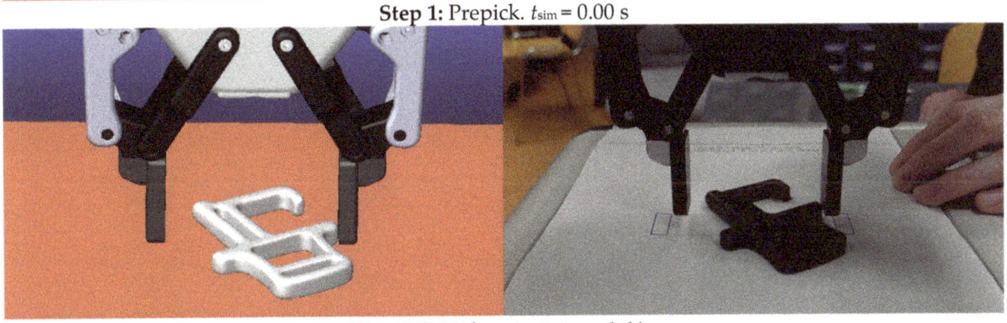

Step 2: Initial contact. $t_{sim} = 1.64$ s

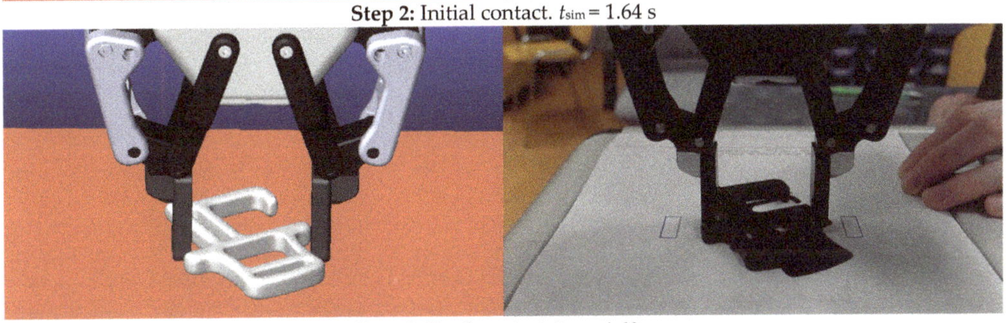

Step 3: Final contact. $t_{sim} = 4.60$ s

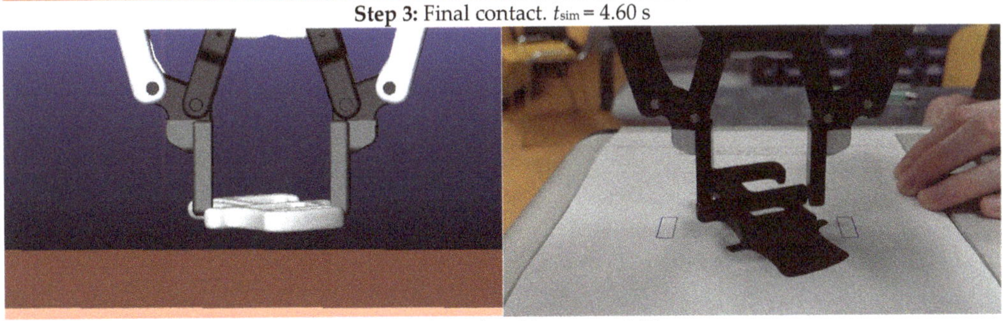

Step 4: Lifting the object. $t_{sim} = 6.48$ s

Figure 16. Graphical analysis of the second pick-point. Source: own.

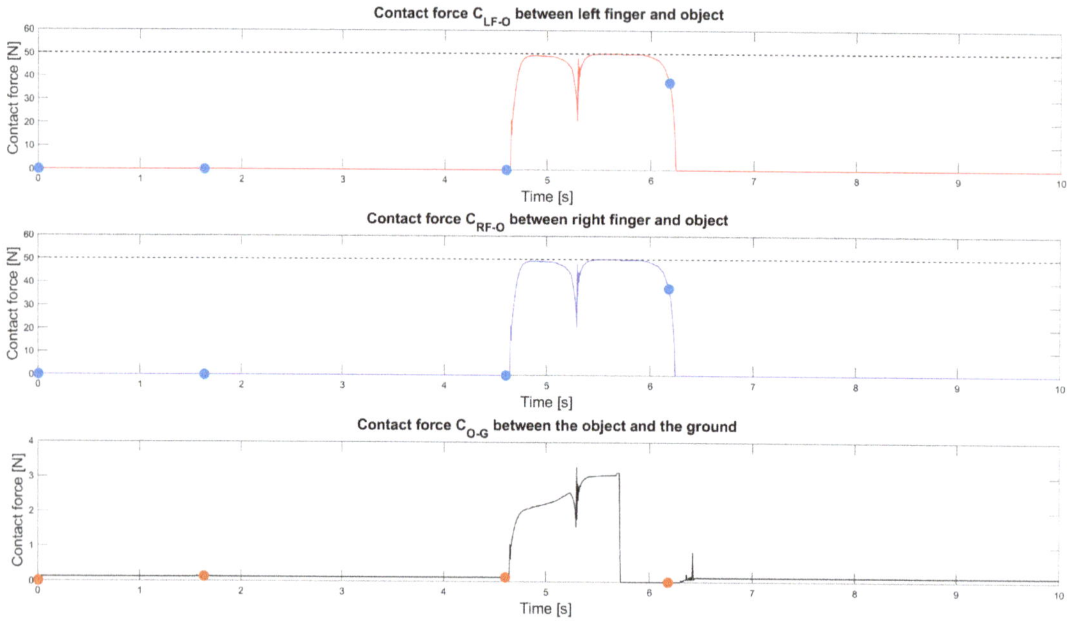

Figure 17. Contact forces between the left and right fingers and the object (red and blue line), and the contact force between the object and the ground (black line) for the second pick-point. The orange and blue circles indicate the events on Figure 16. The dashed line indicates the desired contact force. Source: own.

4.4.3. Analysis of Pick-Point N_3

Figure 18 shows that the gripper can pick the object successfully for only about a portion of the second in the simulation and on the physical test setup, performing nearly identically. The pick-point score for the first pick-point is close compared to the second pick-point, yet a little lower. This analysis shows that this behavior can be closely and precisely simulated, while the slipping problem described above has reasons in the P controllers setting. Figure 19 shows the graphs of contact forces for pick-point N_3.

As can be seen from Figure 17, the robotic gripper initiates the contact at around the t_{sim} = 1.75 s and starts grasping the object at around t_{sim} = 4.50 s. The contact force increases to around 15 N while establishing solid contact. Still, it drops to zero a fraction of a second later. Again, in the next attempt, the contact force increases toward 50 N, as the contact surface is different because the object rotated from its initial position. The object is successfully grasped between time t_{sim} = 4.10 s and t_{sim} = 4.50 s, yet not lifted, as indicated by the nonzero contact force between the object and the surface. At a time around t_{sim} = 4.60 s, the contact force $C_{o\text{-}g}$ becomes zero since the object is lifted from the surface. At around t_{sim} = 6.75 s, the object drops toward the ground due to the lack of proper friction force between the robotic gripper's fingers and the object.

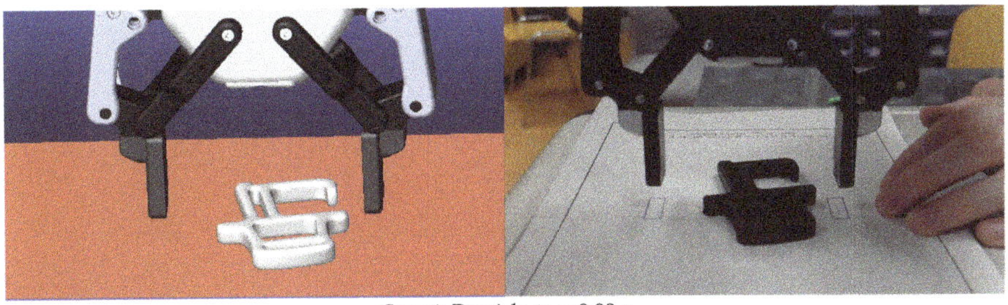

Step 1: Prepick. $t_{sim} = 0.00$ s

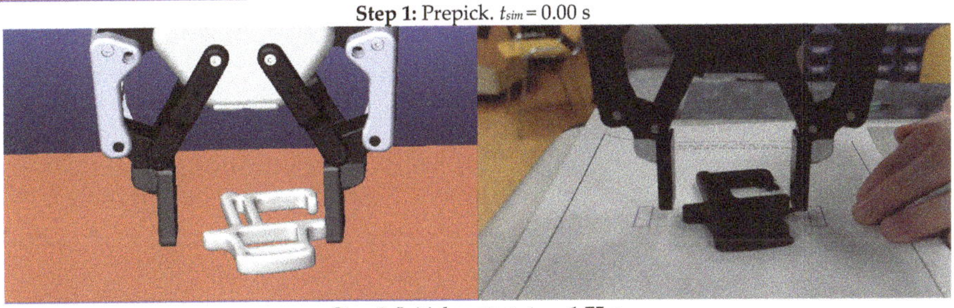

Step 2: Initial contact. $t_{sim} = 1.75$ s

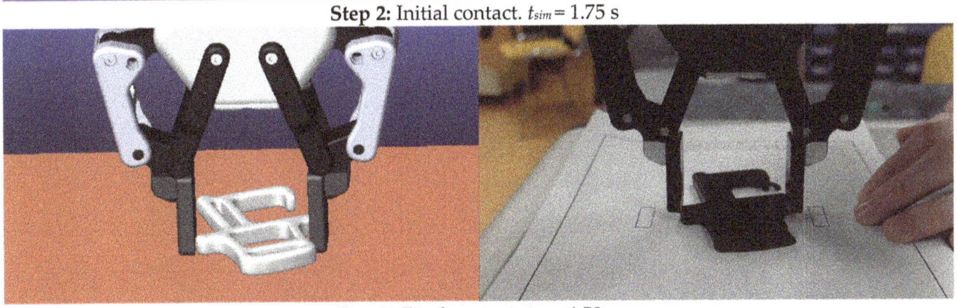

Step 3: Final contact. $t_{sim} = 4.50$ s

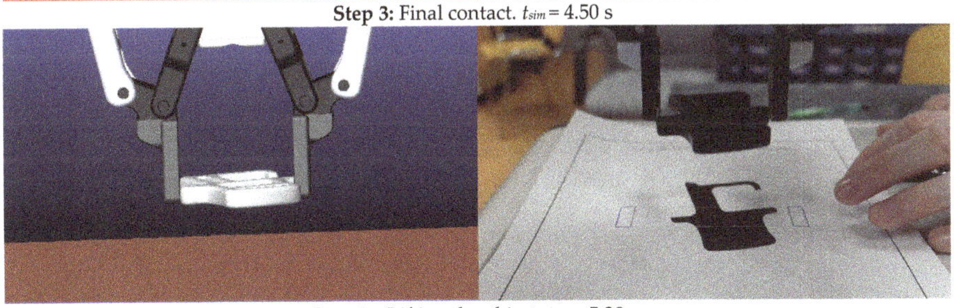

Step 4a: Lifting the object. $t_{sim} = 5.38$ s

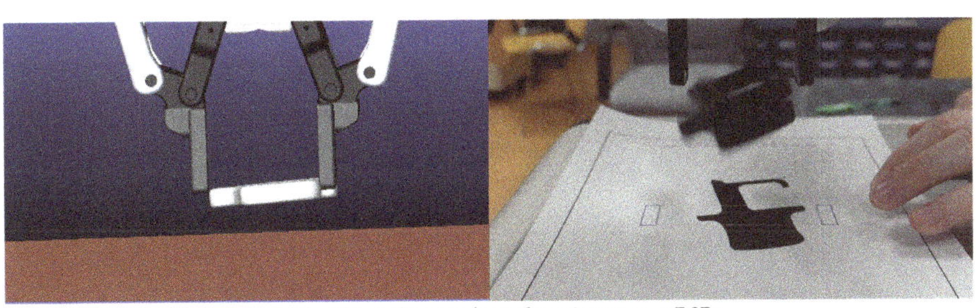

Step 4b: Object slipping from the gripper. $t_{sim} = 7.07$ s

Figure 18. Graphical analysis of the third pick-point. Source: own.

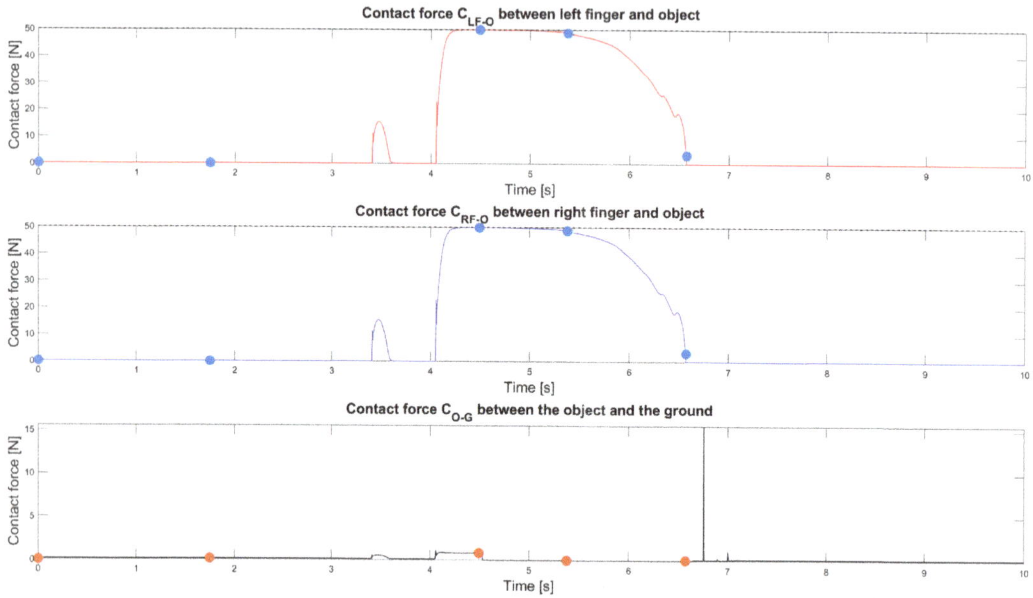

Figure 19. Contact forces between the left and right fingers and the object (red and blue line), and the contact force between the object and the ground (black line) for the third pick-point. The orange and blue circles indicate the events on Figure 18. The dashed line indicates the desired contact force. Source: own.

5. Discussion

From the selected pick-point analysis, we can conclude that the initial simulation model behaves correctly and, in some cases, replicates the actual physical system to a very close extent. The main advantage of using mechanical simulations to determine pick-points is that we can precisely determine an arbitrary number of object pick-points at different parameter settings (e.g., different force settings, different gripper velocity settings, etc.). Additionally, with the ADAMS/MATLAB cosimulation, complex robotic gripper behavior can be modeled along with variants of 2-F robotic grippers (different finger width/depth, stroke width, etc.) since the model is free to be further developed; however, several modifications to the original model are needed in this case. The model's main disadvantage is that these types of simulation require exact contact values and other parameter settings, which are difficult to verify without accurate force/torque sensors. Compared to "deep-learning" approaches, the developed simulation model based on physics simulation achieves lower accuracy. However, it must be noted that the accuracy mainly depends on the accuracy between the set and the actual mechanical parameters of the system. In our paper, we discussed the issue of setting the correct parameters via parallel simulations and comparing the results on the physical system. One must note that even the deep-learning approaches require various parameter settings, which may be even more complex than setting the mechanical model parameters since it is usually clear what those parameters represent. Therefore, if the parameters in the simulation would exactly replicate the actual system, then the accuracy of the model would be significantly higher.

Furthermore, the simulation requires more time to be processed since a simulation accurately simulates real-world physics rather than makes assumptions from the RGB-D data or point cloud. Anyway, in logistics or production companies, where a single type of product is being bin-picked, such analyses are justified to ensure the high performance of the bin-picking operation. In addition, the simulation model's inner working shows transparency since we are not using any "black-box" methods. Of course, determining pick-points must be followed with a quality machine vision system to ensure accurate object (and pick-point) detection.

6. Conclusions

Using a simulation model to evaluate pick-points for a two-finger robotic gripper, we proved that it is possible to systematically check the performance of the selected robotic gripper pick-point using cosimulation with ADAMS and MATLAB/Simulink. Since the model does not yet consider variations of the object depth or allow inner grasping, the proposed object allowed us to test the proof of our concept. The initial results (75% of correctly determined pick-points) proved that the simulation model should be further developed to address the variations of the object depth. Although the simulations do not yet match reality completely, we have successfully laid the foundations for further research. Setting the correct contact parameters requires the most attention. Therefore, we will use machine learning methods to automatically (based on the actual system response) select the most appropriate ones. To further verify the simulated pick-points, we will use force gauges to check the magnitude of the contact forces and the consistency of the set parameters in the simulation and its physical test setup. In the following research, we will first focus on improving the performance of existing simulations and visualization of the pick-points in the 3D model and further reducing simulation time by optimizing the various aspects of the model. Furthermore, to confirm the pick-point resilience toward minor imperfections of the vision system information, we propose that the pick-point evaluation takes place in two steps. First, the pick-point generator should generate points randomly, and the simulation model should determine the most appropriate pick-points. Next, those most appropriate pick-points should be further modified with small perturbations to account for the vision system imperfections. If the original and the perturbed pick-points are still valid, then those pick-points can be considered to lead to reliable picking. Additionally, the limitations on object rotation should be addressed since the possible application of the simulation model is now limited to a few specific applications. Additionally, we would like to use other types of robotic grippers, not limited only to two-fingered robotic grippers, such as vacuum grippers, three-finger grippers, and different special types of robotic grippers.

In combination with a selected 3D vision system, evaluating and setting the pick-points using the developed model could be entirely automated, providing reliable information. In the future, we plan to automate the measurements, which we plan to do using an advanced 3D vision system.

Author Contributions: Conceptualization, P.B., D.H. and T.L.; methodology, P.B.; software, P.B.; validation, P.B., D.H. and T.L.; formal analysis, P.B. and D.H.; investigation, P.B.; resources, T.L.; data curation, T.L.; writing—original draft preparation, P.B., D.H. and T.L.; writing—review and editing, P.B., T.L. and D.H.; visualization, T.L.; supervision, T.L. and D.H.; project administration, T.L.; funding acquisition, T.L. All authors have read and agreed to the published version of the manuscript.

Funding: This research work was supported by the Slovenian Research Agency (ARRS) in ARRS Young Researcher Program (Research activity agreement 2018/2019). T.L. and D.H. were supported by the ARRS Applied Research Project (Research activity agreement 2020/21) entitled: "Warehousing 4.0—Integration model of robotics and warehouse order picking systems"; grant number: L5-2626.

Institutional Review Board Statement: Not applicable.

Informed Consent Statement: Not applicable.

Data Availability Statement: All data obtained in this study are available upon request by contacting the corresponding author.

Acknowledgments: The authors would like to thank the entire team of the Laboratory for Cognitive Systems in Logistics at the Faculty of Logistics for their support, useful comments, and suggestions.

Conflicts of Interest: The authors declare no conflict of interest.

References

1. Lasi, H.; Fettke, P.; Kemper, H.G.; Feld, T.; Hoffmann, M. Industry 4.0. *Bus. Inf. Syst. Eng.* **2014**, *6*, 239–242. [CrossRef]
2. Xu, L.D.; Xu, E.L.; Li, L. Industry 4.0: State of the art and future trends. *Int. J. Prod. Res.* **2018**, *56*, 2941–2962. [CrossRef]
3. Bousdekis, A.; Lepenioti, K.; Apostolou, D.; Mentzas, G. A review of data-driven decision-making methods for industry 4.0 maintenance applications. *Electronics* **2021**, *10*, 828. [CrossRef]
4. Moeuf, A.; Pellerin, R.; Lamouri, S.; Tamayo-Giraldo, S.; Barbaray, R. The industrial management of SMEs in the era of Industry 4.0. *Int. J. Prod. Res.* **2018**, *56*, 1118–1136. [CrossRef]
5. Zhong, R.Y.; Xu, X.; Klotz, E.; Newman, S.T. Intelligent Manufacturing in the Context of Industry 4.0: A Review. *Engineering* **2017**, *3*, 616–630. [CrossRef]
6. Barreto, L.; Amaral, A.; Pereira, T. Industry 4.0 implications in logistics: An overview. *Procedia Manuf.* **2017**, *13*, 1245–1252. [CrossRef]
7. ten Hompel, M.; Kerner, S. Logistics 4.0: The vision of the Internet of Autonomous Things. *Inform. -Spektrum* **2015**, *38*, 176–182. [CrossRef]
8. Boysen, N.; de Koster, R.; Weidinger, F. Warehousing in the e-commerce era: A survey. *Eur. J. Oper. Res.* **2019**, *277*, 396–411. [CrossRef]
9. Calzavara, M.; Battini, D.; Bogataj, D.; Sgarbossa, F.; Zennaro, I. Ageing workforce management in manufacturing systems: State of the art and future research agenda. *Int. J. Prod. Res.* **2020**, *58*, 729–747. [CrossRef]
10. Statista. Logistics Industry-Market Size 2027. Available online: https://www.statista.com/statistics/943517/logistics-industry-global-cagr/ (accessed on 28 December 2022).
11. Statista. Global e-Commerce Market Size 2020–2026. Available online: https://www.statista.com/statistics/1286887/e-commerce-logistics-market-size-worldwide/ (accessed on 28 December 2022).
12. WGTL. Intralogistics—Logistics Journal. Available online: https://www.logistics-journal.de/about/intralogistics (accessed on 28 December 2022).
13. de Koster, R.; Le-Duc, T.; Roodbergen, K.J. Design and control of warehouse order picking: A literature review. *Eur. J. Oper. Res.* **2007**, *182*, 481–501. [CrossRef]
14. Srinivas, S.; Yu, S. Collaborative order picking with multiple pickers and robots: Integrated approach for order batching, sequencing and picker-robot routing. *Int. J. Prod. Econ.* **2022**, *254*, 108634. [CrossRef]
15. D'Souza, F.; Costa, J.; Pires, J.N. Development of a solution for adding a collaborative robot to an industrial AGV. *Ind. Robot* **2020**, *47*, 723–735. [CrossRef]
16. Azadeh, K.; De Koster, R.; Roy, D. Robotized and Automated Warehouse Systems: Review and Recent Developments. *Transp. Sci.* **2019**, *53*, 917–945. [CrossRef]
17. Bormann, R.; de Brito, B.F.; Lindermayr, J.; Omainska, M.; Patel, M. Towards Automated Order Picking Robots for Warehouses and Retail. In Proceedings of the Lecture Notes in Computer Science (including Subseries Lecture Notes in Artificial Intelligence and Lecture Notes in Bioinformatics); Springer: Berlin/Heidelberg, Germany, 2019; pp. 185–198.
18. Du, G.; Wang, K.; Lian, S.; Zhao, K. Vision-based robotic grasping from object localization, object pose estimation to grasp estimation for parallel grippers: A review. *Artif. Intell. Rev.* **2021**, *54*, 1677–1734. [CrossRef]
19. Wang, W.; Liu, W.; Hu, J.; Fang, Y.; Shao, Q.; Qi, J. GraspFusionNet: A two-stage multi-parameter grasp detection network based on RGB–XYZ fusion in dense clutter. *Mach Vis. Appl.* **2020**, *31*, 58. [CrossRef]
20. Jiang, S.; Li, S.; Bai, Q.; Yang, J.; Miao, Y.; Chen, L. Research on generation method of grasp strategy based on deeplab v3+ for three-finger gripper. *Information* **2021**, *12*, 278. [CrossRef]
21. Caldera, S.; Rassau, A.; Chai, D. Review of deep learning methods in robotic grasp detection. *Multimodal Tech. Inter.* **2018**, *2*, 57. [CrossRef]
22. Mahler, J.; Matl, M.; Satish, V.; Danielczuk, M.; DeRose, B.; McKinley, S.; Goldberg, K. Learning ambidextrous robot grasping policies. *Sci Robot* **2019**, *4*. [CrossRef]
23. Wu, B.; Akinola, I.; Gupta, A.; Xu, F.; Varley, J.; Watkins-Valls, D.; Allen, P.K. Generative Attention Learning: A "GenerAL" framework for high-performance multi-fingered grasping in clutter. *Auton. Robot.* **2020**, *44*, 971–990. [CrossRef]
24. Lenz, I.; Lee, H.; Saxena, A. Deep learning for detecting robotic grasps. *Int. J. Robot. Res.* **2015**, *34*, 705–724. [CrossRef]
25. Zechmair, M.; Morel, Y. Penalty-based Numerical Representation of Rigid Body Interactions with Applications to Simulation of Robotic Grasping. In Proceedings of the 2022 International Conference on Electrical, Computer, Communications and Mechatronics Engineering (ICECCME), Maldives, 16–18 November 2022; pp. 1–8.
26. Taylor, J.R.; Drumwright, E.M.; Hsu, J. Analysis of grasping failures in multi-rigid body simulations. In Proceedings of the 2016 IEEE International Conference on Simulation, Modeling, and Programming for Autonomous Robots (SIMPAR), San Francisco, CA, USA, 13–16 December 2016; pp. 295–301.
27. Open Dynamics Engine. Available online: https://www.ode.org/ (accessed on 10 February 2023).
28. Bullet Real-Time Physics Simulation | Home of Bullet and PyBullet: Physics Simulation for Games, Visual Effects, Robotics and Reinforcement Learning. Available online: https://pybullet.org/wordpress/ (accessed on 10 February 2023).
29. DART: Dynamic Animation and Robotics Toolkit. Available online: https://dartsim.github.io/ (accessed on 10 February 2023).
30. SimTK: Simbody: Multi-body Physics API: Project Home. Available online: https://simtk.org/projects/simbody (accessed on 10 February 2023).

31. Erez, T.; Tassa, Y.; Todorov, E. Simulation tools for model-based robotics: Comparison of Bullet, Havok, MuJoCo, ODE and PhysX. In Proceedings of the 2015 IEEE International Conference on Robotics and Automation (ICRA), Seattle, WA, USA, 26–30 May 2015; pp. 4397–4404.
32. GraspIt! Available online: https://graspit-simulator.github.io/ (accessed on 10 February 2023).
33. Malvezzi, M.; Gioioso, G.; Salvietti, G.; Prattichizzo, D. SynGrasp: A MATLAB Toolbox for Underactuated and Compliant Hands. *IEEE Robot. Autom. Mag.* **2015**, *22*, 52–68. [CrossRef]
34. Zechmair, M.; Morel, Y. Assessing Grasp Quality using Local Sensitivity Analysis. In Proceedings of the 2021 IEEE/RSJ International Conference on Intelligent Robots and Systems (IROS), Prague, Czech Republic, 27 September–1 October 2021; pp. 3995–4001.
35. Le, T.N.; Lundell, J.; Abu-Dakka, F.J.; Kyrki, V. Deformation-Aware Data-Driven Grasp Synthesis. *IEEE Robot. Autom. Lett.* **2022**, *7*, 3038–3045. [CrossRef]
36. Huang, I.; Narang, Y.; Eppner, C.; Sundaralingam, B.; Macklin, M.; Bajcsy, R.; Hermans, T.; Fox, D. DefGraspSim: Physics-Based Simulation of Grasp Outcomes for 3D Deformable Objects. *IEEE Robot. Autom. Lett.* **2022**, *7*, 6274–6281. [CrossRef]
37. Li, M.; Ferguson, Z.A.H.; Schneider, T.; Langlois, T.; Zorin, D.; Panozzo, D.; Jiang, C.; Kaufman, D.M. Incremental Potential Contact: Intersection- and Inversion-free, Large-Deformation Dynamics. *ACM Trans Graph.* **2020**, *39*, 49:1–49:20. [CrossRef]
38. Kim, C.M.; Danielczuk, M.; Huang, I.; Goldberg, K. IPC-GraspSim: Reducing the Sim2Real Gap for Parallel-Jaw Grasping with the Incremental Potential Contact Model. In Proceedings of the IEEE International Conference on Robotics and Automation, Philadelphia, PA, USA, 23–27 May 2022; pp. 6180–6187.
39. Bonilla, M.; Farnioli, E.; Piazza, C.; Catalano, M.; Grioli, G.; Garabini, M.; Gabiccini, M.; Bicchi, A. Grasping with Soft Hands. In Proceedings of the IEEE-RAS International Conference on Humanoid Robots, Seoul, Republic of Korea, 3–5 November 2015; pp. 581–587.
40. Utomo, T.W.; Cahyadi, A.I.; Ardiyanto, I. Suction-based Grasp Point Estimation in Cluttered Environment for Robotic Manipulator Using Deep Learning-based Affordance Map. *Int. J. Autom. Comput.* **2021**, *18*, 277–287. [CrossRef]
41. Shukla, P.; Kumar, H.; Nandi, G.C. Robotic grasp manipulation using evolutionary computing and deep reinforcement learning. *Intell. Serv. Rob.* **2021**, *14*, 61–77. [CrossRef]
42. Wang, J.; Li, S. Grasp detection via visual rotation object detection and point cloud spatial feature scoring. *Int. J. Adv. Rob. Syst.* **2021**, *18*. [CrossRef]
43. Vahrenkamp, N.; Westkamp, L.; Yamanobe, N.; Aksoy, E.E.; Asfour, T. Part-based grasp planning for familiar objects. In Proceedings of the 2016 IEEE-RAS 16th International Conference on Humanoid Robots (Humanoids), Cancun, Mexico, 15–17 November 2016; pp. 919–925.
44. Tian, H.; Wang, C.; Manocha, D.; Zhang, X. Transferring Grasp Configurations using Active Learning and Local Replanning. In Proceedings of the 2019 International Conference on Robotics and Automation (ICRA), Montreal, Canada, 20–24 May 2019; pp. 1622–1628.
45. Nechyporenko, N.; Morales, A.; Cervera, E.; del Pobil, A.P. A Practical Approach for Picking Items in an Online Shopping Warehouse. *Appl. Sci.* **2021**, *11*, 5805. [CrossRef]
46. Xu, Z.; Qi, B.; Agrawal, S.; Song, S. AdaGrasp: Learning an Adaptive Gripper-Aware Grasping Policy. In Proceedings of the IEEE International Conference on Robotics and Automation, Xi'an, China, 5 June 2021; pp. 4620–4626.
47. Laili, Y.; Chen, Z.; Ren, L.; Wang, X.; Deen, M.J. Custom Grasping: A Region-Based Robotic Grasping Detection Method in Industrial Cyber-Physical Systems. *IEEE Trans. Autom. Sci. Eng.* **2022**, 1–11. [CrossRef]
48. Cheng, H.; Wang, Y.; Meng, M.Q.H. A Vision-Based Robot Grasping System. *IEEE Sens. J.* **2022**, *22*, 9610–9620. [CrossRef]
49. *Marks' Standard Handbook for Mechanical Engineers*, 12th ed.; McGraw-Hill Education: New York, NY, USA, 2018.

MDPI
St. Alban-Anlage 66
4052 Basel
Switzerland
Tel. +41 61 683 77 34
Fax +41 61 302 89 18
www.mdpi.com

Applied Sciences Editorial Office
E-mail: applsci@mdpi.com
www.mdpi.com/journal/applsci

www.ingramcontent.com/pod-product-compliance
Lightning Source LLC
Chambersburg PA
CBHW061549220926
51850CB00005B/289
* 9 7 8 3 0 3 6 5 7 6 5 9 6 *